T0140416

Security and Quality in Cyber-Physical Systems Engineering

Stefan Biffl • Matthias Eckhart • Arndt Lüder •
Edgar Weippl
Editors

Security and Quality in Cyber-Physical Systems Engineering

With Forewords by
Robert M. Lee and Tom Gilb

Editors
Stefan Biffl
Institute of Information Systems
Engineering
Technische Universität Wien
Vienna, Austria

Matthias Eckhart
Institute of Information Systems
Engineering
Technische Universität Wien
Vienna, Austria

Arndt Lüder
Institute of Ergonomics, Manufacturing
Systems and Automation
Otto von Guericke University Magdeburg
Magdeburg, Germany

Edgar Weippl
Institute of Information Systems
Engineering
Technische Universität Wien
Vienna, Austria

ISBN 978-3-030-25311-0 ISBN 978-3-030-25312-7 (eBook)
https://doi.org/10.1007/978-3-030-25312-7

This Springer imprint is published by the registered company Springer Nature Switzerland AG.
The registered company address is: Gewerbestrasse 11, 6330 Cham, Switzerland

Foreword on Security in Systems Engineering

Fear less, do more. Four simple words that call into focus a reality of our industrial automation field and its nexus with our cybersecurity community. There are real concerns for the cybersecurity of our infrastructure and the industrial systems that they rely upon. There are risks that have been well discussed for the last few decades that are intrinsic to the systems and their designs without outside influence such as faults, errors in system design, and failure of protective and safety functions when they are needed most. There are also risks from outside influences such as malicious adversaries who seek to abuse the systems and their functionality for nefarious purposes. As we learn more about our systems and their evolution as well as our adversaries and their capabilities, it is natural to fear. The consequence can be enormous even though the frequency of impact seems minimal. The entirety of the modern world seems in the balance. So why not gravitate towards fear? Simply put, because our fields of engineering and cybersecurity have done an amazing job and we must appreciate the situation we are in and the advancements that are being made. Yet, we still must realize that the risk is growing, and we must do more to protect our systems. To play to those four opening words I will briefly go through a few case-studies relevant to this collection of manuscripts to ideally set the tone and importance of the works contained in this book.

BTC Pipeline Explosion

In 2008, a Russian cyber-attack pivoted through Internet-connected camera systems running along the Baku-Tbilisi-Ceyhan (BTC) gas pipeline to shut down alarms and over pressurize the pipeline, which resulted in a massive explosion.[1] This story was revealed by Bloomberg news in December 2014 and was on its path of being the first

[1] https://www.bloomberg.com/news/articles/2014-12-10/mysterious-08-turkey-pipeline-blast-opened-new-cyberwar

ever confirmed case of a cyber-attack causing damage or destruction on industrial control systems (ICS). Except, it was not confirmed, nor did it hold up to scrutiny. Almost immediately after the story was published, I started to suspect elements of the story as did a few others in the information security community while the story took on its own life and spread throughout the community.[2] Eventually, Michael Assante, Tim Conway, and I wrote a whitepaper at the SANS Institute noting that many details of the story conflicted with reality and in some ways each other of what such an attack could look like.[3] Later in 2015 a well-researched article in *Sueddeutsche Zeitung* revealed even more of the details about the attack were wrong in comparison to how the pipeline was operated at the time; as an example the camera systems that the adversaries supposedly pivoted through during the attack were not installed until after the attack in response to the explosion.[4]

The BTC pipeline case-study has been captured for years as a real event even though it was debunked by high-profile security experts nearly immediately upon publication. It has been referenced in numerous conferences, academic journals, and even U.S. Congressional testimony. It is otherwise a tantalizing story and to many serves as an example of why we must do more in cybersecurity. The reality of course is that cyber-attacks on gas pipelines are possible. The scenario described is possible in principle. Security is important to safety. However, relying on hyped-up case-studies to make a point is only self-serving. The dedication of resources and studying of attacks to develop defenses and best practices are best suited for real attacks where the study can yield meaningful results. Hyped-up threats only yield results in the wrong direction. Even when there are real threats, we often see the impacts overblown.

Bowman Avenue Dam Infiltration

In 2013, Iranian hackers broke into the Bowman Avenue Dam near Rye Brook, New York. They gained access to a human–machine interface (HMI) and read water levels off the dam.[5] The U.S. Intelligence Community identified the infiltration and informed the U.S. Department of Homeland Security, which kicked off a series of events to identify and respond to the infiltration. The press would later become aware of this infiltration in 2015 and publish on it. Major news outlets covered the story noting that it was a small dam, but significant damage could have been done if only the hackers had the intent. Some argued they had the intent but simply could not manage the attack due to their technical incompetence. Headlines ran with pictures of

[2] https://www.flyingpenguin.com/?p=20958

[3] https://ics.sans.org/media/Media-report-of-the-BTC-pipeline-Cyber-Attack.pdf

[4] https://ics.sans.org/blog/2015/06/19/closing-the-case-on-the-reported-2008-russian-cyber-attack-on-the-btc-pipeline

[5] https://www.cnn.com/2015/12/21/politics/iranian-hackers-new-york-dam/index.html

nuclear meltdowns and major dams breaking with government and industry officials noting this was an example of things to come. Later, the U.S. Department of Justice would indict the hackers and note that they were intending something more nefarious, but the automated controls were down for maintenance that would have allowed them to do their actions. The case was real, there was a lot of hype around it, but many of the technical nuances of the details were not correct.

I closely examined the case and all the details around it including interviews with those involved in New York as well as the U.S. Department of Homeland Security. There were key omissions that were never captured fully. I wrote about the most significant of these in a whitepaper with Michael Assante and Tim Conway, where we covered that the automated controls for the dam were never actually installed by the time the intrusion occurred due to delays in the project.[6] The Iranian hackers had not performed some amazing feat but instead had found an HMI that was remotely connected to the Internet via a cell card with poor authentication. After accessing the system, the Iranians were unable to do anything with it because of the lack of controls, whether they would have intended such, and left. The dam itself was small and handled runover during heavy rains. Damaging it would have resulted in making some people's shoes wet, not the imminent doom and images of nuclear meltdowns that the media captured. Even though the infiltration was real, the details around the case and the potential impact were captured incorrectly. It is important to capture such details properly especially for when attacks are real as such cases can have profound impact on public perception as well as defense lessons learned.

The 2015 and 2016 Cyber-Attack on the Ukraine Electric Grid

In December 2015,a team of adversaries broke into three Ukrainian distribution level power companies. The attack started with malware, known as BlackEnergy3, inside the enterprise information technology (IT) networks but that was just a foothold for the adversary. The real work of the attack was nearly 6 months the adversaries spent inside the operations technology (OT) networks learning the systems and operations of the power companies. The attack was effectively just the adversaries learning how to operate the equipment inappropriately. It was not a focus on exploits and vulnerabilities or malware but instead was the abuse of legitimate functionality and features in the environment for malicious purposes.[7]

The attack ultimately led to around 6 h of outages across 225,000 customers. While this does not seem like a lot, it was the first ever cyber-attack to lead to a power outage. More impactfully, the system operators did not have access to their automation networks and supervisory control and data acquisition (SCADA) systems

---

[6]https://ics.sans.org/media/SANSICS_DUC4_Analysis_of_Attacks_on_US_Infrastructure_V1.1.pdf

[7]https://www.wired.com/2016/03/inside-cunning-unprecedented-hack-ukraines-power-grid/

for nearly a year due to the attack. The recovery was a significant effort. Michael Assante, Tim Conway, and I led an investigation on the attack and in our report to the industry we highlighted that what happened in Ukraine was not unique to Ukraine; the attack could be replicated elsewhere. Moreover, we warned that there were elements of the attack that indicated there could be a follow-on attack.[8]

In December 2016, that follow-on attack manifested in the form of malware known as CRASHOVERRIDE, that led to a loss of power in Kiev, Ukraine, at a transmission level substation. The outage was only about an hour long but the amount of electric power lost was three times the total amount lost in all of the Ukraine 2015 attack due to the differences in distribution versus transmission level substations.[9] My team at Dragos, Inc. wrote the report on this malware drawing specific attention to the fact that the adversary effectively learned from their attack in 2015 and moved the work that took dozens of people into scalable software, CRASHOVERRIDE, that would only take the adversary positioning and activating it.[10] This effort represented a scalable repeatable attack on electric power.

What More Looks Like

There are plenty of high-profile attacks that target OT and ICS environments. It is hard to ignore such attacks especially as we are seeing more that are dangerous such as the TRISIS malware. TRISIS was leveraged in Saudi Arabia to shut down a petrochemical plant while its true design was likely to kill people via disabling the safety system and a follow-on attack against the petrochemical process.[11] The attack failed to kill anyone but managed to shut down the plant costing the site millions. The adversary, code named XENOTIME, remains active as of the time of this publication, targeting other industrial sites around the world.[12]

There are even more incidents that are below such thresholds and thus remain out of the media and remain private to the companies who work those cases. It is easy to see then why people fear. However, I will return to the initial point of this foreword, which is to also highlight that our infrastructure is reliable, our engineering practices sound, and our community is amazing. Far more is done for security than will ever make headlines; the community is quick to notice attacks but slow to see the day-to-day work of defenders around the world. The adversaries are becoming more aggressive, but they must contend with physics and when organizations prepare correctly, they must also contend with defenders who protect our systems.

---

[8] https://ics.sans.org/media/E-ISAC_SANS_Ukraine_DUC_5.pdf

[9] https://www.wired.com/story/crash-override-malware/

[10] https://dragos.com/wp-content/uploads/CrashOverride-01.pdf

[11] https://dragos.com/wp-content/uploads/TRISIS-01.pdf

[12] https://dragos.com/resource/xenotime/

These defenders thrive when the systems are made to be more defensible. If we are not to fear, but instead are to do more, then it is natural to ask what more looks like. To me, it is in contributions such as this book that we find that answer. This book represents an amazing effort by community members, engineers, and academics who are striving to create more defensible systems through better engineering, system design, and implementation of those systems. These quality and security improvements of long-running technical systems such as ICS will yield a safer and more reliable world even in the face of determined adversaries. It is in this type of collaboration that lessons learned can be documented and made available to current and future practitioners.

Gambrills, MD, USA Robert M. Lee

Foreword on Quality in Systems Engineering

Systems Quality Engineering: Some Essential Steps Forward

As citizens in a connected world, the quality of our daily lives depends on critical infrastructure, such as the Internet and energy and transportation networks, and on industrial production systems, the so-called *Cyber-Physical Systems* (CPSs), which we expect to provide high-quality services economically and safely for their environment. In an increasingly networked world, the quality of the services that these systems provide depends on their security against intended and unintended wrongdoing, both during the operation of the system and during the engineering that designs the system.

Therefore, the topic of this book, *Security and Quality in Cyber-Physical Systems Engineering*, is timely and important, but may sound somewhat puzzling, as security is among the qualities of systems engineering. However, the topic makes sense as security in systems engineering is a fundamental quality, since there is no safety of systems operation without information security in systems engineering. Unfortunately, the quality *security* is not well understood and often not well addressed in systems engineering practice. Therefore, I share in this foreword my view on essential steps and principles to improve *systems quality engineering* both in practice and in university teaching.

I will keep my observations to a minimum by sticking to the very basic ideas of *quality engineering of products and services*. References give considerable detail background for the interested reader.

Security and Safety Are Quality Requirements Security and safety are part of the set of system attributes known as *qualities*. Consequently, the systems engineering methods for quality in general help address *Security* and *Safety*. For example, for Usability, Maintainability, Reliability, and Availability, I define *quality* as the attributes that describe *how well* a system functions (Gilb 2005). In my long consulting experience on quality requirements, I have observed that all qualities are variable in their performance levels. Therefore, we can, and must, *quantify all system quality requirements* as a fundamental aspect of any systems engineering methods. While this may sound obvious, even systems engineering subjects taught at university, such as *Quality Function Deployment* (QFD), allow unquantified qualities but should be applied more rigorously by *emphasizing the quantification of quality requirements* (Gilb 2005, 2018b).

Quantification of Quality Requirements All quality attributes vary and can be expressed as a quantified requirement (Gilb 2005). A simple method to see that this is widely understood, and somewhere practiced, is to Google a quality name followed by the term *metrics*, e.g., *Usability Metrics, Security Metrics.*

Quantification of a quality is not the same as measurement, but quantification is the basis for measurement (quality level testing) and other applications. The essential notion of quality quantification is the definition of a *scale of measure*.

For *Security,* an example *Scale* could be *the share (%) of assessed cyber risks with an event likelihood greater than X% and an impact greater than $Y.*

The *Scale* parameter (Gilb 2005) both defines the quality and enables assigning numeric levels to the quality, for a variety of purposes, such as:

1. *Establishing benchmarks* for that quality, in our own system, competitive systems, and past and future performance levels (Gilb 2005)
2. *Establishing scalar constraint* quality requirements, that is, worst acceptable level of the quality for a purpose (Gilb 2005)
3. *Establishing target levels* of the quality dimensions, such as time, space, and conditions
4. *Making estimates* of the impacts of design ideas on the requirement levels
5. *Measuring actual levels* of the quality in real systems
6. *Bidding, costing, and contracting* using these quality levels

Estimation of Design Quality Impact When considering the design for meeting the required quality levels, we must be numeric, logical, and look at the whole picture (all quality requirements, resource budgets, and constraints). Certainly we cannot afford to discuss or evaluate any single design idea solely in a single quantified dimension, such as security or safety. Therefore, the evaluation of the impact of any given design idea, needs to be quantitative as foundation for combining the impact values. Figure 1 illustrates the *Impact Estimation* principle as a conceptual table for assessing the impact of design ideas that, together, have an impact on the project objectives and resources.

1. An estimate of the *expected degree* that the design will satisfy the *constraint* levels, with regard to all concurrent conditions (project deadlines, budgets, constraints, reaching other quality levels, and more);
2. An estimate of the *expected degree* to which the design will satisfy the *target* levels, with regard to all concurrent conditions (project deadlines, budgets, constraints, reaching other quality levels, and more).
3. The estimates need to document the ranges of experience, evidence for the levels, and sources of the levels (*Competitive Engineering* (CE) (Gilb 2005), *Impact Estimation* (Gilb 2008)), for quality control, for responsibility, and for understanding the quality of the information.

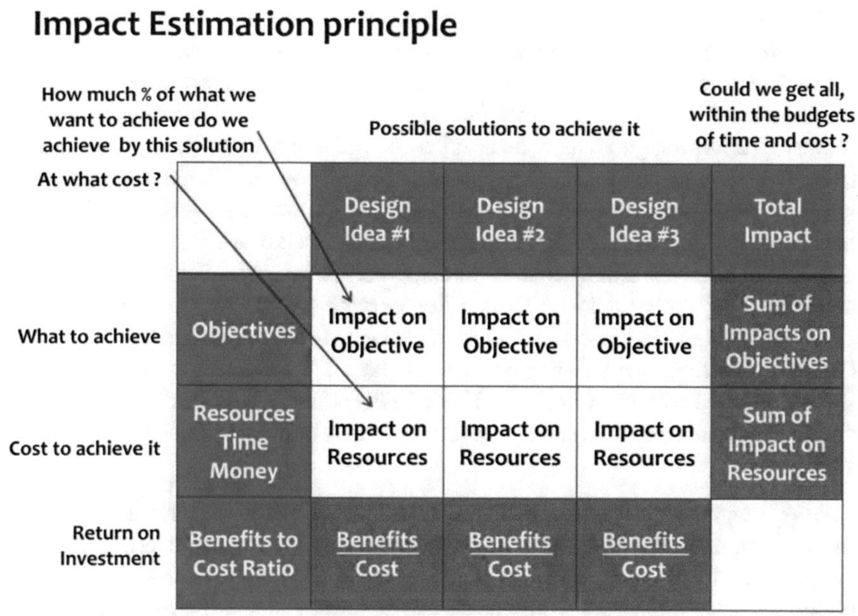

Impact Estimation principle

How much % of what we want to achieve do we achieve by this solution

At what cost ?

Possible solutions to achieve it

Could we get all, within the budgets of time and cost ?

		Design Idea #1	Design Idea #2	Design Idea #3	Total Impact
What to achieve	Objectives	Impact on Objective	Impact on Objective	Impact on Objective	Sum of Impacts on Objectives
Cost to achieve it	Resources Time Money	Impact on Resources	Impact on Resources	Impact on Resources	Sum of Impact on Resources
Return on Investment	Benefits to Cost Ratio	Benefits / Cost	Benefits / Cost	Benefits / Cost	

Fig. 1 A conceptual view of the analysis of quality designs

4. The estimates need to be made in a way that provides at least a rough picture of the *aggregate effects of the designs* (see Fig. 1) (a) together with all other complementary or concurrent designs and (b) with regard to the potential of together meeting the final quality level requirements, preferably with an engineering safety factor.

Measurement of Design Quality, and Correction Once a design is estimated and found to be worthy of at least experimental implementation, we need to measure, at least roughly, not necessarily exactly, how well the design performs, in at least one dimension; and possibly in side effects and cost dimensions as well. If the performance deviates negatively from the expected results, then *root cause analysis* should be used immediately at that increment, and redesign made to get back on track. This process is eminently described by Quinnan in *Cleanroom* (Quinnan 1980) and is an inherent part of my *Evo* approach (Gilb 2005). This is *good engineering agile*, not to be confused with current popular software agile methods, which have no consciousness of qualities, engineering, design, architecture, or dynamic design to cost. When a given *quality constraint* level is reached in the system development process, there is an opportunity to trigger contractual minimum payments. When the *quality target* levels are reached then this can be used to target full payment, and to stop engineering or designing those fully delivered qualities.

Collecting Design Information on Engineering Component Candidates It is a long-standing engineering tradition to organize engineering knowledge about potential design components and processes in engineering handbooks, which include data, tables, etc. regarding their expected attributes. The *World Wide Web* provides convenient means for collecting, accessing, and updating such attribute knowledge. If this knowledge is of good quality, and easily available, then it can be used to make estimates. If it is not good, then the next best thing for those who believe in an idea is to experiment with small-scale incremental implementation and measurement. This might, in fact, provide more useful data on the real system than a general engineering handbook. My father, an engineer and inventor, stressed to me that the engineering tables were not to be blindly trusted, but are likely to be useful in situations without better data. My early books on *Design Engineering* (Gilb1976; Gilb and Weinberg 1977–1978) attempted to show how this might be done for qualities of data structures. One early practical tool-building experiment (later a PhD; see slide 4 in (Gilb 2017) showed how a computer (Apple II, Forth 1979) could automatically pick the best design, if given quantified quality requirements and quantified design component attributes.

Quality Engineering Processes There is a large number of known, and to be invented, engineering processes, which can contribute to the engineering of secure and high-quality products. The important idea regarding a process, as with any system design aspect, are the quality and performance attributes and the resource costs of that process. These quality, performance, and cost attributes could be systematically organized in handbooks, preferably on the Internet for access and updating. Both research and practice could be incorporated. As an example, my *Specification Quality Control* (SQC) process has been studied (Gilb 2005) as a method for measuring the defect quality levels of requirements written in my *Planguage* (Gilb 2005). A key finding by Terzakis from *Intel* (Terzakis 2013) was that *Planguage* together with SQC resulted in a 98% defect reduction in submitted requirements, and in a 233% engineering productivity improvement. Release defects went down to 0.22 defects per 400 (or 600) words, a suitable value for the very high quality levels required for engineering *Intel* chips. Terzakis currently measures the effects of *coaching* on *quality of requirements* (unpublished 2018). My *Planguage* for specification was adopted by over 21,000 *Intel* engineers, demonstrating its viability for embedded hardware logic (Erik Simmons, *Intel*). Therefore, *Competitive Engineering* and *Planguage* (Gilb 2005) should be applied for improving the maturity of engineering *Complex-Cyber-Physical Systems*.

Preventative and Evidence-Based Quality Processes Some of the most interesting engineering processes take a *lean* approach of preventing defects and problems by using common-cause *root-cause analysis* at the grass roots and frequent organization *process and conditions* changes in order to measurably improve organizations' engineering process quality. My favorite generic process is the *Defect Prevention Process* (DPP) developed at IBM (Gilb and Graham 1993; Dion et al. 2018). DPP is a good example of a very generic, organizational engineering improvement, with clear repeatable economic and quality effects that led to lasting results. Unfortunately, the DPP is rarely taught in engineering, or quality engineering, courses, similar to the IBM *Cleanroom* approach (Quinnan 1980). Over the years, I have often seen new ideas advocated without quantitative evidence. I strongly argue that researchers and teachers should build more on proven methods with strong quantitative evidence. Without collecting and sharing data on quality, performance, and cost attributes as evidence on new and traditional processes, systematic scientific progress is not possible. The similar weakness to improve systems engineering practice systematically is reflected in the high project failure rate, which has remained almost constant for decades. Therefore, preventative and evidence-based quality processes should be applied in engineering software-intensive *Complex-Cyber-Physical Systems*.

Integrated Consideration of Requirements I see the integration of quality requirements, security requirements, and all other requirements as essential. It seems risky to limit the engineering evaluation of security and other quality designs to those areas in an isolated way. We cannot allow specialist "security" and "reliability" engineers to make design decisions with impact on a large system, without balanced due regard for other critical factors in the system. Although it is difficult, we must aim at understanding the side effects between all requirements, including example requirements on performance, costs, and non-quality values, such as image and trust or technical debt. In short, security and quality are important parts of the larger systems picture, and have to be properly incorporated in this picture both in practice and in academic research.

Demand for a Stronger Engineering Culture in Software-Intensive Systems Hardware engineering has a mature engineering culture, but is under constant attack from technological change. Systems engineering has also shown engineering culture with good maturity in the face of complexity and change, for example, in space and military application areas. Unfortunately, the newcomer, software, including *logicware*, *dataware*, *peopleware*, *netware*, is, arguably, mostly not engineered in a sufficiently mature *engineering discipline*, but still seems to be all about programming. I have seen this problem for decades.

Hardware experts can engineer a 99.98% available system. In contrast, low-quality software is widely accepted, often with the argument that software can be easily changed. Nobody would buy a machine that seems buggy, but users and lawmakers consider defects in software acceptable if software updates are promised. I see a major reason for this problem in politicians, managers, and researchers not demanding the same quality-of-engineering for IT and software, such as software-intensive systems, as documented in serial aircraft accidents that recently made the news. Therefore, I want to raise awareness for demanding sufficiently high standards for requirements, including security and software quality, for the integrated engineering of software-intensive systems, such as critical infrastructure and industrial systems.

Finally, I summarize my *Basic Principles of Serious Quality Engineering* (Gilb 2018a):

1. All qualities must be treated quantitatively, at all times (Gilb 2005).
2. All other requirements need to be defined rigorously, too.
3. Design options need clear detailed definition, and probably decomposition.

4. All design decisions can, and normally should, be estimated, before selection or prioritization, with regard to their possible and probable impacts on all critical qualities and costs [see Chap. 9 on IET in Gilb (2005)].
5. Incremental quality impacts of designs need to be measured, at that increment (Quinnan 1980).
6. When incremental designs measurably fail to deliver expectations, they need to be immediately replaced or corrected (*Cleanroom*) (Quinnan 1980).
7. Reasonable rigorous quality-control measurements of critical specifications must be carried out with numeric exit levels to determine the economic release level (SQC, Intel; Terzakis 2013).
8. Continuous grassroots analysis of engineering work processes must result in a continuous stream of measurable positive improvement (DPP) (Gilb and Graham 1993; Dion et al. 2018).
9. Technical management (CTO, etc.) need to demand and enforce these principles.
10. Universities need to teach these principles and need to organize the knowledge internationally.

Overall, the topics in this book, *Security and Quality in Cyber-Physical Systems Engineering*, are important and should be complemented with a strong vision on integrated systems engineering that builds on systems quality engineering, according to my *Basic Principles of Serious Quality Engineering* (Gilb 2018a), by quantifying requirements for quality, including security; by collecting and using quantitative evidence on design options to select suitable engineering processes, methods, and tools; and by improving quality based on comparing the quantitative evidence from ongoing projects with the planned constraint and target levels of requirements. My emphasis of quantitative evidence is rooted in empirical scientific principles and has shown to be practical and useful in real-world systems engineering contexts (Terzakis 2013). Therefore, readers of this book can benefit from combining these principles with the lessons on requirements of systems engineering in Part I of this book, on quality improvement approaches in Part II, and on security in engineering in Part III of this book, for improving systems engineering in their academic or practical environment.

Kolbotn, Norway Tom Gilb

References

Dion, K. et al. (2018). https://figshare.com/articles/Raytheon_Electronic_Systems_Experience_in_Software_Process_Improvement/6582863. This gives numeric industrial aspects of the use of *Inspection, and defect prevention processes* over 8 years. They include reducing rework from 43% to under 5%.

Gilb, T. (1976). *Design engineering*. This book was organized to store quantified qualities of data elements and structures. A dataware engineering handbook.

Gilb, T. (2005). Competitive engineering. In *A handbook for systems engineering, requirements engineering, and software engineering using Planguage*. Butterworth: Elsevier. Retrieved from https://www.gilb.com/p/competitive-engineering. This book defines a planning language for quality management, and associated quality processes.

Gilb, T. (2008). *How problems with quality function deployment's (QFD's) house of quality (HoQ) can be addressed by applying some concepts of impact estimation* (IE). http://www.gilb.com/DL119

Gilb, T. (2017). *Ten suggested principles for human factors systems engineering*. http://concepts.gilb.com/dl911. Keynote at WUD (Worldwide Usability Day) Silesia, Katowice Poland, 9 December 2017; Details about the 'Aspect Engine' on slide 4.

Gilb, T. (2018a). *Hundred practical planning principles*. Booklet. https://www.gilb.com/store/4vRbzX6X

Gilb, T. (2018b, December 01). *How to plan for the unknown* about 45 slides keynote at WUD Conference Kawice Poland. Includes Ohno on fake news lean. Americans and happiness index Poland 42, and ten principles of dealing with unknown. http://concepts.gilb.com/dl935

Gilb, T., & Graham, D. (1993). *Software inspection*. See two chapters on *Defect prevention process*. See the many case studies for the industrial attributes of software inspection, and of non-software industrial inspection of specifications.

Gilb, T., & Weinberg, G. (1977–1978). *Humanized input*. This book was organized to store quantified qualities of data elements and structures. A dataware engineering handbook.

Quinnan. (1980). *Mills and Quinnan slides*. http://concepts.gilb.com/dl896

Terzakis, J. (2013). *The impact of requirements on software quality across three product generations*. 21st IEEE international requirements engineering conference (RE), Rio de Janeiro (pp. 284–289). https://www.thinkmind.org/download.php?articleid=iccgi_2013_3_10_10012terzakis

Preface

Sitting at the Berlin Tegel airport, waiting for a flight to Vienna, the preparation of the book at hand provides us with some concerns. Only weeks ago, another *Boeing 373 Max8* airplane had crashed. Following the communication of major air traffic safety organizations, a candidate reason for the accidents was a misleading combination of software and hardware in the airplane, leading to unintended airplane behavior that the crew had not been able to compensate. As a traveler you ask yourself: how can such dangerously misleading combinations occur in a safety-conscious environment?

Modern airplanes (as most large technical systems ranging from trains and airplane systems to power plants and factories) are *complex cyber-physical systems*. They become software-intensive technical systems from combining physical system hardware, such as jet engines, wings, and flaps, with control software assisting the pilots. Such complex cyber-physical systems are developed in large engineering organizations by executing complex engineering processes. Within these processes, several engineering artifacts developed in parallel describe together the architecture and behavior of the intended technical system. In the engineering organization and processes, several engineering disciplines provide their special skills to the overall success of the engineering project.

Even if each involved engineer follows a discipline's best practices, still inconsistencies, incompatibilities, unclear communication, or even errors may occur, may reduce the engineering quality and, in the worst case, may result in an operational disaster, such as the recent *Boeing 373 Max8* incident. Usually, an incident is not intended, but there are cases where malicious acts are performed by individuals, who are interested in causing engineering projects or the developed technical systems to fail.

Do we have a chance to protect engineering organizations against cyber threats and to ensure engineering project quality? Answers to these questions will be given in the book at hand. Therefore, the book contains three parts that logically build up on each other. The first part discusses the structure and behavior of engineering organizations for complex cyber-physical systems. This part provides insights into processes and engineering activities executed and highlights requirements and bordering conditions for secure and high-quality engineering. The second part addresses quality improvements with a focus on engineering data generation, exchange, aggregation, and use within an engineering organization and the need of proper data modeling and engineering result validation. Finally, the third part considers security aspects concerning complex cyber-physical systems engineering. Chapters of the last part cover, for example, security assessments of engineering organizations and their engineering data management (including data exchange), security concepts and technologies that may be leveraged to mitigate the manipulation of engineering data, and discussions of design and run-time aspects of *secure complex cyber-physical systems*.

After reaching Vienna with a safe flight in an *Airbus 319* and sitting in the next *City-Airport-Train*, another *complex cyber-physical system*, we are sure that reading this book can reduce the concerns we had in Berlin and can assist engineers and decision makers, researchers, and practitioners in setting up and improving secure and high-quality engineering processes in appropriate engineering organizations.

Magdeburg, Germany Arndt Lüder

Contents

1 Introduction to Security and Quality Improvement in Complex Cyber-Physical Systems Engineering ... 1
Stefan Biffl, Matthias Eckhart, Arndt Lüder, and Edgar Weippl

Part I Engineering of Complex Cyber-Physical Systems

2 Engineering in an International Context: Risks and Challenges ... 33
Ambra Calà, Jan Vollmar, and Thomas Schäffler

3 Managing Complexity Within the Engineering of Product and Production Systems ... 57
Rostami Mehr and Arndt Lüder

4 Engineering of Signaling Systems ... 81
Johannes Lutz, Kristofer Hell, Ralf Westphal, and Mathias Mühlhause

5 On the Need for Data-Based Model-Driven Engineering ... 103
Alexandra Mazak, Sabine Wolny, and Manuel Wimmer

6 On Testing Data-Intensive Software Systems ... 129
Michael Felderer, Barbara Russo, and Florian Auer

Part II Engineering Quality Improvement

7 Product/ion-Aware Analysis of Collaborative Systems Engineering Processes ... 151
Lukas Kathrein, Arndt Lüder, Kristof Meixner, Dietmar Winkler, and Stefan Biffl

8 Engineering Data Logistics for Agile Automation Systems Engineering ... 187
Stefan Biffl, Arndt Lüder, Felix Rinker, Laura Waltersdorfer, and Dietmar Winkler

9 **Efficient and Flexible Test Automation in Production Systems Engineering** 227
Dietmar Winkler, Kristof Meixner, and Petr Novak

10 **Reengineering Variants of MATLAB/Simulink Software Systems** 267
Alexander Schlie, Christoph Seidl, and Ina Schaefer

Part III Engineering Security Improvement

11 **Security Analysis and Improvement of Data Logistics in AutomationML-Based Engineering Networks** 305
Bernhard Brenner and Edgar Weippl

12 **Securing Information Against Manipulation in the Production Systems Engineering Process** 335
Peter Kieseberg and Edgar Weippl

13 **Design and Run-Time Aspects of Secure Cyber-Physical Systems** 357
Apostolos P. Fournaris, Andreas Komninos, Aris S. Lalos, Athanasios P. Kalogeras, Christos Koulamas, and Dimitrios Serpanos

14 **Digital Twins for Cyber-Physical Systems Security: State of the Art and Outlook** 383
Matthias Eckhart and Andreas Ekelhart

15 **Radio Frequency (RF) Security in Industrial Engineering Processes** 413
Martin Fruhmann and Klaus Gebeshuber

16 **Secure and Safe IIoT Systems via Machine and Deep Learning Approaches** 443
Aris S. Lalos, Athanasios P. Kalogeras, Christos Koulamas, Christos Tselios, Christos Alexakos, and Dimitrios Serpanos

17 **Revisiting Practical Byzantine Fault Tolerance Through Blockchain Technologies** 471
Nicholas Stifter, Aljosha Judmayer, and Edgar Weippl

18 **Conclusion and Outlook on Security and Quality of Complex Cyber-Physical Systems Engineering** 497
Stefan Biffl, Matthias Eckhart, Arndt Lüder, and Edgar Weippl

Contributors

Christos Alexakos Industrial Systems Institute, ATHENA Research Center, Patras, Greece

Florian Auer University of Innsbruck, Innsbruck, Austria

Falko Bendik Otto-v.-Guericke University/IAF, Magdeburg, Germany

Stefan Biffl Institute of Information Systems Engineering, Technische Universität Wien, Vienna, Austria

Bernhard Brenner SBA Research, Vienna, Austria

Ambra Calà Siemens AG, Erlangen, Germany

Violeta Damjanovic-Behrendt Salzburg Research, Salzburg, Austria

Matthias Eckhart Christian Doppler Laboratory for Security and Quality Improvement in the Production System Lifecycle (CDL-SQI), Institute of Information Systems Engineering, Technische Universität Wien, Vienna, AustriaSBA Research, Vienna, Austria

Andreas Ekelhart Christian Doppler Laboratory for Security and Quality Improvement in the Production System Lifecycle (CDL-SQI), Institute of Information Systems Engineering, Technische Universität Wien, Vienna, AustriaSBA Research, Vienna, Austria

Michael Felderer University of Innsbruck, Innsbruck, Austria

Apostolos P. Fournaris Industrial Systems Institute, ATHENA Research Center, Patras, Greece

Martin Fruhmann FH JOANNEUM GmbH, Institute of Internet Technologies & Applications, Kapfenberg, Austria

Klaus Gebeshuber FH JOANNEUM GmbH, Institute of Internet Technologies & Applications, Kapfenberg, Austria

Matthias Gusenbauer SBA Research, Vienna, Austria

Kristofer Hell Siemens Mobility GmbH, Braunschweig, Germany

Aljosha Judmayer SBA Research, Vienna, Austria

Athanasios P. Kalogeras Industrial Systems Institute, ATHENA Research Center, Patras, Greece

Lukas Kathrein Christian Doppler Laboratory for Security and Quality Improvement in the Production System Lifecycle (CDL-SQI), Institute of Information Systems Engineering, Technische Universität Wien, Vienna, Austria

Ismail Khalil Johannes Kepler Universität, Linz, Austria

Peter Kieseberg University of Applied Sciences, St. Poelten, Austria

Andreas Komninos Industrial Systems Institute, ATHENA Research Center, Patras, Greece

Christos Koulamas Industrial Systems Institute, ATHENA Research Center, Patras, Greece

Aris S. Lalos Industrial Systems Institute, ATHENA Research Center, Patras, Greece

Arndt Lüder Otto-v.-Guericke University/IAF, Magdeburg, Germany

Johannes Lutz Siemens Mobility GmbH, Braunschweig, Germany

Alexandra Mazak Christian Doppler Laboratory for Model-Integrated Smart Production (CDL-MINT), WIN-SE, JKU Linz, Linz, Austria

Rostami Mehr Volkswagen AG, Wolfsburg, Germany

Kristof Meixner Christian Doppler Laboratory for Security and Quality Improvement in the Production System Lifecycle (CDL-SQI), Institute of Information Systems Engineering, Technische Universität Wien, Vienna, Austria

Mathias Mühlhause Siemens Mobility GmbH, Braunschweig, Germany

Petr Novak Czech Technical University, Prague, Czech Republic

Johanna Pauly Otto-v.-Guericke University/IAF, Magdeburg, Germany

Felix Rinker Christian Doppler Laboratory for Security and Quality Improvement in the Production System Lifecycle (CDL-SQI), Institute of Information Systems Engineering, Technische Universität Wien, Vienna, Austria

Ronald Rosendahl Otto-v.-Guericke University/IAF, Magdeburg, Germany

Barbara Russo Free University of Bozen-Bolzano, Bolzano, Italy

Marta Sabou Technische Universität Wien, Vienna, Austria

Ina Schaefer Technische Universität Braunschweig, Braunschweig, Germany

Thomas Schäffler Siemens AG, Erlangen, Germany

Alexander Schlie Technische Universität Braunschweig, Braunschweig, Germany

Christoph Seidl Technische Universität Braunschweig, Braunschweig, Germany

Dimitrios Serpanos Industrial Systems Institute, ATHENA Research Center, Patras, Greece

Nicholas Stifter Christian Doppler Laboratory for Security and Quality Improvement in the Production System Lifecycle (CDL-SQI), Institute of Information Systems Engineering, Technische Universität Wien, Vienna, AustriaSBA Research, Vienna, Austria

Christos Tselios Citrix Systems Inc, Patras, Greece

Jan Vollmar Siemens AG, Erlangen, Germany

Laura Waltersdorfer Christian Doppler Laboratory for Security and Quality Improvement in the Production System Lifecycle (CDL-SQI), Institute of Information Systems Engineering, Technische Universität Wien, Vienna, Austria

Edgar Weippl Christian Doppler Laboratory for Security and Quality Improvement in the Production System Lifecycle (CDL-SQI), Institute of Information Systems Engineering, Technische Universität Wien, Vienna, AustriaSBA Research, Vienna, Austria

Ralf Westphal Siemens Mobility GmbH, Braunschweig, Germany

Manuel Wimmer Christian Doppler Laboratory for Model-Integrated Smart Production (CDL-MINT), WIN-SE, JKU Linz, Linz, Austria

Dietmar Winkler Christian Doppler Laboratory for Security and Quality Improvement in the Production System Lifecycle (CDL-SQI), Institute of Information Systems Engineering, Technische Universität Wien, Vienna, Austria

Sabine Wolny Christian Doppler Laboratory for Model-Integrated Smart Production (CDL-MINT), WIN-SE, JKU Linz, Linz, Austria

Ilsun You Soonchunhyang University, Asan-si, South Korea

Chapter 1
Introduction to Security and Quality Improvement in Complex Cyber-Physical Systems Engineering

Stefan Biffl, Matthias Eckhart, Arndt Lüder, and Edgar Weippl

Abstract Providing *Complex Cyber-Physical Systems* (C-CPSs) more efficiently and faster is a goal that requires improvements in engineering process for producing high-quality, advanced engineering artifacts. Furthermore, information security must be a top priority when engineering C-CPSs as the engineering artifacts represent assets of high value.

This chapter overviews the engineering process of C-CPSs, typically long-running technical systems, such as industrial manufacturing systems and continuous processing systems. This chapter also covers major areas of requirements that include: (a) processes with intensive generation of engineering artifacts; (b) challenges regarding dependencies and complexity of engineering artifacts, stemming from variants of a product and the associated production process for a family of products; (c) management of model and consistency rules for dependencies between model parts; (d) the internationalization of the engineering process with partners on different levels of trust; and (e) the security of the engineering processes, such as confidentiality of engineering plans, and the security of the systems to be engineered, such as security aspects in the design phase.

For selected requirement areas, the chapter discusses several approaches for quality improvement from business informatics that addresses important classes of requirements, but introduces new complexity to the engineering process. Therefore, the chapter reviews information security improvement approaches for engineering

S. Biffl (✉)
Institute of Information Systems Engineering, Technische Universität Wien, Vienna, Austria
e-mail: stefan.biffl@tuwien.ac.at

M. Eckhart · E. Weippl
Christian Doppler Laboratory for Security and Quality Improvement in the Production System Lifecycle (CDL-SQI), Institute of Information Systems Engineering, Technische Universität Wien, Vienna, Austria

SBA Research, Vienna, Austria
e-mail: matthias.eckhart@tuwien.ac.at; edgar.weippl@tuwien.ac.at

A. Lüder
Otto-v.-Guericke University/IAF, Magdeburg, Germany
e-mail: arndt.lueder@ovgu.de

S. Biffl et al. (eds.), *Security and Quality in Cyber-Physical Systems Engineering*,
https://doi.org/10.1007/978-3-030-25312-7_1

processes, including the consideration of new security requirements stemming from risks introduced by advanced informatics solutions. Finally, the chapter provides an overview on the book parts and the contributions of the chapters to address advanced engineering process requirements.

Keywords Complex cyber-physical systems · Engineering process · Multidisciplinary engineering · AutomationML · Information security

1.1 Motivation

The engineering of *Complex Cyber-Physical Systems* (C-CPSs), typically long-running and software-intensive technical systems, such as critical infrastructures or industrial production systems and their associated products, is a multidisciplinary, model-driven, and data-driven engineering process that often involves conflicting economic, quality, and security interests, risks, and issues. Quality is a key concern in the engineering process to enable engineers to provide technical systems effectively and efficiently, with a focus on sufficient system quality, particularly on safety and on value to customers. Information security has become increasingly important with growing networking capabilities of technical systems and the rise of the Internet, as the engineering environments and the resulting technical systems are part of a network that allows new kinds of attacks on data and systems. Past cyber-attacks against safety-critical systems, for example, a sewage treatment plant (Slay and Miller 2008) or a steel mill (Lee et al. 2014), demonstrated the devastating consequences that could result from inadequate security measures. Besides potential physical damages, cyber-attacks may also cause silent losses because of intellectual property theft. While practitioners agree that addressing security concerns is crucial for establishing a foundation for system safety and quality, many companies hesitate to introduce sufficient security mechanisms and processes in their environments as well as in their products and engineering processes as they lack methods and information for risk assessment and often cannot relate the benefits to the associated extra cost and reduced usability.

Scope of the Book Providing software-intensive technical systems more efficiently and faster requires improvements of engineering process quality and, often, global cooperation in distributed engineering that requires improvements in security considerations for engineering processes. As a novel research approach, we consider the combination of quality improvement and information security for analyzing and improving engineering processes. Quality improvement contributions tend to make engineering processes faster and more efficient by reducing avoidable rework. On the other hand, even if stakeholders deem security fundamental, they may have difficulties in arguing the considerable extra cost of resources in engineering processes. Therefore, a balance of quality improvement and security would be desirable, a balance that overall reduces the resources required for engineering processes and introduces an adequate level of security, which is necessary for sustainable engineering in a globally distributed environment.

Contributions to Scientific Communities *Automation Systems Engineering*. One family of *Software-Intensive Technical Systems* are *Complex Cyber-Physical Systems* (*C-CPSs*). Following the *Encyclopedia of Business Informatics*,[1] a *Cyber-Physical System* (CPS) is defined as a system of communicating components that have both physical and data processing parts. They usually establish a hierarchy of technical system components that are controlled in a closed-loop structure.

This closed loop is established by measuring the state of the technical system using appropriate sensors, deciding on necessary control actions within an information processing system based on measured state and behavior specification, and executing these control actions by actors (VDI 2206 2004; Lunze 2016). Thus, information processing is essential for the behavior quality of the technical system.

Cyber-Physical Systems range from very simple embedded systems, such as drives or rotary encoders, up to very large systems, such as production systems, power plants, energy transmission systems, air planes, and train systems. In this book, such *Complex Cyber-Physical Systems* (*C-CPSs*) are considered, characterized by the nature of such systems and the fact that

- They require significant effort, material, and financial means.
- Their life cycle is measured in decades.
- Their system behavior is complex.
- Their economic and social impact is significant.

Within the engineering processes of a *C-CPS*, information about the overall structure and the mechanical, electrical, etc., construction of the applicable sensors and actors is required in conjunction with the specification of the intended behavior as input. This information must have sufficiently high quality to finally ensure system safety and economic efficiency (Schnieder 1999). In addition, the engineering results of information processing are required within virtual commissioning and commissioning of the *C-CPS* (Strahilov and Hämmerle 2017). Thus, it is obvious that engineering processes of *C-CPS* require information exchange between engineering disciplines, involved engineers of possibly different legal entities, and involved engineering tools.

Contributions in this book consider requirements, risks, and solutions to guarantee the security and quality of *C-CPS*. Involved engineers and project managers will be enabled to identify possible quality and security challenges they have to cope with. In addition, possible measures are described assisting involved staff to handle the identified challenges.

C-CPS Software and System Quality Analysis and Improvement. Typical assumptions for research in quality assurance and improvement for small software-intensive systems that are safe and easy to reset as well as for business software systems that use standard operating systems and hardware and do not rely on specific real-time hardware may not hold for long-running technical systems, such as critical

[1] http://www.enzyklopaedie-der-wirtschaftsinformatik.de/lexikon/informationssysteme/Sektorspezifische-Anwendungssysteme/cyber-physische-systeme

infrastructure or industrial production systems. Therefore, researchers in quality assurance and improvement can benefit from better understanding challenges on quality assurance and quality improvement coming from the engineering of Complex Cyber-Physical Systems based on the use cases and requirements presented in Part I. The use cases and methods for software and system quality analysis and improvement discussed in Part II, such as engineering process analysis, model-based systems engineering, or test automation, provide researchers with insights from in-depth examples that can be adapted to a range of similar but different applications. Finally, the security threats and countermeasures discussed in Part III provide nonexperts in information security with insights into issues to consider when designing quality improvements in engineering contexts. Therefore, researchers and practitioners can take away requirements and building blocks for future methods and tools from the discussion of quality improvement approaches to address selected challenges from automation systems engineering.

Information Security. Researchers in the information security area gain a comprehensive understanding of the security challenges involved in engineering Complex Cyber-Physical Systems. Furthermore, the proposed concepts for securing the engineering process will allow them to evaluate other approaches in order to determine whether they can be applied to overcome these challenges. Since this book also covers security aspects that go beyond protecting the engineering process, researchers gain insights into how the security of Complex Cyber-Physical Systems can be enhanced by integrating security into the engineering phase. Finally, the discussed open challenges may motivate scholars to develop and pursue new research directions.

The remainder of the chapter is structured as follows: Sect. 1.2 introduces Engineering Processes for Software-Intensive Systems and requirements for quality and security improvement. Section 1.3 discusses business informatics approaches for quality improvement that address classes of requirements from engineering software-intensive systems, but may introduce unwanted IT security risks and requirements. Section 1.4 analyzes potential security issues that may be encountered when engineering software-intensive systems. Based on this analysis, the section defines security requirements that must be met; otherwise, these issues may result in compromised engineering processes, which eventually also affect the security of the systems to be developed. Section 1.5 provides an overview on the book parts and the contributions of the chapters to address advanced engineering process requirements. Section 1.6 suggests relevant contributions in this book for selected reader groups.

1.2 Engineering Processes for Software-Intensive Systems

This section introduces an engineering process view on the engineering of long-running software-intensive technical systems, such as industrial manufacturing systems and continuous processing systems, and derives major areas of requirements.

1.2.1 Background

Our daily life is characterized by technical systems that make our life easier, more comfortable, and safer compared with the lives of our ancestors. These systems create, distribute, maintain, and dispose goods, energy, and information relevant for our lives. Some of these systems are designed for a very long life span, for example, nuclear power plants or production systems in the process industry running several decades. Some of these systems have a medium long life span, for example, production systems, wind mills, or ships running about one decade. And some of them have only a very short lifespan, like rockets.

All of these technical systems (fulfill at least three of the four characteristics of a *C-CPS*) have in common the need to be designed, established, and controlled. Therefore, appropriate information processing systems are required to establish *C-CPS*. Especially, the control of the behavior of these systems is a critical problem requiring high-quality, safe, and secure software systems. Therefore, such systems are considered *cyber-physical systems* combining physical parts (establishing the mechanical, electrical, electronical, etc., construction) and cyber parts (establishing the information processing for control of behavior) (Zanero 2017; Monostori 2014; Lee 2008).

The engineering of such *cyber-physical systems* comprises usually the phases requirement collection, architecture design, implementation, test, deployment, and operation. In all phases, the duality of hardware and software needs to be considered (Gruhn et al. 2017). Considering requirement collection, architecture design, implementation, and operation of these systems is of dual nature. On the one hand, the proper system behavior has to be achieved regarding the reason why the system is designed. In case of a nuclear power plant or a wind mill, proper system behavior is the correct current and voltage intended over time, in case of a ship or a rocket proper system behavior is proper transportation conditions enabling proper transport services; and in case of a production system proper system behavior is the proper product creation. In all cases, the intended output of the system drives the correct behavior.

One special type of long-living *C-CPS* are production systems. These systems possess a duality of products to be produced by executing production processes and the production system executing the production processes. Both system kinds need to be engineered and depend on each other. Here, the so-called PPR concept, concerning product design, production process design, and production resource design, provides the background (Biffl et al. 2017a). Their life cycle and their engineering are detailed in Biffl et al. (2017b). Figure 1.1 illustrates the engineering process steps and selected domain expert roles for engineering a *C-CPS*.

Within such a *C-CPS*, the product definition as the first step of the engineering process provides both the behavior specification as well as technical, economic, environmental, legal, etc., bordering conditions. As an example, a rolling mill shall be considered. Starting point of the rolling mill engineering is the product definition of the rolled steel. Thereby, the production process of the intended steel coils is defined and steel properties, such as steel type and mass, are given. This specification

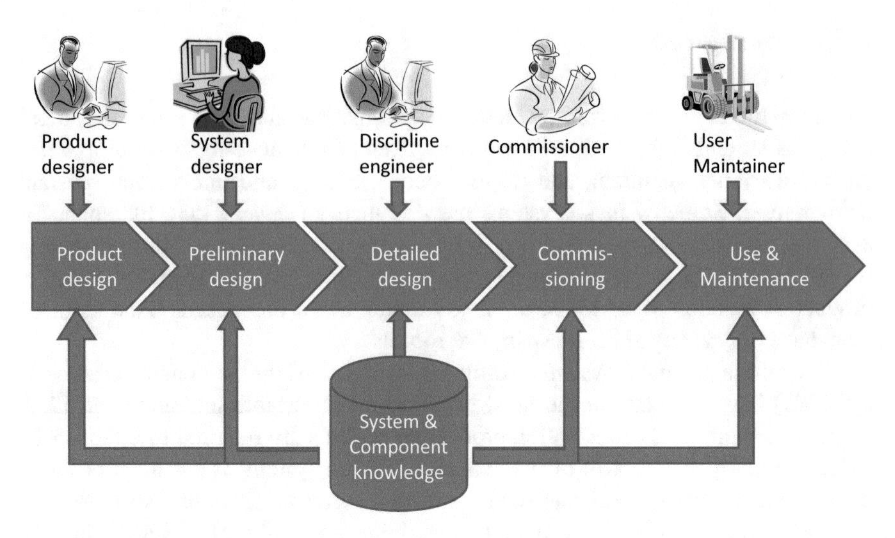

Fig. 1.1 Domain expert roles in the engineering of *C-CPSs*, typically long-running industrial production systems

is from the information point of view similar to the definition of the intended current flow in a power plant or the transport behavior of a rocket.

Usually, the product definition is provided by a product designer as first stakeholder of the engineering process. As indicated by Foehr et al. (2013), the product definition is accompanied by a description of the intended product quality, which is a requirement for the production process execution and, thereby, for the technical and behavioral properties of the production system. For a rolling mill, the product definition is given by the intended material thickness and material properties resulting in a set of necessary milling steps with milling pressure and cooling properties.

The product definition is succeeded by the preliminary design of the production system. Within this second phase, the production system designer as second involved stakeholder develops the overall system structure by assigning manufacturing resources to the different required production process steps of the product. In a steel mill, this overall system structure defines for example the number of mating roles, the length of the transport system (and thereby the number of transport roles), and the number of cooling units.

Exploiting the developed overall system, design groups of discipline-specific engineers covering mechanical, electrical, automation, etc. engineering act as the next stakeholders in the third production system engineering (PSE) process phase. They detail the production system design in the different engineering disciplines in a parallel and round-trip-driven way leading to a system specification in such detail that it can be physically established. For a rolling mill for example, the mechanical engineering defines type and location of drives and roles of the different required types, the electrical engineering defines their wiring related to electrical power transmission and control signal exchange, and the automation system engineering implements the necessary control system parts required to drive the roles dependent on the intended steel quality.

The next phase in the production system engineering process contains its physical realization based on the developed engineering artifacts by several involved installation and ramp-up specialists as last set of stakeholders. As indicated by Lüder et al. (2017a, b), the involved stakeholders exchange engineering information along the complete life cycle of the production system. Therefore, they require standardized information exchange technologies as indicated by Lüder et al. (2017c). The intended increasing digitalization of a wide range of technical systems following approaches like *Industry 4.0* or *Industrial Internet of Things* (IIoT) establishes additional intentions to the engineering. The different components of technical systems shall be accompanied by their digital representation realized as the *management shell* of the *Industry 4.0 component* or the *digital shadow* in the IoT world. Using this digital representation, necessary activities like system maintenance and system optimization can be simplified and improved, leading to safer, more environmental-protective, and more economic systems. Thus, also stakeholders like users, maintainers, etc., come into play for the high-quality and secure engineering process.

Another main development direction in the engineering of production systems is the increasing use of system knowledge. This covers for example the reuse of existing artifacts, the system-wide standardization of components, the application of system family architectures, etc. This trend is mainly addressed by engineering organization improvement approaches as for example in VDI Guideline 3695 (2009). Comparing the engineering and use of other types of *C-CPS* with the engineering and use of production systems, it is easy to see that they usually involve similar life cycle phases and sets of stakeholders (Lindemann 2007).

1.2.2 Research Questions

The intended quality and security improvement of engineering processes of *Complex Cyber-Physical Systems* (*C-CPS*) initially requires an improvement of the understanding of these processes to enable the identification of possible impacts of quality and security improvement approaches that have been developed in information sciences. Thus, the initial research question considered in this book is related to a detailed knowledge about structure, behavior, and further characteristics of engineering processes for *C-CPS*.

RQ1a: What Are Typical Characteristics of Engineering Processes for Long-Running Software-Intensive Technical Systems? As indicated earlier, an engineering process of *C-CPS* establishes a network of engineering activities connected by information exchange. The different involved engineering disciplines step-by-step enrich the overall model system of the intended technical system. Each discipline works with its own models, while all related models need to be kept consistent, that is, there are model-crossing consistency rules to be considered. In addition, models of early engineering phases are applied to generate further models of higher level of detail.

Information science consists of different software-related disciplines. In addition, model-driven software engineering (Brambilla et al. 2017) is well known. But software engineering (in the context of engineering a *C-CPS*) is of limited discipline-related variance. Thus, the applicability of security and quality improvement measures from information sciences for improvement of *C-CPS* is unclear. The applicability is likely to strongly depend on the characteristics of the model creation and use within the established engineering network.

Within engineering networks of *C-CPS*, not only cyclic information exchange following round-trip engineering, merging of engineering information, and handling intended and unintended (not to be disclosed) but also incomplete engineering information occurs. In addition, the involved engineering disciplines might have model elements with similar semantic but different syntactic representations leading to the misinterpretation of exchanged information.

To understand the characteristics of engineering processes for *C-CPS* and to derive requirements related to quality and security improvement in Part I of this book, three engineering processes coming from different industries are described in detail. In one chapter, the topic of interrelation of model-driven *C-CPS* engineering with the models of other life cycle phases of *C-CPS* is discussed in order to derive specific data quality requirements and dependencies as inputs to Part II and Part III of this book.

RQ1b: What Are the Requirement Areas from Engineering Processes for Long-Running Software-Intensive Technical Systems That Require Informatics Contributions? Mostly, engineering processes are considered similar to business processes. Therefore, the accumulated requirements usually focus on economic issues. However, engineering processes of *C-CPS* are based on a detailed engineering data logistics, which has to ensure the availability of engineering information at the right point at the right moment with the right costs and (most important) with the right quality. Quality in this case means as correct as necessary at a point in the engineering process.

Correctness can be impacted in different ways: models may be inadequate, incomplete, or contradicting, etc. This incorrectness can be a result of unintended behavior (like erroneous tool use) but also from intended behavior (like maliciously changed data). *Availability of engineering information* can be considered as a straightforward process management issue. However, in parallel engineering, it might be intentional to provide incomplete data temporarily to shorten the overall project duration. In this case, the manager needs to identify at which point in time and in the process the engineering data providers have to provide information at which degree of completeness.

Both problems can be summarized under the term engineering data quality. Requirement areas for quality improvement include (QI1) processes with iterative generation of engineering artifacts (like *round-trip engineering*); (QI2) challenges regarding dependencies and complexity of engineering artifacts stemming from variants of intended system behaviors; and (QI3) management of model and consistency rules for dependencies between model parts. Requirements areas for IT security include (Sect. 1.1) the internationalization of the engineering process with partners on different levels of trust; and (Sect. 1.2) security, such as the confidentiality of engineering plans and traceable changes to engineering artifacts.

1.3 Requirements for Quality Improvement

For selected requirement areas, this section discusses approaches for quality improvement from business informatics that can address classes of requirements. However, some quality improvement approaches tend to introduce new complexity to the engineering process and security risks that require further research to address.

1.3.1 Background

Companies that engineer *Complex Cyber-Physical Systems* (*C-CPS*), such as industrial production systems based on third-party components, aim at developing systems of sufficient quality in an efficient manner. These systems typically are part of a family of similar but different systems and require for their efficient engineering, capabilities for modeling, refactoring, and reusing system variants, ideally incorporating in new designs the lessons learned from the operation and maintenance of systems in the family.

In this context, several trends in engineering *C-CPS* require capabilities for quality improvement.

System Aspects The Industry 4.0 vision aims at developing flexible *C-CPSs* that require stronger capabilities for engineering more complex systems. These systems may have capabilities for self-adaptation making their validation more difficult. The engineering of families of *C-CPS* requires mastering the dependencies between complex engineering artifacts both between engineering discipline model views and between variants of the system in the family and the associated production process. These dependencies complicate the management of versioning and consistency checking.

Process Aspects The development of *C-CPSs* in short innovation cycles supports the trend of moving from traditional paper-based waterfall process to a modern agile iterative software- and data-based engineering process. To shorten the project duration, parallel engineering requires automating the information exchange following round-trip engineering and advanced change management capabilities for merging of engineering information coming from several disciplines at different levels of maturity and completeness. To make engineering more efficient, the automated engineering requires capabilities for automated checking of model-crossing consistency rules, for generating engineering artifacts, and engineering organization improvement (VDI 3695 2009) for better cooperation of engineering disciplines.

To make engineering cheaper, companies cooperate globally requiring capabilities for the internationalization of the engineering process and associated information security. Unfortunately, the applicability of security and quality improvement measures from information sciences and business informatics remains unclear due to important differences in the industrial context, such as limitations from regulations and technology.

Business Requirements Requirements for quality definition, assurance, and improvement come from stakeholders (Biffl et al. 2017a). System operators and maintainers require *C-CPSs* that enable providing services, for example, for production, effectively and efficiently; systems that are easy to set up, train, maintain, extend, and analyze for service optimization. Technology providers of components and tools for engineering require the *C-CPS* to follow established system architecture standards and influence the solution space in engineering projects considerably. Regulators require the *C-CPS* to follow established validation standards to ensure surpassing minimal quality levels for safety and the environment. Domain experts and system engineers from several discipline-specific workgroups require an engineering process that allows them to work effectively and efficiently in parallel on enriching the engineering plans and configuration as design information becomes more specific and valid along the development process. Project and quality managers require an overview on the discipline contributions to the engineering process to allow assessing whether the engineering data providers have provided their information on a sufficient degree of completeness, accuracy, and validity.

These system and process aspects as well as stakeholder requirements concern quality aspects and capabilities for quality assurance and improvement.

Quality According to IEEE (ISO Square 2014), software and systems quality refers to two aspects: (a) the degree to which a system, component, or process meets specified requirements; and (b) the degree to which a system, component, or process meets customer or user needs or expectations. While the first aspect refers to the verification of software and systems (Kaner et al. 1999), the latter aspect refers to validation of software and systems (Myers and Sandler 1979). Gilb (2005) introduced the *Planguage* to define quality requirements for improving systems engineering in quickly changing environments. Assessing the quality of a *C-CPS*, for example, an industrial production system, concerns system capabilities and technical parameters, and requires capabilities for system validation regarding well-defined quality requirements. Assessing the quality of processes, for example, engineering and production processes, such as process capabilities, effectiveness, efficiency, duration, cost, or risk, requires capabilities for engineering process analysis as well as for representing and processing knowledge on the system, the processes, and their relationships during engineering and operation (Schleipen et al. 2015). Assessing the quality of data, for example, engineering artifacts, data, and knowledge, such as completeness, correctness, adequacy, consistency, accuracy, validity, and understandability for humans and machines (as foundation for Industry 4.0 functions) requires capabilities for advanced engineering data management, such as the data logistics between workgroups coming from different disciplines.

Software Engineering Challenges in a *C-CPS* Engineering Context Software tools and data processing are a core foundation for automating engineering process. Unfortunately, software tools are often designed for addressing the requirements of one engineering discipline and do not know or care about partner disciplines or project aspects. Over the last decades, domain experts with systems engineering background have moved from using software tools toward configuring software tools, and toward writing software scripts and home-grown tools of increasing size, complexity, and importance to running an engineering project.

Unfortunately, many domain experts do not have the software engineering experience to make informed software design decisions, leading to considerable technical debt in data models and in software, for example, scripts that exchange engineering artifacts and data or generate engineering artifacts. The symptoms of technical debt can be found easily when analyzing engineering processes and project experiences in a typical medium-to-large engineering company regarding the modeling and management of dependencies, such as model-crossing consistency rules, between engineering artifacts coming from different disciplines, cyclic information exchange between engineering workgroups, merging of engineering information coming from several data providers, and handling of incomplete engineering information. Therefore, engineering processes of *C-CPS* that depend on a detailed engineering data logistics to ensure the availability of engineering information at the right point at the right moment with the right costs and (most important) with the right quality, tend in practice to be error-prone and costly, and require strong capabilities for quality assurance and improvement.

Quality Assurance This is a way of preventing mistakes and defects in manufactured products and avoiding problems when delivering solutions or services to customers, which ISO 9000 (2015) defines as "part of quality management focused on providing confidence that quality requirements will be fulfilled." Quality assurance (QA) includes in the software and systems engineering lifecycle capabilities for constructive QA, to make sufficient quality more likely to achieve, and for analytic QA, such as system validation and process analysis. *Constructive quality assurance* (Tian 2005; Laporte and April 2018) includes processes to plan and control a software or systems engineering project, that is, by defining individual steps within the system life cycle and methods that support individual steps along the engineering project. In the context of software engineering, examples include *model-driven engineering* (Borky and Bradley 2018; Whittle et al. 2019) and *test-driven engineering* aiming at enabling the construction of high-quality products based on defined processes with integrated quality assurance approaches. *Analytic quality assurance* (Wagner 2007) refers to methods and tools aiming at identifying defects effectively and efficiently along the system life cycle. While static quality assurance approaches, such as reviewing (Zhu 2016), are applicable to non-executable artifacts, such as models or specification documents including (software) code documents, dynamic quality assurance approaches (Kenett et al. 2018), such as software and system tests, require executable code and/or models within an engineering and test environment.

Quality Improvement According to the *Shewhart cycle for learning and improvement* (PDSA; Deming 1986, 1993) or the Japanese *Plan-Do-Check-Act* (PDCA; Sokovic et al. 2010) cycle, quality improvement refers to analyzing the state of the practice, such as quality improvement in the software and systems engineering lifecycles; identifying and implementing improvement actions; evaluating implemented improvements; and adjusting knowledge for further improvements in the context of a continuous improvement strategy (Ning et al. 2010). Often, quality assurance and improvement is organized in a quality assurance ecosystem (Bosch 2009; Axelsson and Skoglund 2016). Quality improvement approaches typically focus on the improvement of processes and organizations in the context of standards, such as ISO 9001 (Hoyle 2017), CMMI (Beth et al. 2011), and SPICE (Abowd et al. 2018), and methods and tools, typically following the PDSA (Deming 1986, 1993) or the PDCA approach (Sokovic et al. 2010).

While these approaches for quality assurance and improvement have been used successfully in business software engineering contexts, their application to the engineering processes of *C-CPS* requires considerable adaptations that require empirical validation.

1.3.2 Research Questions

Business informatics has successfully improved and automated business processes with software in diverse areas, such as business administration, government, and health care. Business informatics approaches, such as agile software development, test-driven development, and DevOps, have the potential to significantly improve the capabilities of software engineering teams to deliver software solutions faster, more efficiently, more flexibly, and with higher quality. Therefore, the adaptation of successful business informatics approaches to the engineering of *Complex Cyber-Physical Systems* (*C-CPS*) could be promising to address requirements of stakeholders that look similar in both areas.

However, the engineering of *C-CPS* has characteristics that differ in important ways from typical business software engineering contexts, such as real-time hardware capabilities, the cooperation of several engineering disciplines with widely varying models that depend on each other, and requirements for safety that limit the direct application of business informatics approaches. Therefore, we look in Part II of this book at the following overarching research questions.

RQ2: How Can Proven Business Informatics Approaches Be Adapted for Improving the Engineering of Complex Cyber-Physical Systems? From this research question, we derive the following research questions that consider both (a) the potential of better quality improvement capabilities for engineering *C-CPS*, such as more effective dependency management between engineering disciplines, and (b) security implications that may come from interventions of quality improvement approaches, such as centralizing and integrating engineering knowledge.

RQ2a: How Can Approaches Adapted from Business Informatics Address Quality Improvement Requirements Coming from Characteristics of C-CPS Engineering? This research question concerns first the identification of promising business informatics approaches, such as agile software development, business process analysis, knowledge representation, model-based engineering, test-driven development, and variability management for addressing selected trends in the engineering of *C-CPS*. For the selected business informatics approaches, the book chapters will discuss how these approaches are likely to address selected requirements coming from the engineering of *C-CPS*, in particular, different characteristics of engineering business software solutions and *C-CPS* that require the adaptation of business informatics approaches. Finally, the book chapters in Part II will identify interventions to the engineering of *C-CPS*, such as changes to the engineering process, models, methods, or mechanisms to facilitate quality improvement and addressing important requirements of *C-CPS* stakeholders.

RQ2b: Which Interventions of Quality Improvement Approaches Are Likely to Introduce or Increase Information Security Requirements and Risks? Following the goal of the book, toward combining approaches from quality improvement and information security improvement, we will analyze the interventions to the engineering of *C-CPS* identified when answering RQ2a, such as centrally accessible data repositories and the collaboration with partially trusted and untrusted parties, regarding their impact on information security requirements, capabilities, and risks. For example, an open question is how a secure round-trip engineering process can be implemented and applied in automation systems development—a major management goal of PSE is the reduction of PSE resource needs, while securing any process tends to add complexity and resource requirements. As a result, we will introduce a model on how selected factors from the domains of *C-CPS* engineering, quality improvement, and security improvement influence each other as foundation for analyzing where the combination of approaches from quality improvement and information security improvement is most relevant to enable the comparison of options that consider both quality and security solution aspects. Finally, we will discuss options for combining security with quality improvement approaches, such as the early consideration of typical security requirements and contributions in adapted quality assurance and improvement methods, for example, when applying the VDI guideline 3695 for engineering organization improvement or for representing security as a system quality in variability handling approaches.

1.4 Security Considerations for the Engineering of Complex Cyber-Physical Systems

This section discusses security concerns that arise in the context of engineering *Complex Cyber-Physical Systems (C-CPSs)*. Based on this, we derive requirements and, in further consequence, research questions that aim to address these security concerns.

1.4.1 Background

Similar to the part that discusses security, this section covers security issues related to engineering C-CPSs from two different points of view. First, we provide background information for the chapters on securing the engineering process and then give context to the chapters that deal with the aspects to be considered for developing more secure C-CPSs.

1.4.1.1 Securing the Engineering Process

As introduced in Sect. 1.2.1, the engineering of C-CPSs represents a multidisciplinary process that involves the exchange of engineering-related information among stakeholders. As part of each engineering step, artifacts are created, managed, and edited by engineers who make use of domain-specific tools. These engineering artifacts may then also serve as an input for subsequent engineering steps. As a result, a data logistics solution that is seamlessly integrated into the engineering workflow, coupled with the use of a common data exchange format (e.g., AutomationML), may improve the engineering efficiency and, in further consequence, increase the overall quality of the engineering process. However, considering that the exchanged data not only includes valuable know-how about the system to be developed but also poses a severe security and safety threat if altered with malicious intent, protecting the engineering data logistics is paramount. For instance, adversaries may attempt to obtain the exchanged engineering data for the purpose of industrial espionage or even tamper with blueprints to implant flaws into the system's design, allowing them to exploit these vulnerabilities during operation (Weippl and Kieseberg 2017, 2018). Thus, the data logistics system is an attractive target for cyber-attacks, especially if it is similar to a data repository that provides a central point of access to *all* engineering artifacts. This issue is exacerbated by the fact that engineering projects may be executed by globally distributed teams, some of which may even belong to external companies (i.e., subcontractors) (Weippl and Kieseberg 2017). Consequently, engineering data must be protected against external threat actors (i.e., individuals who are not involved in engineering projects and launch attacks via the

Internet or Intranet) and insider threats (i.e., employees who sabotage engineering activities or leak data).

Challenges and Requirements There are several research challenges that must be overcome to work toward a secure engineering process while keeping engineering efficiency and quality high.

First, a lack of understanding about potential *attack vectors and the threats to assets of the data logistics solution* may result in inaccurate risk assessments, which would affect the treatment of risks. In other words, if security risks are not thoroughly analyzed, the data logistics system may not be adequately protected against cyber threats, or the implemented countermeasures may undermine its value for improving the engineering workflow. To give a concrete example, current tools within the engineering toolchain lack a fine-grained access control mechanism (Weippl and Kieseberg 2017), meaning that engineering artifacts may be disclosed to subjects that do not even require access to these resources. On the other hand, if standardized access control solutions are introduced without considering the peculiarities of engineering projects, a decrease in productivity is inevitable. Thus, research is required to study how existing security approaches can be tailored to fit the needs of engineering projects while paying attention to achieving a fair balance between meeting engineering quality and security requirements. Still, research in this area would benefit from insights gained from prior works that deal with threat modeling and risk assessment related to the exchange of engineering information.

Second, novel security methods that guarantee the *confidentiality*, *integrity*, and *availability* of engineering data at rest as well as at transit are required. The primary challenge in this regard is to provide seamless integration of these security methods into engineering toolchains of different kinds. According to Hundt and Lüder (2012), there are three main types of engineering toolchains, viz., "One Tool for All," "Best of Breed," and "Integration Framework." As the authors point out, these approaches vary in terms of the achieved level of integration into engineering activities, which in turn drives the complexity of information models and flows. Thus, security methods that attempt to accomplish the aforementioned security goals in engineering processes must be able to cope with the complexity that each tool integration approach entails. Considering that the tools used within the engineering process are often proprietary, equipping existing closed-source engineering tools with security features represents an additional issue. In this context, developing secure interfaces to engineering tools that lack security features may be a promising approach to tackle this challenge. Yet, it is also worth mentioning that establishing a secure engineering workflow is not only a technical but also an organizational task. Identifying the security gaps in the underlying business processes of the engineering lifecycle lays the foundation for setting up organizational security policies in order to comply with governance and legal requirements.

Third, ensuring the *traceability* of processed and shared engineering data represents an essential requirement for data logistics solutions. In particular, information about the origins of data, who modified what, when, and how is vital for creating an audit trail. The recorded events can then be monitored to detect suspicious activity from engineers, enabling security analysts to resolve potential issues in a timely manner. Additionally, reviewing the audit trail for further investigations of security incidents may assist them in taking reactive security measures. In this way, secure collaboration between engineers may be fostered. Besides protecting valuable engineering artifacts from security threats, meeting this requirement may also prevent data inconsistencies and thereby provide round-trip engineering support. However, thus far, it remains unclear how such traceability features can be implemented in data logistics systems.

Fourth, countering industrial espionage in the context of engineering C-CPSs deserves more attention from both academia and industry. Intellectual property theft represents a severe threat for vendors and systems integrators, especially if collaborating with third parties, such as local suppliers, is required (Weippl and Kieseberg 2017). Thus, novel techniques for *protecting know-how* are needed to prevent adversaries from stealing knowledge or at least being able to hold them legally accountable for their wrongdoing. According to Kieseberg and Weippl (2018), there are two concepts that may be adopted to address this challenge, viz., obfuscation (i.e., modifying data in a way that it is no longer useful for adversaries) and watermarking (i.e., hiding additional information for the purpose of detecting data leaks and tracing misuse). Although methods from both categories have been extensively researched (e.g., code obfuscation, privacy, digital watermarking), it is still unknown how these methods can be applied to engineering artifacts.

1.4.1.2 Considering Security Aspects Along the Engineering Phase

The realization of the Industry 4.0 vision (Kagermann et al. 2013) demands extensive connectivity capabilities of technical systems. In fact, ubiquitous connectivity as an enabler for efficiency improvements and cost reductions can be considered as one of the driving forces behind the *Information Technology* (IT) and *Operational Technology* (OT) convergence (Hahn 2016). However, with the ever-increasing need to interconnect OT systems, which are clearly related to C-CPSs, the number of points of attacks grows likewise, causing the risk of cyber-attacks to rise. This issue is aggravated by the fact that inherently insecure OT systems become exposed to cyber threats. The reason for the increased susceptibility of OT systems is that stakeholders of engineering projects tend to give more weight to availability than other security goals, such as confidentiality and integrity (Ullrich et al. 2016). However, with the IT/OT convergence, which signifies the end of air-gapped OT systems, information security becomes a pressing issue. Moreover, considering that the lifecycle of technical systems is approx. 15–30 years (Macaulay and Singer 2016), legacy systems represent a long-lasting risk factor. Legacy systems often lack fundamental security features and may also not be updatable because of (1) the lack of patches, (2) missing testing capabilities, (3) inefficient patch deployment techniques, and (4) the necessity of undergoing recertification upon patching (Weippl and Kieseberg 2017, 2018). However, even if patching would be a viable option, flaws in the design of systems may render security updates ineffective. According to a recent report published by Dragos, Inc. (2018), this seems to be the case for 64% of the vulnerability patches for OT components that have been released in 2017. To resolve this problem, engineers must consider security aspects already early on in the systems' lifecycle, that is, the engineering phase. While in the software industry this is a long-established practice, as for example, McGraw (2004, 2006) and Microsoft (Howard and Lipner 2006) proposed their concepts for the development of secure software more than a decade ago, the industrial informatics community took the first steps in this direction just a couple of years ago (Schmittner et al. 2015).

Challenges and Requirements The combined safety and security development lifecycle proposed by Schmittner et al. (2015) represents a valuable contribution that may be used as a starting point. Yet, given its generic nature, it may be too coarse to assist stakeholders of C-CPSs engineering projects in their day-to-day work. In particular, engineers may benefit from a variant of the lifecycle discussed by Schmittner et al. (2015) that is tailored to the subprocesses typically involved in the engineering of C-CPSs (e.g., plant design, electrical planning). By rigorously following such a *security development lifecycle for complex cyber-physical systems*, information security may be established as an integral part of the engineering process and thereby foster a security-aware culture in each engineering discipline. In this way, the industrial systems' architecture in its entirety would be designed with security in mind, covering the (1) hardware, (2) firmware, (3) software, (4) network, and (5) process layer (McLaughlin et al. 2016). However, the key challenge in this context is to identify the specific security activities that either fit within the realm of each engineering subprocess or that do not entail a radical restructuring of the engineering workflow. Furthermore, engineers would benefit from *tool-supported* security activities or practices that augment the traditional C-CPSs engineering process. Thus, the development of tools that aim to improve the efficiency and effectiveness of the security development lifecycle for complex cyber-physical systems poses a further research challenge. One emerging research area that may prove to be beneficial in overcoming this challenge is the concept of digital twins. For instance, Eckhart and Ekelhart (2018a, b) introduced security-related use cases of digital twins, such as penetration testing. Although these use cases may be applicable in the engineering phase, Eckhart and Ekelhart (2018a, b) barely touch on the specific applications within the engineering process, leaving great room for exploring this subject in more detail.

Furthermore, given that the connectivity of C-CPSs significantly increases in light of Industry 4.0, *securing communications technologies* deserves more focus in the engineering process. As a matter of fact, wireless connectivity seems to be one of the building blocks of the Industrial Internet of Things (IIoT), meaning that its wide adoption can be expected in the years to come. Yet, from a security perspective, this may increase the risk of a cyber-attack, as adversaries no longer need to work their way forward to the (wired) control network, nor do they require (on-site) physical access; instead, they can launch attacks against control devices if they are within reach of wireless networks (Stouffer et al. 2015). In response to the increased level of risk that systems with wireless network capabilities may pose, the NIST SP 800-82 guideline (Stouffer et al. 2015) recommends that they should only be deployed in industrial settings that have a low impact. Still, given their susceptibility to cyber-attacks,[2] research in the areas of *wireless network security* and *IIoT security*, in general, with an emphasis on engineering and design aspects, is required.

On a final note, it must be ensured that the C-CPSs to be engineered operate not only in a secure but also *safe* manner. As Ullrich et al. (2016) correctly point out, this can be achieved by considering security and safety aspects jointly in the engineering process, such as defined in the lifecycle proposed by Schmittner et al. (2015).

1.4.2 Research Questions

As indicated in Sect. 1.4.1, improving the security of C-CPSs requires securing the engineering process, on the one hand, and considering security aspects when engineering these systems, on the other hand. In this way, cyber threats targeting the engineering process (e.g., with the objective to implant vulnerabilities into the systems' design) can be mitigated while ensuring that these systems have been engineered with security in mind. Since Part III of this book aims to help organizations in achieving these security improvements by addressing some of the challenges discussed in the previous subsection, we have defined the following two research questions:

RQ3a: Which Security Concepts Mitigate Cyber Threats Targeting the Engineering Process of Complex Cyber-Physical Systems? This research question deals with the security challenges and requirements presented in the first part of Sect. 1.4.1. In essence, the book chapters that attempt to answer this research question will provide a guide to hardening a data logistics solution and discuss novel methods to reduce the risk of tampering with engineering artifacts and information disclosure. Answering this research question requires the introduction of a risk-based approach to exchanging engineering information by means of centrally accessible data repositories. Furthermore, an investigation of data integrity threats and information theft is needed to fend off targeted attacks against data repositories that aim to sabotage systems' engineering or steal know-how of high value. This will allow readers to gain an in-depth understanding of the threats pertaining to engineering data logistics and how they can be mitigated with the security concepts presented.

RQ3b: How Can the Security of Complex Cyber-Physical Systems Be Enhanced by Considering Security Aspects During the Engineering Phase? This research question focuses on the security challenges and requirements discussed in the second part of Sect. 1.4.1. In particular, selected security improvement approaches, which aim at supporting engineers to develop C-CPSs with security in mind, will be

[2]See, e.g., Radmand et al. (2010) for a taxonomy of attacks against wireless sensor networks that are used in industrial environments.

discussed. To answer this research question, the book chapters will examine (1) novel concepts that may support security activities within the lifecycle of C-CPSs, (2) measures to protect the connectivity layer of these systems, and (3) recent approaches to designing them in a way that they are able to tolerate arbitrary faults, which is especially important if they are deployed in safety-critical environments. In this way, readers will be equipped with novel concepts that can be applied during the engineering phase in order to develop more secure C-CPSs.

1.5 Book Structure

This book discusses challenges and solutions for quality and security improvements for design processes of long-running technical systems, such as design process integrity, that differ from well-known processes, methods, and mechanisms from business software engineering. The authors consider how a combination of quality and security viewpoints can help to improve engineering processes of long-running technical systems in a way that allows practitioners to introduce sufficient security considerations. An example for such a combination would be a data logistics system that improves the effectiveness and efficiency of work groups in industrial engineering, such as mechanical, electrical, and software engineering. The introduction of such a data logistics system raises new information security risks as the integrated engineering models provide more information for a hacker on the planned product that would be interesting intellectual property to leak to competitors and would provide intelligence knowledge to plan systematic attacks on the resulting product in operation.

Therefore, the book is structured in three parts to cover security and quality improvements in the engineering of long-running technical systems. Application examples for long-running technical systems will come mainly from the area of industrial production systems such as automotive, steel works, power plant, or oil production (including the vision of Industry 4.0), but the considerations will be applicable to the engineering of a wide range of traditional and future technical systems.

1.5.1 *Part I: Product Engineering of Complex Cyber-Physical Systems*

Part I on *Engineering Networks* as application context discusses challenges and approaches for the *C-CPS* engineering processes including key stakeholders, structured engineering methods, engineering data quality, reuse of engineering know-how, and data security in engineering. During the *C-CPS* lifecycle and the installation of the system, information processing concerns the creation, change, exchange, and use of engineering data and artifacts in order to characterize, design, configure, and verify the future *C-CPS* and its parts. *C-CPSs*, such as production systems as manufacturing cells or steel mills, power generation and distribution systems like wind mills and HVAC stations, and transportation systems like trains control powerful and risky physical processes and must meet domain-specific safety, environmental, and quality standards. Meeting these standards is challenging for traditional technical systems and even more challenging for Complex Cyber-Physical Systems according to the Industry 4.0 vision. Thus Part I collects challenges and requirements regarding information management and exchange to ensure sufficient quality and security of engineering data.

Chapters with their contributions to the overall book in Part I are the following.

Chapter 2: Engineering in an International Context—Risks and Challenges This chapter discusses risks and challenges based on the increased globalization of engineering of *C-CPS*. This chapter discusses possible organizational structures and reachable performance resulting from increasing globalization and digitalization on systems engineering. It describes best practice for internationalization strategies and collects requirement to information management resulting from these strategies.

Chapter 3: Managing Complexity Within the Engineering of Product and Production Systems This chapter displays the complexity problem emerging from the required parallelization of engineering of products (i.e., cars) and production systems (i.e., car production) in automotive industry. Starting from a detailed presentation of the engineering process the relevance of engineering data within the complexity management is considered and challenges to information management to be solved to improve process management, change management (required to cope with late product changes within the production system engineering), and knowledge management are named.

Chapter 4: Engineering of Signaling Systems This chapter discusses the engineering within a strongly regulated field of rail-based public transport systems that is an example of very safety critical systems. Thus, their engineering is very quality and security sensitive. The chapter considers the general structure of the engineering of rail systems focusing on the engineering of signaling systems within them. Based on this process discussion the chapter gives challenges on the engineering tools (and tool chains), engineering processes, and engineering data.

Chapter 5: On the Need for Data-Driven Model-Based Engineering This chapter motivates requirements for linking system design-time models to run-time models as foundation for improving the design-time models based on the analysis of data from operation. The chapter considers the evolutionary aspects of the engineering data creation process and derives challenges to be tackled when integrating run-time and design-time models and processes. As a theoretical solution a Temporal Model Framework is presented. This framework is applied in three illustrative use cases that discuss technology/methodology candidates to be used within a technical architecture for connecting design-time models and run-time models to improve design-time models.

Chapter 6: On Testing Data-Intensive Software Systems This chapter motivates with illustrating use cases the need for novel testing approaches beyond purely functional behavior, in particular, as data foundations increasingly become a significant part of the system under test, beyond input and context parameters. The chapter discusses the role of data in testing systems and requirements for appropriate testing approaches for data-intensive software. The chapter provides a state-of-the-art survey on testing of data-intensive software. The chapter extends the testing dimensions of software systems based on the observation that data quality and systems of systems thinking become more relevant in data-intensive software systems.

In total, the collected challenges in the chapters of Part I can be categorized as follows:

- Challenges related to the increasing digitalization of CPS in general and *C-CPS* in special including challenges of data/model integration
- Challenges related to the required guarantee of high-quality, consistent, and fault-free engineering information within and crossing involved engineering disciplines and lifecycle phases
- Challenges related to information management within engineering networks like version and responsibility management
- Challenges related to knowledge use and protection like variant management and run-time data application
- Challenges related to the confidentiality of engineering information
- Challenges related to the required business processes and their secure execution

1.5.2 Part II: Engineering Quality Improvement

Part II on *Engineering Quality Improvement* discusses processes, methods, models, and mechanisms for quality assurance, analysis, and improvement in the engineering of long-running technical systems with consideration of supporting sufficient IT security. Quality improvement aspects in production system families include more efficient testing, more efficient reuse, efficient refactoring, and efficient data/model exchange between work groups in multidisciplinary engineering. This part will point out target qualities that IT security should defend and methods and mechanisms that introduce new IT security risks.

First, Chaps. 7 and 8 discuss the engineering process improvement in general. Chapter 7 provides a method for product/ion-aware analysis of production system engineering (PSE) processes to reduce risks coming from the insufficient representation of knowledge on the product to be produced and on the planned production process during PSE and operation. Chapter 8 introduces a method for efficient Engineering Data Exchange in PSE to enable the frequent synchronization of parallel workgroups in production system engineering (PSE) to reduce the risk of diverging local data views and avoidable rework. Second, Chap. 9 considers how to improve testing of software-intensive systems. Chapter 9 focuses on the reuse of test artifacts in similar but different test environments. Finally, Chap. 10 discusses a method for comprehensive analysis of product variants in a family of production systems for collapsing redundant parts and identifying reusable parts in order to reduce the size of the system variants and to facilitate better quality assurance.

Therefore, readers gain insights into a variety of approaches to improve engineering quality for addressing requirements raised in Part I of this book for making better informed decisions in the design and improvement of PSE processes. Further, Part II brings up security threats as input to the security methods addressed in Part III of this book.

Chapter 7: Product/ion-Aware Analysis of Multidisciplinary Systems Engineering Processes This chapter discusses use cases that show how insufficient explicit representation of knowledge on product and production process characteristics can raise risks in later PSE stages and operation. Reporting on a case study, the chapter discusses strengths and limitations of traditional business and engineering process analysis methods. Based on these findings, it introduces a *product/ion-aware engineering process analysis* (PPR EPA) method and the notation for a *product/ion-aware data processing map* (PPR DPM) that enables advanced analyses of PPR knowledge needs and gaps. The chapter considers requirements for storing PPR knowledge as foundation for PPR knowledge retrieval during the PSE process. Finally, the chapter discusses security issues that PPR EPA can identify as input to a subsequent security analysis.

Chapter 8: Engineering Data Logistics for Agile Automation Systems Engineering This chapter motivates requirements of efficient *Engineering Data Exchange* (EDEx) as foundation for the frequent synchronization of parallel workgroups in *production system engineering* (PSE) to reduce the risk of diverging local data views, effort for rework, and unclear project progress assessment. The chapter introduces the illustrating use case *round-trip engineering* (RTE) and derives requirements for (a) a process for negotiating data elements requested by data consumers and matching to data elements coming from data providers; and (b) a data exchange method and mechanism for executing the agreed data exchanges between domain experts and their data sources and sinks. The chapter introduces the *Engineering Data Exchange* (EDEx) process facilitated by an *EDEx information system* (EDExIS). As there is no suitable out-of-the-box technology to link discipline-specific views on data, the chapter introduces a software architecture based on AutomationML data models that address these challenges. It reports on a case study to evaluate the data negotiation process and the data exchange software architecture with representative use cases from real-world engineering environments. Finally, the chapter discusses security issues that an EDExIS can help address and identifies new security threats as input to a subsequent security analysis.

Chapter 9: Efficient and Flexible Test Automation in Production Systems Engineering This chapter motivates the need for test automation in *production system engineering* (PSE) and discusses strengths and limitations of traditional testing approaches in PSE with illustrating use cases. The chapter focuses on the reuse of testing artifacts in similar but different PSE environments. Based on the *Behavior Driven Development* (BDD) approach, the chapter introduces a *BDD-to-test-code weaving* approach that allows representing BDD concepts in a language-independent model for the efficient generation of test source code in different test frameworks. The chapter demonstrates the feasibility of the key capabilities of the *BDD-to-test-code weaving* approach based on a prototype and discusses practical application cases. The chapter discusses security issues that reusing test automation constructs can help address. Further, the chapter identifies new security threats as input to a subsequent security analysis.

Chapter 10: Reengineering Variants of Matlab/Simulink Software Systems This chapter motivates the need for a comprehensive analysis of product variants in a family of production systems in order to identify the systems variability and their relations. It provides illustrative use cases that demonstrate the impact of variability on system quality and risk. The chapter introduces a technique to capture course-grained variability as foundation for the identification of similar and redundant parts. By collapsing redundant parts and identifying reusable parts, the approach allows reducing the size of the system variants and facilitates better quality assurance. Based on knowledge about the relations between system variants from the portfolio analysis, affected software systems can be identified across variant boundaries, mitigating security concerns for the entire product portfolio. Finally, the chapter discusses security issues that the variant analysis of MATLAB/Simulink system portfolios can help address and identifies new security threats as input to a subsequent security analysis.

1.5.3 Part III: Engineering Security Improvement

Part III on *Engineering Security Improvement* discusses novel methods to enhance the security of C-CPSs. More specifically, the methods proposed in this part will not only deal with the protection of the engineering process itself but also address security aspects that ought to be considered when designing these systems. Thus, the objective of this part of the book is twofold. First, Chap. 11 provides a security assessment of a data repository used to exchange engineering artifacts, followed by Chap. 12, which examines security concepts and technologies that may be leveraged to mitigate industrial espionage and the manipulation of engineering data. Second, Chap. 13 discusses the design and run-time aspects of secure C-CPSs, and thereby lays the foundation for subsequent chapters that touch on the security-by-design principle for C-CPSs. After that, Chap. 14 explains how the concept of digital twins can be used to improve the security of C-CPSs throughout their lifecycle, starting from the engineering to the operation phase. Furthermore, Chaps. 15 and 16 address security concerns that emerge due to the increasing connectivity of these systems. Finally, Chap. 17 provides valuable insights into Byzantine fault tolerance (BFT) and how safety-critical distributed systems can be designed to achieve this property.

In this way, readers gain insights into security threats and adequate countermeasures concerning C-CPSs, allowing them to integrate profound scientific knowledge in their decision-making process during engineering activities.

Chapter 11: Security Analysis and Improvement of Data Logistics in AutomationML-Based Engineering Networks This chapter investigates the security flaws of the state of practice regarding the exchange of engineering data. Furthermore, it provides a systematic approach to designing secure central data repositories, specifically tailored to engineering environments. To ensure that threats pertaining to these central data repositories are adequately considered, the chapter demonstrates how threat modeling can be applied in the context of engineering data exchange. Based on this, the chapter proposes practical countermeasures that mitigate the identified threats while maintaining the efficiency of engineering workflows. Finally, the chapter provides an overview of counterexamples and mistakes to avoid, which further supports the risk treatment process.

Chapter 12: Securing Information Against Manipulation in the Production Systems Engineering Process This chapter discusses methods that aim to ensure the integrity of engineering data, and to identify leakages to unauthorized parties, which may also provide the means for holding them legally accountable for their malicious actions. The security concepts proposed in this chapter supplement Chap. 11, as protecting the data exchange does not prevent insiders (e.g., employees) or third parties that are involved in the engineering process (e.g., contractors) from manipulating or stealing engineering artifacts. In particular, the chapter covers concepts that provide proof of ownership, proof of correctness, and the ability to track leaked engineering data. In this way, readers develop a thorough understanding of methods that can be used to protect engineering knowledge against manipulation or theft.

Chapter 13: Design and Run-Time Aspects of Secure Cyber-Physical Systems This chapter provides an introduction to the principle of security by design for CPSs and associated run-time aspects. Based on a discussion of the CPSs threat landscape, the chapter reviews methods for detecting and preventing attacks, which may be worth implementing when engineering these systems. The chapter then delves into the monitoring of run-time behavior of CPSs for security purposes. In particular, the authors discuss key aspects and requirements for implementing run-time security monitoring in CPSs based on existing work in this field. Owing to the use case presented, readers will be able to understand the concrete security challenges and issues that emerge in an industrial context. Finally, the chapter discusses the implications of the findings and suggests directions for further research in this area.

Chapter 14: Digital Twins for Cyber-Physical Systems Security—State of the Art and Outlook This chapter explores security-specific use cases of the concept of digital twins and provides suggestions on how they may be put into practice. The chapter aims to demonstrate that digital twins can holistically improve the security of CPSs by applying security measures in various phases of their lifecycle. In particular, the chapter shows that the simulation aspects of digital twins can be used to conduct security tests in the design, testing, and commissioning phases, while the replication of states to digital twins, so that they follow the states of their physical counterparts, may reveal intrusions. The chapter completes with an extensive analysis of challenges and open questions that are worth studying in future work.

Chapter 15: Radio Frequency Security in Industrial Engineering Processes This chapter addresses the need for securing wireless communication technologies of C-CPSs. First it outlines radio frequency technologies that are used in an industrial setting and then points out security shortcomings. Based on this analysis, practical countermeasures against the identified attack vectors are proposed. Due to the fact that security flaws that have its origins in the design are costly to fix, the chapter also shows how approaches to security testing can be applied in the engineering phase of wireless systems. As strategic initiatives, such as Industry 4.0, motivate the adoption of wireless technologies, an in-depth analysis of the security of radio frequency communication represents a valuable contribution to the book at hand.

Chapter 16: Secure and Safe IIoT Systems via Machine and Deep Learning Approaches This chapter reviews security and safety challenges for *Internet of Things* (IoT) applications in industrial environments. Similar to Chap. 15, this work addresses both the security and safety concerns emerging from connectivity capabilities, which have to be considered jointly during the engineering phase, as both influence each other. On the one hand, security concerns arise from the expanding attack surface of C-CPSs due to the increasing connectivity on all levels of the industrial automation pyramid. On the other hand, safety concerns magnify the consequences of traditional security attacks. Based on the thorough analysis of potential security and safety issues of *Industrial IoT* (IIoT) systems, the chapter surveys *machine and deep learning* (ML/DL) methods that can be applied to counter the security and safety threats that emerge in this context. In particular, the chapter explores how ML/DL methods can be leveraged in the engineering phase for designing more secure and safe IoT-enabled C-CPSs. However, the peculiarities of IoT environments (e.g., resource-constrained devices with limited memory, energy, and computational capabilities) still represent a barrier to the adoption of these methods. Thus, this chapter also discusses the limitations of ML/DL methods for IoT security and how they might be overcome in future work by pursuing the suggested research directions.

Chapter 17: Revisiting Practical Byzantine Fault Tolerance Through the Lens of Blockchains This chapter discusses how blockchain technologies have rekindled an interest in the topic of Byzantine fault tolerance (BFT) that reaches well beyond the hype behind cryptocurrencies. This topic is highly relevant for the engineering of safety-critical systems, especially if they constitute distributed systems. Being able to tolerate not just crash or omission failures, but potentially arbitrary and malicious (i.e., Byzantine) behavior of a subset of a system's components can greatly improve its resilience and security. The vast improvement in computing power, networking, and hardware costs over the last 20 years has greatly diminished the impact of the overhead that is incurred for achieving Byzantine fault tolerance (BFT) and it should no longer be perceived as impractical or infeasible for real-world applications. It is essential that safety-critical systems are engineered in a way that allows them to tolerate Byzantine failures, and Blockchain technologies introduce many new paradigms and techniques that can be seen as complementary to classical BFT approaches. This chapter examines how blockchain technologies relate to classical Byzantine fault tolerance and outlines which aspects need to be considered when making design decisions.

1.6 Who Shall Read This Book?

This book is intended for several target groups. Computer science researchers will be enabled to identify research issues related to the development of new methods, architectures, and technologies for quality and security improvements in multidisciplinary engineering, pushing forward the current state of the art. Researchers on the engineering of Complex Cyber-Physical Systems will get a better understanding of the challenges and requirements of multidisciplinary engineering that will guide them in future research and development activities on quality and security improvements. Engineers and managers with engineering background will be able to get a better understanding of the benefits and limitations of applicable methods, architectures, and technologies for selected use cases.

Acknowledgments The financial support by the Christian Doppler Research Association, the Austrian Federal Ministry for Digital and Economic Affairs, and the National Foundation for Research, Technology, and Development is gratefully acknowledged.

References

Abowd, P., Hörmann, K., Vanamali, B., Wall, D., & Schnetzer, S. (2018). *Automotive spice essentials: Automotive spice v3.1 – at a glance*. Kugler Maag.

Axelsson, J., & Skoglund, M. (2016). Quality assurance in software ecosystems: A systematic literature mapping and research agenda. *JSS, 114*, 69–81.

Beth, M., Chrissis, B., & Konrad, M. (2011). *CMMI for development: Guidelines for process integration and product improvement*. Boston, MA: Addison Wesley.

Biffl, S., Gerhard, D., & Lüder, A. (2017a). Introduction to the multi-disciplinary engineering for cyber-physical production systems. In *Multi-disciplinary engineering for cyber-physical production systems* (pp. 1–24). Cham: Springer.

Biffl, S., Lüder, A., & Gerhard, D. (Eds.). (2017b). *Multi-disciplinary engineering for cyber-physical production systems – Data models and software solutions for handling complex engineering projects*. Cham: Springer.

Borky, J. M., & Bradley, T. H. (2018). *Effective model based systems engineering*. Cham: Springer.

Bosch, J. (2009, August). From software product lines to software ecosystems. In *Proceedings of the 13th international software product line conference* (pp. 111–119). Pittsburgh: Carnegie Mellon University.

Brambilla, M., Cabot, J., & Wimmer, M. (2017). Model-driven software engineering in practice. *Synthesis Lectures on Software Engineering, 3*(1), 1–207.
Deming, W. E. (1986). *Out of the crisis*. Cambridge, MA: MIT Press.
Deming, W. E. (1993). *The new economics*. Cambridge, MA: MIT Press.
Dragos, Inc. (2018). *Industrial control vulnerabilities: 2017 in review* (Technical report). Hanover, MD: Dragos. https://dragos.com/media/2017-Review-Industrial-Control-Vulnerabilities.pdf.
Eckhart, M., & Ekelhart, A. (2018a, May). Towards security-aware virtual environments for digital twins. In *Proceedings of the 4th ACM workshop on cyber-physical system security* (pp. 61–72). ACM.
Eckhart, M., & Ekelhart, A. (2018b). Securing cyber-physical systems through digital twins. *ERCIM NEWS, 115*, 22–23.
Foehr, M., Jäger, T., Turrin, C., Petrali, P., & Pagani, A. (2013). Methodology for consideration of product quality within factory automation engineering. In *2013 IEEE international conference on industrial technology (ICIT)* (pp. 1333–1338). Cape Town.
Gilb, T. (2005). *Competitive engineering: A handbook for systems engineering, requirements engineering, and software engineering using Planguage*. Amsterdam: Elsevier.
Gruhn, V., Gries, S., Hesenius, M., Ollesch, J., Ur Rehmann, S., Schwenzfeier, N., Wahl, C., & Wessling, F. (2017). Engineering cyber-physical systems, within H. Fujita, A. Selamat, S. Omatu, new trends in intelligent software – Methodologies, tools, and techniques. In *Proceedings of 16th SoMeT*. Amsterdam: IOS Press.
Hahn, A. (2016). Operational technology and information technology in industrial control systems. In *Cyber-security of SCADA and other industrial control systems* (pp. 51–68). Cham: Springer.
Howard, M., & Lipner, S. (2006). *The security development lifecycle* (Vol. 8). Redmond: Microsoft Press.
Hoyle, D. (2017). ISO 9000 quality systems handbook-updated for the ISO 9001:2015 standard. In *Increasing the quality of an organization's outputs*. Abingdon: Taylor & Francis.
Hundt, L., & Lüder, A. (2012, September). Development of a method for the implementation of interoperable tool chains applying mechatronical thinking—use case engineering of logic control. In *Emerging technologies & factory automation (ETFA), 2012 IEEE 17th conference* (pp. 1–8). IEEE.
ISO 9000:2015. (2015). *Quality management systems – Fundamentals and vocabulary*.
ISO/IEC 25000:2014. (2014). *Systems and software engineering – Systems and software quality requirements and evaluation (SQuaRE) – Guide to SQuaRE*.
Kagermann, H., Helbig, J., Hellinger, A., &Wahlster, W. (2013). *Recommendations for implementing the strategic initiative INDUSTRIE 4.0: Securing the future of German manufacturing industry*. Final report of the Industrie 4.0 working group, Forschungsunion.
Kaner, C., Falk, J., & Nguyen, H. Q. (1999). *Testing computer software*. Hoboken, NJ: Wiley.
Kenett, R. S., Ruggeri, F., & Faltin, F. W. (2018). *Analytic methods in systems and software testing*. Hoboken, NJ: Wiley.
Kieseberg, P., & Weippl, E. (2018). Security challenges in cyber-physical production systems. In *International conference on software quality* (pp. 3–16). Cham: Springer.
Laporte, C. Y., & April, A. (2018). *Software quality assurance*. Hoboken, NJ: Wiley.
Lee, E. A. (2008). Cyber physical systems: Design challenges. In *11th IEEE symposium on object oriented real-time distributed computing (ISORC)* (pp. 363–369). IEEE.
Lee, R. M., Assante, M. J., & Conway, T. (2014). German steel mill cyber attack. *Industrial Control Systems, 30*, 62.
Lindemann, U. (2007). *Methodische Entwicklung technischer Produkte*. Berlin: Springer.
Lüder, A., Schmidt, N., Hell, K., Röpke, H., & Zawisza, J. (2017a). Fundamentals of artifact reuse in CPPS. In *Multi-disciplinary engineering for cyber-physical production systems: Data models and software solutions for handling complex engineering projects* (pp. S113–S138). Cham: Springer.
Lüder, A., Schmidt, N., Hell, K., Röpke, H., & Zawisza, J. (2017b). Identification of artifacts in life cycle phases of CPPS. In *Multi-disciplinary engineering for cyber-physical production systems: Data models and software solutions for handling complex engineering projects* (pp. S139–S167). Cham: Springer.

Lüder, A., Schmidt, N., Hell, K., Röpke, H., & Zawisza, J. (2017c). Description means for information artifacts throughout the life cycle of CPPS. In *Multi-disciplinary engineering for cyber-physical production systems: Data models and software solutions for handling complex engineering projects* (pp. S169–S183). Cham: Springer.

Lunze, J. (2016). *Automatisierungstechnik – Methoden für die Überwachung und Steuerung kontinuierlicher und ereignisdiskreter Systeme*. De Gruyter Studium.

Macaulay, T., & Singer, B. L. (2016). *Cybersecurity for industrial control systems: SCADA, DCS, PLC, HMI, and SIS*. Abingdon: Auerbach.

McGraw, G. (2004). Software security. *IEEE Security & Privacy, 2*(2), 80–83.

McGraw, G. (2006). *Software security: Building security in* (Vol. 1). Boston, MA: Addison-Wesley.

McLaughlin, S., Konstantinou, C., Wang, X., Davi, L., Sadeghi, A. R., Maniatakos, M., & Karri, R. (2016). The cybersecurity landscape in industrial control systems. *Proceedings of the IEEE, 104*(5), 1039–1057.

Monostori, L. (2014). Cyber-physical production systems: Roots, expectations and R&D challenges. In *Proceedings of the 47th CIRP conference on manufacturing; systems, procedia CIRP* (Vol. 17, pp. 9–13).

Myers, G. J., & Sandler, C. (1979). *The art of software testing*. Hoboken, NJ: Wiley.

Ning, J., Chen, Z., & Liu, G. (2010, August). PDCA process application in the continuous improvement of software quality. In *Computer, mechatronics, control and electronic engineering (CMCE), 2010 international conference* (Vol. 1, pp. 61–65). IEEE.

Radmand, P., Talevski, A., Petersen, S., & Carlsen, S. (2010). Taxonomy of wireless sensor network cyber security attacks in the oil and gas industries. In *Advanced information networking and applications (AINA), 2010 24th IEEE international conference* (pp. 949–957). IEEE.

Schleipen, M., Lüder, A., Sauer, O., Flatt, H., & Jasperneite, J. (2015). Requirements and concept for plug-and-work. *Automatisierungstechnik, 63*(10), 801–820.

Schmittner, C., Ma, Z., & Schoitsch, E. (2015). Combined safety and security development lifecycle. In *Industrial informatics (INDIN), 2015 IEEE 13th international conference* (pp. 1408–1415). IEEE.

Schnieder, E. (1999). *Methoden der Automatisierung*. Braunschweig: Vieweg.

Slay, J., & Miller, M. (2008). Lessons learned from the Maroochywater breach. In *International conference on critical infrastructure protection* (pp. 73–82). Boston, MA: Springer.

Sokovic, M., Pavletic, D., & Pipan, K. K. (2010). Quality improvement methodologies–PDCA cycle, RADAR matrix, DMAIC and DFSS. *Journal of Achievements in Materials and Manufacturing Engineering, 43*(1), 476–483.

Stouffer, K., Pillitteri, V., Lightman, S., Abrams, M., & Hahn, A. (2015). Guide to Industrial Control Systems (ICS) security. *NIST Special Publication, 800*(82). http://dx.doi.org/10.6028/NIST.SP.800-82r2.

Strahilov, A., & Hämmerle, H. (2017). Engineering workflow and software tool chains of automated production systems. In S. Biffl, A. Lüder, & D. Gerhard (Eds.), *Multi-disciplinary engineering for cyber-physical production systems – Data models and software solutions for handling complex engineering projects* (pp. 207–234). Cham: Springer.

Tian, J. (2005). *Software quality engineering: Testing, quality assurance, and quantifiable improvement*. Hoboken, NJ: Wiley.

Ullrich, J., Voyiatzis, A. G., & Weippl, E. R. (2016). Secure cyber-physical production systems: Solid steps towards realization. In *Cyber-physical production systems (CPPS), 2016 1st international workshop* (pp. 1–4). IEEE.

VDI Richtlinie 3695. (2009). *Engineering von Anlagen – Evaluieren und optimieren des Engineerings*. Berlin: Beuth.

VereinDeutscherIngenieure. (2004). *VDI-Richtlinie 2206 – Entwicklungsmethodik für mechatronische Systeme*. Düsseldorf: Beuth.

Wagner, S. (2007). *Cost-optimisation of analytical software quality assurance*. Munich: Technical University Munich.

Weippl, E., & Kieseberg, P. (2017). Security in cyber-physical production systems: A roadmap to improving IT-security in the production system lifecycle. In *AEIT international annual conference* (pp. 1–6). IEEE.

Whittle, J., Hutchinson, J., & Rouncefield, M. (2019). *Model-driven development – A practical approach*. Abingdon: Routledge.

Zanero, S. (2017, April). Cyber-physical systems. *Computer, 50*(4), 14–16.

Zhu, Y.-M. (2016). *Software reading techniques: Twenty techniques for more effective software review and inspection*. New York: Apress.

Part I
Engineering of Complex Cyber-Physical Systems

Chapter 2
Engineering in an International Context: Risks and Challenges

Ambra Calà, Jan Vollmar, and Thomas Schäffler

Abstract The increasing globalization has had a positive impact on multinational companies. Their customers are often doing their business globally and require the availability of technical services around the world. If on the one hand companies can have access to new markets, on the other hand globalization gives companies access to labor at cheaper prices. Outsourcing and offshoring of business activities to countries such as India or China represent significant potential savings of labor costs. Therefore, multinational companies should consider also some changes in the way of working since coordinating the tasks of teams located thousands of kilometers away or adapting to each country's regulations and procedures represent big challenges for the companies' engineering.

Today, engineering and manufacturing projects are usually executed across two or more geographically dispersed units or departments, research centers, or companies. This requires a shift of the whole or partial engineering value chain to other countries, a process typically referred to as "internationalization." Based on the experience of a multinational company, active in various technical domains, the authors aim at identifying the challenges of internationalization in project business for the engineering of software-intensive technical systems.

This chapter intends to provide a definition of internationalization in engineering, pointing out its main characteristics. Challenges and impacts that industries face in the organization and performance of internationalization in a multinational company are discussed. Moreover, engineering best practices gathered so far to manage internationalization in engineering are presented and further envisioned research steps are shown.

Keywords Internationalization · Engineering · Project business · Readiness check · Co-operation

A. Calà (✉) · J. Vollmar · T. Schäffler
Siemens CT, Erlangen, Germany
e-mail: ambra.cala@siemens.com

S. Biffl et al. (eds.), *Security and Quality in Cyber-Physical Systems Engineering*,
https://doi.org/10.1007/978-3-030-25312-7_2

2.1 Internationalization

Globalization is a key trend in the business world today (Khikhadze 2019). Over the past decades, globalization has been accelerated by the development and diffusion of new communication, information, and work-sharing technologies (Myers and Smith 1999), which make national boundaries less important in terms of political, cultural, technological, financial, environmental, and national security issues (Wust 2011).

Some aspects of globalization, like the removal of trade barriers among trading countries, have brought both advantages and disadvantages to industries and companies, particularly in developed countries. On the one hand, customers from all over the world can be reached more easily and resources can be sourced from a global supplier market. On the other hand, customers need local presence of suppliers, such as Engineering, Procurement, and Constructions (EPCs), for technical services and support. Moreover, even more companies are expanding their business worldwide, striving to increase their global competitiveness.

To achieve global competitiveness, new ways of doing business have been generated, for example, outsourcing, offshoring, and in-sourcing (Friedman 2005). The business strategy of relocating internal services of a company to a company internal partner or affiliated company across national or international borders is here called regionalization (Schaeffler et al. 2014) or internationalization.

The reasons for internationalization of production projects are numerous. Sometimes, the imposition of restrictions on imports by the foreign countries forces the establishment of manufacturing facilities in other countries. Foreign countries may have higher availability of high-quality materials and low-cost inputs. Moreover, the relocation of production activities lowers the cost of transportation and the complexity of logistic management resulting in a reduction of the overall costs and project timelines (Nekpuri 2011).

The idea of lowering costs is the initial impetus for companies to outsource, for example, businesses with overhead costs can have the excess cost cut down in countries that have relatively deflated currencies as well as low cost of living (Apex 2007).

Moreover, by fragmenting technical operations internationally, the work load can be shared by employees in several countries who work together on a project on the basis of their individual background, knowledge, and experience, transmitting ideas for new products and new ways of making goods around the world (Schaeffler et al. 2014). The location of operations and services within a geographic area has a major impact on inventory levels in terms of speed of production and work within the area.

Besides cost-related opportunities, there are various motivational factors for the relocation of technical services. Dunning (1993) summarizes the benefits of internationalization for a company in the following three aspects: resource advantages; gaining new customers; and improving efficiency. Fletcher (2001) proposes a different classification of these factors and reviews them into four categories: management characteristics; organizational characteristics; external impediments; and external incentives. Other classifications can be found in Sachse (2002) and Deloitte (2005).

To summarize, the opportunities listed below are taken from the general relocation literature but can be applied to engineering.

Cost Savings As stated above, lower wages and costs in the destination countries are attractive for the management. But costs can be also reduced considering the specialization in particular functions of the different company affiliations. Therefore, fixed costs can be converted into variable costs by eliminating the need to reserve capacity at any time and retrieving it when needed. The more the demand for a service fluctuates, the greater are the savings (Pisani and Ricart 2008).

Performance Improvements By using dislocated services, a company can react more quickly to current market development. The easy access to qualified personnel can also avoid bottlenecks by requesting resources from the headquarters (HQ) or other affiliated companies. Additional capacities and processing of tasks by specialists can lead to a faster result, reducing the time to provide a service. Moreover, the specialization of functions in different areas enhances the focus on core competencies with strategic importance and makes the structure of the company leaner (Azuayi 2016).

Country-Related Opportunities In some cases, the creation on site is caused by the task itself, for example, if the creation of a sales order can be physically created only at the project site. A practical example is the construction industry, in which it is necessary to make use of engineering services on site (Apex 2007). Besides the availability of better quality resources and materials, tax advantages may also arise in the provision of services abroad.

The trend of nextshoring has caused manufacturers and their local partners within the supply chain to adapt and prepare for the changing nature of manufacturing, leveraging the impact they are making on the area where they are located.

This chapter addresses the problem of adapting a project business process that has to be executed by a company across different countries. Main attention is paid on the relocation of engineering activities, highlighting the risks and challenges for international technical companies. The following section describes what characterizes internationalization in project business in terms of enablers for the engineering of software-intensive technical systems. The third section uses an example of the plant industry to represent the typical challenges in internationalization of engineering. The fourth and fifth sections provide some practical suggestions to deal with the issues described in the previous section. The chapter concludes with a summary and outlook.

2.2 Characteristics of Internationalization in Engineering Projects

The international trade of technological services results also in an increasing demand for the processing of engineering activities on a global scale. For companies active in technical domains, engineering is one of the most important disciplines. In project business (Artto and Wilkström 2005), to realize this requirement, engineering activities must be relocated from the original place of performance to abroad to ensure performance and/or cost benefits. For example, the engineering of components may take place where the headquarters is located, while manufacturing and testing occur where the factory plant is built (Schaeffler et al. 2014).

In general, relocation projects are very complex and usually take a period of several months to years to be completed. Especially in plant engineering, there is little experience with the relocation of engineering activities, so the risk of project failure is high.

Engineering work in plant construction companies has always been considered a core activity that had to remain tightly controlled within the home country location of a company and conducted within the boundaries of the company. However, in case of large international projects, companies can take advantage of transferring some engineering activities to affiliated companies in other countries to cope with changing market demands such as faster time-to-market for products, lower design engineering costs, and higher product quality. This results in a shift of the engineering value chain to other countries and, therefore, requires some specific project execution measures to maximize the likelihood of success of such a relocation project.

Figure 2.1 represents the engineering reference process. An engineering process is defined in the work by Artto and Wilkström (2005) as "a sequence of activities of creative application of scientific principles to design or develop structures, machines, apparatus or manufacturing processes with respect to their intended function and economic and safe operations," An engineering process in project business mainly consists of four phases: concept engineering, basic engineering, detail engineering, and installation and commissioning (Engineering Council 1941).

The concept engineering concerns the definition of a technically feasible design. The basic engineering results in a technically verifiable design. The detail engineering results in a design ready for implementation. Finally, installation and commissioning concern site engineering with the definition of the final operational design. In project business, engineering plays a fundamental and complex role since it designs a solution out of predefined products and systems (Schaeffler et al. 2013).

Fig. 2.1 Engineering reference process

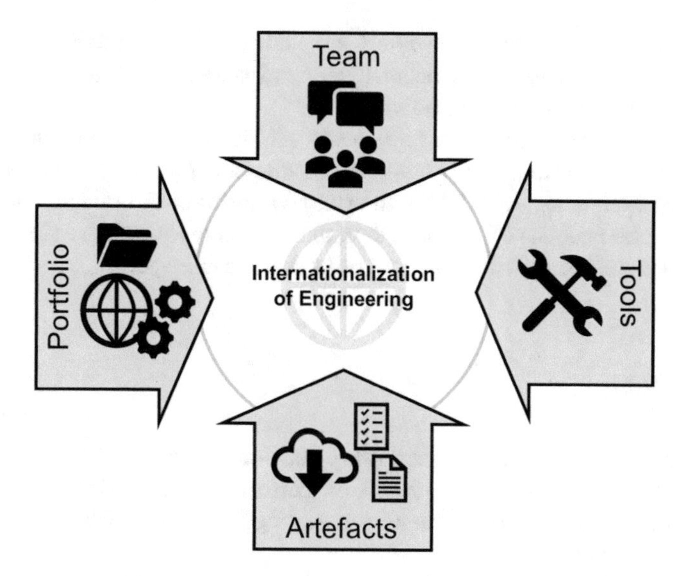

Fig. 2.2 Enablers for the internationalization of engineering

Of course, engineering processes can be performed differently among companies depending on the considered business model. More generally, the phases described in the engineering reference process describe the main engineering activities required in project business and are here described only as a reference.

To enable the execution of these engineering activities in international project business a set of enablers should be considered, ranging from the availability of adequate engineering expertise and required information to the use of appropriate engineering tools. The enablers depicted in Fig. 2.2 are common for all international plant construction projects in engineering-intensive businesses.

2.2.1 Solution Portfolio

In principle it is possible to outsource individual functional areas of engineering or entire engineering processes for the solution considered, ranging from single engineering activity (e.g., finite element method calculation) up to engineering phases (e.g., conceptional design of a system). In any case, it is necessary to determine which areas are suitable for relocation. Therefore, a solution portfolio has to be defined. This portfolio collects the elements of the company that can be internationalized in terms of products, systems, and modules, defining also the related engineering activities.

Portfolio elements include engineering solutions, product-related services and personnel, namely, technologies, processes, and competencies. Specifications and characteristics of the portfolio elements must be clearly described in terms of core components, required or available interfaces, and complexity in their structure. The better an activity is specified, documented, and standardized, the easier it is to outsource complex engineering activities.

Usually, it is recommended to relocate activities that are not internal core competencies of the company. In contrast, activities and processes that are uncritical of business operations are good for an external relocation. A building construction work would be necessarily performed on the destination country, while basic and detail civil engineering activities could be performed elsewhere.

2.2.2 Team

The internationalization of engineering projects is characterized by the engineering team responsible for the activities to be performed across different countries. Typically, the size of an international engineering team ranges from few engineers up to several hundred engineers.

Employees should be selected based on the required knowledge and skills their intended role in the engineering project demands, allocated, recruited or delegated, inserted into the local organization and trained in order to become valuable resources. They should be able in the end to execute the engineering activities, which create information reflecting the taken design decision for a specific solution.

To ensure the success of the international project, the working team should be composed by engineers fulfilling different roles in terms of skills, competences and physical abilities to perform a set of certain engineering activities.

Managers of global technology teams need to understand how cross-cultural differences influence technology development. Cross-boundary skills are the most needed since the communication across disciplinary, organizational, and cultural boundaries is the core of engineering in international projects.

Typically, technologies and tools differ from country to country, as well as the engineers' working approach to different domains. Collaborative working hence involves not only the experts and engineering team but also the project manager and department managers that define the team structure, roles, and responsibilities and coordinate the team during the project.

2.2.3 Artifacts

Engineers require the exchange of information of various types and data formats among different stakeholders involved in the international project. Data are usually related to the design specifications of the technical systems, at different granularity of details, but also to the description of the engineered system and the documentation of the activities conducted during the project phases.

The exchange of information is characterized by the different sources and targets of the information itself, according to the engineering phase considered, and by the communication sequence resulting from the engineering steps.

Also, engineers should provide a clear documentation of all the engineering activities in an understandable way, taking into account that people from different countries and languages will have access to it. Guidelines are strongly recommended but should be agreed upon at global and local levels. In addition, the exchange of know-how across several locations of the project must be protected. To this end, adequate measures to ensure the intellectual property of engineering activities related to the different location units are also necessary.

2.2.4 Tools

Within an internationalization project where different engineering activities are dislocated among different teams, only the use of appropriate tools in a defined tool chain can ensure the correct exchange of information. Usually, highly specialized software tools are used to support different engineering activities, ranging from computer-aided design (CAD) tools, programmable logic controller (PLC) programming tools, product lifecycle management (PLM) tools to vendor-specific commissioning tools. Each of these engineering tools creates or consumes input/output artifacts with its own semantic and specific data format (Steinmann et al. 2014).

Clearly defined tool input and output within the entire tool chain will avoid information loss over borders. Different teams may use different tools depending on their engineering activity and location in which they operate. To enable a smooth interplay across organizational units and international borders, a properly integrated tool landscape for the different engineering activities is ensured by adequate tool interfaces. The technology transfer is enhanced by the harmonization of the tools and the integration of the toolchain within engineering processes, avoiding the use of different tools by different teams in the same project that block flexibility in project execution.

2.3 Challenges of Internationalization of Engineering in Industry

The challenges presented in this chapter have been identified based on analysis and interviews with stakeholders from various projects and involved organizational units. The more generic challenges will be explained with examples from a specific use case. This use case is based on an organizational unit that is active in project business in process industry. The main scope of supply consists of electrification and

Fig. 2.3 Severed industries of use case

automation equipment and related services. The organization is active in different industries represented in Fig. 2.3 (e.g., marine, fiber, minerals). Customers are global players as well as regional companies. The engineering organization is organized in global engineering hubs and regional engineering centers.

Projects are always executed in a global setup, staffed with engineers from different locations. The projects vary in terms of scope and volume. In some projects (e.g., modernization) only small parts of the system are within the scope of supply, whereas in green-field projects the complete system is in the scope of supply.

One essential prerequisite to successfully deliver international engineering projects is an adequate organization of collaboration, which ranges across organizational setup, engineering disciplines, and system architecture levels.

The challenges for collaboration can be clustered in different aspects:

- Workflow management
- Resource management
- Information exchange and engineering change management
- System architecture

2.3.1 Workflow Management

The systems in the use case are often complex systems with a high degree of uncertainty (e.g., varying scope of supply, new sub-supplier, unclear customer requirements). Therefore, constant planning and rescheduling, as well as controlling, is needed.

However, the projects face various challenges in planning and rescheduling. The planning of workflows and related execution steps (tasks) is often complicated and inflexible and should consider non-transparent interdependencies between tasks. In an environment that is characterized by a high degree of uncertainty (e.g., changes and errors), this results in a high amount of effort for regular updates and rescheduling. As the management of workflows is implemented with a very low degree of automation (mostly manual handover and resulting waiting time), the challenges are increased even more.

When considering resource management, it is very challenging to define needed skills/competences in global setup without clearly defined skill/competence profiles for different engineering tasks. The planning tools often lack the support of the multi-project environment (e.g., in terms of hours and periods for various project at the same time).

Also, the project controlling can be challenging as the evaluation of the actual status of the workflow progress needs a high amount of effort. It is often not possible to predict or to identify bottlenecks in workflow execution.

2.3.2 Resource Management

When the workflow has been defined and the resource need is identified, it has to be matched with actual resources considering the needed skills, needed qualification, and expected experience level. If these demands cannot be matched, alternative scenarios must be considered. Additional external resources can be acquired, internal resources can be transferred from other project (multi-project management) or additional workflow steps must be added (e.g., additional reviews) or design lead time adapted (e.g., lower experience level). There is a constant adaptation of resource needs based on changes in project schedule (e.g., offer is placed later than planned), workflow and design (refer to engineering change management).

In the use case, the organization has regional engineering hubs that use one or more regional Engineering Centers, which provide engineering services for certain parts of the engineering value chain. Collaboration between the regional engineering hub and engineering center is managed directly by the regional hub, as depicted in Fig. 2.4. The engineering hubs are located in different countries or regions to address engineering for customer projects in these countries or regions for specific domains (e.g., in Norway for oil and gas). Engineering centers are usually located in low-wage countries to achieve cost benefits.

Managing competences of engineers is a crucial activity to enable a smooth execution of the global projects. To ensure that engineers with the right skills are assigned to the projects, a set of competences are defined. There are over 600 competences, clustered in engineering discipline specific groups (e.g., for automation, electrical, mechanical), cross-discipline groups (e.g., project management, quality management) and tooling related competencies (e.g., COMOS, 3D CAD). Additionally, different skill levels are defined (basic, advanced, expert). In this way, it is possible to create skill profile for all engineers.

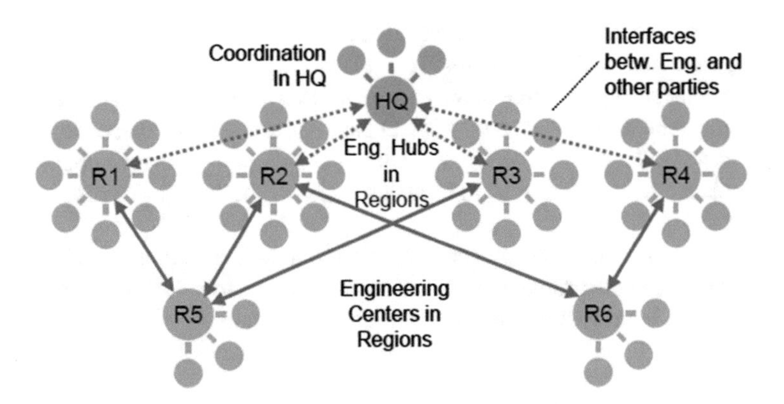

Fig. 2.4 Setup of engineering hubs and engineering centers

2.3.3 *Information Exchange and Engineering Change Management*

The information exchange logistic should ensure lossless information transmission and use. The right information should be transferred in the adequate format to the right human resource at the right time to interpret them and ensure interoperability (Hundt et al. 2010). The VDI (association of German engineers) states that seamless engineering is characterized by the facts, that a result of an engineering activity can be used for another step in the value chain, ensuring consistency of information with less additional effort and redundant work as possible (Foehr et al. 2013).

When considering engineering resources located across international borders, the collaboration process and respective workflows represent a critical aspect in terms of exchanging engineering relevant information among the different involved stakeholders. This exchange is characterized by stakeholders' needs and capabilities. Moreover, different engineering activities are related to different engineering disciplines that are supported by highly specialized software tools. As in the use case most customers demand the use of specific tools for each engineering discipline, (for example, NX, Inventor for mechanical engineering or PCS7, PowerCC, STEP7 for automation engineering), it is difficult to limit the number of tools for the engineering company. A high number of additional tools, for example, for parameterization and configuration of devices or for document management and application for specialized tasks (e.g., net studies) increase the complexity of tools. Thereby, all of them have a specialized way to describe the engineering information with its own syntactical and semantic domain, resulting sometimes in different notations for the same objects (Steinmann et al. 2014).

Another challenge is the adequate handling of engineering-related changes. As with complex projects the degree of uncertainty is quite high (especially in early phases), changes are inevitable. An adequate support to analyze the impact of changes is often missing (e.g., impact on schedule, which other systems or components need to be changed). In case of changes, normally previous engineering steps or phases (e.g., from detail engineering to basic engineering) must be repeated. This kind of round-trip engineering is only weakly supported by the tool landscape,which is mostly designed as a "one way" tool chain. In the use case the number of changes per day ranges from 2 to 4 for highly standardized portfolio elements up to 100 for more complex portfolio elements.

When working in a global setup and exchanging data with various customers, suppliers and authority data security is another big challenge. These partners expect confidentiality of project information (e.g., encrypted data transmission, access control), integrity of the delivered solutions (e.g., prevention of software manipulation, prevent manipulation of data) and transparency of changes (e.g., audit trails).

2.3.4 System Architecture

Also, the system architecture must support engineering collaboration. State-of-the-art systems engineering demand for a system of system approach (INCOSE 2014). But often an overall system architecture is not defined or there are different engineering discipline specific architectures in place that must be integrated in every project (Vollmar et al. 2017).

As in the use case, usually only a part of the overall solution is in the scope of supply (e.g., only automation and electrification) so the system architecture and implementation concepts that are defined during conceptual engineering cannot be influenced by the engineering organization. In this case it is difficult to use a predefined architecture and implementation concept.

Reuse concepts are often based on copy and paste and not on modularized reference solution with reusable modules, which results in low degree of modularization and standardization, ranging from normally 20% up to 80% for few highly standardized products portfolios elements.

From the organizational point of view, internationalization will challenge the staffing, style, and formal and informal information systems of the organization.

Human interactions as well as IT systems and applications will need to enable the dynamic allocation of workflow and support quality assurance.

2.4 Internationalization Process

This section describes a process showing how such internationalization efforts can be effectively executed and managed. The process has been obtained by literature study in combination with expert interviews and it has been validated by applying it to real internationalization projects in a multinational company.

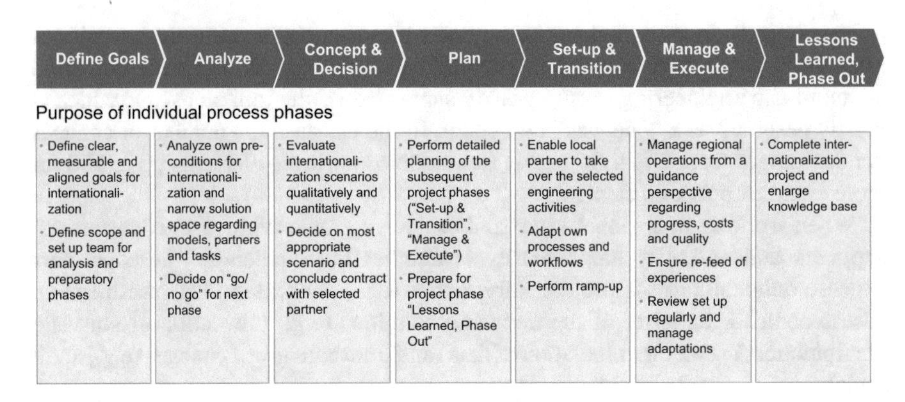

Fig. 2.5 Internationalization process

The process for internationalization of engineering is depicted in Fig. 2.5 and comprises seven process steps. Its purpose is an effective preparation, planning, and execution of internationalization of engineering.

The process starts with the "Define Goals" phase in which clear, measurable, and aligned goals for internationalization are defined. The purpose of internationalization might be to extend the business to emerging countries, or to use the advantage of lower local labor costs. In this phase, the interests of the various stakeholders need to be balanced, taking into account cultural aspects within the different countries involved, in order to guarantee the success of the internationalization. Also, countries identified for regionalization should be chosen alongside the business responsibility assigned to them. Moreover, the persons involved should define duties and schedules required for the internationalization project.

This phase results in the definition of the scope (and non-scope) of internationalization, thus possible elements of the solution portfolio, and sets up the team for the analysis and preparatory phases that follow. The goals are defined taking into account the initial situation of the engineering entities involved, concerned disciplines and activities, possible international and local partners, and the impact on the headquarter organization.

The second phase is the "Analyze" phase, in which the preconditions for internationalization are analyzed based on the organization's solution portfolio in order to narrow the solution space and to identify models, partners, and tasks accordingly. Especially, the models for internationalization include the setup or extension of an engineering unit in an already existing one, the setup of a newly incorporated subsidiary, the acquisition of a local company, or the co-operation with an existing company. Based on the suitability of the identified model, potential partners can be investigated. Special attention in this phase has to be dedicated to the required permissions and certificates based on local law of the partners identified.

With the prerequisites of the first two phases, in the "Concept and Decision" phase the analyses are combined to build possible internationalization scenarios. The purpose of this phase is to evaluate these scenarios qualitatively and quantitatively and to identify the most appropriate one. A scenario consists of a set of selected portfolio elements, tasks, partners, and models. Within this analysis phase it is crucial to involve engineering experts. Subjects to be agreed upon are prices and costs, responsibilities, delivery times, quality criteria, etc. The phase ends with a signed contract or agreement among partners involved.

The following "Plan" phase performs detailed planning of the subsequent project phases: "Set-up and Transition" and "Manage and Execute". The plan includes a definition of the processes and workflows between headquarter and local entities that need to be aligned. Furthermore, the required manpower and ramp-up in the local entities have to be planned. The know-how transfer has to be organized in line with ramp-up and training for newly hired staff. Next to this, also procurement, transfer and installation of infrastructure such as automation components, computers, software and testing devices need to be identified and planned.

The objective of "Set-up and Transition" is to enable or set-up the local entities to take over the previously agreed upon engineering steps. The emphasis in this phase is on employees that should be selected based on the knowledge and skills required for their intended role. These depend on the engineering project requirements. People need to be allocated, recruited or delegated, inserted into the local entities' organization and trained in order to become valuable resources. In general, the plans established in the previous phase are now carried out starting with a pilot projects, whose results are agreed upon within this phase.

"Manage and Execute" is the phase in which the internationalized operations of the engineering project are managed according to the perspectives of progress, costs, and quality. With increasing knowledge and experience, work load and responsibilities in the local entities, the resolution of unforeseen situations becomes easier. Regular meetings are recommended in this phase to build up relationships among entities to better monitor the work execution and check and track the fulfillment of the objectives and the achievement of milestones of the project.

In order to manage the quality of the engineering artifacts produced, suitable quality checks and reviews are called for. All experiences gained within projects in an internationalized environment should be collected and introduced as a part of planned continuous improvement. This is the purpose of the "Lessons Learned and Phase Out" phase. A phase out may be decided upon after a predefined duration in an internationalized environment or upon termination of an underlying project.

The internationalization process iteratively defines the enablers for the internationalization of engineering identified in Sect. 2.2 during all seven phases: from the initial definition of the scope of internationalization, thus the elements of the solution portfolio, and the team involved, up to the identification and redefinition (with the lessons learned) of the most suitable artifacts and tools.

2.5 Best Practices for the Internationalization of Engineering

Internationalization of engineering is an initiative with high immanent risk. In order to show further basic internationalization concepts, the example of an electrical substation project (see Fig. 2.6) is used in this chapter. Substation projects require large civil and mechanical installations (VDI 2016). These projects can be performed on the basis of the project process shown in Fig. 2.1. Figure 2.6 shows a typical substation. In order to fulfill the customer contract to build a substation, the commercial regulations stipulated have to be abided by. These often require that a certain portion of the contract value has to originate from or has to be purchased in the country where the substation will be installed [so-called local content rules (Gremmel 2001; Bundesministerium 2013)]. For this reason, international projects are often particularly interesting for value chain optimizations. An internationalization initiative is a typical way to perform such an optimization.

In this example, the headquarters' (HQ) perspective of a strategic business unit (SBU) of a multinational company is adopted (Sarraf et al. 2012). The basic case under consideration here is that this SBU has the intention to regionalize its business to one or more local entities (LEs).

A value chain element of interest in the context of internationalization could be, for example, the engineering of protection and substation control. The fact that in electrical substations all equipment must be intensively checked and tested before putting it into operation is one important boundary condition. Such testing will naturally take place right before energizing the substation, that is, directly on site. Engineering tasks such as designing the general layout of the substation control

Fig. 2.6 HV substation in gas insulated technology (GIS)

system, configuring automation devices, or calculating the tripping schemes of the protection devices can basically take place anywhere. On the other hand, performing these tasks may require certain skills, which are not available locally; they may furthermore be dependent on other engineering tasks such as civil design or design of high-voltage components. In addition, some tasks may be considered "core" in the sense that the HQ wishes to keep control of these tasks because of safety, confidentiality, intellectual property, or other considerations.

These reflections show some important aspects relevant to internationalization of engineering. The fundamental question the SBU has to answer is to which extent it is ready for internationalization of engineering and which actions are needed to be performed for doing it efficiently. In the case of an ongoing internationalization initiative, management may wish to get information about possible blocking points or critical issues. This need is especially addressed by an internationalization readiness check, which is subsequently described.

2.5.1 *Readiness Check: Structure*

The structure of the readiness check follows the topics of internationalization of engineering (IoE) as presented in Fig. 2.7. These topics are characterized by their applicability to several IoE phases as shown in Fig. 2.5. It is suitable to organize them in fivecategories characterized by typical questions an organization comes across during IoE. These categories are explained in brief, subsequently.

"Why and What for" stands for the reasons behind internationalization. Expected cost benefits can be supposed to be key drivers; extension of capacity or customer proximity may be others. IoE requires that significant costs are to be deployed; risks

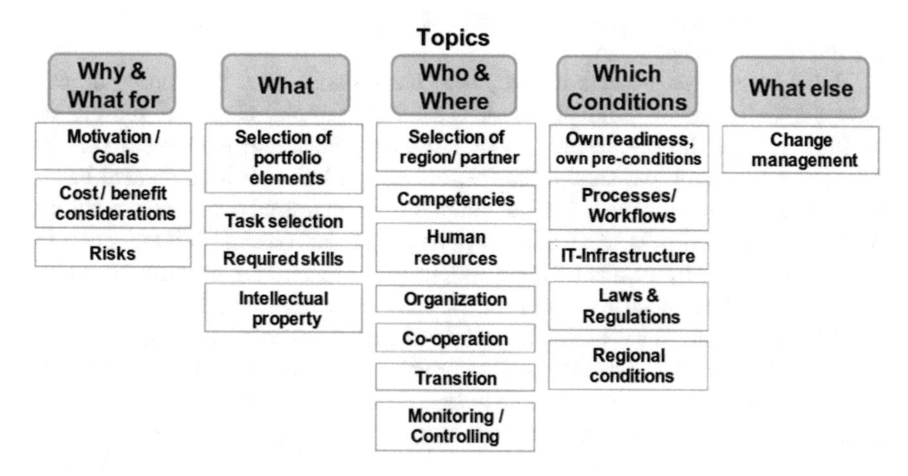

Fig. 2.7 Internationalization topics

have to be identified and monitored in order to proceed successfully. "What" is a question about the portfolio elements and engineering tasks that can be regionalized. Skills required and experiences as well as intellectual property (IP) aspects have to be considered.

"Who and Where" are questions concerning possible partners and concerned people. Organizational as well as operational structures of the selected partner have to be agreed upon by share of workload in order to perform engineering tasks. Questions concerning hiring and subsequent training of people in the LE's organization are dealt with in this category, as well as monitoring and controlling, which ensure that internationalization effort is on track. The category "Which Conditions" deals with important boundary conditions, such as own readiness, processes and workflows that have to be adapted, and infrastructure issues. Local laws and regulations, concerning labor or trade and commerce have to be taken care of, as well as other regional conditions, which are generally different to the company's HQ. The aspect of change management deals with the influences IoE has on the people involved and their situation. This is addressed by the category "What else."

The readiness check should be completed at the latest by the end of the IoE phase "Plan" (see Fig. 2.5). At this time there should not be any unidentified need for action as significant means for IoE are going to be deployed in the phases "Setup and Transition" and "Manage andExecute."

It is, of course, possible and even encouraged to start with the IoE readiness check right from the beginning of the IoE initiative. Thus, progress can be traced and monitored; moreover, management can be informed in appropriate detail.

In Fig. 2.8, a high-level view of the IoE readiness check is given, showing the first two structure levels "categories" and "topics." The two subordinated structure levels, "subtopics" and "questions," are explained in extracts in the next section by means of an actually performed IoE project in a big multinational company. This validation example has been modified and anonymized for reasons of confidentiality, but nevertheless shows important aspects of structure, contents, and utilization of the IoE readiness check.

Typical customer projects in the power transmission and distribution domain have the objective to build plants such as, for example, electrical substations, or high voltage direct current (HVDC) stations, including equipment such as transformers, circuit breakers, disconnectors, etc. All electrical equipment has to be protected from negative influences coming from undesired operating conditions. Moreover, operators need to have access to important data (such as voltage or current measurands) and, by means of interventions, need to be in a position to modify certain states of the electrical equipment, such as "open/close" operations on circuit breakers or similar. This is taken care of by means of a control and protection (C&P) system. Such a system is composed of programmable logic controllers, human–machine interface equipment, and dedicated devices for special purposes, such as measuring or closed-loop control in certain applications (see VDI (2016) for further information).

The objective of the SBU considered here is to internationalize a part of its control and protection engineering from HQ to another country for the sake of

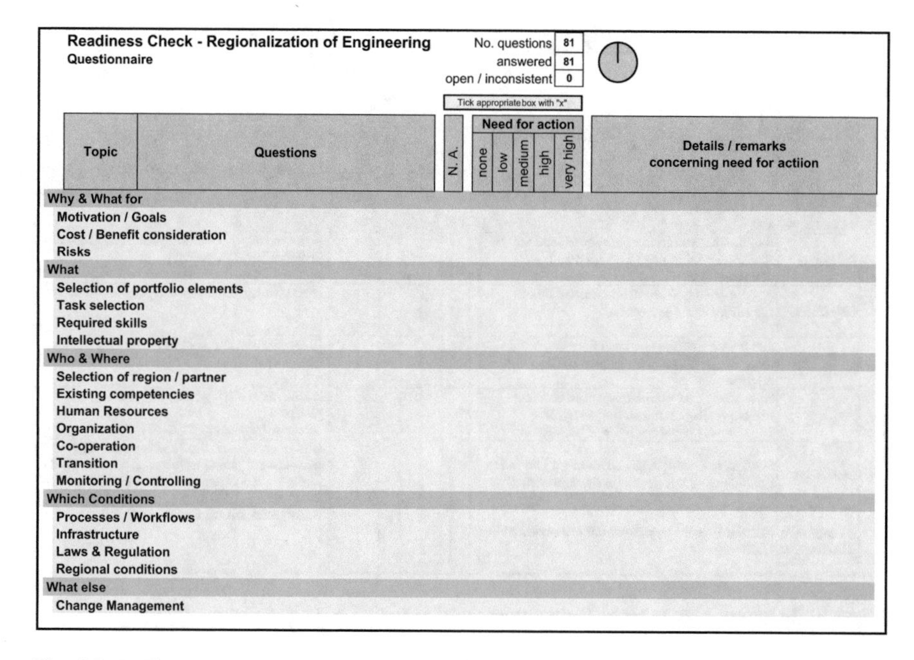

Readiness Check - Regionalization of Engineering
Questionnaire

No. questions 81
answered 81
open / inconsistent 0

Tick appropriate box with "x"

Topic | Questions | N. A. | Need for action: none, low, medium, high, very high | Details / remarks concerning need for actiion

Why & What for
- Motivation / Goals
- Cost / Benefit consideration
- Risks

What
- Selection of portfolio elements
- Task selection
- Required skills
- Intellectual property

Who & Where
- Selection of region / partner
- Existing competencies
- Human Resources
- Organization
- Co-operation
- Transition
- Monitoring / Controlling

Which Conditions
- Processes / Workflows
- Infrastructure
- Laws & Regulation
- Regional conditions

What else
- Change Management

Fig. 2.8 Readiness check for IoE—overview

increased manpower and cost advantages. At the time this readiness check has been performed, the first pilot projects were already running. Hence, from the SBU engineering management's perspective, the main purposes in performing an IoE readiness check was firstly, to verify whether all important aspects for IoE have already been considered appropriately and secondly, to identify action items if appropriate.

The readiness check has been performed with the responsible Engineering Manager of this SBU. In the next sections, findings of selected IoE topics are given. The selection is such that one topic with "very high," one topic with "high," and one topic with "medium" need for action have been chosen, taking confidentiality requirements into account. A detailed presentation of IoE topics will be given in an upcoming publication.

2.5.2 *Readiness Check: Topic "Motivation/Goals"*

The inquiry into this validation example starts with the consideration of motivations and goals. Figure 2.9 gives an overview of the parts of the IoE readiness check relevant here. Cost reduction and capacity increase have been named as reasons for the internationalization initiative. They are relatively clear, apart from their insufficient documentation. Based on the general reasons, concrete goals have been

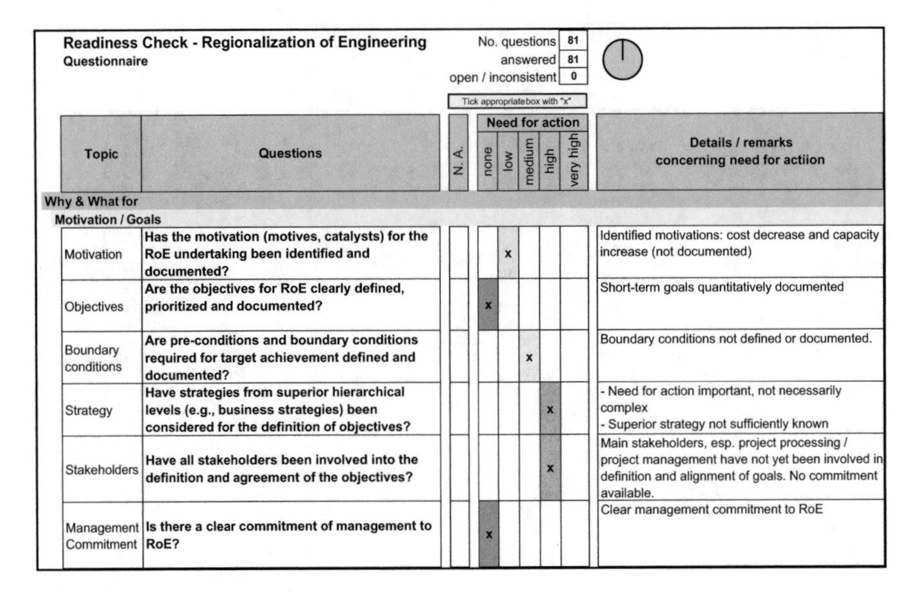

Readiness Check - Regionalization of Engineering
Questionnaire

No. questions 81
answered 81
open / inconsistent 0

Tick appropriatebox with "x"

Topic	Questions	N. A.	Need for action: none	low	medium	high	very high	Details / remarks concerning need for actiion
Why & What for								
Motivation / Goals								
Motivation	**Has the motivation (motives, catalysts) for the RoE undertaking been identified and documented?**			x				Identified motivations: cost decrease and capacity increase (not documented)
Objectives	**Are the objectives for RoE clearly defined, prioritized and documented?**		x					Short-term goals quantitatively documented
Boundary conditions	**Are pre-conditions and boundary conditions required for target achievement defined and documented?**				x			Boundary conditions not defined or documented.
Strategy	**Have strategies from superior hierarchical levels (e.g., business strategies) been considered for the definition of objectives?**					x		- Need for action important, not necessarily complex - Superior strategy not sufficiently known
Stakeholders	**Have all stakeholders been involved into the definition and agreement of the objectives?**					x		Main stakeholders, esp. project processing / project management have not yet been involved in definition and alignment of goals. No commitment available.
Management Commitment	**Is there a clear commitment of management to RoE?**		x					Clear management commitment to RoE

Fig. 2.9 Readiness check: topic "motivation/goals"

defined and documented. Hence, there is no need for action regarding goals, at least not in the short term.

The situation concerning boundary conditions is less clear, as these have been neither defined nor documented. Such boundary conditions include the availability of suitable customer projects in number and volume or the commitment of concerned project managers to support the IoE initiative. A further boundary condition might be that underlying customer contracts allow enough degree of freedom for the engineering department. Such contracts may set limits regarding allowed sub-suppliers or allowed portion of project volume that can be allotted to third parties.

A "high" need for action has been identified with respect to the business strategy which, at the time of the interview, was not sufficiently known by engineering management. Therefore, the goals for IoE could not be derived from, aligned with, or checked against the business strategy.

There is further need for action concerning different stakeholders and their intentions. Project managers, for example, have not been sufficiently involved and hence, their commitment is missing. On the other hand, there is a clear management commitment to IoE. This may be helpful in case frictions occur.

2.5.3 Readiness Check: Topic "Task Selection"

For all questions in this topic, a "high" need for action has been identified. An overview is given in Fig. 2.10.

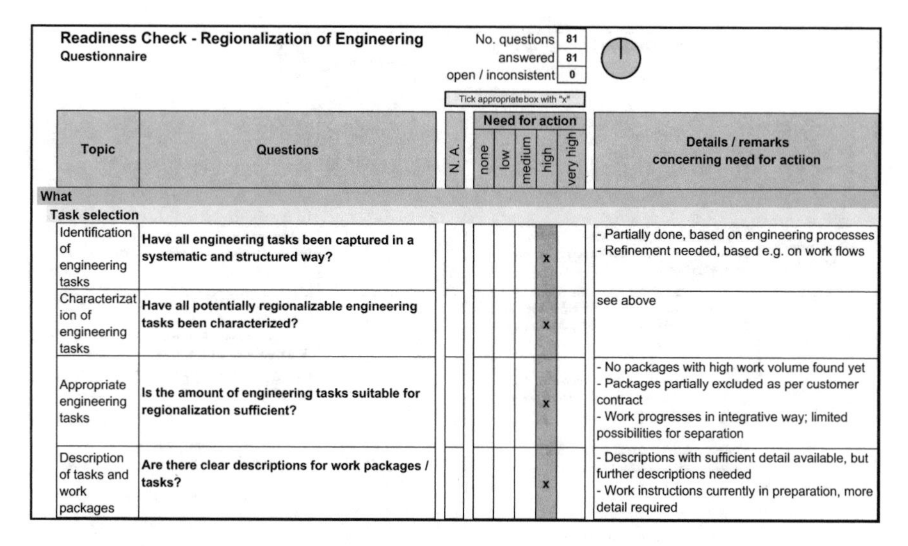

Readiness Check - Regionalization of Engineering
Questionnaire

No. questions 81
answered 81
open / inconsistent 0

Tick appropriate box with "x"

Topic	Questions	N. A.	Need for action: none	low	medium	high	very high	Details / remarks concerning need for actiion
What								
Task selection								
Identification of engineering tasks	**Have all engineering tasks been captured in a systematic and structured way?**					x		- Partially done, based on engineering processes - Refinement needed, based e.g. on work flows
Characterization of engineering tasks	**Have all potentially regionalizable engineering tasks been characterized?**					x		see above
Appropriate engineering tasks	**Is the amount of engineering tasks suitable for regionalization sufficient?**					x		- No packages with high work volume found yet - Packages partially excluded as per customer contract - Work progresses in integrative way; limited possibilities for separation
Description of tasks and work packages	**Are there clear descriptions for work packages / tasks?**					x		- Descriptions with sufficient detail available, but further descriptions needed - Work instructions currently in preparation, more detail required

Fig. 2.10 Readiness check: topic "task selection"

The first question deals with methods or structures to capture engineering tasks. This has only been done partially on the basis of available process descriptions. Further detailing with respect to the engineering value chain is required in order to identify more tasks suitable for internationalization. This is in line with the observation that in order to regionalize more of work packages, these have to be identified and described in detail (Cole 2003). Engineering in the given organization proceeds in work streams according to today's process; these have to be analyzed in detail in order to find ways to split them up. Such an analysis is needed to leverage internationalization. It gives insight into the interdependencies between engineering tasks, shows delimitations and interfaces between these tasks and gives hints where splits (and therefore, internationalization) can be performed. Further detail is given in Cole (2003). It is important to consider that engineering is an integrative approach with several activities in parallel which involve people from various disciplines, such as control, automation, human–machine interface, measuring, communication, etc. Therefore, new interfaces between process steps may have to be created in certain situations. This may have consequences to lead time and effort needed.

2.5.4 Readiness Check: Topic "Co-operation"

The details of this IoE topic are given in Fig. 2.11.

The first question of the topic "Co-operation" deals with an internationalization scenario. Such a scenario is a set of instantiations (shown in square brackets in the enumeration below) of the following parameters (Schaeffler et al. 2014):

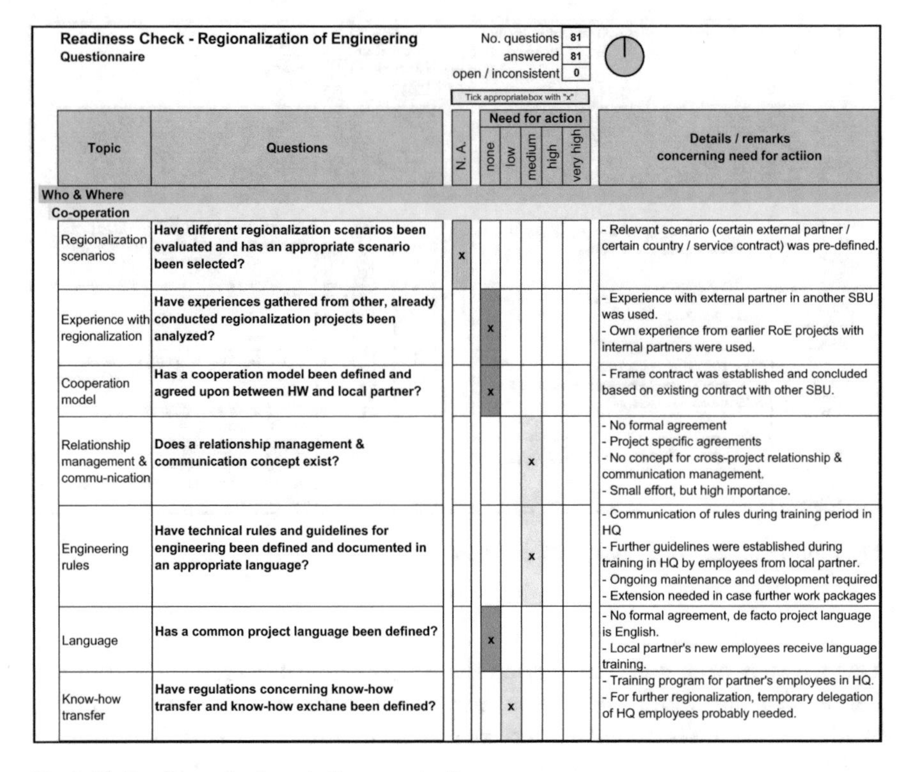

Readiness Check - Regionalization of Engineering
Questionnaire

No. questions: 81
answered: 81
open / inconsistent: 0

Tick appropriatebox with "x"

Topic	Questions	N. A.	Need for action: none	low	medium	high	very high	Details / remarks concerning need for actiion
Who & Where								
Co-operation								
Regionalization scenarios	**Have different regionalization scenarios been evaluated and has an appropriate scenario been selected?**	x						- Relevant scenario (certain external partner / certain country / service contract) was pre-defined.
Experience with regionalization	**Have experiences gathered from other, already conducted regionalization projects been analyzed?**		x					- Experience with external partner in another SBU was used. - Own experience from earlier RoE projects with internal partners were used.
Cooperation model	**Has a cooperation model been defined and agreed upon between HW and local partner?**		x					- Frame contract was established and concluded based on existing contract with other SBU.
Relationship management & commu-nication	**Does a relationship management & communication concept exist?**				x			- No formal agreement - Project specific agreements - No concept for cross-project relationship & communication management. - Small effort, but high importance.
Engineering rules	**Have technical rules and guidelines for engineering been defined and documented in an appropriate language?**				x			- Communication of rules during training period in HQ - Further guidelines were established during training in HQ by employees from local partner. - Ongoing maintenance and development required - Extension needed in case further work packages
Language	**Has a common project language been defined?**		x					- No formal agreement, de facto project language is English. - Local partner's new employees receive language training.
Know-how transfer	**Have regulations concerning know-how transfer and know-how exchane been defined?**			x				- Training program for partner's employees in HQ. - For further regionalization, temporary delegation of HQ employees probably needed.

Fig. 2.11 Readiness check: topic "co-operation"

- Selected portfolio elements [C&P engineering]
- Tasks [not detailed here—see previous section]
- Region or partner [external partner in pre-defined country]
- Internationalization model [service contract]

In the present case, as the parameters have already been set (with the exception of "tasks"), there is no further need for action. Moreover, experience from other SBUs in the given company has been made available and could already be used for the IoE initiative. For example, important clauses of an already existing framework agreement could be reused and are the basis for the co-operation model used in the present case. This includes work split, applicable rules and regulations, and mutual support under given circumstances.

Need for action has been identified with respect to relationship/communication and technical regulations. Relationship and communication management includes items such as contact persons, communication paths, and escalation routines. These are only available on a project-specific basis. Therefore, there is a need to define cross-project approaches. Engineering regulations set the basis for the technical performance of tasks, limit the possible choices and document the state of the technology in the given domain. These have been established partially during common trainings of local staff in HQ, but need to be further detailed, especially in the case that new work packages are intended to be regionalized.

2.5.5 *Readiness Check: Evaluation*

As the readiness check has the purpose not only to ask detailed questions about the IoE initiative, but also to give a status overview to management, a special focus has been put on evaluation functionality. The evaluation of the IoE scenario explained in the earlier is shown in Fig. 2.12.

The percentage value of answers with respect to the underlying scale is given for every category and for every topic. Topics with medium to very high need for action are indicated in a separate column and the pie charts give details per category and overall.

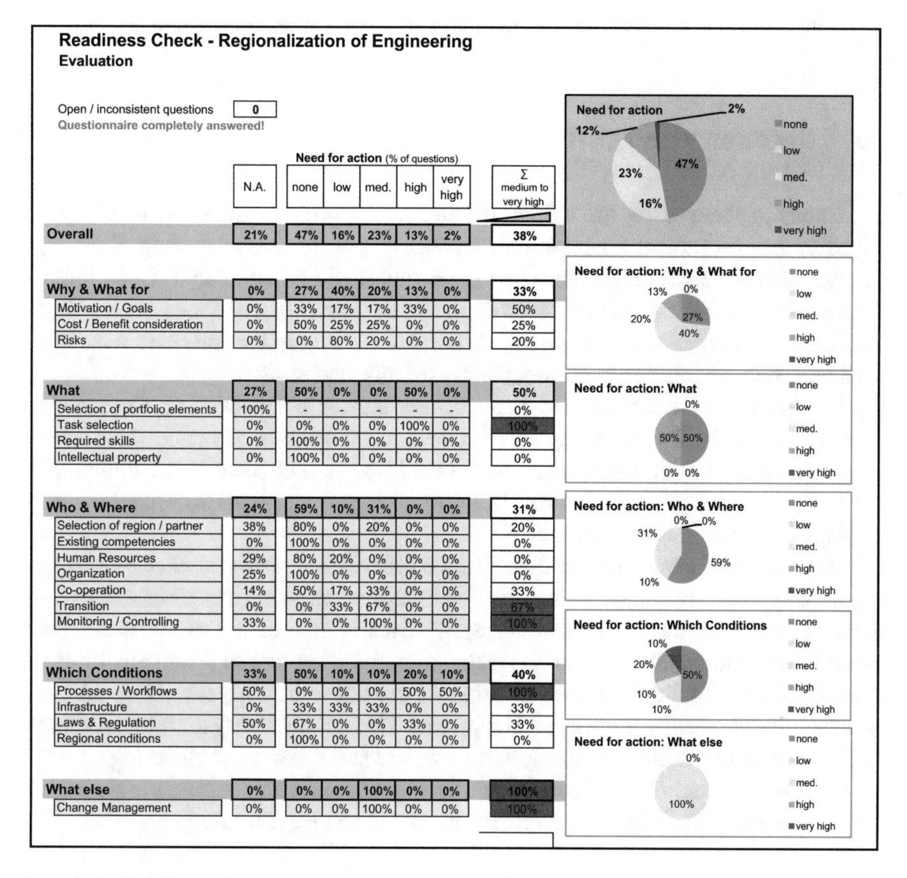

Readiness Check - Regionalization of Engineering
Evaluation

Open / inconsistent questions: 0
Questionnaire completely answered!

	N.A.	Need for action (% of questions): none	low	med.	high	very high	Σ medium to very high
Overall	21%	47%	16%	23%	13%	2%	38%
Why & What for	0%	27%	40%	20%	13%	0%	33%
Motivation / Goals	0%	33%	17%	17%	33%	0%	50%
Cost / Benefit consideration	0%	50%	25%	25%	0%	0%	25%
Risks	0%	0%	80%	20%	0%	0%	20%
What	27%	50%	0%	0%	50%	0%	50%
Selection of portfolio elements	100%	-	-	-	-	-	0%
Task selection	0%	0%	0%	0%	100%	0%	100%
Required skills	0%	100%	0%	0%	0%	0%	0%
Intellectual property	0%	100%	0%	0%	0%	0%	0%
Who & Where	24%	59%	10%	31%	0%	0%	31%
Selection of region / partner	38%	80%	0%	20%	0%	0%	20%
Existing competencies	0%	100%	0%	0%	0%	0%	0%
Human Resources	29%	80%	20%	0%	0%	0%	0%
Organization	25%	100%	0%	0%	0%	0%	0%
Co-operation	14%	50%	17%	33%	0%	0%	33%
Transition	0%	0%	33%	67%	0%	0%	67%
Monitoring / Controlling	33%	0%	0%	100%	0%	0%	100%
Which Conditions	33%	50%	10%	10%	20%	10%	40%
Processes / Workflows	50%	0%	0%	0%	50%	50%	100%
Infrastructure	0%	33%	33%	33%	0%	0%	33%
Laws & Regulation	50%	67%	0%	0%	33%	0%	33%
Regional conditions	0%	100%	0%	0%	0%	0%	0%
What else	0%	0%	0%	100%	0%	0%	100%
Change Management	0%	0%	0%	100%	0%	0%	100%

Fig. 2.12 Readiness check: evaluation

In the given case, all questions have been answered in a consistent way. For 47% of the questions, no further need for action has been identified. However, there is need for action in the topics “Motivation/Goals” and “Task selection.” Further topics with need for action, but not detailed here for lack of space, are:

- “Transition” concerning further training and infrastructure measures
- “Monitoring/Controlling,” that is, certain elements of quality management
- “Processes/Workflows” with respect to a modified engineering process in HQ needed for further internationalization
- The aspect of “Change Management,” dealing with the influence IoE has on the people involved.

The user feedback gathered with the IoE readiness check has been unanimously positive. By answering 81 questions in roughly 4 hours, the user is given detailed insight into his/her IoE initiative. Users have pointed out so far that from their point of view, all important aspects of IoE are addressed. The readiness check has not only proven its usefulness by confirming points the users were aware of, it has also helped to address unknown issues deemed critical enough to evoke subsequent actions.

2.6 Summary and Outlook

This chapter focused on the effects the increasing globalization and internationalization has on project business for software and systems engineering. Table 2.1 summarizes the most important challenges multinational companies intending to internationalize all or part of their engineering activities are faced with. These challenges are organized in readiness clusters according to the internationalization enablers, such as solution portfolio, team, artifacts, and tools.

A central approach presented here to face these challenges is an internationalization process. By following this process, a company can tackle the task of internationalization in a structured way. One important element is the company’s own internationalization readiness. This can be measured by performing the check explained in detail in this chapter.

These two elements—internationalization process and readiness check—support a company to perform internationalization by asking valuable questions and giving hints to possible solutions. Nevertheless, by virtue of their nature and on account of the inherent complexity of internationalization, the level of detail remains limited. Moreover, some of the challenges in Table 2.1 are not sufficiently taken into account yet, such as data security and data management. This paves the way for future research in the domain of internationalization of hardware and software engineering.

Solutions related to engineering quality improvement and engineering security improvement can be found, respectively, in Part II and Part III of this book.

Table 2.1 Identified challenges of internationalization in engineering

Clusters	Challenges
Solution portfolio	• Low degree of modularity • Unclear or nonstandard interfaces • Missing system-of-systems approach • Missing adequate data model
Team	• Find suitable and available engineers • Ensure qualification of available engineers • Discipline-specific system models • Working and environmental conditions • Safety and security • Stability of political conditions
Artifacts	• Capture/distribute knowledge • Strategy at local/global level • Law and regulation (ECC) • Ensure the use of data models across value chain steps • Ensure a common syntax and semantic • Ensure linking data/data models, e.g., define relationships and interdependencies between information/data • Consistent and scalable data exchange structures and formats to ensure collaboration within international engineering projects • Ensure consistency in engineering change management • Ensure data quality across different locations • Ensure security of engineering data (confidentiality, integrity, authenticity)
Tools	• Low degree of automation of workflows • Tool support • Interoperability • Missing connection of tools (tool chains), no possibility for round-trip engineering

References

Apex Engineering Solutions. (2007). The globalization of engineering. In *White paper*. Miami Beach, FL: Apex CoVantage.

Artto, K. A., & Wilkström, K. (2005, July). What is project business? *International Journal of Project Management, 23*(5), 343–353.

Azuayi, R. (2016). Internationalization strategies for global companies: A case study of Arla foods, Denmark. *Journal of Accounting & Marketing, 05*.

Bundesministerium für Wirtschaft und Technologie. (2013, May). Bestehende "Local-Content"-Regelungen.

Cole, G. A. (2003). *Strategic management: Theory and practice*. Stanford: Cengage Learning.

Deloitte Touche Tohmatsu. (2005). *Calling a change in the outsourcing market: The realities for the World's largest organizations.*

Dunning, J. H. (1993). *The globalization of business: The challenge of the 1990s*. Abingdon: Routledge.

Engineering Council for Professional Development. (1941). *Science*.

Fletcher, R. (2001). A holistic approach to internationalisation. *International Business Review, 10*, 25–49.

Foehr, M., Köhlein, A., Elger, J., Schaeffler, T., & Lüder, A. (2013, April). Optimization of the information chain within engineering process of production systems. In *IEEE international systems conference (SysCon)*. Orlando, FL.
Friedman, T. L. (2005). *The world is flat: A brief history of the twenty-first century*. New York: Farrar, Straus and Giroux.
Gremmel, H. (2001). *Switchgear manual*. Berlin: Cornelsen.
Hundt, L., Lüder, A., & Estévez-Estévez, E. (2010, September). Engineering of manufacturing systems within engineering networks. In *15th IEEE international conference on emerging technologies and factory automation (ETFA 2010)*. Bilbao.
INCOSE. (2014). *A world in motion: Systems engineering vision 2025*.
Khikhadze, L. (2019). Modern trends of the development of economic and cultural globalism. *Ecoforum Journal, 8*(1).
Myers, M., & Smith, K. (1999). Xerox: The global market and technology innovator. In R. Boutellier, O. Gassmann, & M. von Zedtwitz (Eds.), *Managing global innovation: Uncovering the secrets of future competitiveness* (pp. 299–315). Berlin: Springer.
Nekpuri, A. (2011). *Drivers, globalization of market, production, investment, technology*. Jabalpur: Xavier Institute.
Pisani, N., & Ricart, J. (2008). Offshoring and the global sourcing of talent: Understanding the new frontier of internationalization. In *Proceedings of 2nd annual offshoring research network conference and workshop*. Temple University.
Sachse, U. (2002). *Internationalisation of medium-sized enterprises: An integrated approach to management consulting*. Sternenfels, Germany: Verlag Wissenschaft & Praxis.
Sarraf, G., Elborai, S., & Kombargi, R. (2012, September). *Traditional approach to local content growth no longer reaping full benefits for developing resource-rich nations*. BusinessIntelligence Middle East.
Schaeffler, T., Foehr, M., Kodes, R., & Lüder, A. (2014). Regionalization of engineering. In *IEEE international conference on engineering, technology and innovation (ICE)*.
Schaeffler, T., Foehr, M., Kodes, R., Müller-Martin, A., & Lüder, A. (2014). A process for regionalization of engineering. In *Proceeding of the 2014 industrial and systems engineering research conference*.
Schaeffler, T., Foehr, M., Lüder, A., & Supke, K. (2013). Engineering process evaluation. In *Proceeding of of 22nd IEEE international symposium on industrial electronics (ISIE 2013)*. Taipei.
Steinmann, F., Voigt, K., Schaeffler, T., & Vollmar, J. (2014). Challenges in procurement of engineering services in project business. In *2014 proceedings of PICMET'14: Infrastructure and service integration* (pp. 2538–2549). Kanazawa.
VDI Statusreport. (2016). *Durchgängiges Engineering in Industrie 4.0-Wertschöpfungsketten*.
Vollmar, J., Gepp, M., Palm, H., & Calà, A. (2017). Engineering framework for the future – Cynefin for engineers. In *IEEE international symposium on systems engineering*. Wien.
Wust, E. L. (2011). *The effects of globalization on the civil engineering profession*. www.marquette.edu.

Chapter 3
Managing Complexity Within the Engineering of Product and Production Systems

Rostami Mehr and Arndt Lüder

Abstract Changing conditions on costumer, material, and technology market force producing companies to decrease duration of product and production system development. Especially in case of complex products like cars, this reduction leads to a strategic need for parallel development of products and production systems. Thus automotive industry organizes a complex interplay between product engineering and production system engineering within the *new product and production system development processes* (NPPDP).

This chapter discusses complexity challenges from automobile manufacturing that NPPDP have to cope with, and surveys strengths and limitations of complexity management methods for production system development. As no complexity management method can fully address the NPPDP challenges, the chapter derives types of NPPDP requirements and discusses a future framework for managing the complexity in NPPDP.

Keywords Complexity management · Interlinked product and production system engineering · Requirements for engineering processes

3.1 Introduction

Due to trends such as globalization and increased digitalization, manufacturing companies today operate in an environment very different from that a few decades ago. On the one hand, globalization has provided unique opportunities to companies to attract customers, to split up the work among specialized contributors, and to purchase services and materials from all over the world. On the other hand, it has increased competition as a result of the increasing number of international players. Hence, today customers have more choices and act in a buyers' market.

R. Mehr · A. Lüder (✉)
Otto-v.-Guericke University/IAF, Magdeburg, Germany
e-mail: rostami.mehr@st.ovgu.de; arndt.lueder@ovgu.de

S. Biffl et al. (eds.), *Security and Quality in Cyber-Physical Systems Engineering*,
https://doi.org/10.1007/978-3-030-25312-7_3

To respond to such global opportunities, companies must adapt their product and/or service portfolio to be able to satisfy various customer requirements, leading to the consideration of different market niches. The resulting market segmentation, with sharply delineated products tailored to local needs, increases the required amount of product/service variations and adaptations. This problem is intensified in companies and domains that basically have a high number of product variants like the automotive industry or home appliances.

In addition to the increasing number of products and product variants, the duration of product lifecycles is decreasing rapidly (MirRashed et al. 2016). Apart from this, the speed of technological changes has dramatically increased in the last few decades. Such rapid changes have increased the time pressure on companies to develop and introduce new products.

Furthermore, the development of new products takes place in an increasingly international environment. Several companies and several departments collaborate in a big network, an engineering organization (VDI 2010), to develop a product or to adjust existing products to new markets or to new customer requirements (Lindemann et al. 2006). The number of connected and parallelized processes in development procedures is rising (MirRashed et al. 2016). Products are getting more complex. The interlinking of products on production processes, their mutual impact, and the required resources for them are growing (Lindemann et al. 2006).

In summary, market and environment conditions of manufacturing companies are becoming increasingly complex, and, consequently, these complexities affect the manufacturer as an organization. As a result, the complexity of *new product and production system development processes* (NPPDP) increases. In this chapter, the NPPDP is defined as the complete network of engineering activities (taking design decisions based on available engineering information and skills and knowledge of engineers and using appropriate tools) that are required to design a new product (or set of product variants) and the production system required to produce this (these) product(s). In the automotive industry, the NPPDP covers the design of a car (with car body, power train, and all internal technical and other elements) and the different production systems to create them. This chapter illustrates this complexity and draws conclusions on the engineering process embedded in the NPPDP.

The dilemma of complexity has attracted growing attention as the effects on different parts of organizations became increasingly evident. In recent decades, an increasing number of researchers have attempted to respond to the following *Complexity-related Research Questions* (CrRQ):

CrRQ1: How can an organization deal with the growing complexity within new product and production system development processes (NPPDP)?

To answer this question, three main research activities have been combined: literature research, collecting evidence from practice, and experiments. Together they all have enabled the identification of available complexity management approaches, their evaluation with respect to their applicability within the NPPDP in the automotive industry, and the identification of required improvements leading to the development of a new complexity management framework.

This complexity management requires appropriate methodological and technical support within the engineering organization. This support strongly affects the quality of the management and, finally, the quality of the engineered systems themselves leading to the following question.

CrRQ2: What are requirements to the engineering organizations intending to integrate stronger complexity management within new product and production system development processes (NPPDP)?

The answer to this question is based on the consideration of a new complexity management framework. In this chapter, a future complexity management framework is drafted by identifying its main building blocks and sketching required IT technologies needed within.

While answering the "Complexity Related Research Questions," this chapter will contribute to the research questions*RQ1a*: "What are typical characteristics of engineering processes for long-running software-intensive technical systems?" and *RQ1b*: "What are requirement areas from engineering processes for long-running software-intensive technical systems that require informatics contributions?" mentioned in Chap. 1.

Based on the background of the authors, the automotive industry will be applied as running example within this chapter. In this industry, complexity management has a significant impact on the efficiency and quality of engineering and helps reduce the cost and risk of NPPDPs.

To answer the CrRQ in relation to engineering within the automotive industry, this chapter is structured as follows.

Section 3.1 introduces as focus of research new product and production system development processes (NPPDP) that address the strategic need for parallel development of products and process variants.

Section 3.2 discusses complexity challenges from automobile manufacturing that NPPDP have to cope with.

Section 3.3 surveys complexity management methods for production system development.

Section 3.4 summarizes NPPDP types of requirements.

Section 3.5 discusses a framework for managing the complexity in NPPDP.

3.2 Complexity Challenges

The most important driver of complexity in industry is the product (Schoeller 2009). The product is formed on the basis of customer needs. Customer needs are affected by trends like regionalization, fragmentation, and saturation (Maune 2011).

Regionalization can be illustrated in the example of the automobile variants sold in different areas. The station wagons are the models that are in highest demand in Europe, but in Asia, sedans sell much better than station wagons. Fragmentation becomes visible in the increasing electrification of cars leading to similar cars with different drive chain concepts, going beyond different fuels, such as petrol

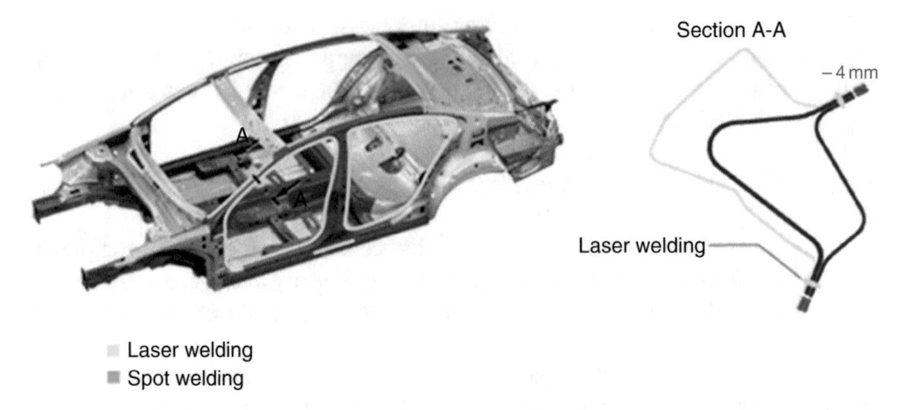

Fig. 3.1 Example of laser welding impacting product and production system engineering

and diesel, now also including completely electrical-driven cars, hybrid cars, and hydrogen or natural-gas-driven cars. The third trend that impacts customer demand—saturation—is a trend most prominent in west-European countries, North America, and Japan. Here the number of licensed cars is stagnating. It requires automobile manufacturers to differentiate and individualize their products forcing both other trends. Premium car manufacturers could benefit from this trend (Wemhoener 2005), but for high-volume manufacturers and manufacturers of commercial vehicles, the trend of saturation can be very challenging.

Another factor that has a significant impact on products is regulation. As automobiles significantly affect the environment and accordingly influence society, the legislators stipulate the requirements that must be followed by manufacturers. Such legislation could vary from country to country (Maune 2011). Most prominent legislation trend is related to reducing emissions. There are two fields considered. On the one hand, the emission of the product (the car) shall be reduced based on improved drive chains or reduced overall mass. On the other hand, the energy consumption of the production system needs to be reduced resulting in considerations related to production technologies, such as low-energy car body welding or even replacing welding by gluing. Figure 3.1 depicts an example affecting both the product and the production system. Here new joining methods in car body production, which contribute to reducing the weight of cars, are presented. As is obvious, legislation is one of the important drivers for developing and deploying new technology that affects products and production systems.

The third factor refers to the changes in technologies, such as the introduction of a new material or new production techniques. The applications of new materials like high-strength steel, hot-formed parts, aluminum, sandwich materials, fiber composites, and so on, attract increasing attraction. Although the applications of these materials have been common in the premium car segment for some time, recently these materials have become more common in high-volume cars. The increasing number of materials in the body of cars generates challenges in selecting

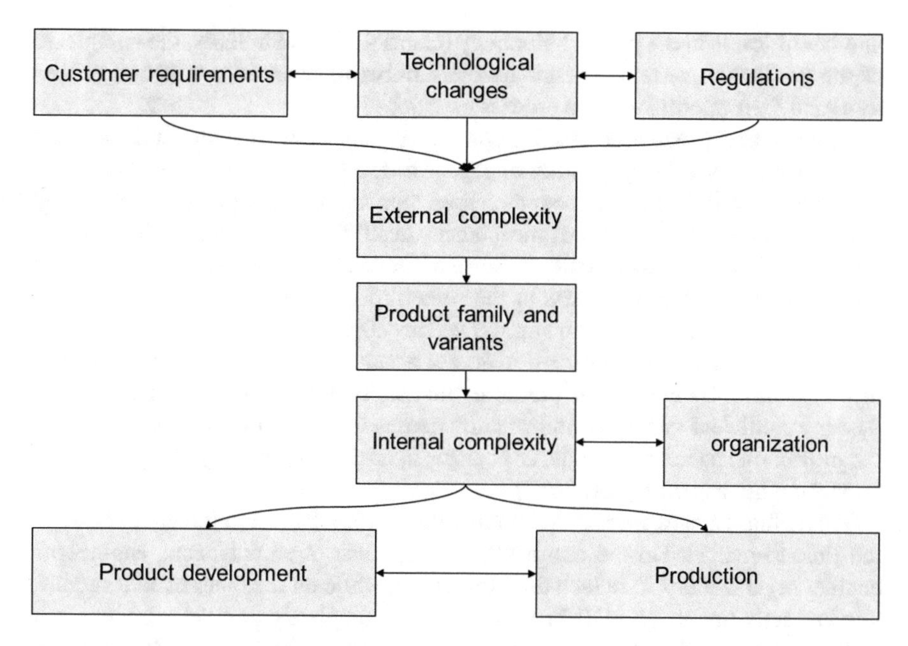

Fig. 3.2 Internal and external complexity in automobile manufacturing (Brosch 2014)

and adjusting the production process and, thereby, the selection of appropriate production resources within the production system engineering.

These three factors—customer requirements, rapid technological changes, and regulations—can be categorized as factors shaping *external complexity*. External complexity has a direct impact on product characterization. As a consequence, the product shapes the organization and exerts an enormous influence on *internal complexity* as depicted in Fig. 3.2. Transferring all three influencing factors to the product and shaping the product in a way that responds exactly to all requirements of these external factors, is challenging. The exact matching of product structure and external requirements could create an important competitive advantage for the organization through cost and risk reduction and efficiency increase.

3.2.1 General Problem in Practice

New product and production system development processes (NPPDP) are considered complex processes, especially in the automobile industry. The basis for this consideration is, on the one hand, the complexity of the product and, on the other hand, the complexity of the production system required for the product.

Cars are complex products due to the high number of components used in the depth and breadth of the product. Product structure breadth is defined by the number of components used for the parent item and the depth of the product is defined by the number of levels in the product hierarchy (Gabriel 2007). Similarly, the complexity of the production system depends on the number of required production process steps and their interlinking in a process network.

Product and production system complexity can have different influences on different parts of an organization engaged in development processes. The trends of individualization have led to an increased number of product variants, and a high number of variants intensify the time intensity in NPPDP. Time intensity in a complex project intensifies the complexity because of the increased number of activities that must be completed more or less at the same time. It also causes increases in the received information at the same time (DeVries 2005).

Increased amounts of information do not necessarily represent the same quality of information. Owing to time pressure, the quality of information could decrease. Again, insufficient quality of information causes extra complexity. This is akin to the project management triangle, as changes in the time of activity completion may impact quality and cost (PMI 2017).

Following the increased market competition, NPPDP are accelerated to decrease the time-to-market. On the one hand, the approach of simultaneous engineering enables organizations to reduce time-to-market, while on the other hand it requires starting activities in NPPDP in a partially or completely parallel manner. This, again, results in an increasing information flow between the departments engaged in development.

Since, in the early phase of NPPDP projects, car designs are mostly conceptual and not completed, the required information for production system engineering is not presented in detail. Nevertheless, production planning engineers start their task in an early phase parallel to the designers. Therefore, the number of changes during the development phase of the production system could rise, following the changes in car design. The increased number of changes under time pressure could again intensify time pressure and increase complexity levels more than before (Benedikt et al. 2012).

Another effect of the high number of variants is the so-called *recourse leveling*. In projects like the development of cars with a high number of variants and models, companies first develop the main variant of the car—that is, the model sold the most—and then they develop next variants and models, in sequence. This strategy enables them to first stabilize the production line for the first variant, and then adjust the production line for the next upcoming variants. This strategy also provides them the ability to manage fluctuations of the required work for the development of high variant cars. Unfortunately, this approach causes a production system planning without having the exact information of the upcoming variants. The phenomenon of incomplete information in production development processes generates more uncertainty and, accordingly, more complexity (MirRashed et al. 2016).

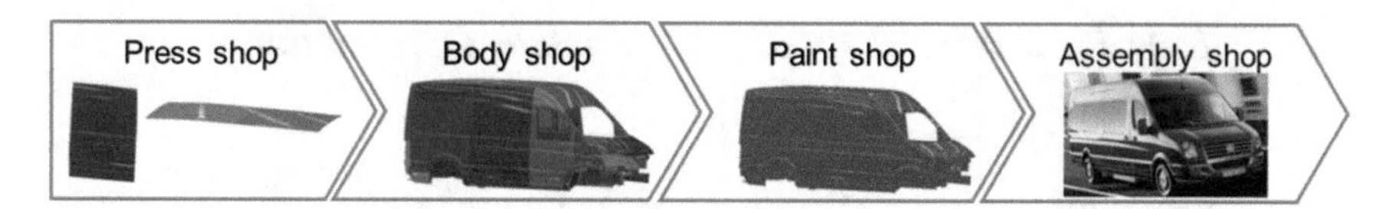

Fig. 3.3 Four main fields of car manufacturing

3.2.2 Products and Production Systems

As mentioned earlier, the overall complexity of the NPPDP comes from the complexity of the product, the complexity of the production system, and their dependencies. To make this complexity more visible in the following the car production process is reviewed.

Usually, the overall car production process is chronologically divided into four fields: press shop, body shop, paint shop, and assembly shop (see Fig. 3.3).

This division of labor supports the car production companies to manage their internal organization and processes within and between these fields.

The production starts with the press shop where the necessary metal sheets for a car are stamped out of steel coils. A typical body of car includes approximately 200 sheet metal parts for passenger cars and almost 250 sheet metal parts for light commercial vehicles.

In the subsequent body shop, these metal sheets are joined together. The most common joining methods include spot welding, stud welding, weld bonding (combination of adhesive and spot welding), clinching, MIG arc welding, and laser welding.

Next, the created car body is coated within the paint shop. Here, different layers of different materials are applied and dried in coating and oven lines. Finally, within the assembly shop, the car body is assembled with all necessary further car parts, including car wire harness, seats, power trains, lights, and windows. Up to 10,000 additional parts may have to be mounted to the car.

It shall not be neglected that the parts to be mounted and their production are also complex, especially for the power train.

The above-mentioned four fields of production constitute more detailed organizational divisions, for instance, the body shop is divided into four segments for platform, side walls, main body, and hang-on parts, while the assembly shop is subdivided into suspension system, engine, gearbox, seats, glasses, and plastic parts.

This division of the manufacturing process activities also forms the development processes and finally impacts the design of the organizational units for production development. Thus, the production development processes are segmented similar to the production processes.

3.2.3 New Product and Production System Development Process

The NPPDP in the automotive industry, designed for the joint development of car and production system, is established by six teams of engineers:

- The *project management team* has the job of coordinating and controlling the whole planning processes including the support of other planning team during the product development through the construction of prototypes, feasibility studies, and tests of the production technologies and processes.
- The *car engineering team* is responsible for the development of the product, that is, the car including the design and the evaluation of required manufacturing processes.
- The *production systemengineering teams of press shop, body shop, paint shop, and assembly line* are responsible for determining the required tools, machinery, technologies, and processes of manufacturing for the product developed. They also influence products in order to ensure the manufacturability, reduce the total project cost, and achieve the desired quality of the product.

The inner part of Fig. 3.4 depicts the work of these engineering teams.

Usually, the *car engineering team* starts the product engineering with design and construction of the product, which is accompanied by building first prototypes. Mostly at the same time *production system engineering teams* of the four different production system fields start to plan production facilities. When the *car engineering team* has reached the product release state, the *production system engineering teams* can finish the production system planning and start with supplier acquisition, detailed engineering, building and commissioning the production system. After product

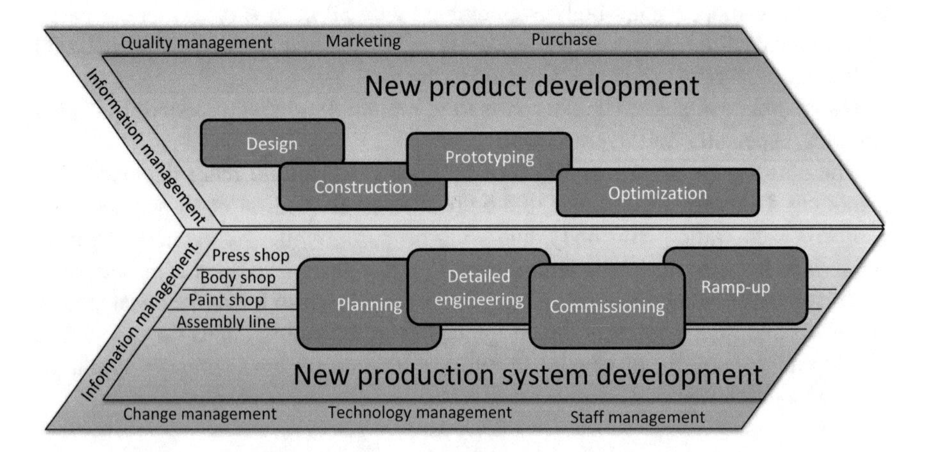

Fig. 3.4 General structure of NPPDP

release and in parallel to production system detailed engineering and commissioning, the prototyping and testing of product will take place by the *car engineering team*.

It becomes clear that the engineering activities of the different engineering teams run in parallel and constitute sequences of activities to conceptualize, construct, and commercialize a product. These activities are mostly mental and organizational instead of physical (Oyama et al. 2015), and, in addition, interlinked with each other. Two examples are the identification of problems or optimization possibilities within product engineering that change the product (like reinforcement of car body for optimization of crash test behavior) and lead to necessary changes in production system detailed engineering and, consequently, the identification of a problem regarding manufacturability by a supplier (like discovering the collision of manufacturing tools with the product) leading to a product change within the product construction. In addition, the named engineering processes are linked with additional functions of the overall company like quality management, purchase, and marketing on the product side and change management, technology management, and human resource management on the production system side.

It shall not be neglected that the described process is in some sort iterative. By quasi-parallel product and production system release prototyping and testing of product will take place. Thereafter, design engineer can start to discover problems or optimization possibility related to the product. The emerging product changes can have impact on the production system design. For example, the reinforcement of car body for optimization of crash test behavior can lead to additional handlings in car body welding. In addition, also on the production system side, improvement possibilities related to manufacturability or economic issues can be identified, possibly leading to requests for product change. An often relevant example is the identification of collisions of the welding tool with the car body requiring a shift of the welding spot location.

Beyond these overall company function, the information management has an important impact on the overall engineering organization, as it is responsible for the creation, exchange and storing of engineering information along the complete life cycles of product and production system. Information management covers all IT hardware and software such as Product Life Cycle Management (PLM) systems, databases, and servers.

The number of individual activities related to each named organizational unit might be very high. In automobile industries, this number can exceed a thousand activities (Kirchhof 2003). This results in enormous flow of information. One illustration for this fact could be the engineering of the body shop. The VDI Guideline 4499 (VDI 05/2011) introduces more detailed subphases for the planning phase mentioned in Fig. 3.4. This guideline defines the *concept planning phase* covering activities like finding production concepts, joining sequence planning, and geometrical validation, all intending to detail the production process to be executed. The next subphase is the *detailed planning* targeting the discussion of required production resources covering, for example, jigs and fixtures planning, material flow planning, and ergonomics analysis. This phase is closed by cost calculation and offline programming (see Fig. 3.5).

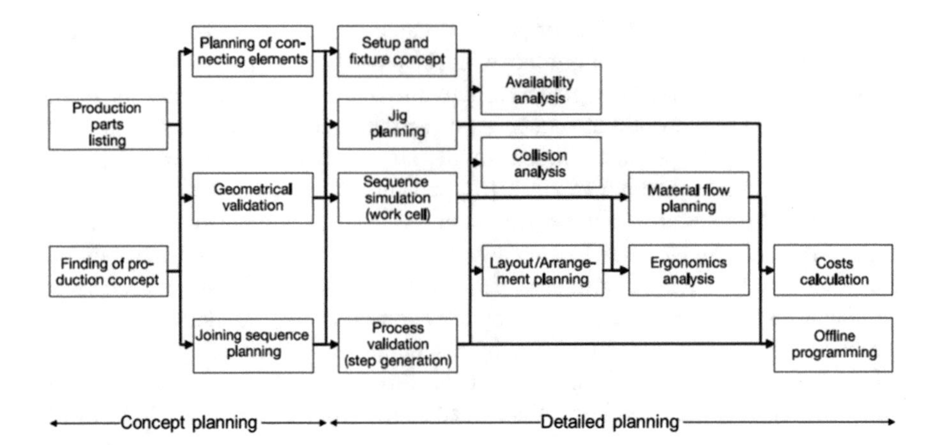

Fig. 3.5 Body shop production development (VDI 05/2011)

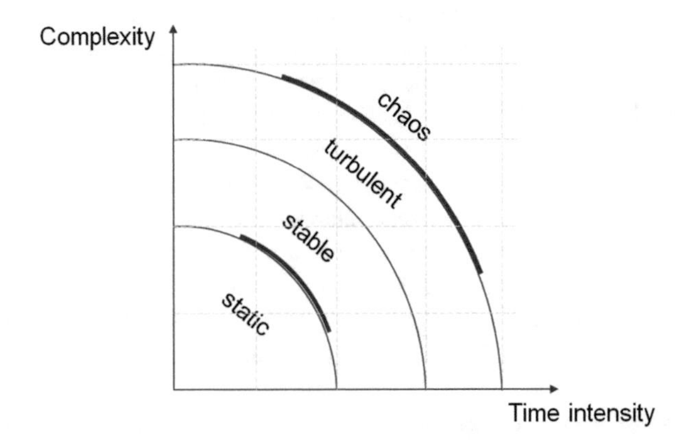

Fig. 3.6 Dynamic complexity of NPPDP

Originating from the changing market conditions, the complexity of the NPPDP has additionally increased by uncertainties regarding time to market, market demand, fluctuation of demand, available technology, speed of development of new technologies, and required resources such as human resources and capital (Grussenmeyer and Blecker 2013). With the presence of features, like a high number of involved engineering decisions (diversity) that are strongly interrelated by required information exchange (connectivity), and frequent changing bordering conditions (uncertainty), NPPDP can be categorized as a complex. In addition, their duration, dynamics, and human labor-intensiveness makes NPPDP time intensive. NPPDP are usually turbulent following the high dynamics and the strong dependencies between engineering activities and increasingly show a tendency to become chaotic (see Fig. 3.6). Thus, there is a strong need for management mechanisms to handle complexity before becoming chaotic.

3.3 Complexity Management for Production System Development

Several research studies have worked on complexity and approaches to deal with complexity. In the following, some studies shall be discussed opening up the existing broad scope and summarizing the main features applied in these approaches.

In complexity management sciences, there are approaches based on a holistic view of a system/organizationthat recommend steps to manage complexity. As these approaches are based on the strategy of a comprehensive view of systems, the applications of these approaches in praxis are very challenging. Two examples in this field are the work of Vogel (2017) and the work of Budde (2016).

Vogel (2017) introduced a *comprehensive complexity management* approach for resource planning, which has the following four steps: complexity analysis, complexity evaluation, application of complexity strategies, and complexity planning and control. The application of such approaches in praxis requires supporting tools to evaluate or identify a suitable strategy of complexity, which could vary drastically from case to case.

Budde (2016) introduced a parameter that provides better visualized understanding of complexity named *Complexity Value Level* (CVL). This parameter delivers a quantitative value that shows whether an organization is a complexity master or outperformer, market performer, and complexity underperformer. Based on this complexity index, an organization can define actions to improve its CVL and improve complexity management. Even though providing a quantitative index for complexity is very helpful, still this index could not help in defining strategies for specific complexity drivers to manage or reduce complexity.

On the other hand, there are several approaches addressing predefined complexity drivers and providing methodologies to handle them. For instance, the *variant management strategy* is a common approach to solve the problem of complexity caused by variants (Thiebes and Plankert 2014). Since these strategies are based on single complexity drivers, usually the overall complexity of the production system will remain unaffected. This is known as remaining complexity. The limitation of variant management in solving complexity in the early phase of production development processes is an example of this deficit for such approaches.

Finally, there are approaches addressing sets of complexity drives. Schuh (2005) and Schuh et al. (2011, 2015, 2016) introduced a framework to evaluate the complexity in NPPDP. This framework includes three main sections: evaluating complexity by using complexity drivers, analyzing the interdependencies between drivers, and segmenting or rating the drivers. This evaluation of complexity drivers serves as a basis to define proper methods to manage complexity by dealing with its causes and origin.

Meier et al. (2005) suggested a comprehensive approach for managing the complexity of products by means of, inter alia, variant management and managing the complexity of processes through lean management.

Daryani and Amini (2016) drew a five-step process for complex organization management by understanding the complexity type, investigating the causes of complexity, identifying solutions, selecting an effective solution, and implementing and evaluating the selected solution. They intended to provide decision-making assistance in complex situations.

Lasch and Gießmann (2009) introduced a complexity management method enhancing the well-known PCDA cycle. They aggregated existing engineering methodologies like variant management, ABC analysis, failure modes and effect analysis (FMEA), and impact matrices to one larger methodology.

For production organizations, Marti (2007) introduced a complexity management model with three steps: strategy and product lifecycle assessment, product complexity management, and driving guidelines for action. The first step includes analyzing the strategy of the company and product positioning in the market and its lifecycle. The second step analyzes product details for optimization of its architecture. And the last step provides guidelines for action in accordance with the findings in the first two steps.

In total, there is no approach available that addresses the intensified dynamic complexity caused by time pressure, incomplete information, and strong linking of engineering activities as given in the NPPDP.

To gain an essence of the available approaches in complexity management science, the next section investigates the common features of these complexity management approaches.

3.3.1 System Thinking

The system engineering sciences have strong interrelations with complexity management sciences. Within systems engineering, a system is defined as a collection of elements that are interrelated and distinct from the environment (Chen et al. 2009). A system is designed for a purpose, that is, predefined goals or objectives (Fuchs 2018). The term "element" is applied to any part of a system (e.g., organizational units, employees, documents) and, therefore, can also be used for subsystems (Checkland 1999). A system may have interconnections with its environment.

As mentioned earlier, the definition of complexity has strong interconnection with system definition. Complexity of systems is defined as a high number of interconnected elements that have a variable status. This variation can include but is not limited to the number of elements, their interaction within the system, and their interaction with the environment. To better understand the complexity of a system, it is strongly recommended to consider the system within its interacting elements and its environment, that is, to consider complexity from the systems thinking side.

Especially Maurer (2007) emphasizes a systematic approach toward complexity management to resolve the challenge of missing clarity. The identification of the system with elements and interactions as well as its environment helps to define the origins of complexity and their impact.

Thus, understanding the NPPDP as a system with its structure and behavior is the first main requirement within complexity analysis and management.

3.3.2 *System Analysis*

System analysis provides tangible fundaments to identify the origin of complexity.

The structure of interactions among system elements can be, for example, more modular oriented, forming clusters of elements that are stronger linked while interactions among clusters are only sparse. Another example is the integral structure where all elements interact with each other more or less equally strong. It is obvious that changes in the behavior of one element will have different impacts on the behavior of the other elements in both structures (Fuchs 2018). Therefore, the structures may provide information on complexity.

Analyzing the system reveals also the type of interactions between the elements. Elements can be disconnected from other elements, have one-way connection, or can be connected which each other in both directions. These interactions can be represented by dependency structure matrices (Jacob and Paul 2016) or by graph-based methods (Maurer 2007). Different interaction types will impact complexity in a different way.

Nevertheless, it is difficult, up to impossible, to identify all interactions among the elements of an NPPDP (Malik 2016). Therefore, a method has to be considered to collect relevant information or sort gathered information in an effective manner.

In addition to the static analysis of the system also the dynamic features of a system (especially in the case of an NPPDP) are relevant. During the system lifetime, the status of system elements as well as interactions of system elements may change. The speed of changes is one of the determining factors in complexity. In some works, this factor is called "fast flux," expressing the transient nature of the organization and its environment (Schwandt 2009).

System analysis by itself does not result in complexity management, but represents the fundament of the system under consideration, enabling better interpretation, management, and (finally) improvements (Maurer 2007). System analysis also reveals potentials in two ways: optimization of the system structures such as eliminating redundant elements or dependencies, clustering of elements to build modules, and building a path to trace complexity effects on the system.

3.3.3 *Drivers and Effects Analysis*

As mentioned earlier, the system analysis shall be the foundation for an analysis of the complexity drivers and their impact on the system or subsystems. Successful analysis of complexity drivers and effects is a mandatory enabler for appropriate complexity management methods. Indeed, an initial concern regarding the system

Fig. 3.7 General complexity management process for NPPDP

definition is the origin of complexity. Sources of complexity and their impact can be located in different system parts (Brosch 2014; Maurer 2007; Velte et al. 2017).

Several methods have been developed to find the origin of complexity. The first step of all methods is collecting information. Based on information gathered in the system analysis [from available documents and experts (Weber et al. 2014)] methods from Cause and Effect Analysis or Root Cause Analysis (RCA) can be applied (Lee et al. 2018).

Beyond the application-case-related, cause-and-effect analysis, there are also generalizations of complexity causes. For example Velte et al. (2017), surveyed complexity management and categorized complexity drivers. Thereby, three groups of complexity drivers related to production systems have been identified: *internal complexity* including product, organization, process, order fulfillment; *interface complexity* including purchase, communication, and sales; and *external complexity* covering customer, competition, and legislation.

Wildemann (2012) has also categorized the complexity drivers in three groups: company structure based drivers, information systems based drivers, and communication system based drivers.

Similar to complexity causes the complexity impacts are various. Most often, high-level impacts are related to project costs (Sinha 2014), production (Kieviet 2014), and quality (Lasch and Gießmann 2009).

3.3.4 Summary

Summarizing the considerations in this section, there are two main types of methods relevant within complexity management for NPPDP: holistic methods and special driver related methods. These methods all share the same basic structure depicted in Fig. 3.7. Usually, they start with modeling and analyzing the system of interest. Based on this analysis, complexity drivers and complexity effects are identified and used for the definition of complexity management measures.

However, both approaches mentioned have their specific drawbacks. The holistic approach lacks details for industrial usage. These methods are too general and face many unclear situations, making them hard to apply in industrial practice. The methods targeting special complexity drivers might not take care of relevant complexity drivers for the system of interest. Thus, they may be too detailed and miss a comprehensive overview. This can be considered as a research gap that must be investigated. To close this gap for NPDDP a two-step approach is applied, where the two steps reflect requirement modeling and complexity management framework configuration.

3.4 Requirements

The *classical requirements* of NPPDP are related to criteria like production cost, investment, cycle time, optimal layout, flexibility, and meeting general project goals (time, cost, and quality) (Schady 2008). As shown, besides these classical requirements, it is vital to define the *complexity management related requirement.* They are categorized in three main groups.

Process-related requirements are related to the engineering process execution. Here transparency, modularity, reusability, adaptability, and finally standardization of processes and their corresponding outputs as well as their predictability are relevant.

Interconnectivity-related requirements focus on the information exchange along the engineering process chains. The quality of the information exchange related to correctness, completeness, and appropriateness shall be ensured. Thus, misinterpretation, or extensive retreatment of information shall be prevented along the engineering chain.

Dynamics-related requirements face the dynamics within the engineering chain and its volatility. They include NPPDP monitoring and agile change management.

3.5 Complexity Management Framework

Figure 3.8 illustrates the main parts of a framework for NPPDP complexity management that shall address all characteristics of complexity. As shown earlier, the interaction of and dependencies between the involved engineering steps and their volatility cause high information flow and are the main complexity drivers (MirRashed et al. 2016). These information flows have two mainstreams: information flows within product engineering and information flows within production system engineering. They require a detailed understanding of the engineering chains (that may be based on an appropriate process model) (Gadatsch 2015). In addition, both engineering chains are accompanied by knowledge management activities to ensure effective reuse of engineering data.

As product engineering results in changing product design, change management is required to ensure the transmission of relevant impacts to production system engineering and back. The availability and completeness of product data strongly depend on the number and position of quality gates and product-release strategies defined within the project management as quality assurance mechanisms.

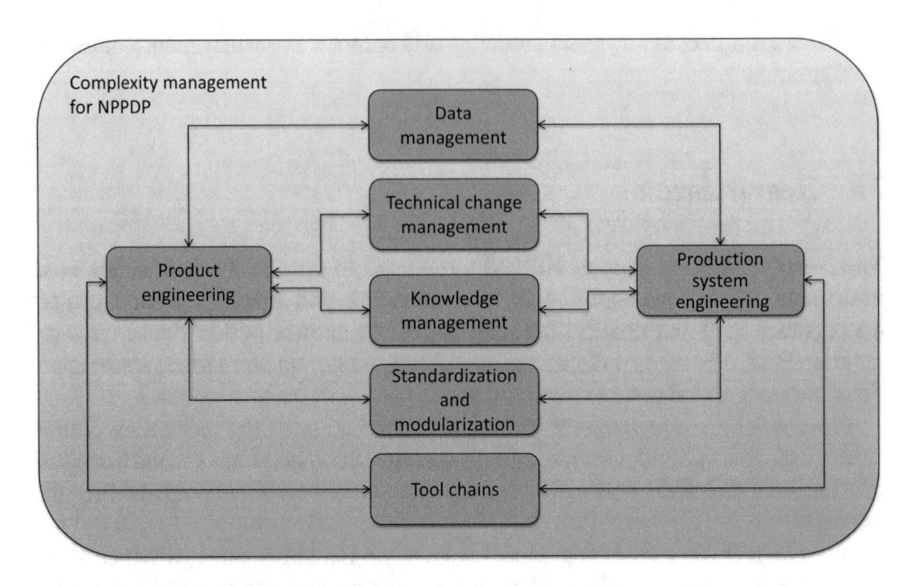

Fig. 3.8 General complexity management framework for NPPDP

The tool chains applied within the engineering processes and their volatility are another complexity driver. Especially the increasing digitalization leads to relevant impacts on engineering quality and efficiency (Biffl et al. 2017).

Finally, standardization and modularization is an important driver in NPPDP. It affects both the modularization and standardization of the product as well as the production system leading to component catalogues to be applied by engineers.

3.5.1 Engineering Processes

Gadatsch (2015) defines a process as a regular repeating activity set that has determined starting and end points. This activity set processes predefined input information to provide output information. There are many methods for modeling engineering processes such as flowcharts, RACI (Responsible, Accountable, Consultedand Informed) charts, activity diagrams, and SIPOC (Supplier, Inputs, Process, Output, Customer) diagrams. These methods provide an appropriate overview of the whole system and the interconnection between system elements. Thus, they help to obtain transparency.

Within the proposed NPPDP complexity management approaches, the first step shall be related to modeling the engineering processes. As the model and the illustration of whole engineering processes in NPPDP must be transparent and understandable to ensure clarity and easy inside, the level of details must be corresponding.

Even if a swim-lane diagram may initially seem appropriate for a large-scale project like NPPDP, where milestones and responsible organizational units can be added to the diagram to increase perception of the whole system, this kind of diagram is only one option. It is up to further research to find appropriate modeling mechanisms for reflecting the necessary information exchange within the engineering chain covering required engineering data quality.

3.5.2 Data Management

Product engineering data and production system engineering data are both main pillars of the NPPDP and main complexity drivers. They must be timely exchanged within NPPDP owing to the gradual development of product and production system variants, simultaneous engineering, and technical changes in the development phase. Incomplete information or time-delayed information could mislead and intensify complexity.

The issue of information management is addressed in many studies. A common approach for information management in NPPDP processes is the use of Product Lifecycle Management (PLM) tools interconnected with the different tools within the engineering tool chain (Sindermann 2014). In the PLM tools, the product data act as core model including direct and indirect relations to process and production system information. Recently PLM data tend to be enriched to overall system engineering models covering product, production process, and production system engineering information (Biffl et al. 2017). An example of such information are product data related to the car body components that include main assembly, subassembly structures, geometries, and joining information as basis for process and for production system engineering.

In multivariant NPPDP used in the automotive industry, the completeness of required engineering information is essential for the work of all production system engineers. Hence, a dedicated product data release strategy is required to manage the complexity coming from the interconnectivity of planning processes. These release strategies are mostly compatible with approaches in the ramp-up process. Therefore, recent ramp-up strategies (Schuh et al. 2015) can be adjusted for product release. In the early phase of NPPDP proper product data release supports to avoid unnecessary complexity in the production system engineering by eliminating uncertainty due to incomplete or time-delayed information. The quality and accuracy of product data help to reduce misunderstandings and increase transparency.

Quality gates (QG) assure the quality of delivered data from product engineering by means of a set of measurable criteria that were initially agreed upon within the complete NPPDP (Richter and Walther 2017). The challenge of defining quality gates (QG) for product data is characterized by three factors: a set of measurable criteria, placing the QG in the proper phase of NPPDP, and the frequency of QG. QG help to identify and correct errors in product data in the early phase and to avoid costly changes later on.

Due to the increasing number of variants handled within NPPDP, quality criteria are increasingly relevant issues within the definition of QG. These quality criteria may range from simple completeness check-ups to detailed reasonability criteria for the consistency of engineering information within the multidisciplinary engineering of an NPPDP. Here, appropriate means for modeling, integration, and evaluation of complex and discipline-crossing consistency rules within PLM systems are required and still not well addressed.

3.5.3 Technical Change Management

According to DIN 69901, change management includes five activities: record, evaluate, decision making, documentation, and implementation of changes. Following the increasing complexity of products and the fast evolution of customer and technology markets, it is almost impossible to avoid technical changes in NPPDP. A research study revealed that approximately 20% of the product engineering and 40% of the production system engineering efforts were dedicated to technical change management (Köhler 2009). As presented in the sections above, the strong interconnectivity of engineering activities within NPPDP requires a detailed change propagation making changes to complexity drivers.

The cost of technical changes, depending on the phase of occurrence, could vary and affect the cost of NPPDP. Therefore, it is very important to distinguish and communicate changes. The faster the change identification and change information propagation process, the more cost-effective is the technical change management. Monitoring and tracing changes in the overall set of engineering information, therefore, provides a reasonable basis. Hence, the engineering data need to be enriched by appropriate data management information covering thinks like version and revision management information, data owner information, etc. This is still an open issue within engineering data management systems.

3.5.4 Knowledge Management

Knowledge management refers to the ability of identifying, collecting, sorting, storing, and retrieving a set of scientific and technical information (Carayannis 2013), which can be reused. Within NPPDP that can cover product and production system data from previous projects, which can be combined with production system component data from suppliers. In the traditional approach, thisinformation is not applied in the early phases of NPDP. The early engagement of this knowledge in NPPDP can reduce complexity.

However, the collection and quick evaluation of such information is a challenging task due to the big information flow (big data), missing structure, and the corresponding IT system (Olsen 2017). Thus, structured information in combination with agile data management shall be considered within the integrated information management system of NPPDP in order to avoid and manage complexity.

3.5.5 Tool Chains

Digital engineering tools are the foundation of NPPDP (Biffl et al. 2017). Providing engineering data in an overall engineering data logistics that extends PLM systems helps to reduce the development time and provides the essential data for the work of engineers within all engineering phases. Especially, the engineering data logistics enables the connection between product engineering and production system engineering (Bracht et al. 2017). Simplifying the communication between product and production system engineering enables the production system engineers to be involved in very early phases of NPPDP, giving bordering conditions also for the product engineering.

NPPDP operate in a more agile manner and engineers analyze numerous scenarios in a very short period by means of simulation. In the automobile industry, an engineer spends approximately 60% of their capacity to obtain information (Reijers and Mendling 2008). Digital engineering tools provide the possibility of perceiving the data in a way that would be suitable for further processing and save the time of engineers for engineering activities.

Nevertheless, these tool chains can only appropriately interact if they all support the interaction with the envisioned engineering data logistics based on appropriate engineering data exchange technologies (Biffl et al. 2017).

3.5.6 Standardization and Modularization

The last part of the framework, like many other approaches in complexity management, recommends the standardization and modularization of the product, production processes, and production systems, especially for multivariant production as in the automotive industry. Using the standard and modular elements in NPPDP generally speeds up the development and ensures the termination of activities within planned time (Reijers and Mendling 2008).

Standardization and modularization approaches reduce the possible variability within the objects to be engineered (product and production system) by enabling the definition of generic system architectures and system components (VDI 2010).

However, standardization and modularization require a modeling methodology for system components, in case of NPPDP components of products and production systems with their interrelation over processes. This modeling methodology shall go beyond currently existing methodologies defined, for example, in PLM tools or ISA 95 standard.

3.6 Summary

Increasing globalization and technological improvement have resulted in a change to customer markets also in the automotive industry. This has a major impact on the joined engineering of products and production system within this industry as it enforces system complexity increase und development time reduction. It can be stated that the joined engineering of products and production system tends to become dynamically complex with three main characteristics of complexity: diversity, connectivity, and uncertainty.

Currently, there is no holistic complexity management framework available applicable for the joined engineering of products and production system within the automotive industry. Thus, this chapter has addressed *Complexity-related Research Questions* (CrRQ) that shall assist the development of such a framework.

The first research question (CrRQ1) has addressed means for complexity management, enabling organizations to deal with the growing complexity within *new product and production system development processes* (NPPDP) while the second one (CrRQ2) has concentrated on requirements to engineering organizations intending to integrate an increased complexity management NPPDP.

To address these questions, Sect. 3.2 collected complexity challenges arising in the joined engineering of products and production systems in the automotive industry within an NPPDP, characterized as dynamic complexity. Section 3.3 reviewed existing complexity management methods and evaluated how they can contribute to a holistic NPPDP complexity management. Section 3.4 discussed requirement groups for such a holistic complexity management. Finally, Sect. 3.5 presented the first ideas for such a framework, which consists of the six main pillars of defining and modeling engineering processes, data management, technical change management, knowledge management, standardization, and modularization, and tool chain management. The modeling of engineering processes provides a clear overview of the NPPDP as a system and creates a better understanding. The combination of product and production system engineering data management and knowledge management supported by appropriate engineering tools helps to resolve the main part of complexity by providing integrated information management. This combination provides transparency and eliminates unnecessary dependencies between two NPPDP engineering activities. Finally, standardization and modularization again reduce dependencies by means of proven and commonly known approaches. All these parts of the framework follow the common approach to avoid, reduce, and manage complexity.

Nevertheless, there are still some challenges to tackle by research and development to make this framework applicable. This chapter has identified the following challenges:

- Providing appropriate modeling mechanisms for engineering chain modeling and analysis reflecting the necessary information exchange within the engineering chain covering required engineering data quality

- Providing modeling, integration, and evaluation means for complex and discipline-crossing consistency rules applicable in multidisciplinary engineering data management systems
- Enhancing multidisciplinary engineering data management systems by appropriate data management information modeling means covering thinks like version and revision management information, data owner information, etc.
- Integrating multidisciplinary engineering data management systems with agile data management
- Providing appropriate engineering data exchange technologies for multidisciplinary engineering data management systems
- Providing modeling methodologies for system components applicable in multidisciplinary engineering data management systems

The subsequent chapters of this book take up most of these challenges and provide means for their solution.

By discussing the named CrRQs, this chapter has contributed to the RQ1a: "What are typical characteristics of engineering processes for long-running software-intensive technical systems?" and RQ1b: "What are requirement areas from engineering processes for long-running software-intensive technical systems that require informatics contributions?" named in Chap. 1 discussing especially challenges related to complexity management.

References

Biffl, S., Gerhard, D., & Lüder, A. (2017). *Multi-disciplinary engineering for cyber-physical production systems*. Cham: Springer.

Bracht, U., Geckler, D., & Wenzel, S. (2017). *Digital Fabrik, Methoden und Praxisbeispiele*. Singapore: Springer.

Brosch, M. (2014). *eine Methode zur Reduzierung der produktvarianteninduzierten Komplexität (German)*. Hamburg: TuTech.

Budde, L. (2016). *Integriertes Komplexitätsmanagement in produzierenden Unternehmen, Ein Modell zur Bewertung von Komplexität*. Dissertation, University of St. Gallen.

Carayannis, E. G. (2013). *Encyclopedia of creativity, invention, innovation and entrepreneurship*. New York: Springer.

Checkland, P. (1999). *Systems thinking, systems practice*. New York: Wiley.

Chen, C. C., Nagl, S. B., & Clack, C. D. (2009). *Complexity and emergence in engineering systems*. Berlin: Springer.

Daryani, M. S., & Amini, A. (2016). Management and organizational complexity. *Procedia- Social and Behavioural Sciences, 230*, 359–366.

DeVries, E. (2005). *How assortment variety affects assortment attractiveness: A consumer perspective*. PhD Thesis, Erasmus University Rotterdam, Erasmus Research Institute of Management.

Dellaert, B. G. C., Donkers, B., & Van Soest, A. (2012). Complexity effects in choice experiment–based models. *Journal of Marketing Research, 49*(3), 424–434.

Fuchs, C. (2018). *Mastering disruption and innovation in product management*. Cham: Springer.

Gabriel, A. (2007). *The effect of internal static manufacturing complexity on manufacturing performance*. Dissertation, Clemson University.

Gadatsch, A. (2015). *Geschäftsprozesse analysieren und optimieren, Praxistools zur Analyse, Optimierung und Controlling von Arbeitsabläufe (German)*. Cham: Springer.
Grussenmeyer, R., & Blecker, T. (2013). Requirement for the design of a complexity management method in new product development of internal and modular products. *International Journal of Engineering, Science and Technology, 5*(2), 132–149.
Jacob, J., & Paul, A. (2016). Business service integration using dependency structure matrix in view of e-government system. *International Journal of Engineering Research & Technology (IJERT), 5*, 5.
Kieviet, A. (2014). *Implications of additive manufacturing on complexity management within supply chains in a production environment*. PhD Thesis, University of Louisville.
Kirchhof, R. (2003). *Ganzheitliches Komplexitätsmanagement: Grundlage und Methodik des Umgangs mit Komplexität im Unternehmen (German)*. Wiesbaden: Springer.
Köhler, C. (2009). *Technische Produktänderungen – Analyse und Beurteilung von Lösungsmöglichkeiten auf Basis einer Erweiterung des CPM/PDD-Ansatzes (German)*. Dissertation, University of Saarland.
Lasch, R., & Gießmann, M. (2009). *Qualität- und Komplexitätsmanagement- Parallelitäten und Interaktionen zweier Managementdisziplinen* (pp. 93–124). Wiesbaden: Gabler, GWV Fachverlage.
Lee, M. G., Chechurin, L., & Lenyashin, V. (2018). Introduction to cause-effect chain analysis plus with an application in solving manufacturing problems. *The International Journal of Advanced Manufacturing Technology, 99*(9–12), 2159–2169.
Lindemann, U., Reichwald, R., & Zäh, M. F. (2006). *Individualisierte Produkte – Komplexität beherrschen in Entwicklung und Produktion (German)*. Berlin: Springer.
Malik, F. (2016). *Strategy for managing complex systems, a contribution to management cybernetics for evolutionary systems* (11th ed.). New York: Campus.
Marti, M. (2007). *Complexity management-optimizing product architecture of industrial products*. Wiesbaden: Deutsche Universität.
Maune, G. (2011). *Möglichkeiten des Komplexitätsmanagements für Automobilhersteller auf Basis IT-gestützter durchgängiger Systeme (German)*. Dissertation, Universität-GH-Paderborn.
Maurer, M. S. (2007). *Structural awareness in complex product design*. PhD Thesis, TU München.
Meier, H., Hanenkamp, N., & Mattern, C. (2005). *Komplexitätsmanagement im Lebenszyklus individualisierter Produkte im Maschinen- und Anlagebau*. Wiesbaden: Springer.
MirRashed, A., Rostami Mehr, M., Mißler-Behr, M., & Lüder, A. (2016). *Analysing the causes and effects of complexity on different levels of automobile manufacturing systems*. IEEE.
Olson, D. L. (2017). *Descriptive data mining*. Singapore: Springer.
Oyama, K., Learmonth, G., & Chao, R. (2015). Analyzing complexity science to new product development: Modelling considerations, extensions. *Journal of Engineering and Technology Management, 35*, 1–24.
Project Management Institute (PMI). (2017). *A guide to the project management body of knowledge (Pmbok Guide)* (6th ed.). Project Management Institute.
Reijers, H. A., & Mendling, J. (2008). *Modularity in process models: Review and effects*. Berlin: Springer.
Richter, K., & Walther, J. (2017). *Supply chain integration challenges in commercial aerospace, a comprehensive perspective on the aviation value chain*. Cham: Springer.
Schady, R. (2008). *Methode und Anwendungen einer wissensorientierten Fabrikmodellierung (German)*. Dissertation, Otto-von-Guericke Universität Magdeburg.
Schoeller, N. (2009). *Internationales Komplexitätsmanagement am Beispiel der Automobilindustrie (German)*. Dissertation, Rheinisch-Westfälischen Technischen Hochschule Aachen.
Schuh, G. (2005). *Produktkomplexität managen: Strategien- Methoden- Tools (German)*. München: CarlsHanser.
Schuh, G., Gartzen, T., & Wagner, J. (2015). Complexity-oriented ramp-up of assembly systems. *Journal of Manufacturing Science and Technology, 10*, 1–15.
Schuh, G., Kamoker, A., & Wesch-Potente, C. (2011). Condition based factory planning. *Production Engineering Research Development, 5*, 89–94.

Schuh, G., Riesener, M., & Mattern, C. (2016). Approaches to evaluate complexity in newproduct development projects. *International Journal of Design & Nature and Ecodynamics, 11*(4), 573–583.

Schwandt, A. (2009). *Measuring organizational complexity and its impact on organizational performance – A comprehensive conceptual model and empirical study*. Dissertation, Technical university of Munich.

Sindermann, S. (2014). *Modellbasierte virtuelle Produktentwicklung, Schnittstellen und Datenaustauschformate (German)*. Berlin: Springer.

Sinha, K. (2014). *Structural complexity and its implications for design of cyber-physical systems*. PhD thesis, Massachusetts Institute of Technology.

Thiebes, F., & Plankert, P. (2014). *Komplexitätsmanagement in Unternehmen (German)*. Wiesbaden: Springer.

Velte, C. J., Wilfahrt, A., Müller, R., & Steinhilper, R. (2017). Complexity in a life cycle perspective. *Procedia CIRP, 61*, 104–109.

VDI 3695. (2010). *Engineering of industrial plants, evaluation and optimization, Part 1*. Berlin: BeuthVerlag.

Verein Deutscher Ingenieure (VDI). (2011, May). *VDI guideline 4499 – Digital factory operations*. Berlin: Beuth.

Vogel, W. (2017). *Complexity management approach for resource planning in variant-rich product development*. Wiesbaden: Springer.

Weber, W., et al. (2014). *Einführung in die Betriebswirtschaftslehre (German)*. Wiesbaden: Springer.

Wemhoener, N. (2005). *Flexibility optimization for automotive body shops*. Dissertation, Techn. Hochsch. Aachen.

Wildemann, H. (2012). *Komplexitätsmanagement in Vertrieb, Beschaffung, Produkt, Entwicklung und Produktion (German)* (13th ed.). München: TCW.

Chapter 4
Engineering of Signaling Systems

Johannes Lutz, Kristofer Hell, Ralf Westphal, and Mathias Mühlhause

Abstract Rail system products demand high standards on safety, security, and quality. To guarantee this, the design of such a system including its processes, utilized engineering tools, and produced data must be approved to meet the strict requirements. Engineering of signaling systems can be utilized as a use case to derive industries' safety and quality challenges regarding engineering tools and data flows.

To understand these challenges in a comprehensive way, this chapter gives an overview about the general signaling business, its engineering processes, and its engineering tools including aspects of data flows and semantics. Based on the author's findings in these subjects, requirements and respectively challenges are derived and summarized in the last section of this chapter. Each list entry is classified as quality and/or safety challenge. Furthermore, this chapter enables the link between industrial needs regarding engineering tool chains and current research in this field.

Keywords Rail automation · Signaling system · Rail system engineering · Integrated data management · Quality-safety and security management

4.1 Introduction: Signaling Industry

The transport of people and goods can be accomplished in various ways. One efficient option is traveling by train. On September 27, 1825, the world's first public steam-powered locomotive, the Locomotion No. 1 began a new era of carrying people and goods. Just 5 years later, also the world's first intercity railway line connecting Liverpool and Manchester was established (Jeans 1875). Since then, the evolution of railway industries continued from mechanical and later electrical systems into fully integrated digital solutions. Expensive construction costs but also increasing

J. Lutz (✉) · K. Hell · R. Westphal · M. Mühlhause
Siemens Mobility GmbH, Braunschweig, Germany
e-mail: johannes.lutz@siemens.com; kristofer.hell@siemens.com; ralf.westphal@siemens.com; mathias.muehlhause@siemens.com

S. Biffl et al. (eds.), *Security and Quality in Cyber-Physical Systems Engineering*,
https://doi.org/10.1007/978-3-030-25312-7_4

popularity and increasing demand of railway transports pushed developments for signaling systems to make traveling by train safer and more efficient.

In this chapter, the engineering process of these signaling systems is presented with focus on quality, security, and safety challenges regarding engineering data in this domain. To understand the field of signaling system engineering, it is necessary to introduce the main components of a rail system as well as its stakeholder and processes. A detailed overview about rail systems is given in Pachl (2018) and Theeg and Vlasenko (2017). While the first section introduces the basic terms in signaling systems, Sect. 4.2 describes the general engineering process of rail systems with a specific focus on signaling systems. Section 4.3 describes the engineering tool types and their relations to each other in establishing the engineering tool chain. Section 4.4 gives an overview on data flows and explains the different roles. Finally, Sect. 4.5 summarizes the domain-specific challenges on safety, security, and quality.

4.1.1 Classification of Signaling Systems

A signaling system can be distinguished into its life cycle phases and target markets based on the considered track distances and transportation type which were commissioned (see Fig. 4.1).

A signaling system has different life cycle phases: the phase of planning and engineering and the phase of operation. In the phase of operation, there are two stakeholders involved: on the one hand, the rail operator (e.g., *DB, SNCF*), which runs trains and transports passengers and goods and, on the other hand, the rail network operator (e.g., *DB Netz, Network Rail*), which provides the infrastructure and defines and ensures the overall operating procedure.

In the life cycle phase of engineering, there are two groups of players: the contractors/manufacturers (e.g., *Siemens Mobility GmbH, Bombardier Transportation, Thales*) that deliver trains to rail operator or trackside equipment

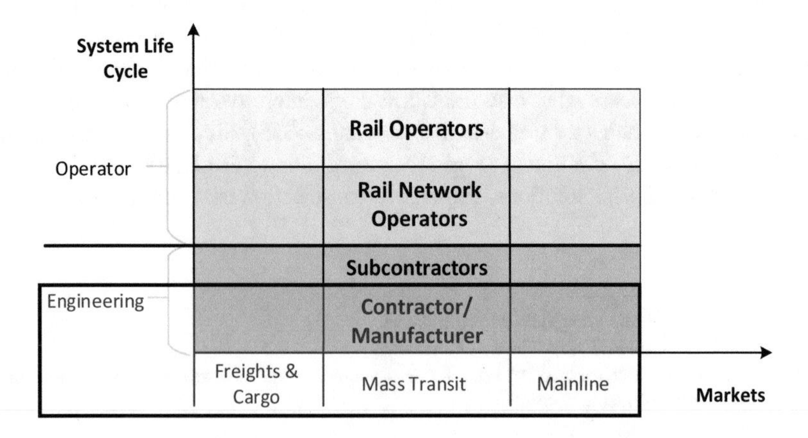

Fig. 4.1 Analysis of signaling systems by life cycle phases and target markets

like electrification, civil works, or the signaling system. Besides, there are subcontractors/submanufacturers who supply parts, modules, or engineering services to contractors/manufacturers. The content of this chapter is limited to the engineering phase of signaling systems and will be described from a contractor's point of view, as in Fig. 4.1.

A further dimension to classify signaling systems is the target market. Thus, it can be distinguished between freight traffic, urban metro lines, and mainline traffic, due to specific product requirements. While mainline products focus on moving people as well as goods on long distances with high speed and various train types, mass transit provides signaling systems for cities and regional customers on short distances like metro or light rail with special focus on high operational capacity. Freight traffic covers particularly highly automated freight terminals with special components, for example, weight and speed detections and rail brakes as well as special signaling systems, such as for mines. By means of typical engineering processes for mainline signaling products, this chapter will present challenges regarding quality and security in signaling engineering data flows. Even if metro or freight systems have specific requirements and dedicated products, the principles of the mainline examples can be transferred into these application areas.

4.1.2 Signaling Products and Systems

Even if the typical lifecycle of a signaling system is longer than in other domains of automation, the impact of digital products and services has changed business rapidly in the last decade. Starting in England with the first mechanical interlocking patented in 1856 (Theeg and Vlasenko 2017) continuously new interlocking types were developed over decades according to the latest technological standards. Nevertheless, digitization also started here with electronic and digital interlockings that were already developed. However, a huge amount of older interlockings is still in deployment and moreover, new products like track vacancy detection systems or automatic train control systems were introduced. Thus, signaling systems became more and more complex. But a modernization of the whole railway infrastructure to the latest state-of-the-art products is very cost-intensive. This gives a first idea of the complexity of the interaction of older mechanical and electromechanical products with new software-controlled and microprocessor-equipped products and components. Depending on its concrete function, the safety integrity level of those products may vary from Safety Integrity Level 1 (SIL1) to SIL4 according to CENELEC standards. Within the control level, interlocking systems as well as train control systems can be used.

ZVEI (2010) compares different domains of automation systems and summarizes that signaling systems in average have longer lifecycles than systems, for example, in the process or automotive industry. The overall life time of equipment is different based on its purpose. It can be determined in outdoor equipment (e.g. signals) and indoor systems like interlockings. Furthermore, the demands from various

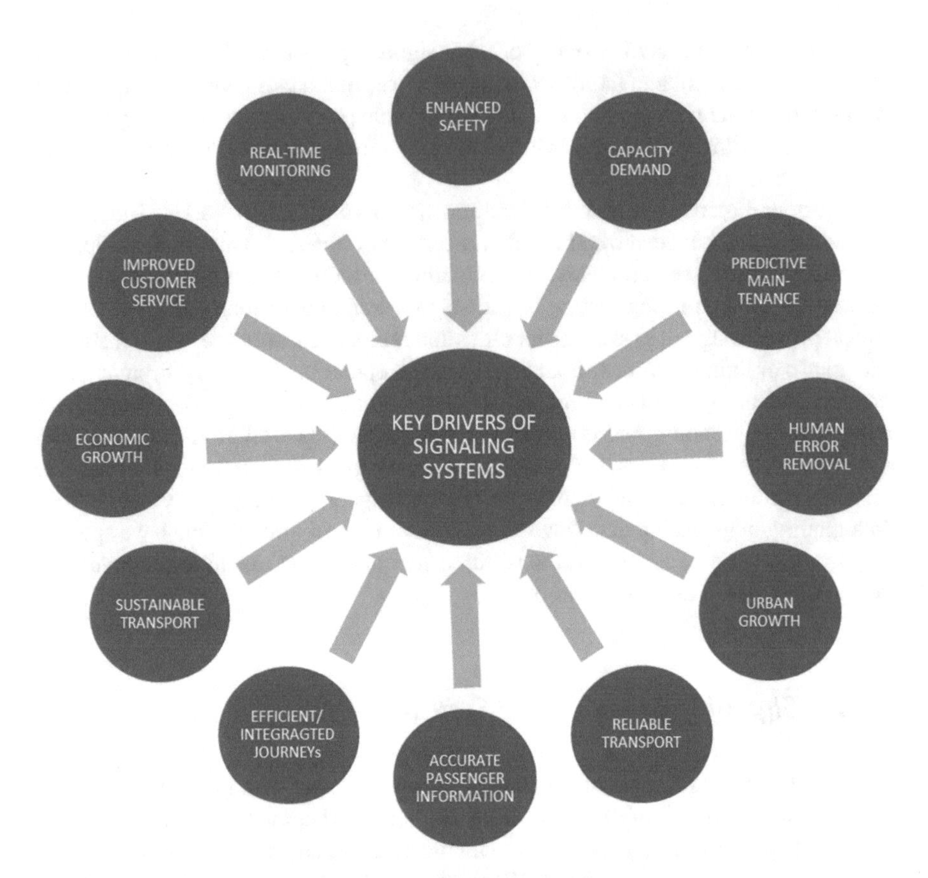

Fig. 4.2 Key drivers of signaling systems (TUEV-SUED 2017)

stakeholders on rail transportation systems, mostly achievable by signaling products, have also become more complex, as shown in Fig. 4.2.

These general demands on key drivers of signaling products are independent of the target market of the products. Several products relating to infrastructure are required to provide a modern signaling system that meets the customer expectations (see Fig. 4.3). Here, only main product groups are shown as every manufacturer has an own portfolio belonging to each product group. The classification will help to understand the general correlation between the functionality of signaling systems and their quality, safety, and security requirements. A detailed view about demands and principles is given in Theeg and Vlasenko (2017).

Operations Control Center Overall control of railway operations in the assigned railway network section. These functionalities cover the allocation of trains as well as its timetables, optimization of operational capacity of railway networks, intervention in failure scenarios like blocking routes, real-time monitoring, and providing passenger information at stations.

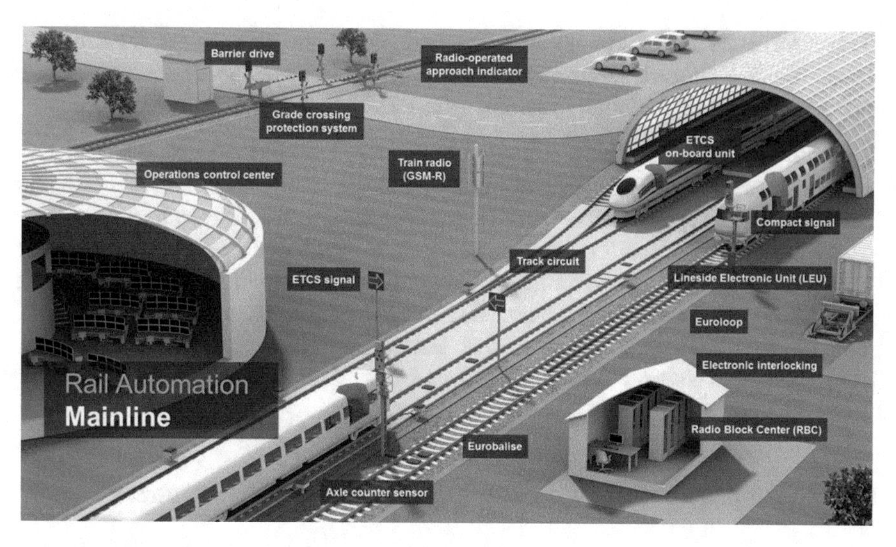

Fig. 4.3 Main products of a mainline system (Siemens AG, 2015)

Interlocking System Setting of safe routes based on logical rules. Therefore, the interlocking controls points and signals that are linked by fixed logical rules to provide a safe route, that is, guarantee route locking, flank protection, opposing route locking, track vacancy. Due to its safety-critical functionality this product group is typically classified SIL4.

Automatic Train Control System Safety system to decrease the risk of accidents made by human factors. Therefore, the automatic train control can supervise the train and automatically reduce the speed or follow a speed profile as well as stop the train. Towards a higher realized automation level (e.g. automatic train protection up to automatic train operation), the number of functionalities as well as the required equipment is increasing. The figure shows the automation control system called European Train Control System (ETCS), which consists of signals, axle counter sensors, or track circuits to detect the track vacancy, Eurobalises, and according to the ETCS level it will be complemented by radio block centers (RBC) and train radios (Global System for Mobile Communications—Rail(way), GSM-R) for continuous wireless communication and/or signals. Train control systems allow a higher degree of automatic train control up to driverless operation that is already in use for urban metro systems and expected in the upcoming decade also for mainline systems.

The field level can be distinguished between sensors to detect trains on a track and actors to control movement of trains.

Level Crossing Safe crossing of railway tracks and motorways. A level needs to detect an approaching train and needs to lock the motorway to ensure that nobody enters the railway tracks until the train has passed. The approach detection can be realized with different technical methods and data. The level crossing can be enhanced with further safety devices like radar.

Track Vacancy Detection System that evaluates the track vacancy, for example, based on axle counter sensors or based on track circuits. The track vacancy is a necessary information for an interlocking to lock/unlock a certain safe route.

Components Devices that are directly built on or next to the track. Thus, signal, axle counter sensor, balises are examples of single components that are essential parts of superior systems like interlockings or automatic train controls.

Additionally, it is required to interact within the trackside control system and the train. For this, positions are transferred via balises to the train and commands from the Radio Block Center (RBC) with GSM-R technology to the train (ETCS Level 2). Besides, the train itself is equipped with sensors like odometrical pulse generators, radars, etc., and controlled by an On-Board-Unit.

4.1.3 Requirements and Challenges of Signaling Products

In this section, general requirements and challenges for engineering rail industry products are listed. This collection will give the reader an understanding of this industry to further derive demands toward safety and quality of engineering data in signaling (see also (Falkner and Schreiner 2014) or (ZVEI 2010)).

4.1.3.1 Challenge 1: Product Variety and High Degree of Product Customization

While it looks easy to point out the overall product groups, it is much harder to get an overview about certain products that belong to a product group. Since every country in the world and even sometimes the different signaling systems within a country have their own regulations, a wide range of products emerged often specially adapted to the customer needs of the rail operator or rail network operator.

4.1.3.2 Challenge 2: Very Long Lifecycles of Products

One additional typical characteristic is the very long product lifecycle of the products in the rail domain. Product lifecycles of signaling systems are typically up to at least 30 years; especially outdoor equipment is used longer. Common demands for new products are coexistence, compatibility, migration, or interoperation to existing installed products, which will remain. Thus, realizations of technology leaps are highly challenging due to high amount of brown field projects. Availability, timeliness and digitalization of engineering data have to be considered in brown field projects and especially against the background of very long product lifecycles. Concepts regarding handling, archiving, and updating engineering data including its configuration tools are also needed to fulfill long-term service and maintenance contracts according to products' lifecycles with respect to future-proof data format and data update capability.

4.1.3.3 Challenge 3: Large-Scale Railway Network

A crucial reason why the rail industry is such a complex field is the open cross-border railway network. But with respect to the passenger needs, the transnational railway network allows fast and comfortable journeys through different countries. If a train crosses a country border it still must fulfill all requirements on mechanics, electrical, control logic, regulations, etc. That means that rail systems are not closed systems and need to be interoperational or at least a consideration of system transitions is necessary, which applies to both passenger as well as freight transport.

4.1.3.4 Challenges 4: Country-Specific Laws, Regulations, Operational Guidelines, Authorities, and Infrastructure

Considering this background, signaling products need to be developed and adapted for cross-national railway regularities, for long-term reliability as well as high-level safety integrity level. Regarding the long-term use, products need to be developed and built as simple and robust as possible to guarantee long-term maintenance and availability. Due to the long service demands as well as high demands on safety, spare parts based on commercial off-the-shelf (COTS) are typically limited on nonsafety parts and thus they need to be developed as generic as possible as well to fulfill these requirements.

As mentioned above, the diversity of country-specific regularities is a huge challenge for the manufacturers and their engineering departments as well as for the manufacturing process. Although every rail network of several countries need the same product portfolio of special product types out of the portfolio of a single country, they often cannot be reused in other countries and need to be reengineered or at least adapted to the new customer requirements. As a result of individual country-specific requirements, usually, the manufacturer deals with hundreds and up to thousands of different products—each with a very small manufacturing lot size. This leads to a high complex engineering and manufacturing process, which needs to be coordinated and synchronized very efficiently and effectively.

4.1.3.5 Challenge 5: Diversity of Rolling Stock

A characteristic of mainline railway networks is the occurrence of different train types in mixed traffic operations that use the same infrastructure and need to interact with the same signaling systems. Examples are high speed trains, local trains, freight trains or special trains for shunting, construction, etc. Parameters to consider are among other things platform height & length, maximal speed, train length, communication and signaling equipment onboard the trains, acceleration & braking characteristics, route gradients, technique for completeness check of trains to avoid loss of wagons.

4.1.3.6 Challenge 6: Data Security in Networks

Compared to terminated area of automotive or process sites, large-scaled signaling systems have advanced demands on IT-security and data integrity. With respect to an increasingly connected world with strong impacts of network effects and its unauthorized manipulation, data security of all assets in the field are of the highest priority and are in focus right from scratch of product development (IEC 62443 3-3 SL 3, 2013). The continuous process of product improvement guarantees also the latest security standards for older installed products.

4.1.3.7 Challenge 7: Safety of Engineering and Run-Time Data

At the very end, all efforts guarantee the required level of integrity, availability, traceability, and confidentiality. The field applications, products, systems, and services are developed and used in accordance with the applicable legal and normative standards meeting highest standards in all relevant phases of the lifecycle (see Fig. 4.4).

Fulfilling highest safety requirements (Safety Integrity Level, SIL) and specific customer demands because of different regulatory requirements of each country, a seamless collaboration of each participating project partner is strictly necessary. Thus, all engineering tools and generators of run-time data have to be assessed and if necessary approved against the assigned safety level. Thus, components of the shelf cannot be simply used or integrated in the engineering workflow. Extensive safety and quality assessments and approvals are mandatory for tools as well as for generated data.

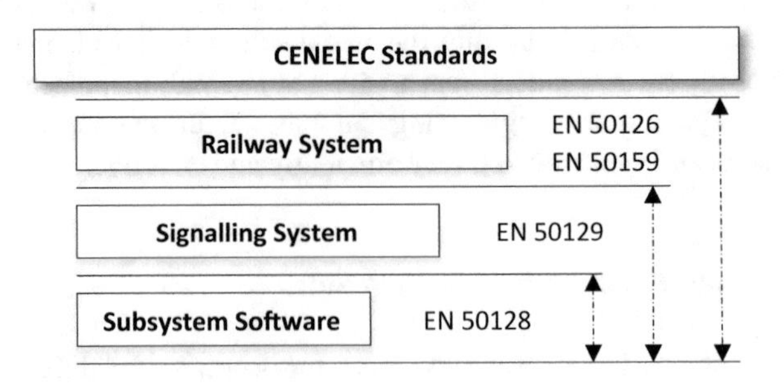

Fig. 4.4 Main products of a mainline system

4.2 Engineering Process of Signaling Systems

A typical engineering process for a system can be divided into four phases: project kick-off and bid phase, preliminary design, civil works, operation and service(see Fig. 4.5). Civil works can be refined into subdisciplines like the installation of tracks, concrete works, electrification, or the signaling system that is analyzed in this chapter. The concrete set of process steps as well as required roles must be tailored depending on systems newly built up (green field projects) and extension/refurbishment of existing sites (brown field projects).

Furthermore, the size or complexity of the project influences the chosen process approach. The bandwidth of project duration differs from a few months for small updates or installations with less complexity up to frameworks with several years. In terms of the engineering processes, small projects requires fewer engineering steps whereas large projects must be orchestrated into different engineering and commissioning stages as well as much more synchronization steps in between the involved disciplines.

On an abstract level there are different parties involved: customer (rail operator), engineering offices/subcontractors in charge for the preliminary planning, and the contractor. Depending on the project, contractors may be distinguished into further

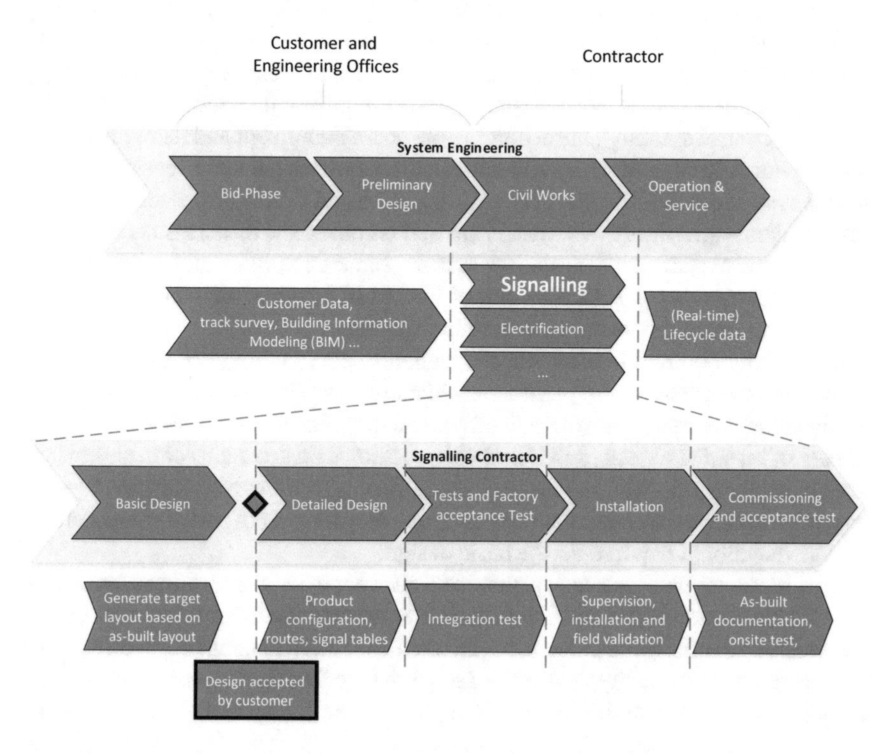

Fig. 4.5 Phases of the engineering process of signaling systems

roles, for example, engineering, construction, or commissioning. Of course, each main discipline may be refined to its organization specific roles.

During the bid phase, the customer defines requirements to be fulfilled with the new or refurbished project. The main drivers are:

- The improvement of the systems performance (increase the capacity to transport passengers and goods)
- Improvement of systems efficiency (e.g., reduction of energy consumption, resources, maintained assets)
- Improvement of services and customer or internal processes

The preliminary design includes the basic characteristics of the system like basic characteristics of the railway topology (geographic, gradients, location of points, etc.). It is the basic input for the disciplines summarized in the civil works.

From the signaling contractors' point of view, the first planning steps are initiated in the bid phase. Depending on the size of the project this phase can last between few weeks up to years. The railway companies publish a bid invitation including basic information about the project, requirements on the solution as well as basic information about topology and—in case of brownfield projects—about existing installations.

Bidding signaling companies create a first architecture proposal how to fulfill the customers' requirements. This submits an initial proposal how to equip the railway track as well as corresponding installation locations, e.g. for the main hardware (HW) assets (signals, point machines etc.) including the cabling concept for power supply and data communication. For signaling systems requiring interaction with the train (e.g. train control systems), the according on-board equipment and its requirements must be considered as well. Additionally, simulations are done to ensure that the offered system can fulfill requirements on capacity and timetable or—especially for urban systems—the headway which means in this context the actual distance between two consecutive trains on the same track in the same direction. Among commercial aspects, technical aspects are also part of the bid like system architecture in which the performance of the technical system is described.

In case the contract is awarded, the signaling contractors need more accurate input data than given in the preliminary planning. Among the requirements itself the signaling contractor gets the refined preliminary planning data, such as type and number of installed objects, and if available position data and status of these objects, latest as built maps as well as schematic track layouts. If necessary, track surveys are done additionally to increase data quality (e.g., detailed lists of installed equipment, accurate locations) required for the basic design.

One trend to be observed is that data exchange between railway company, measurement campaigns, and contractor is more and more supported by digital interfaces. Concerning the process, the interaction of data exchange between contract partners is not clarified. Typically, the data exchange format only covers syntax and semantics. Compared to the paper-driven process, the sender of data is responsible for their correctness, which is specified in the associated regulation processes. Thus, digital data exchange must also guarantee the data quality. A possible approach is defined by Buder (2017) to embed the data exchange format PlanPro into an exchange process.

The basic design phase refines the architecture that was created for the bid. As one example, the preliminary track layout includes only basic signaling elements like points, signals, axle counters, as well as important installation locations like stations, etc. The complete track layout includes every element with impact on the signaling system, for example, emergency plungers, platform screen doors, etc. (see Fig. 4.6).

Furthermore, the system architecture is extended, for example, to determine borders of the interlocking system, communication concept, etc. as well as safety/RAMS (Reliability, Availability, Maintainability, Safety) evaluations. The basic design phase is finished if the railway company accepts the contractor's proposal.

The design phase can be subdivided into product design processes (see Fig. 4.7). Depending on the contractor's setup the team is organized in different engineering groups contributing to the overall systems solution.

Focusing on the detail design of interlockings, here, two technical disciplines can be distinguished: interlocking hardware design and software design. Tasks in hardware design are, among others, the configuration/engineering of various electrical cabinets, related circuit diagrams and other technical drawings, bill of material to order the equipment, and the hardware interconnection between the various cabinets.

Typical tasks in interlocking software design are definition of routes, of logical dependencies, signal tables, and enhancement of track layouts with attributes needed for interlocking software, for example, directions, neighbor objects. Finally, the software or its configuration files are generated to integrate and commission the system. The detailed design phase is followed by the factory acceptance test phase, which contains extensive integration tests.

The last phases are the construction phase, in which the signaling system is installed in the field, and the commissioning phase. Here, the engineering outputs of the detailed design phase are integrated. They must interact together and meet all the customer requirements. Finally, the system must be validated by the railway regulatory authority to be handed over to the customer and start the operation phase.

4.3 Engineering Tools and the Engineering Tool Chain

To finish the design of complex systems, engineers must be supported by tools in order to ensure the quality of the solution as well as to fulfill the different engineering activities efficiently. Thus, they support the assembling of all single products into a system that meets all the customer requirements (see Fig. 4.8).

Among other industries, the tool support during the engineering process of signaling systems is a mandatory measure to get rid of paper-driven work and processes—of course, various engineering activities still require documentation as an output, for example, to fulfill customer requirements or regulations. Nevertheless,

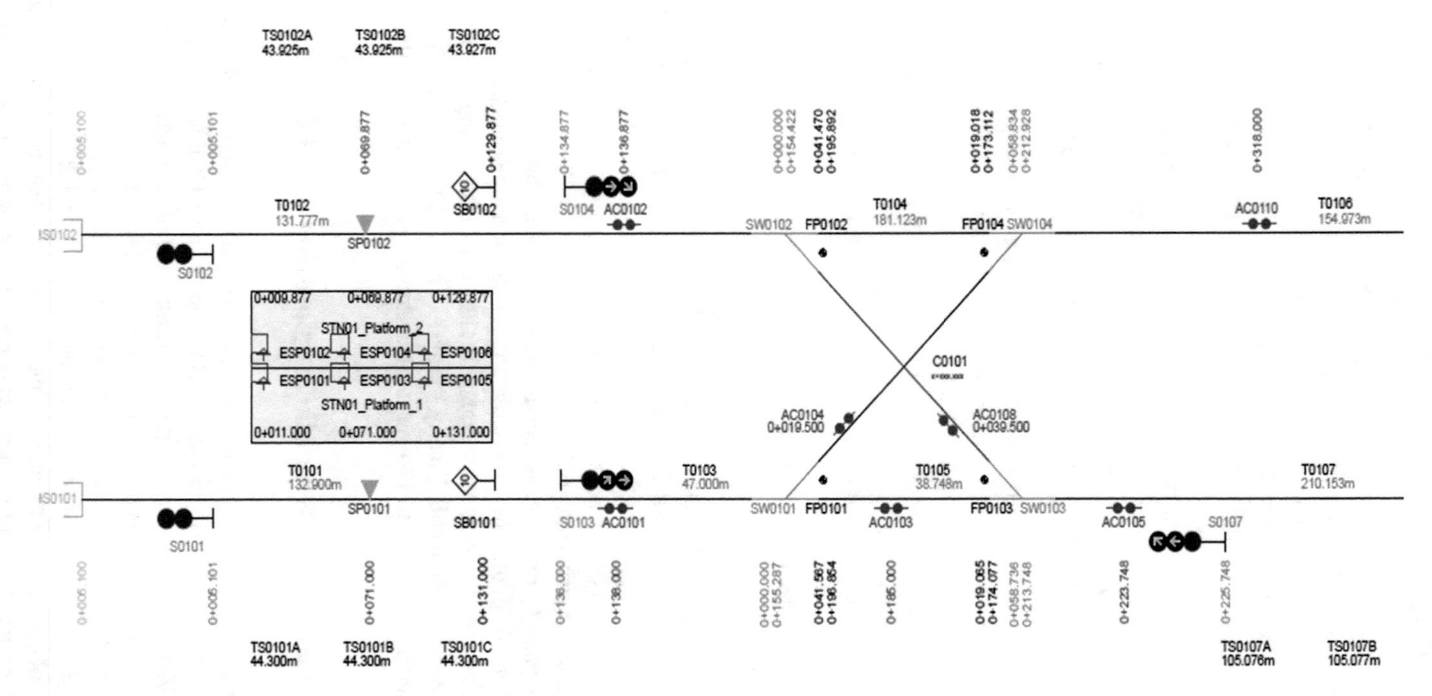

Fig. 4.6 Exemplary track layout in the basic design phase

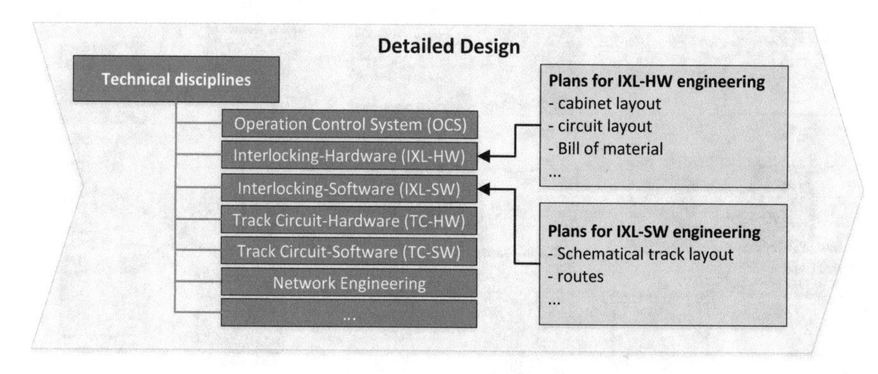

Fig. 4.7 Phase of detailed design with its subdisciplines and input data

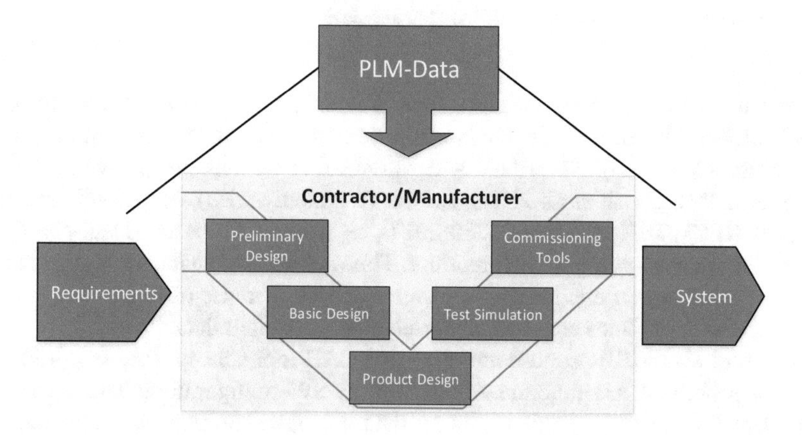

Fig. 4.8 General overview of signaling system engineering

the tool support has to be automated as much as possible to reduce manual extra work and to decrease process duration.

The engineering tool enables an activity or user-centered view to design the model of the system. Thus, different tools' aspects can be identified:

- CAE/CAD-centered tools to document the design
- Software (SW) configuration tools to generate the files for the run-time system

The variety of solutions and their architecture depend on the organization needs. For standard CAD-design tasks often the COTS solutions on the market are used. The data exchange with those interfaces is typically document driven. This is accompanied by a loss of meta data, besides, dependencies within engineering data may not be transferred. This missing information has to be potentially restored in the subsequent tool chain. Within the engineering process this class of tools is typically part of a tool chain—data is exchanged sequentially. Depending on the engineering activity this tool chain is combined with SW-configuration tools. The data exchange

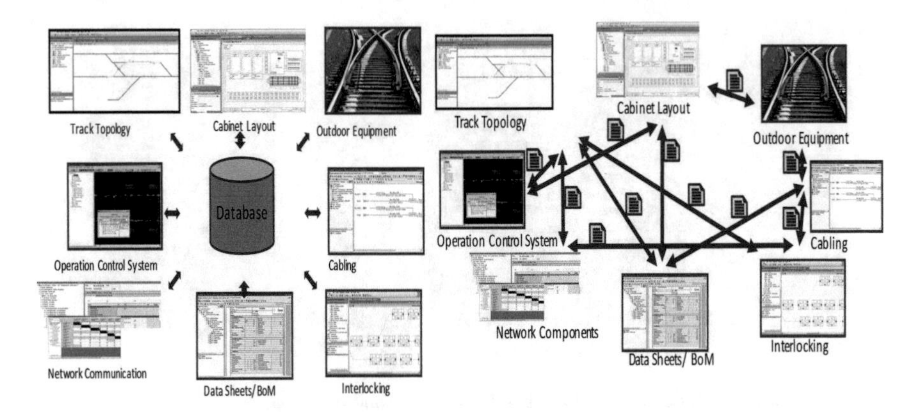

Fig. 4.9 Centralized vs. distributed engineering approach

is organized by means of interface files. For this, proprietary formats are used, frequently combined with domain-specific content (see Linder and Grimm (2012) and Maschek (2018)). Nevertheless, harmonization on domain-specific semantics such as railML (Nash et al. 2004), Eulynx (www.eulynx.eu), Industry Foundation Classes (IFC) (DIN EN ISO 16739 2017), as well as 3D-based (Deutsche Bahn 2018) formats, is getting more important. This architecture enables a higher degree of standardization of engineering semantics as well as reuse of tool assets, that is, the different disciplines are coupled closely and share their data directly.

Holm et al. (2012) propose an integrated CAE tool concept (see Fig. 4.9) that combines both—CAD-functionality as well as SW-configuration. This tool class organizes its semantics explicitly in its data management, that is, documents are treated as "intelligent view" on engineering semantics like a table-based graphical user interface (GUI) (see Fig. 4.10) (Holm et al. 2012). Furthermore, they stated that this approach causes higher requirements on data integrity as well as version and workflow management. Decoupled, single tools manage their own particular—and most times—terminated scope on one engineering discipline. Data integrity and version management are allocated externally. Compared to this, integrated tool approaches have a high demand to have the manage functionality internally. Seidel et al. (2017) propose how to extend the integrated approach with workflow management systems known, for example, from commercial market platforms to guide customers through the order process. Bigvand and Fay (2017) verify this strategy for the domain of process automation.

Another characteristic of engineering tools used in this domain is the high level of data integrity checks as well as automatic derivation to ensure data integrity to fulfill safety-relevant needs and foster efficient engineering. Falkner and Schreiner (2014) and Haselboeck and Schenner (2014) define an expert system—as known from other domains like medical processes or predictive maintenance systems. Engineering knowledge is formalized in a data model as well as constraints to check data correctness and rules to derive new data automatically. Duggan and Boraelv (2015) describe a mathematical proof of an automated environment that ensures data

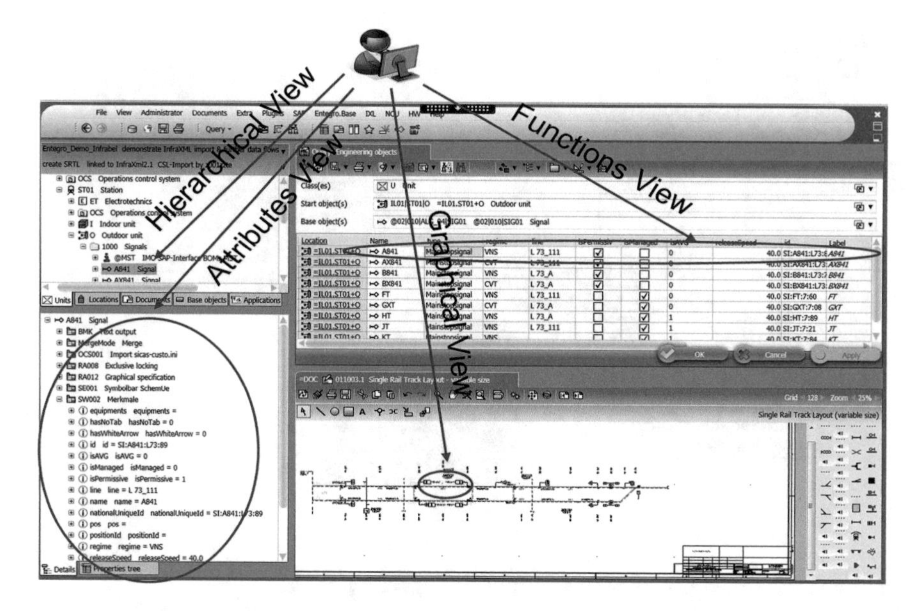

Fig. 4.10 Data views based on engineering discipline and role

integrity for interlocking systems. The precondition in both approaches is the ability to describe the engineering domain in a formal and all-embracing model. It ensures data correctness but restricts flexibility in case of customer-specific needs. From the authors' point of view, this is partially achieved in the engineering process if model and logic is standardized and repeatable. The unwanted side effect is shrinking responsiveness of engineering activities with higher needs on flexibility.

To reflect the engineering process the scope of engineering tools can be classified in general within the main process steps (see Fig. 4.11).

Basic design tools enable the design of the railway target topology based on the preliminary design data. The level of detail is broken down so that both customer and involved disciplines are able to understand the main installations on the railway track. Furthermore, tools are used to define the system architecture and document basic characteristic of the systems (train control, interlocking) as well as its interconnections. Finally, simulation tools are used to ensure that the design fulfills the customer requirements, for example, for customer capacities or headways.

If the basic design is accepted, the product design tools are used to configure and document its specifics. Depending on the product type and its functions this tool class requires a higher level on safety critical development specified in the tool classification of (DIN 50128 2012) typical outputs of product design tools are:

- Refinement of the basic design architecture
- Documentation of HW for the construction and commissioning staff as well as for the customer
- Bill of materials to initiate order processes
- Data configuration to generate target files

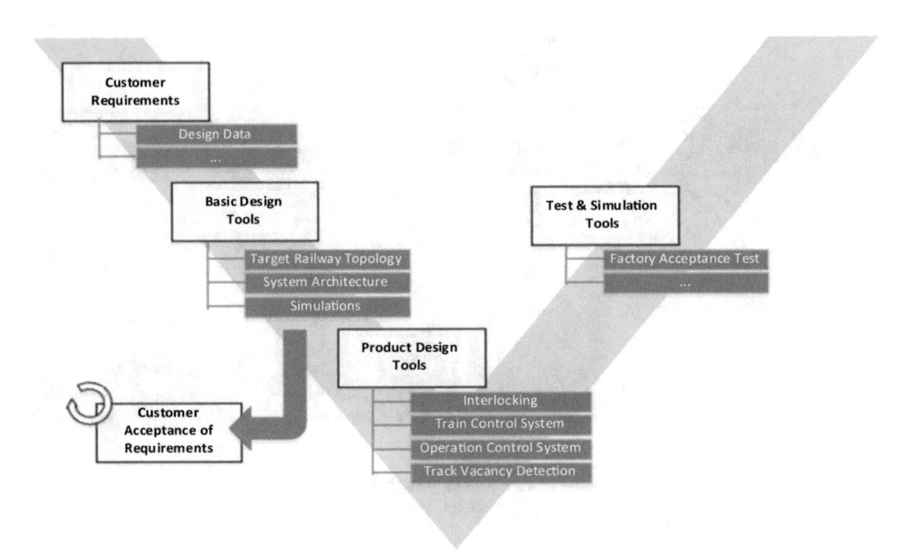

Fig. 4.11 Generated engineering data in different engineering phases

Test and simulation tools are mainly used in factory acceptance tests. These tests must ensure the correct data configuration in the product design process. This includes virtual commissioning of the configured software and HW-tests, for example, to approve the correct assembly of HW-modules in the cabinets.

Depending on the used product, the data quality must meet the requirements up to SIL 4. This is ensured either by process-oriented measures like 4-eyereviews or the engineering tooling being designed based on the required tool classification level (DIN 50128 2012) to guarantee that the data is not corrupted. Further measure to ensure correctness on approved data is the usage, for example, of checksums for data integrity.

Furthermore, the impact on IT-security is rising due to the fact that organizations, including their IT-systems, are getting more and more global. Also, modern tool architectures are being changed from local file-based installations to client-server architectures up to cloud solutions with global usage. This leads to the demand to protect engineering tools, its data, and of course the embedded IT-environment against unauthorized access.

4.4 Engineering Model

It is important to define required engineering data, its flows, as well as the different roles of stakeholders dealing with the data. The previous chapters introduced the process aspects including tool support to design complex signaling systems. With respect to the semiotic triangle of the model theory (Odgen and Richards 1923),

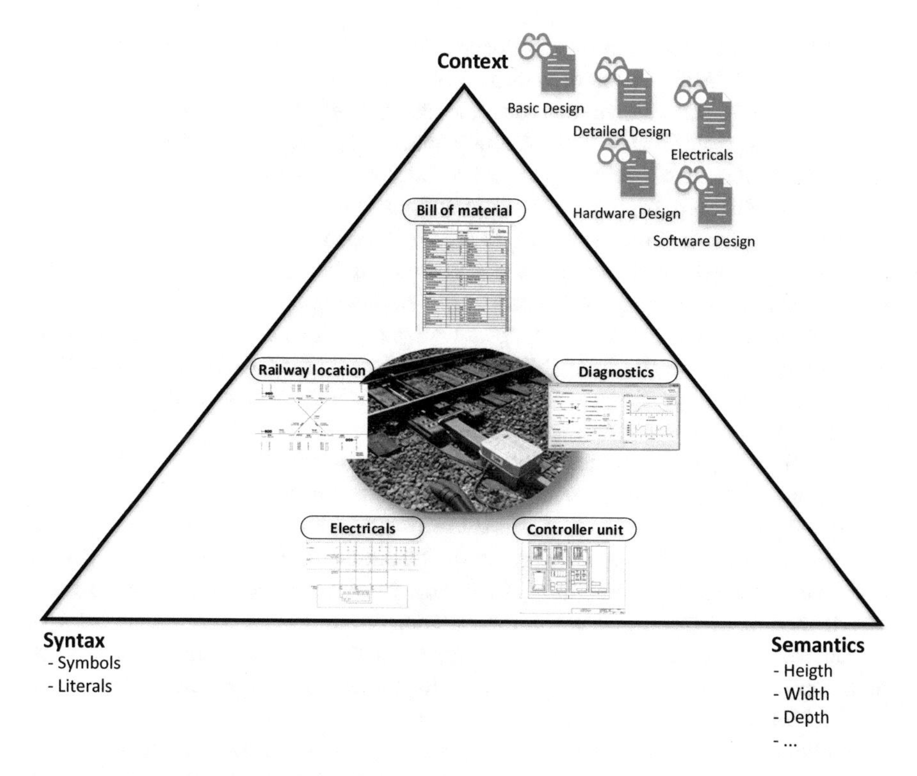

Fig. 4.12 Requirements on data models for signaling engineering

the process as well as the stakeholder of a process step reflect the pragmatics (see Fig. 4.12). The engineering tool is an instance of a semiotic triangle because it:

- Defines the basic of available signs and symbols that can be used to describe an engineering artifact (syntax)
- Structures the aspects of each engineering entity and its relations based on a model (semantics)
- Transfers and visualizes relevant semantics in the graphical user interface for the different engineering roles and enables interaction to design the system (pragmatics)

Each facet of the system design needs to be specified in the engineering model. Similar to DIN EN 81346-1 (2013), it is possible to distinguish each entity in the categories of function, location, and product, which is used in various classification systems, for example, power generation or process automation. Functional entities specify the signaling system in a generic and product-independent manner. Typical examples are elements on railway track layouts like signals, switches, buffer stops, or axle counters (see also Fig. 4.12). The location is an aspect to define a position of an entity, that is, it is another product-independent perspective on the railway design and thus important, for example, for construction or ensuring that distances between elements fulfill the safety requirements.

The product perspective includes the semantics of each asset installed in the signaling system and is typically captured by the life cycle data (PLM data). Also, the PLM perspective is typically aggregated by sets of data, for example, electricals, mechanicals, SW-configuration, commercials, or graphical representation for documents as well as 3D-applications.

The coherency of these aspects is ensured within the design process. Depending on the engineering activity the according project discipline is interested in a subset of design data. The basic designer's perspective typically covers functional as well as locational aspects by defining the generic elements on the railway layout as well as the overall systems architecture including its interfaces. The succeeding disciplines specify the concrete products and their detailed data.

The following example illustrates the refinement of data based on a signal. The functional or safety-related demand to control start/stop positions of trains is typically part of the preliminary design (e.g., start/stop in railway stations, control the sequence of train runs within the signaling system). Depending on the signaling companies system it may be possible that this information is refined in the basic design process and depending on the customers and safety acceptance.

The basic designer creates the railway track layout, the control table defining start and stop points, as well as possible paths for trains and assigns the signal to the interlocking system from where it is controlled.

Within the product design, the HW-designer defines the mechanical and electrical relevant data by assigning the concrete trackside signaling equipment as well as the indoor equipment to be installed in the cabinet and its wiring in between. The view of the SW-designer is to set the data required to control the signal, for example, setting of the IO-configuration to display the correct signal aspect. The designer of the control center uses the set of data to display it for the dispatcher, set up the time table system or interpretation of the diagnostic messages, etc.

4.5 Conclusion and Challenges on Engineering Tools, Processes, and Data

In Sect. 4.1.3, challenges of the business and therefore on the domain-specific systems are summarized. Of course, they also have an impact on the engineering process as well as on its tools and data. With respect to the engineering of signaling systems, the following challenges or main drivers on quality, safety, and security need to be considered.

4.5.1 Challenge: Long Lifecycles of Signaling Systems

Since signaling systems are in operation over several decades, also the accessibility of its engineering data must be ensured. This demands high requirements on archiving as well as on the ability to open and update data at each point in the lifecycle. For this, either the original tools must still be available—that is challenging in rapidly changing IT-infrastructures—or a safe and reliable data migration strategy is used (see also Kuny 1997).

4.5.2 Challenge: Safety

Safety needs to be proven for final products, thus, tools providing engineering or run-time data have to be approved according to safety requirements. To ensure this, the safety standards accordingly describe technologies and processes for its development and maintenance. A possible solution to ease the development of safety-critical tool environments is a modular approach to distinguish between safety and non-safety-related parts.

Furthermore, the design of a safety-critical signaling system must follow rules to ensure the correctness of data. For this, the process and its engineering tools must follow principles of user roles, process milestones, and approved data integrity.

4.5.3 Challenge: Digital Twin

Railway and rail operators shall be able to access and reuse up-to-date operational as well as engineering data through the whole product life cycle of each product at any time, for example, for optimization or maintenance scenarios. Thus, the engineering data generated by tools has to be maintainable and accessible over the very long product life cycle by several different stakeholders. Furthermore, they must be capable of being integrated and to combine with other data sets, for example, operational data, asset management data to derive new inferences.

4.5.4 Challenge: Variant Management

Signaling projects must ensure a high degree of flexibility on country or customer needs, that is, the main systems are not COTS and information to be engineered often differs. For this, the engineering tools must also support a high flexibility to configure and integrate product variants for engineering tools and keep the high degree regarding process automation. The basis for this is a generic metamodel on data representation and semantics like, for example, railML or eCl@ss (Nash et al. 2004).

4.5.5 Challenge: Reuse of Data

Tools shall support and enable the reuse of existing engineering data in upcoming engineering processes, for example, for refurbishments of systems, for new products or for other life cycle phases of the same product. Thus, the engineering processes and used tools shall provide premises for reuse like product-independent engineering processes; processes and tools to find, access, and select existing engineering data; and processes to ensure data quality and validity (see VDI 3695 2010).

4.5.6 Challenge: Integrated Data Flow

Engineering data shall be consistent at any time through the tool chain/in different tools. Thus, the tools must be integrated to enable the consistency requirements without process interruptions or manual processes in-between. Possible solutions are: a monolith system architecture or a common data exchange format that must be able to represent all data types required in the signaling industry.

4.5.7 Challenge: Version Management and Updates

Parallel working within a discipline as well as interacting with other stakeholders shall be enabled. Thus, the tools themselves require functionality to support the processes (Holm et al. 2012) or require embedded measures like workflow management systems (Seidel et al. 2017) or an engineering service bus (Sudindyo et al. 2013) within processes that could be used. This data management contains archiving of outdated data, traceability of changes, inconsistency checks especially for parallel working, rule out data changes by third-party disciplines which are not writer of this particular data, display of updated data made by others and notify users about impact for tasks dedicated to the responsible engineer.

4.5.8 Challenge: Global Business and Collaboration

Due to increasing global supply chains, availability and accessibility became important drivers. The possibility to access the engineering data from multiple destinations, multiple parties with different roles and to create an underlying process to ensure the quality, traceability, and especially security is a steady growing demand.

Table 4.1 Impact matrix on quality, safety, and security

Challenge	Quality	Safety	Security
Long lifecycles of signaling systems	X		X
Safety		X	
Digital twin	X		X
Variant management	X	X	
Reuse of data	X		
Integrated data flow	X	X	X
Version management and updates	X	X	
Global business and collaboration	X		X
Data integrity between railway operator and constructor		X	X
Security on access of tools and data			X

4.5.9 Challenge: Data Integrity Between Railway Operator and Constructor

Exchange of digital engineering data between the railway operator and contractors must ensure integrity. One challenge is the commitment on common data semantics like railML. Furthermore, the sender must ensure correctness of data to be transferred as it is guaranteed in a paper-driven process (see also Buder 2017).

4.5.10 Challenge: Security on Access of Tools and Data

Due to the increasing availability of engineering tools and their data, the protection against unauthorized access becomes a higher demand compared to paper documentation that is only available locally. Encryption technologies and methods to restrict access are possible solutions.

A summary of the impact of the mentioned challenges with respect to quality, safety, and security an overview is shown in Table 4.1.

References

Bigvand, P. G., & Fay, A. (2017). *A workflow support system for the process and automation engineering of production plants*. Toronto: IEEE ICIT.

Buder, J. (2017). *Neues Planungsverfahren für Anlagen der Leit- und Siche-ungstechnik auf Basis durchgängiger elektronischer Datenhaltung*. Dissertation, Saechsische Landesbibliothek-Staats- und Universitaetsbibliothek Dresden: Technische Universität Dresden.

DB Station&Service AG & DB Netz AG. (2018). *BIM-Vorgaben, BIM-Methodik, Digitaltes Planen und Bauen*. Berlin.

DIN EN 50128. (2012). *Railway applications – Communication, signaling and processing systems – Software for railway control and protection systems*. German version EN 50128:2011. Beuth: Berlin.

DIN EN 81346-1. (2013). *Industrial systems, installations and equipment and industrial products – Structuring principles and reference designations – Part 1: Basic rules (IEC 81346-1:2009)*. German version EN 81346-1:2009. Beuth: Berlin.

DIN EN ISO 16739:2017-04. (2017). Industry Foundation Classes (IFC) for data sharing in the construction and facility management industries (ISO 16739:2013); English version EN ISO 16739:2016.

Duggan, P., & Boraelv, A. (2015). *Mathematical proof in an automated environment for railway interlockings*. Technical paper in IRSE presidential program presented at IRSE NEWS 217, London.

Falkner, A., & Schreiner, H. (2014). Configuration and reconfiguration in industry. In A. Felfernig, L. Hotz, C. Bagley, & J. Tiihonen (Eds.), *Knowledge-based configuration – From research to business cases* (pp. 199–210). Waltham, MA: Morgan Kaufmann. (chapter 16).

Haselboeck, A., & Schenner, G. (2014). S'UPREME. In A. Felfernig, L. Hotz, C. Bagley, & J. Tiihonen (Eds.), *Knowledge-based configuration – From research to business cases* (pp. 263–269). Waltham, MA: Morgan Kaufmann. (chapter 22).

Holm, T., Horn, S., Lehmann, O., & Seidel, H. (2012). Reference model based design of tool landscapes for rail infrastructure engineering. In B. Katalinic (Ed.), *DAAAM international scientific book 2012* (11th ed., pp. 267–276). DAAAM International Vienna.

IEC 62443-3-3. (2013). *Industrial communication networks – Network and system security – Part 3-3: System security requirements and security levels*, IEC2013.

Jeans, J. S. (1875). *Jubilee memorial of the railway system: A history of the Stockton and Darlington railway and a record of results*. Michigan: Longmans, Green, and Company.

Kuny, T. (1997). The digital dark ages? Challenges in the preservation of electronic information. In *Paper presented at the 63rd International Federation of Library Associations and Institutions (IFLA) council and general conference*. Copenhagen, 31 August–5 September 1997. Retrieved December 17, 2018, from https://archive.ifla.org/IV/ifla63/63kuny1.pdf.

Linder, C., & Grimm, M. (2012). Datenmodellanalyse zum Austausch von Projektierungsdaten für Stellwerkssysteme in INESS. *Journal of Signal and Draht, 104*(9), 16–21.

Maschek, U. (2018). *Sicherung des Schienenverkehrs. Grundlagen und Planung der Leit- und Sicherungstechnik* (4. überarbeitete und erweiterte Auflage). Wiesbaden: Springer.

Nash, A., Huerlimann, D., Schuette, J., & Krauss, V. P. (2004). *RailML – A standard interface for railroad applications*. In COMPRAIL 2004, Dresden.

Odgen, C. K., & Richards, I. A. (1923). *The meaning of meaning*. London: Kegan Paul, Trench, Trubner.

Pachl, J. (2018). *Railway operation and control* (4th ed.). Mountlake Terrace, WA: VTD Rail.

Seidel, H., Mühlhause, M., Jaeger, T., Fay, A., & Diedrich, C. (2017). Automatic workflow generation in engineering processes. *Automatisierungstechnik, 65*(1), 37–48. https://doi.org/10.1515/auto-2016-0096.

Sunindyo, W., Moser, T., Winkler, D., & Mordinyi, R. (2013). Project progress and risk monitoring in automation systems engineering. In D. Winkler, S. Biffl, & J. Bergsmann (Eds.), *Software quality – Increasing value in software and systems development* (pp. 30–54). Berlin: Springer.

Theeg, G., & Vlasenko, S. (2017). *Railway signalling& interlocking: International compendium* (2nd revised ed.). Bingen: PMC Media House.

TÜV-SÜD. (2017). *The future of rail automation*. Retrieved October 25, 2018 from https://www.tuev-sued.de/rail-en/the-future-of-rail-automation.

Verein Deutscher Ingenieure, VDI 3695, Part 3. (2010). *Engineering of industrial plants; Evaluation and optimization; subject methods*. Beuth: Berlin.

ZVEI – Zentralverband Elektronik- und Elektronikindustrie e.V. (2010). *Leitfaden Life-Cycle-Management für Produkte und System der Automation*. Frankfurt am Main.

Chapter 5
On the Need for Data-Based Model-Driven Engineering

Alexandra Mazak, Sabine Wolny, and Manuel Wimmer

Abstract In order to deal with the increasing complexity of modern systems such as in software-intensive environments, models are used in many research fields as abstract descriptions of reality. On the one side, a model serves as an abstraction for a specific purpose, as a kind of "blueprint" of a system, describing a system's structure and desired behavior in the design phase. On the other side, there are so-called runtime models providing real abstractions of systems during runtime, for example, to monitor runtime behavior. Today, we recognize a discrepancy between the early snapshots and their real-world correspondents. To overcome this discrepancy, we propose to fully integrate models from the very beginning within the life cycle of a system. As a first step in this direction, we introduce a data-based model-driven engineering approach where we provide a unifying framework to combine downstream information from the model-driven engineering process with upstream information gathered during a system's operation at runtime, by explicitly considering also a timing component. We present this temporal model framework step-by-step by selected use cases with increasing complexity.

Keywords Model-driven engineering · Data-driven engineering · Model repositories · Model profiling · Sequence mining

5.1 Introduction

The evolutionary aspect of engineering artifacts refers to the fact that they change over time. Models in engineering processes are usually developed from initial ideas to first drafts. They are then continuously revised, often by taking into account feedback from engineers, until they are finally released. However, the feedback after the release from the ongoing operation should also be reflected in those models to

---

A. Mazak (✉) · S. Wolny · M. Wimmer
Christian Doppler Laboratory for Model-Integrated Smart Production (CDL-MINT), WIN-SE, JKU Linz, Linz, Austria
e-mail: alexandra.mazak-huemer@jku.at; sabine.wolny@jku.at; manuel.wimmer@jku.at

S. Biffl et al. (eds.), *Security and Quality in Cyber-Physical Systems Engineering*,
https://doi.org/10.1007/978-3-030-25312-7_5

cover the complete life cycle of a system, meaning from design through runtime and backward. This means that models should be used throughout and beyond the life cycle of a system. For this purpose, we envision a new kind of adaptive model, which we call "liquid model." This term, firstly introduced in (Mazak and Wimmer 2016b), is used to stress that models as any kind of engineering artifact should not be isolated and static but cooperative and evolutionary.

To realize this vision, models should be hosted in enhanced model repositories. In such repositories, models are storable as integrated artifacts leading to a set of interconnected models. Having this as a baseline, services for cooperation and reactivity can be directly implemented and offered within the repository and, thus, are reusable for all modeling tools that provide connections to the repository. The same is true for collecting runtime information. Instead of building bilateral solutions between each engineering tool and runtime environment, the central repository acts as a mediator between the engineering tools and runtime environments. Overall, the model repositories are becoming an engineering and operation backbone for software-intensive systems, such as Cyber-Physical Systems (CPS), and may present the next generation of modelware platforms in general.

In particular, the focus is on connecting runtime environments, such as different Internet-of-Things (IoT) platforms, to model repositories to extract operational models (i.e., models derived from data gathered during a system's operation) and to enable their connection to design models (i.e., models built during the design phase of a system). Thus, specific services are needed to connect to runtime environments and to deal with data streams, for example, in order to be able to efficiently react to events occurring in highly distributed systems, however, not at data level but at model level. This enables runtime models to be compared to design time models, and if required to enhance the latter one with meta-information collected during a system's operation, which may be useful for decision-making. Therefore, the aim of our approach of liquid models is to combine model-based downstream information derived from the design phase with measurement-based upstream information of an operating system by combining model-driven and data-driven techniques. However, the implementation of such liquid models is not straightforward. The implementation of this kind of models is highly interdisciplinary and can only succeed if different techniques, methods, and approaches from various research disciplines are interweaved.

The intention of this chapter is to present parts of our work, which we have conducted within the scope of our 2-year research activities in the module Reactive Model Repositories of the Christian Doppler Laboratory for Model-Integrated Smart Production (CDL-MINT). In Sect. 5.2, we discuss related work and present research challenges in each of the presented research fields. In Sect. 5.3, we present a unifying framework to combine data-driven and model-driven techniques. For demonstration purposes, Sect. 5.4 presents selected use cases with increasing complexity for illustrating the implementation of this framework. Finally, Sect. 5.5 concludes this chapter and outlines future work.

5.2 Related Work and Research Challenges

Several research directions are relevant to realize our goal to combine downstream information from the model-driven engineering process with upstream information gathered during a software system's operation. This section summarizes approaches and discusses challenges in the scientific communities related to model repositories, runtime models, as well as process mining and data analytics. Firstly, we discuss emerging model repositories as well as modeling approaches considering runtime aspects in addition to model-driven design. Secondly, we survey techniques to deal with operational data and to turn it into abstracted model representations. In this context, we specifically highlight process mining and runtime models and give insights in new directions of data analytics.

5.2.1 Model Repositories

Research concerning model repositories comprises mainly two areas: concurrent modeling using a central repository to coordinate the editing of models (Brosch et al. 2010; Koegel and Helming 2010) and scalability in storing and retrieving models (Kolovos et al. 2013). The general services offered by a model repository is to load a complete model from a repository and to store a complete model to a repository. Other services, such as more fine-grained model loading or manipulation, are often missing in most repositories (Basciani et al. 2014). Hartmann et al. (2014) tackle this challenge by a versioning system for model elements, but the versions must be explicitly introduced and managed in the versioning systems.

The scalability problems of loading large models represented by XML-based documents into memory have been already recognized several years ago. One of the first improved solutions for models is the Connected Data Objects[1] (CDO) model repository, which enables to store models in all kinds of database back ends, such as traditional relational databases as well as not only Structured Query Language (NoSQL) (Davoudian et al. 2018) databases. CDO supports the ability to store and access large-sized models due to the transparent loading of single objects on demand and caching them. If objects are no longer referenced, they are automatically garbage collected.

There are also several projects for storing very large Eclipse Modeling Framework (EMF) models such as MongoEMF[2] or Morsa (Espinazo Pagan and Garcia Molina 2014). Both approaches are built on top of MongoDB.[3] Furthermore, graph-based databases and map-based databases are also exploited for model storage, such as

[1] http://projects.eclipse.org/projects/modeling.emf.cdo.

[2] http://code.google.com/a/eclipselabs.org/p/mongo-emf.

[3] https://www.mongodb.com/de.

done in Neo4EMF (Benelallam et al. 2014; Gómez et al. 2015) where also different unloading strategies for partial models are explored (Daniel et al. 2014). Clasen et al. (2012) elaborate on strategies for storing models in a distributed manner by horizontal and vertical partitioning in Cloud environments. A similar idea is explored in (Deak et al. 2013) where different automatic partitioning algorithms are discussed for graph-based models. In addition to these research works, (Bergmayr et al. 2018) conducted a systematic review on cloud modeling languages.

Challenge 5.2.1: Temporal Model Repositories

All mentioned approaches have in common that models are residing behind the walls of the model repository without an appropriate connection to the environment, as it is required, for example, for enabling to immediately react on anomalies within a system's operation during runtime. Thus, temporal model repositories are needed to shift "static" models to evolutionary artifacts, where the focus is not only on the current state to steer the system but also on the history of changes. Such temporal model repositories are highly needed in application fields where models have to be used throughout the complete system life cycle.

5.2.2 Runtime Models

There are several different approaches for runtime modeling. Blair et al. (2009) show the importance of supporting runtime adaptations to extend the use of model-driven engineering (MDE). They propose models that provide abstractions of systems during runtime, the so-called operational models. These models are an abstraction of runtime behavior. Due to this abstraction, different stakeholders can use the models in various ways for runtime monitoring. Hartmann et al. (2015) combine the concept of runtime models with reactive programming. Reactive programming aims on enabling support for interactive applications, which react on events by focusing on real-time data streams. For this purpose, a typical publish/subscribe pattern, well known as observer pattern in software engineering (Vlissides et al. 1995), is used. Khare et al. (2015) show the application of such an approach in the IoT domain.

Hartmann et al. (2015) define runtime models as a stream of model chunks, as it is common in reactive programming. The models are continuously updated during runtime; therefore, they grow indefinitely. With their interpretation that every chunk has the data of one model element, they process them piecewise without looking at the total size. In order to prevent the exchange of full runtime models, peer-to-peer distribution is used between nodes to exchange model chunks. In addition, automatic reloading mechanism is used to respond on events for enabling reactive modeling. As the models are distributed, operations like transformations have to be adapted. For this purpose, transformations on streams as proposed by Cuadrado and de Lara (2013) as well as Dávid et al. (2018) can be used.

Challenge 5.2.2: Hosting Runtime Monitoring

The paradigm Models@run.time refers to the runtime adaptation mechanisms that leverage software models to dynamically change the behavior of the system based on a set of predefined conditions. While these approaches provide a model infrastructure to instantiate models, they are not focusing on the history of changes (e.g., state changes, value changes) during the system's lifetime. For this purpose, a framework is required that enables an automatic monitoring of systems by using the aforementioned temporal model repositories.

5.2.3 Process Mining

Process mining (PM) techniques (van der Aalst 2012; van der Aalst et al. 2011) analyze processes based on events stored in the so-called event logs. Therefore, these techniques involve both event data and process models. In PM, logs are sequential events recorded at runtime by an information system (Dumas et al. 2005). In this context, Mannhardt et al. (2018) address the problem that low-level events recorded by such information systems may not directly match high-level activities, which build the basis of process owners for decision-making activities. Therefore, the authors present a so-called guided Process Discovery Method (GPD) based on behavioral activity patterns to translate low-level events into high-level ones. This approach seems very similar to Complex Event Processing (CEP) introduced by Luckham (2001). Generally, approaches based on CEP assume a stream of events over where queries are evaluated. Whenever such a query matches a higher level, an activity is detected. However, CEP does not consider the notion of process instance (i.e., case) like GPD, and in case of overlapping queries (e.g., shared functionalities), both high-level activities would be detected (Cugola and Margara 2012). In addition, most of the approaches based on CEP (e.g., in Bülow et al. 2013; Hallé and Varvaressos 2014) do not provide a complete discovery model as output like GDP.

For an efficient capturing of behavioral patterns, van der Aalst and his research group introduce specialized algorithms (e.g., alpha-algorithm, inductive miner) to extract knowledge from event logs (e.g., in van der Aalst 2012; van der Aalst et al. 2004). These algorithms produce outputs, for example, in the form of a Petri net, which can then be easily converted into an appropriate process model such as a BPMN model, UML activity diagram, etc. To automatically learn a control-flow model from event data is challenging, but in recent years, many powerful discovery techniques have been developed, as presented and discussed, for example, in (van der Aalst 2018; van der Aalst et al. 2011). However, PM is not limited to a control-flow perspective. There are further perspectives as introduced in (van der Aalst 2012) and again summarized in (van der Aalst 2018), namely, (1) the organizational perspective focusing on information about resources (e.g., people, systems, departments) hidden in event logs, (2) the case perspective focusing on properties of cases, and (3) the time perspective concerned with the timing and frequency of events.

PM operates on the base of events that belong to cases (i.e., process instances) that are already completed (van Dongen and van der Aalst 2005). Since such an off-line analysis is not suitable for cases which are still in the pipeline, in (van der Aalst 2012), the author proposes to support both on-line and off-line analyses by mixing current data with historic data.

Challenge 5.2.3: Execution-Based Model Profiling

In PM, each event should have a case identifier (case ID), an activity name, and a time stamp as a bare minimum. Unfortunately, logs generated during monitoring automation systems such as software controllers or log documentations of model executions do not meet this required minimum. Thus, it is hard to determine a case ID for sensor data streams. Therefore, logs or data streams have to be preprocessed in order to usefully apply PM algorithms appropriately for analyzing purposes. In addition, current event processing technologies usually monitor only a single stream of events at a time. Even if users monitor multiple streams from different sources, they often end up with multiple "silo" views. A more unified view is required that correlates with events from multiple streams of different sources as well as in different formats.

5.2.4 Data Analytics

The term "data analytics" subsumes techniques examining big datasets to uncover hidden patterns, unknown correlations, and other useful information that is used to make better decisions. This motivates a technology shift to data-centric architectures and operational models (Demchenko et al. 2014).

The steps that data passes through in most applications are the following: acquiring, cleansing, storing, exploring, learning, and predicting (Pedersen 2017). Mostly, the acquisition of information that derive from a big amount of data gathered from heterogeneous sources is challenging. Since, there are very large data volumes (Volume), data arrives very fast in the form of data streams (Velocity), and data has varied and complex formats, types, and meanings (Variety) (Anagnostopoulos et al. 2016; Laney 2001). These three big data properties (i.e., the 3Vs) put high requirements on data storage (e.g., NoSQL, RDBMS), data processing (e.g., batch, incremental, interactive), orchestration (scheduling, provisioning), and interfaces for offering access points (e.g., SQL, script, graphical), to mention only four of six pillars of the big data analytic ecosystem as introduced in (Khalifa et al. 2016).

As discussed in (Pedersen 2017), new data cycles are needed to handle big multidimensional data, such as merging steps (doing several steps in combination), hierarchical steps (steps inside steps), models in all steps (models for data acquisition, storage, etc.) as well as a step for combining prediction and optimization, and the new step prescribe. The goal of this latter step is to prescribe the right course of action for optimizing a given goal (Pedersen 2017). All these steps are combined in an iterative manner.

For analysis purposes, the fulfilment of the 3Vs common data processing technologies are not sufficient (Yaqoob et al. 2016). Therefore, there are a multitude of techniques such as data mining firstly introduced in (Fayyad et al. 1996), machine learning (Bishop 2007), anomaly detection (Chandola et al. 2009), predictive analytics (Shmueli and Koppius 2011), and prescriptive analytics (Pedersen 2017). At the end, the aim is to recognize new data models within big data, for example, to recover existing patterns and to discover new ones (Domingos and Hulten 2001).

Challenge 5.2.4: Data Model Integration

An important challenge when linking data to runtime models is how to integrate data from various sources with different formats (Microsoft Excel, XML, databases) to get a useful representation. For this purpose, it is important to combine data analytic technologies with model-driven technologies, such as model transformation engines. For instance, the transformation of log files to models enables appropriate analysis of statistical information such as anomaly detection or pattern mining.

Synopsis
One important open challenge, for linking design models to runtime models, is how to combine advanced data analytics technologies and modeling technologies, such as model transformation engines or model checkers, in order to get the best of both worlds. This is a particular challenge toward a moving target, since there are still some unsolved problems in these areas to be combined.

5.3 Approach for Enabling Data-Based Model-Driven Engineering

In this section, we present from a theoretical point of view how to connect runtime environments, for example, a traffic light system to temporal model repositories. Additionally, we present specific techniques needed for dealing with data streams to react efficiently to events occurring in physically distributed data sources at runtime (see Sect. 5.4). We have investigated the research issues about the extended usage of models also during operation to enable temporal model repositories for allowing runtime monitoring, execution-based model mining, as well as design model enhancement in order to address the challenges discussed in Sect. 5.2. The approach is divided into two main contributions.

Contribution 1: Managing Temporal Models

For managing temporal models, we need a framework that deals with the challenges of temporal model repositories and hosting runtime monitoring (see Challenge 5.2.1 and Challenge 5.2.2 in Sect. 5.2). To tackle these challenges, we consider them from two different perspectives: (1) a model-driven perspective and (2) a data-driven one. In addition, communication from models to the environment should be enabled at runtime. For instance, events can be used to establish a reactive exchange of messages.

Table 5.1 Assignment of the contributions to the challenges including the associated use cases

	Contribution 1	Contribution 2	Use case 1	Use case 2	Use case 3
Challenge 5.2.1	✓		✓		
Challenge 5.2.2	✓		✓	✓	✓
Challenge 5.2.3		✓	✓	✓	
Challenge 5.2.4		✓			✓

Contribution 2: Evolutionary Model Mining

Evolutionary model mining is an umbrella term covering the evolutionary aspect of engineering artifacts (i.e., class diagram, state machine) for describing their changes over time (e.g., state changes, value changes). Monitoring this evolution bases on data collected over time from executions. The integration and unification of this observed data is a cross-disciplinary challenge, where MDE techniques have to be combined with mining techniques, such as process mining and multivariate statistics, for enabling an incremental evolutionary model mining process. Thereby, the so-called model profiles are automatically generated from execution logs of running systems (see Challenge 5.2.3 in Sect. 5.2). Using this execution-based model profiling as transformation of log files to models enables appropriate model mining, for example, to check whether the code generator works as intended (see Challenge 5.2.4 in Sect. 5.2). Table 5.1 shows the assignment of our contributions to the identified challenges and the selected use cases.

It should be mentioned that for the implementation of our contributions, basic technologies are applied to collect data from different data sources. These data streams are combined into topics, which are used for further investigations and analyses. All contributions therefore work with data from the extended system.

5.3.1 Temporal Model Framework

First, we analyzed research directions in the field of temporal models. Current evolving models are stored in model repositories, which often rely on existing versioning systems or standard database technologies. These approaches are sufficient for hosting different versions of models, which are stored separately. However, the time dimension is often not explicitly represented and accessible (Bill et al. 2017a). In order to support use cases where models are used not only in the system design phase but also in simulation and runtime environments, a more explicit representation of time is needed in order to reason about temporal aspects such as model element changes (Bill et al. 2017a). Thus, a framework is needed where such temporal aspects are explicitly taken into account. Thereby, there are certain challenges to be dealt with such as storage, consistency, access, and visualization (Bill et al. 2017a,b). The need for an explicit time component is also identified by standardization bodies such as the Object Management Group (OMG).

Currently, the OMG is working on a new version of the Systems Modeling Language (SysML) where enhancements are planned to have a timing component in models, which is an important issue, for example, for modeling continuous and discrete systems.

This development coincides with the findings from our systematic mapping study (Wolny et al. 2019). On the basis of these findings, we develop a temporal model framework dealing with downstream information from the MDE process with upstream information gathered at runtime through system executions (Mazak et al. 2016). Figure 5.1 presents this framework starting with an MDE perspective and ending up in a data-driven one.

Starting with an MDE perspective at design time, we choose a modeling language such as UML, SysML, or a domain-specific language (DSL) for describing a domain of interest, for example, a software controller for a multipurpose machine. The output of this model engineering process is a model that has to conform to this language. This means that all elements of this model can be instantiated from the corresponding classes of the metamodel (see Fig. 5.1 on the left hand side, the relation between modeling and metamodeling level) (Brambilla et al. 2017). In a next step, this design model acts as input for a code generator by which an artifact, for example, code in case of software, is automatically generated. For such a vertical transformation from the modeling to the realization level, we employ a Model-to-Text (M2T) transformation (see Fig. 5.1 on the left hand side, the arrow from automation level to realization level). By deploying this M2T transformation, model elements are automatically transformed to corresponding code statements (Brambilla et al. 2017).

In order to be able to access data logs from a currently running code at an execution platform, we need a special language for specifying specific runtime models over time. We call this language "runtime history language" (see Fig. 5.1 in the middle, data-driven perspective). Since we are interested in all model elements

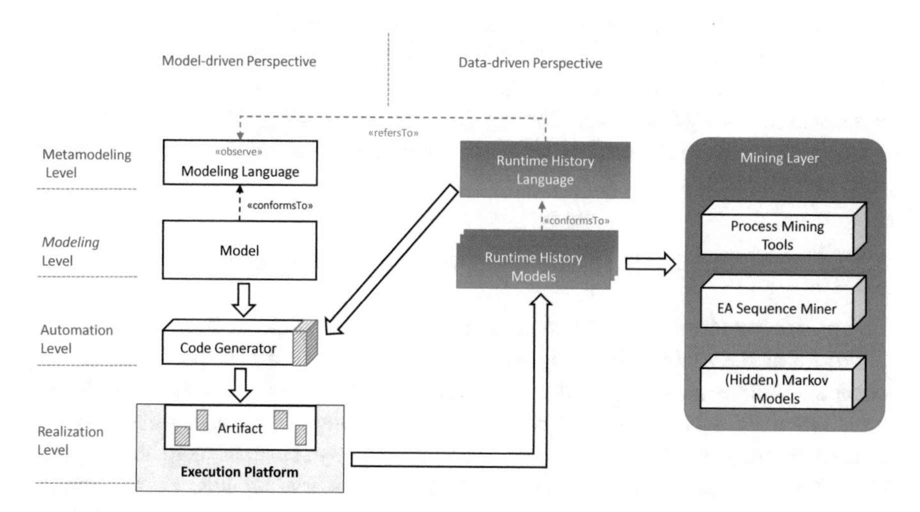

Fig. 5.1 Overview of the temporal model framework

that could change during runtime, this language has to refer to the operational semantics of the used modeling language (see Fig. 5.1, «refersTo» relation between runtime history language and modeling language). As discussed in (Brambilla et al. 2017), the operational semantics defines the meaning of the used modeling language. In particular, running systems can only be generated from executable models. A model is executable if its operational semantics is fully specified (Brambilla et al. 2017). This means that operational semantics defines everything that is changing in a system at runtime, for example, attribute values, or the current state of the system or its components. Therefore, we have to define a so-called «observe» stereotype for annotating selected model elements to ensure that these elements are automatically recognized by the code generator as additional log information. Thus, the runtime history language determines which runtime changes should be logged (e.g., state changes, attribute value changes, etc.) by influencing the logging of the code generator as a kind of meta-logging language (see Fig. 5.1, the arrow from the runtime history language to the extended code generator). In addition to operational semantics, the approach is based on translational semantics in the form of code generators to produce code for a concrete platform to realize the software system (see Fig. 5.1 on the left hand side, automation level to realization level). By this, it is feasible to continuously monitor system changes at the model layer than as usually common at the data layer so far.

This modus operandi enables to generate specified model profiles already at design time (Kadam et al. 2017; Mazak and Wimmer 2016a) and could be seen as a kind of embedded data wrangling. Thus, we do not need time-consuming algorithms in the preprocessing phase of data analytics for extracting profiles afterward at the mining layer from massive datasets (see Fig. 5.1 on the right hand side, mining layer). This means that the generated runtime history models have already a specific meaning when used as input at this layer. For instance, these models can be used as input (1) for PM tools to discover process models out of the running code (Mazak and Wimmer 2016a); (2) for an object-oriented visualization of communication among systems components by employing the Enterprise Architect Sequence Miner, a plug-in of the Enterprise Architect that we developed and implemented in the course of our research work (Mazak and Wimmer 2017); or (3) for computing Hidden Markov Models to analyze information, which is not directly observable or measurable at runtime (Mazak et al. 2017).

5.3.1.1 First Realization of the Temporal Model Framework

Figure 5.2 presents a first version of realizing our temporal model framework approach at the metamodeling level (see Fig. 5.1, model-driven perspective). We choose a simple subset of UML for modeling class diagrams and state machines, which we call CSC (Class State Chart) modeling language. The upper part of Fig. 5.2 defines the structural as well as behavioral aspects of a system to be modeled at design time. The lower part of Fig. 5.2 shows the runtime history language, corresponding to the CSC language, which defines the structure of the model profiles to be logged at runtime.

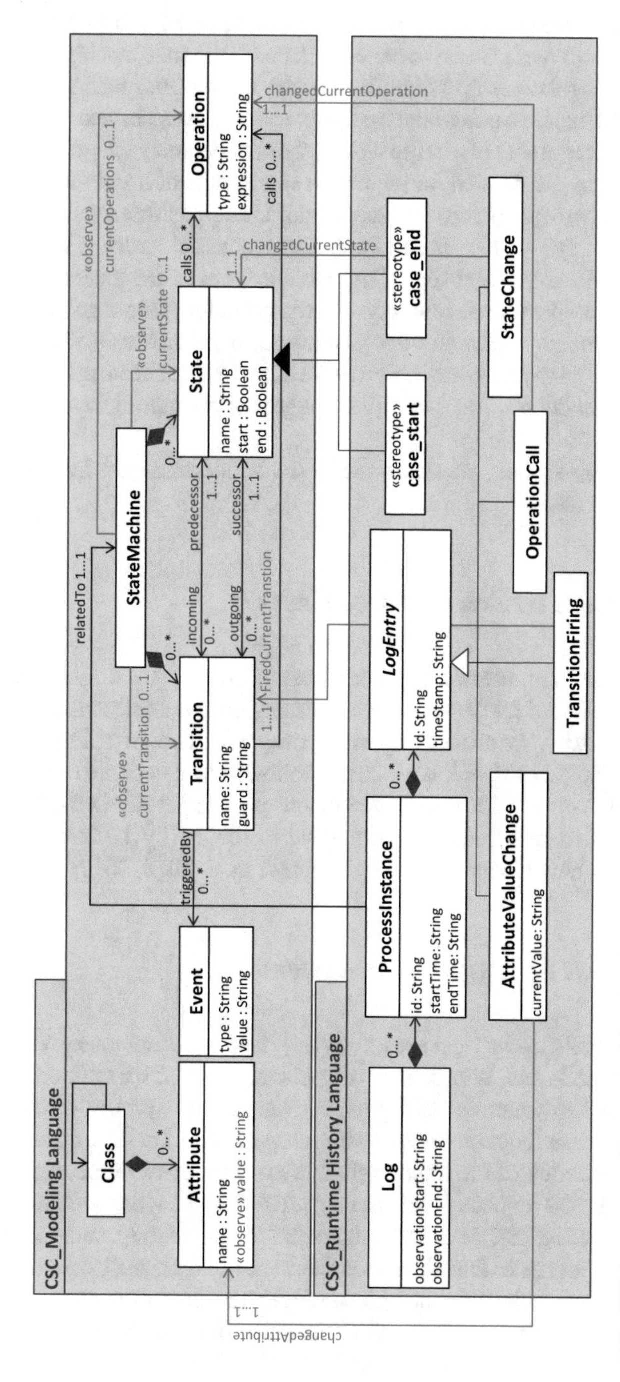

Fig. 5.2 CSC language at the metamodeling level (Mazak et al. 2016)

The class LOG represents a logging session of a running software system with a registered start and end (see Fig. 5.2, observationStart, observationEnd). A log is a composition of process instances related to the state machine (see Fig. 5.2, the relationTo relationship symbolized by the arrow from ProcessInstance to StateMachine). In cases where we do not have a clearly defined start/end, as it is the case in our first example of a traffic light system, the stereotypes «case_start» and «case_end» are defined for capturing single runs of a state machine during operation. A ProcessInstance represents a composition of log entries (see Fig. 5.2, LogEntry) with a time stamp for sequential order. The time stamp indicates when the entry is registered. The specific types of the generalized log entry has to refer to the operational semantics of the modeling language (see Sect. 5.2). Through the annotated «observe» stereotype, AttributeValueChanges, TransitionFirings, OperationCalls, and StateChanges are automatically logged as model profiles for further analysis at the mining layer (see Fig. 5.1, mining layer).

In the following section, we demonstrate the temporal model framework based on selected use cases.

5.4 Realization by Selected Use Cases

In this section, we demonstrate the usefulness of the temporal model framework for enabling data-based model-driven engineering on three selected use cases with increasing complexity. We are starting with a simple use case of a traffic light system for pedestrians and cars (Mazak et al. 2016), followed by a use case of a self-driving car (Mazak and Wimmer 2017) provided by our project partner in the CDL-MINT, and ending up with a lab-sized production system provided by the Otto von Guericke University Magdeburg (Artner et al. 2017; Mazak et al. 2018, 2017).

5.4.1 Use Case 1: Traffic Light System

The first use case of a traffic light system mainly bases on preliminary research work presented in (Mazak and Wimmer 2016a; Mazak et al. 2016). Figure 5.3 shows a picture of the hardware and two perspectives on this system, which are early blueprints. The class diagram contains the class TrafficLightController describing the structure of the traffic light system. This controller consists of lights for pedestrians (pedG for pedestrian green, pedGR for pedestrian red), lights for cars (carG for car green, carR for car red, and carY for car yellow), and a blink counter (bc). The state chart describes the states the TrafficLightController can take during runtime with different light settings. The controller starts in an initial SafetyState, which is only reached during initialization, and then switches through different states based on which the lights should be switched on and off during

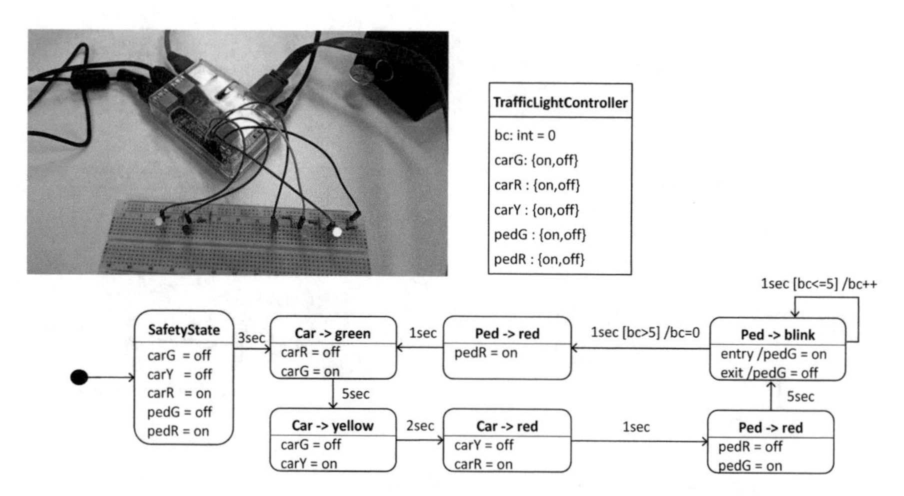

Fig. 5.3 CSC model: class diagram and state machine of the traffic light system (Mazak et al. 2016)

operation. Both models conform to the CSC metamodeling language presented in Sect. 5.3.1.1.

Based on this CSC metamodeling language and the extended CSC runtime history language (both presented in Fig. 5.2), we develop an M2T transformation as code generator to transform CSC models to executable Python code. This enables us to automatically logging `AttributeValueChanges` and `StateChanges` during the system's operation by the «observe» stereotype (see Sect. 5.3). Figure 5.4 shows the implementation workflow for realizing the data-based model-driven engineering approach for this use case.

Firstly, the CSC model of the traffic light system is used as input for the code generator (`CSC2Python`) to produce automatically Python code for the Raspberry Pi execution platform (see Fig. 5.4, vertical transformation on the left hand side from the CSC model to the Python code running at Raspberry Pi). Secondly, during execution, the Raspberry Pi generates JSON[4] logs which are further processed in a log recording microservice, which we have implemented as data collector (see Fig. 5.4, in the lower part). This service stores the log information as runtime history models in the model repository Neo4EMF,[5] a NoSQL database. Thirdly, for analyzing these runtime history models, selected as model profiles, we use ProM Lite[6] (a PM tool) at the mining layer (see Fig. 5.1, mining layer). For loading the model profiles in ProM, we have to transform them to a specific workflow format basing on an XSD schema. For this purpose, we employ a Model-to-Model (M2M) transformation (Brambilla et al. 2017) for transforming the model profiles to appropriate workflow instances. Based on this M2M transformation, we are able to

[4]https://www.json.org/.

[5]https://www.neoemf.com/.

[6]http://www.promtools.org/doku.php.

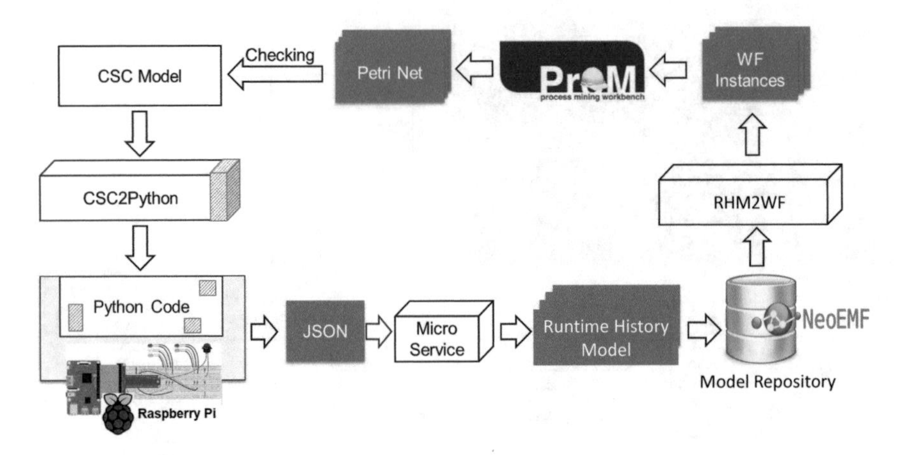

Fig. 5.4 Deployed traffic light system (Mazak et al. 2016)

discover different kinds of PM models for analyzing structural as well as behavioral aspects of the traffic light system at the mining layer. For more insights into PM and its techniques and algorithms, we refer the interested reader to (van der Aalst 2012, 2018; van der Aalst et al. 2004, 2011).

Figure 5.5 shows a Petri net as output generated by ProM Lite based on the alpha++ algorithm. Through a manually performed alignment, we could check if the model profile of state changes corresponds to specified parts of the state chart (see Fig. 5.3, at the bottom). In addition, this enables us to validate the implemented `CSC2Python` code generator for correctness. In a further example, we checked if the condition (i.e., invariant) holds by logging the `AttributeValueChanges` (see Fig. 5.2, CSC_Runtime History Language) of the blink counter of the pedestrian light (see Fig. 5.3, `Ped→blink`). This enables us to detect, for example, if there were any unexpected events during runtime, such as failures of the blink counter, etc.

5.4.2 Use Case 2: Self-Driving Car—PiCar

The second use case of a self-driving car, the so-called PiCar, mainly bases on our research work presented in (Mazak and Wimmer 2017). PiCar is a simple automated system equipped with a software controller, sensors, and two actuators `MotorControl` and `ServoControl`. Figure 5.6 shows the modeled behavior of this car described by a state machine at design time using SysML as modeling language.

In this use case, from a mining layer perspective, we are interested in the frequency of actual executed operations of the two actuators, `MotorControl` and `ServoControl` (see Fig. 5.2, `OperationCalls`). The state machine

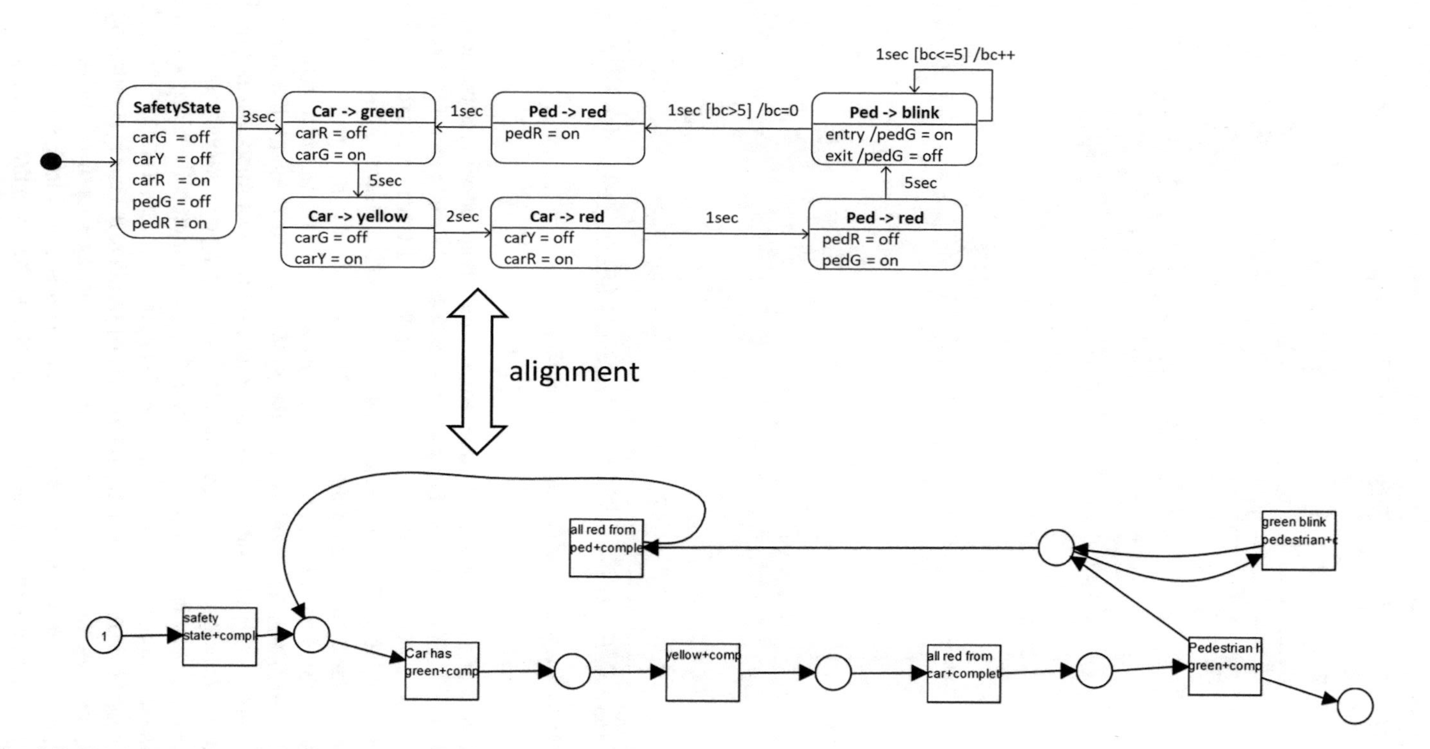

Fig. 5.5 Validation of the CSCPython code generator

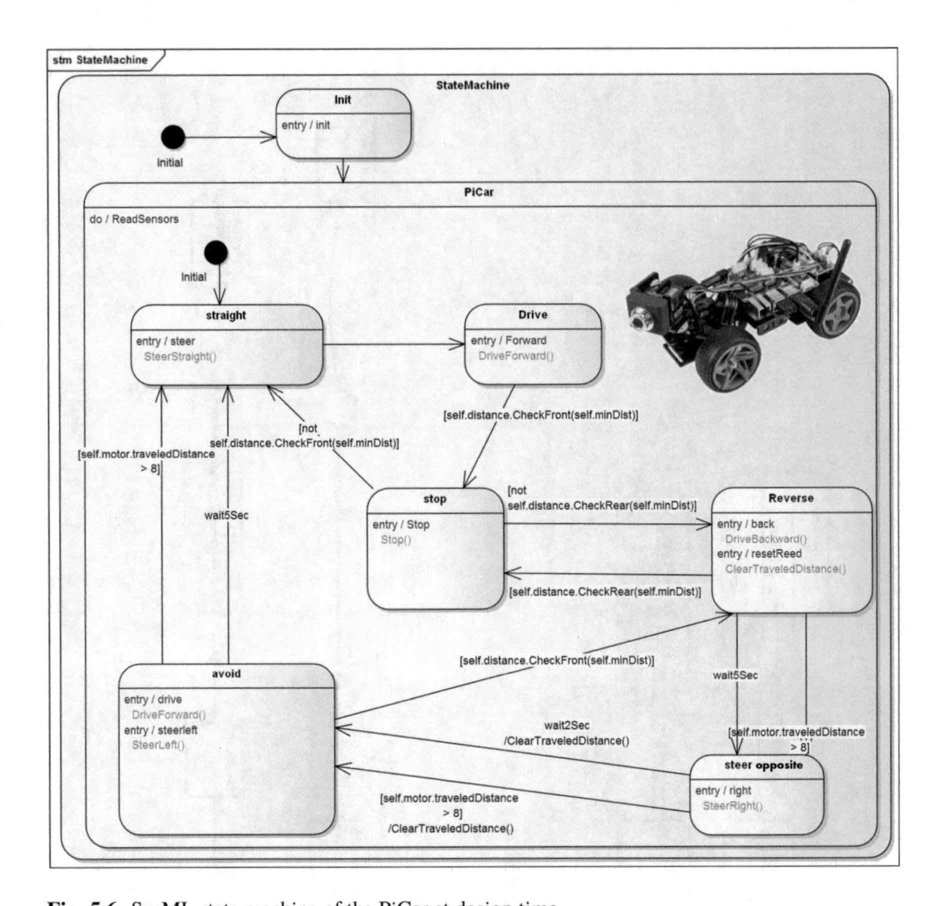

Fig. 5.6 SysML state machine of the PiCar at design time

presented in Fig. 5.6 shows that there are in total nine possible sequences of operation calls such as `drive_forward` → `steer_left`, `drive_forward` → `steer_straight`, and `drive_forward` → `steer_right`, to show exemplary three of it. We call these sequences "episodes" according to common sequential mining approaches (Han et al. 2007; Tax et al. 2016).

Similar to the traffic light example, we present in Fig. 5.7 the technical realization of the PiCar use case. As input, we take the SysML state machine for the so-called "VanillaSource" code generator, a product provided by our project partner in the CDL-MINT. We extended this code generator by a logging component based on the «observe» stereotype in order to log automatically operation calls during runtime. The VanillaSource code generator generates the Python execution code of the PiCar-software controller running on a Raspberry Pi. At runtime, the car is autonomously driving in an environment and stops, reverses, and changes its direction whenever barriers are detected by the sensors. Based on the predefined logging structure in the code generator, we get the data of realized episodes between the two actuators in

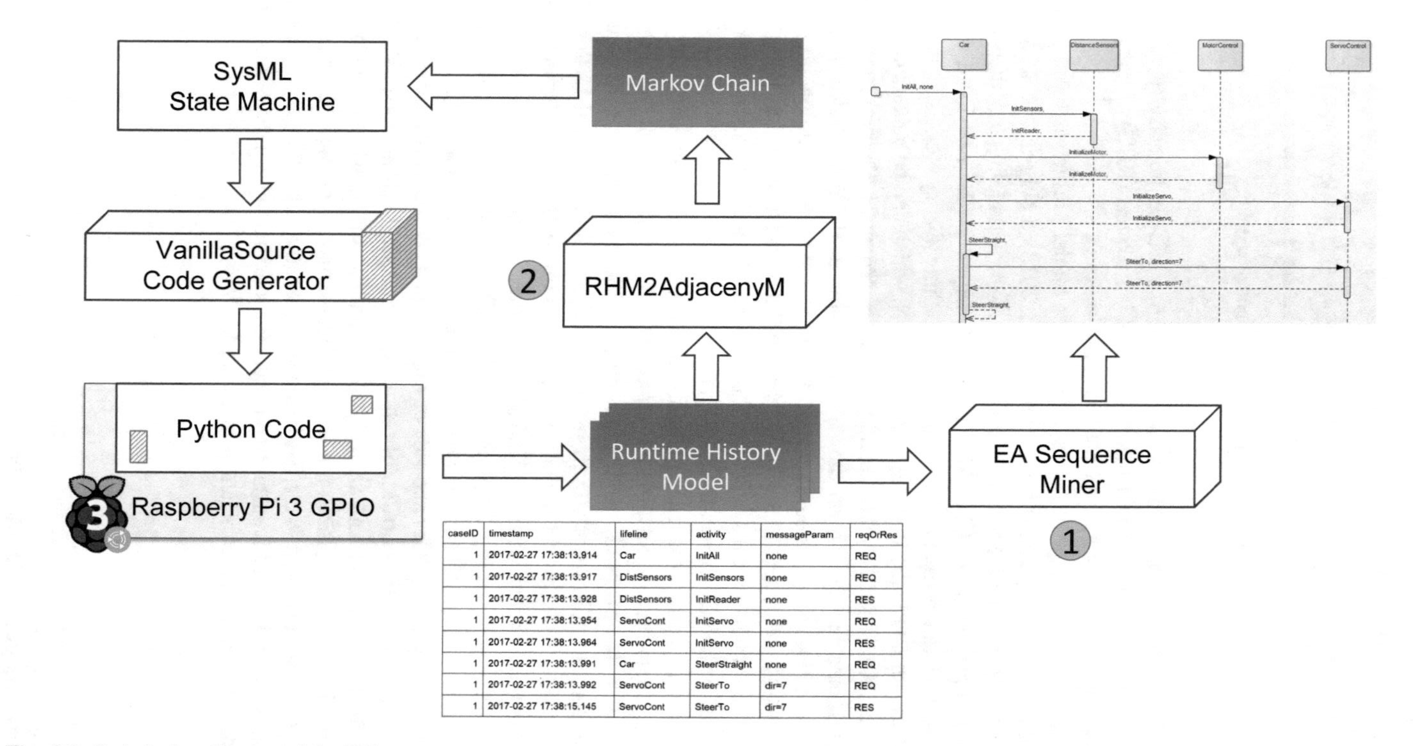

caseID	timestamp	lifeline	activity	messageParam	reqOrRes
1	2017-02-27 17:38:13.914	Car	InitAll	none	REQ
1	2017-02-27 17:38:13.917	DistSensors	InitSensors	none	REQ
1	2017-02-27 17:38:13.928	DistSensors	InitReader	none	RES
1	2017-02-27 17:38:13.954	ServoCont	InitServo	none	REQ
1	2017-02-27 17:38:13.964	ServoCont	InitServo	none	RES
1	2017-02-27 17:38:13.991	Car	SteerStraight	none	REQ
1	2017-02-27 17:38:13.992	ServoCont	SteerTo	dir=7	REQ
1	2017-02-27 17:38:15.145	ServoCont	SteerTo	dir=7	RES

Fig. 5.7 Technical realization of the PiCar

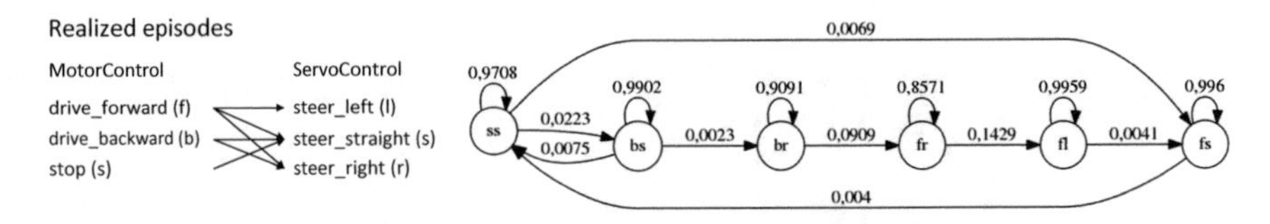

Fig. 5.8 Realized episodes of the PiCar in a certain environment setting

a CSV format, as following: a caseID, a time stamp, a lifeline, an activity, message parameters, and the information if the method call is a request (REQ) or a response (RES) (see Fig. 5.7, excerpt of the CSV file at the bottom).

In a first variant (see Fig. 5.7, circled 1), we use model profiles as input for our Enterprise Architect Sequence Miner (Hafner et al. 2018), a plug-in which we have developed for the Enterprise Architect[7] (EA). The EA is a modeling tool distributed by our project partner of the CDL-MINT. The EA sequence miner performs a Text-to-Model (T2M) transformation by transforming runtime history models, in the form of CSV files, automatically into sequence diagrams, which can be visualized in the EA. An excerpt of such an automatically generated sequence diagram for our use case is presented in Fig. 5.7, on the right hand side at the top. Such a diagram visualizes an object-oriented communication among the system components of the car. The software controller (see Fig. 5.7, first lifeline) acts as a kind of interaction manager. It continuously reads data from the sensors and sends operation calls to the two actuators, which then execute these operations.

Unfortunately, such a sequence diagram can become very long, confusing, and unreadable. Therefore, we adapted our approach in a second variant (see Fig. 5.7, circled 2). For this purpose, we use the runtime history models as input of an M2M transformation to transform them into an adjacency matrix and further into a Markov chain. This second variant mainly bases on preliminary research work presented in (Mazak and Wimmer 2017; Mazak et al. 2017). The Markov chain in the form of a digraph (see Fig. 5.8, at the right hand side) shows the realized episodes between `MotorControl` and `ServoControl`. For better readability, we present a textual summary on the left hand side of Fig. 5.8. In a certain environment setting (we let the car drive in the corridor of our institute and set up some barriers), six of nine possible episodes were actually realized. In addition, the transition probabilities on the digraph may be used, for example, to predict the driving behavior of the car for certain settings. As opposed to the EA-based sequence diagram, this graph provides a clear overview and makes it easier to compare the outcome with parts of the initial SysML state machine (target/actual comparison).

[7]https://www.sparxsystems.de/uml/neweditions/.

5.4.3 Use Case 3: Lab-Sized Production System

In this last use case, we present a reverse engineering approach where we gather logs from machines during operation in the form of raw sensor data. As case study example, we employ the lab-sized production system hosted at IAF[8] of the Otto von Guericke University Magdeburg. This use case is based on our research work started in (Artner et al. 2017; Wimmer et al. 2017) and continued in (Mazak et al. 2017). In this third use case, we are interested on properties which cannot be directly measured during operation, such as workload and performance bottlenecks. These properties are also known as nonfunctional properties. For this purpose, we collect data through sensors attached on machines, turntables, and conveyor belts in each area of a lab-sized production system (see Fig. 5.9). In this example, we focus on the resources of area 2 as follows: one multipurpose machine, four turntables, and three conveyor belts. In order to compute certain workload characteristics, we are analyzing the utilization of these resources based on the duration of executed operations, the so-called operation durations.

By logging and recording these operation durations per resource in a time series database (TSDB), we are able to compute the utilization of each resource as hidden states by a Hidden Markov Model. Based on this model, we are able to automatically generate a workload matrix for all resources of area 2 (Artner et al. 2017; Mazak et al. 2017). We use InfluxDB[9] as TSDB to handle the massive amount of time-stamped data generated in this use case. Figure 5.10 presents the outcome for which we build a model editor[10] to get a probabilistic finite state automaton for enabling a better overview as well as decision support. The figure shows this automaton for an exemplary manufacturing process, which we have really executed in area 2. We routed workpieces through this area to process them on the turntables (`T1` to `T4`), the conveyor belts (`C1` to `C3`), and the multipurpose machine (`M1`). Thereby, each workpiece coming from area 1 as input was routed and processed through the individual work units before they were transported to area 3 as new input (see Fig. 5.9).

The think time (Fig. 5.10, `thinkT`) is the average time of requests (start of an operation) and responses (completion of an operation) for a single workpiece. The total time (Fig. 5.10, `totalT`) is the sum of think time, this means the time all workpieces are processed at a specific resource (e.g., `Turntable_T2`) in a certain scenario. The total time is taken for approximating the utilization of each specific resource of area 2.

As shown in Fig. 5.10, the conveyor belt `C1` and the turntables `T2` and `T3` have a high processing frequency, which indicates a longer waiting time for workpieces to be processed. This fact may increases the need for a higher buffer size. Additionally,

---

[8]http://www.iaf-bg.ovgu.de/en/lueder.html.

[9]https://www.influxdata.com/.

[10]https://cdl-mint.big.tuwien.ac.at/case-study-artefacts-for-case-2017/.

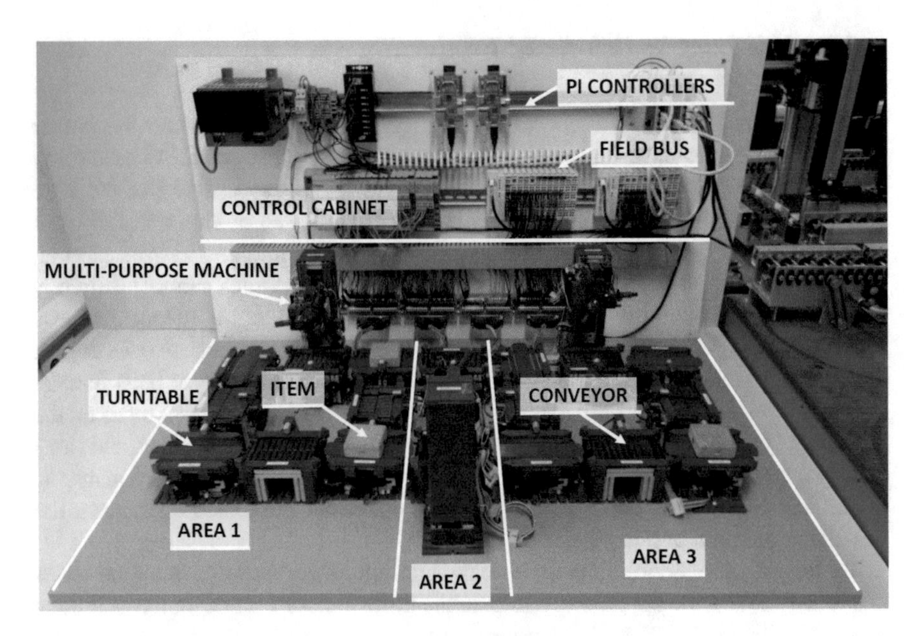

Fig. 5.9 Lab-sized production system hosted at IAF of the Otto von Guericke University Magdeburg http://www.iaf-bg.ovgu.de/en/technische_ausstattung_cvs.html

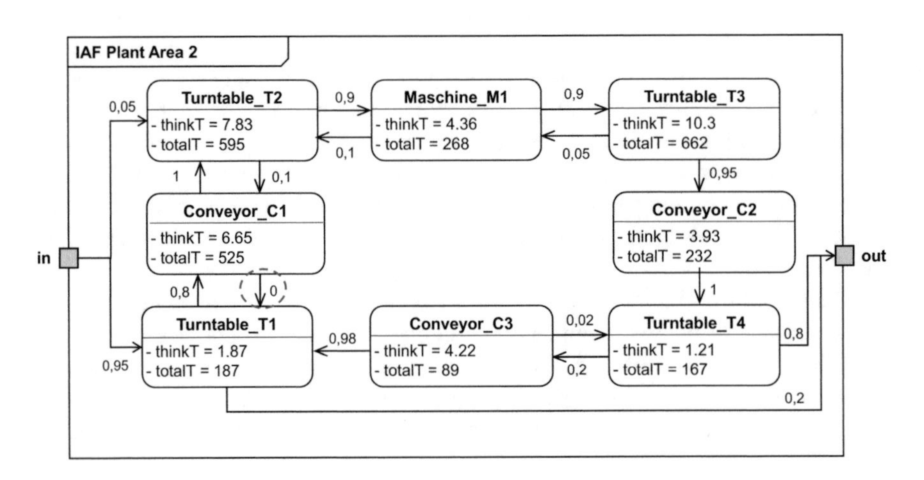

Fig. 5.10 Finite state automaton of area 2 of the lab-sized production system (Artner et al. 2017)

the figure shows one bottleneck. There is a backward routing from turntable T3 to the multipurpose machine M1 and further to turntable T2 and the conveyor belt C1. However, it does not make sense, since there is no further backward routing possible, only forward routing (see Fig. 5.10, red circle).

In an additional scenario of the lab-sized production system, firstly presented in (Mazak et al. 2018), we implemented the t-digest algorithm introduced by Dunning (2014). This algorithm is a kind of probabilistic data structure especially used for anomaly detection. This means to detect unexpected behavior or outliers. In this scenario, we use it for detecting performance bottlenecks. For this purpose, we approximate a so-called perfect timing for processing workpieces made of various materials (wood, aluminum, and titanium) for the process of transportation as well as the machining operations drilling and milling. These approximations are based on SPS cycle time and domain expert knowledge. For the computations, we used each with a dataset of 50.

In a next step, we applied the t-digest algorithm to calculate an anomaly detector. Based on this anomaly detector and the datasets, we additionally calculate acceptable deviations as upper and lower bounds for each material and working process (transportation, drilling, and milling). We used the anomaly detector as well as the lower and upper bounds for defining monitoring alarms by Grafana[11] for our TSDB. Based on these benchmarks, an alert is given whenever one of the thresholds will be exceeded or fallen below. In one of the scenarios, alerts occurred for workpieces of aluminum with a thickness of 125 mm. We found out that this alarm appeared due to a delay sometimes occurred during transportation at the conveyor belt C2, caused by a jamming of the workpiece.

5.5 Conclusion and Outlook

In order to meet the challenges we discussed in Sect. 5.2, a well-defined mix of MDE, data-driven engineering, reverse engineering, PM, performance engineering, and data analytics is required. We presented that there is currently emerging research work in progress focusing on runtime phenomena, runtime monitoring, and runtime analytics for enabling a high assurance in software and systems engineering. All these techniques, methods, and approaches aim at better understanding the concrete data and events used in or by a system as well as to be able to focusing on particular aspects whenever needed.

In this context, we have shown that we could further advance the research work in this direction by introducing a unifying temporal model framework dealing with downstream information from the model-driven engineering process and upstream information gathered during a system's operation at runtime. For realizing such a framework that combines two perspectives with each other, one at design time and

[11] https://grafana.com/.

one at runtime, we presented two contributions: (1) managing temporal models and (2) evolutionary model mining. In the presented use cases, we were focusing on the difference between design models, needed for realizing a system, and runtime models, used for describing how a (software) system is actually realized and how it is operating in certain environments. Thereby, model profiles are generated automatically from execution logs of software systems during operation. First outcomes of our different use cases showed the feasibility and usefulness of a temporal model framework as well as the analyzed data of operating systems at model level instead of data level.

In future work, we will extend this framework by query components based on, for example, Complex Event Processing (CEP), firstly introduced in (Luckham 2001) for implementing a continuous behavior mining during runtime. Thereby, event queries are used to automatically validate states of continuous systems by producing time series for their observable variables based on sensor value streams. This further approach allows us to identify and verify system states/events based on various raw sensor data. A first promising approach in this direction is presented in (Wolny et al. 2017).

Acknowledgements This work has been supported by the Austrian Federal Ministry for Digital and Economic Affairs; by the National Foundation for Research, Technology and Development; and by the FWF in the Project TETRABox under the grant number P28519-N31.

References

Anagnostopoulos, I., Zeadally, S., & Exposito, E. (2016). Handling big data: Research challenges and future directions. *The Journal of Supercomputing, 72*(4), 1494–1516.

Artner, J., Mazak, A., & Wimmer, M. (2017). Towards stochastic performance models for web 2.0 applications. In J. Cabot, R. D. Virgilio, & R. Torlone (Eds.), *Proceedings of the 17th International Conference on Web Engineering (ICWE 2017). Lecture Notes in Computer Science* (Vol. 10360, pp. 360–369). Berlin: Springer.

Basciani, F., Rocco, J. D., Ruscio, D. D., Salle, A. D., Iovino, L., & Pierantonio, A. (2014). MDEForge: An extensible web-based modeling platform. In *Proceedings of the 2nd International Workshop on Model-Driven Engineering on and for the Cloud (CloudMDE) Co-located with the 17th International Conference on Model Driven Engineering Languages and Systems (MoDELS)* (pp. 66–75). CEUR-WS.org

Benelallam, A., Gómez, A., Sunyé, G., Tisi, M., & Launay, D. (2014). Neo4EMF, a scalable persistence layer for EMF models. In J. Cabot & J. Rubin, (Eds.), *Proceedings of the 10th European Conference on Modelling Foundations and Applications, ECMFA 2014. Lecture Notes in Computer Science* (Vol. 8569, pp. 230–241). Berlin: Springer.

Bergmayr, A., Breitenbücher, U., Ferry, N., Rossini, A., Solberg, A., Wimmer, M., et al. (2018). A systematic review of cloud modeling languages. *ACM Computing Surveys, 51*(1), 22.

Bill, R., Mazak, A., Wimmer, M., & Vogel-Heuser, B. (2017a). On the need for temporal model repositories. In M. Seidl & S. Zschaler (Eds.), *2017 Collocated Workshops on Software Technologies: Applications and Foundations, STAF, Revised Selected Papers. Lecture Notes in Computer Science* (Vol. 10748, pp. 136–145). Berlin: Springer.

Bill, R., Neubauer, P., & Wimmer, M. (2017b). Virtual textual model composition for supporting versioning and aspect-orientation. In *Proceedings of the 10th ACM SIGPLAN International Conference on Software Language Engineering, SLE 2017* (pp. 67–78). New York, NY: ACM.

Bishop, C. M. (2007). *Pattern recognition and machine learning. Information science and statistics* (5th ed.). Berlin: Springer.

Blair, G., Bencomo, N., & France, R. (2009). Models@ run.time. *Computer, 42*(10), 22–27.

Brambilla, M., Cabot, J., & Wimmer, M. (2017). *Model-driven software engineering in practice. Synthesis lectures on software engineering* (2nd ed.). Morgan & Claypool Publishers.

Brosch, P., Kappel, G., Seidl, M., Wieland, K., Wimmer, M., Kargl, H., & Langer, P. (2010). Adaptable model versioning in action. In *Proceedings of the German Modellierung Conference* (pp. 221–236). GI.

Bülow, S., Backmann, M., Herzberg, N., Hille, T., Meyer, A., Ulm, B., et al. (2013). Monitoring of business processes with complex event processing. In N. Lohmann, M. Song, & P. Wohed (Eds.), *2013 International Workshops on Business Process Management Workshops - BPM, Revised Papers. Lecture Notes in Business Information Processing* (Vol. 171, pp. 277–290). Berlin: Springer.

Chandola, V., Banerjee, A., & Kumar, V. (2009). Anomaly detection: A survey. *ACM Computing Surveys, 41*(3), 15:1–15:58.

Clasen, C., Didonet Del Fabro, M., & Tisi, M. (2012). Transforming very large models in the cloud: A research roadmap. In *Proceedings of the 1st International Workshop on Model-Driven Engineering on and for the Cloud (CloudMDE) Co-located with the 8th European Conference on Modelling Foundations and Applications (ECMFA)* (pp. 1–10). HAL.

Cuadrado, J. S., & de Lara, J. (2013). Streaming model transformations: Scenarios, challenges and initial solutions. In *Proceedings of the 6th International Conference on Theory and Practice of Model Transformations (ICMT)* (pp. 1–16). Berlin: Springer.

Cugola, G., & Margara, A. (2012). Processing flows of information: From data stream to complex event processing. *ACM Computing Surveys, 44*(3), 15:1–15:62.

Daniel, G., Sunyé, G., Benelallam, A., & Tisi, M. (2014). Improving memory efficiency for processing large-scale models. In *Proceedings of the 2nd Workshop on Scalability in Model Driven Engineering (BigMDE)* (pp. 31–39). CEUR-WS.org.

Dávid, I., Ráth, I., & Varró, D. (2018). Foundations for streaming model transformations by complex event processing. *Software and Systems Modeling, 17*(1), 135–162.

Davoudian, A., Chen, L., & Liu, M. (2018). A survey on NoSQL stores. *ACM Computing Surveys, 51*(2), 40:1–40:43.

Deak, L., Mezei, G., Vajk, T., & Fekete, K. (2013). Graph partitioning algorithm for model transformation frameworks. In *Proceedings of the International Conference on Computer as a Tool (EUROCON)* (pp. 475–481). Piscataway, NJ: IEEE.

Demchenko, Y., de Laat, C., & Membrey, P. (2014). Defining architecture components of the big data ecosystem. In *2014 International Conference on Collaboration Technologies and Systems, CTS* (pp. 104–112). Piscataway, NJ: IEEE.

Domingos, P. M., & Hulten, G. (2001). Catching up with the data: Research issues in mining data streams. In *Proceedings of the 6th International Workshop on Research Issues in Data Mining and Knowledge Discovery (DMKD)*. cs.cornell.edu.

Dumas, M., van der Aalst, W. M. P., & ter Hofstede, A. H. M. (2005). *Process-aware information systems: Bridging people and software through process technology*. London: Wiley.

Dunning, T. (2014). *Practical machine learning: A new look at anomaly detection* (1st ed.) . Sebastopol, CA: O'Reilly Media.

Espinazo Pagan, J., & Garcia Molina, J. (2014). Querying large models efficiently. *Information and Software Technology, 56*(6), 586–622.

Fayyad, U. M., Piatetsky-Shapiro, G., & Smyth, P. (1996). From data mining to knowledge discovery: An overview. In *Advances in Knowledge Discovery and Data Mining* (pp. 1–34). Menlo Park, CA: AAAI.

Gómez, A., Tisi, M., Sunyé, G., & Cabot, J. (2015). Map-based transparent persistence for very large models. In *Proceedings of the 18th International Conference on Fundamental Approaches to Software Engineering (FASE)* (pp. 19–34). Berlin: Springer.

Hafner, C., Medetz, M., & Wapp, M. (2018). *Enterprise Architect Sequence Miner*. Technical report, TU Wien.

Hallé, S., & Varvaressos, S. (2014). A formalization of complex event stream processing. In M. Reichert, S. Rinderle-Ma, & G. Grossmann (Eds.), *18th IEEE International Enterprise Distributed Object Computing Conference, EDOC 2014* (pp. 2–11). Washington, DC: IEEE Computer Society.

Han, J., Cheng, H., Xin, D., & Yan, X. (2007). Frequent pattern mining: current status and future directions. *Data Mining and Knowledge Discovery, 15*(1), 55–86.

Hartmann, T., Fouquet, F., Nain, G., Morin, B., Klein, J., Barais, O., et al. (2014). A native versioning concept to support historized models at runtime. In J. Dingel, W. Schulte, I. Ramos, S. Abrahão, & E. Insfrán (Eds.), *Proceedings of the 17th International Conference on Model-Driven Engineering Languages and Systems, MODELS 2014. Lecture Notes in Computer Science* (Vol. 8767, pp. 252–268). Berlin: Springer.

Hartmann, T., Moawad, A., Fouquet, F., Nain, G., Klein, J., & Traon, Y. L. (2015). Stream my models: Reactive peer-to-peer distributed models@run.time. In *Proceedings of the 18th International Conference on Model Driven Engineering Languages and Systems (MoDELS)*. ACM/IEEE.

Kadam, S., Maltsev, A., Patsuk-Bösch, P. (2017). *Model Profiling*. Technical report, TU Wien.

Khalifa, S., Elshater, Y., Sundaravarathan, K., Bhat, A., Martin, P., Imam, F., et al. (2016). The six pillars for building big data analytics ecosystems. *ACM Computing Surveys, 49*(2), 33:1–33:36.

Khare, S., An, K., Gokhale, A. S., Tambe, S., & Meena, A. (2015). Reactive stream processing for data-centric publish/subscribe. In *Proceedings of the 9th International Conference on Distributed Event-Based Systems (DEBS)*, (pp. 234–245). New York, NY: ACM.

Koegel, M., & Helming, J. (2010). EMFStore: A model repository for EMF models. In J. Kramer, J. Bishop, P. T. Devanbu, & S. Uchitel (Eds.), *Proceedings of the 32nd ACM/IEEE International Conference on Software Engineering, ICSE 2010* (Vol. 2, pp. 307–308). New York, NY: ACM.

Kolovos, D. S., Rose, L. M., Matragkas, N., Paige, R. F., Guerra, E., Cuadrado, J. S., et al. (2013). A research roadmap towards achieving scalability in model driven engineering. In *Proceedings of the Workshop on Scalability in Model Driven Engineering (BigMDE)* (pp. 2:1–2:10). New York, NY: ACM.

Laney, D. (2001). *3-D Data Management: Controlling Data Volume, Velocity, and Variety*. Technical report, META Group.

Luckham, D. C. (2001). *The power of events: An introduction to complex event processing in distributed enterprise systems*. Reading, MA: Addison-Wesley.

Mannhardt, F., de Leoni, M., Reijers, H. A., van der Aalst, W. M., & Toussaint, P. J. (2018). Guided process discovery—a pattern-based approach. *Information Systems, 76*, 1–18.

Mazak, A., Lüder, A., Wolny, S., Wimmer, M., Winkler, D., Kirchheim, K., et al. (2018). Model-based generation of run-time data collection systems exploiting automationml. *Automatisierungstechnik, 66*(10), 819–833.

Mazak, A., & Wimmer, M. (2016a). On marrying model-driven engineering and process mining: A case study in execution-based model profiling. In P. Ceravolo, C. Guetl, & S. Rinderle-Ma (Eds.), *Proceedings of the 6th International Symposium on Data-driven Process Discovery and Analysis (SIMPDA 2016), CEUR Workshop Proceedings* (Vol. 1757, pp. 78–88). CEUR-WS.org

Mazak, A., & Wimmer, M. (2016b). Towards liquid models: An evolutionary modeling approach. In *18th IEEE Conference on Business Informatics, CBI 2016*, E. Kornyshova, G. Poels, C. Huemer, I. Wattiau, F. Matthes, & J. L. C. Sanz (Eds.) (pp. 104–112). Piscataway, NJ: IEEE.

Mazak, A., & Wimmer, M. (2017). Sequence pattern mining: Automatisches erkennen und auswerten von interaktionsmustern zwischen technischen assets basierend auf sysml-sequenzdiagrammen. In *Tag des Systems Engineering 2017, TdSE 2017* (pp. 145–156). Munich: Carl Hanser Verlag GmbH. KG.

Mazak, A., M. Wimmer, & P. Patsuk-Boesch (2017). Reverse engineering of production processes based on Markov chains. In *13th IEEE Conference on Automation Science and Engineering, CASE 2017* (pp. 680–686). Piscataway, NJ: IEEE.

Mazak, A., Wimmer, M., & Patsuk-Bösch, P. (2016). Execution-based model profiling. In P. Ceravolo, C. Guetl, & S. Rinderle-Ma (Eds.), *Data-Driven Process Discovery and Analysis - 6th IFIP WG 2.6 International Symposium, SIMPDA 2016, Revised Selected Papers. Lecture Notes in Business Information Processing* (Vol. 307, pp. 37–52). Berlin: Springer.

Pedersen, T. B. (2017). Managing big multidimensional data: A journey from acquisition to prescriptive analytics. In J. Bernardino, C. Quix, & J. Filipe (Eds.), *Proceedings of the 6th International Conference on Data Science, Technology and Applications, DATA 2017* (p. 5). SciTePress.

Shmueli, G., & Koppius, O. R. (2011). Predictive analytics in information systems research. *MIS Quarterly, 35*(3), 553–572.

Tax, N., Sidorova, N., Haakma, R., & van der Aalst, W. M. (2016). Mining local process models. *Journal of Innovation in Digital Ecosystem, 3*(2), 183–196.

van der Aalst, W. M. P. (2012). Process mining. *Communications of the ACM, 55*(8), 76–83.

van der Aalst, W. M. P. (2018). Process discovery from event data: Relating models and logs through abstractions. *Wiley Interdisciplinary Reviews: Data Mining and Knowledge Discovery, 8*(3), e1244.

van der Aalst, W. M. P., Adriansyah, A., de Medeiros, A. K. A., Arcieri, F., Baier, T., Blickle, T., et al. (2011). Process mining manifesto. In *Proceedings of the Business Process Management Workshops (BPM)* (pp. 169–194). Berlin: Springer.

van der Aalst, W. M. P., Weijters, T., & Maruster, L. (2004). Workflow mining: Discovering process models from event logs. *IEEE Transactions on Knowledge and Data Engineering, 16*(9), 1128–1142.

van Dongen, B. F., van der Aalst, W. M. P. (2005). A meta model for process mining data. In *Proceedings of the International Workshop on Enterprise Modelling and Ontologies for Interoperability (EMOI) Co-located with the 17th Conference on Advanced Information Systems Engineering (CAiSE)*. CEUR-WS.org.

Vlissides, J., Helm, R., Johnson, R., & Gamma, E. (1995). *Design patterns: Elements of reusable object-oriented software*. Reading, MA: Addison-Wesley.

Wimmer, M., Garrigós, I., & Firmenich, S. (2017). Towards automatic generation of web-based modeling editors. In J. Cabot, R. De Virgilio, & R. Torlone (Eds.), *Proceedings of the 17th International Conference on Web Engineering (ICWE 2017). Lecture Notes in Computer Science* (Vol. 10360, pp. 446–454). Berlin: Springer.

Wolny, S., Mazak, A., Carpella, C., Geist, V., & Wimmer, M. (2019). Thirteen years of SysML: A systematic mapping study. *Software and System Modeling*. 10.1007/s10270-019-00735-y

Wolny, S., Mazak, A., Konlechner, R., & Wimmer, M. (2017). Towards continuous behavior mining. In P. Ceravolo, M. van Keulen, & K. Stoffel (Eds.), *Proceedings of the 7th International Symposium on Data-driven Process Discovery and Analysis (SIMPDA 2017). CEURWorkshop Proceedings* (Vol. 2016, pp. 149–150). CEUR-WS.org.

Yaqoob, I., Hashem, I. A. T., Gani, A., Mokhtar, S., Ahmed, E., Anuar, N. B., & Vasilakos, A. V. (2016). Big data: From beginning to future. *International Journal of Information Management, 36*(6), 1231–1247.

Chapter 6
On Testing Data-Intensive Software Systems

Michael Felderer, Barbara Russo, and Florian Auer

Abstract Today's software systems like cyber-physical production systems or big data systems have to process large volumes and diverse types of data which heavily influences the quality of these so-called data-intensive systems. However, traditional software testing approaches rather focus on functional behavior than on data aspects. Therefore, the role of data in testing has to be rethought, and specific testing approaches for data-intensive software systems are required. Thus, the aim of this chapter is to contribute to this area by (1) providing basic terminology and background on data-intensive software systems and their testing and (2) presenting the state of the research and the hot topics in the area. Finally, the directions of research and the new frontiers on testing data-intensive software systems are discussed.

Keywords Data-intensive systems · Data engineering · Cyber-physical production systems · Software testing · Data quality

6.1 Introduction

Information technology is evolving toward a kind of software-intensive systems that, more and more extensively, collect, integrate, process, and analyze large data volumes that have to be processed with high velocity coming from a number of diverse data sources, the so-called big data. Examples for such data-intensive systems can be found in all domains (e.g., finance, automotive, production) and all types of systems (e.g., information systems and cyber-physical systems). Especially also, large and long-running cyber-physical systems that can be found in manufacturing plants are today

M. Felderer (✉) · F. Auer
University of Innsbruck, Innsbruck, Austria
e-mail: michael.felderer@uibk.ac.at; florian.auer@uibk.ac.at

B. Russo
Free University of Bozen-Bolzano, Bolzano, Italy
e-mail: barbara.russo@unibz.it

S. Biffl et al. (eds.), *Security and Quality in Cyber-Physical Systems Engineering*, https://doi.org/10.1007/978-3-030-25312-7_6

data- and software-intensive as they are controlled by software and have to process large volumes and different types of data in realtime (see Sect. 6.4). The rapid growth of such systems is generating a paradigm shift in the field of software engineering. We are moving from traditional software engineering, which is functionality-centric, toward modern software and data engineering, which is data-centric and where the functionality is driven by the availability of data. Modern systems collect raw data from the users and their personal devices (like smart watches) and from the environment and its smart objects (like sensors in industrial plants), as well as higher-level data coming from information providers like social platforms, open-data sites, and data silos. Also, different types of data like image, sound, video, or textual data have to be taken into account. Beyond functionality, the success (and added value) of such systems is tied to the availability of the data that is processed as well as its quality.

Data-intensive (software) systems are therefore becoming increasingly prominent in today's world, where the collection, processing, and dissemination of ever-larger volumes of data have become a driving force behind innovation. However, data-intensive systems also pose new challenges to quality assurance (Hummel et al. 2018) and especially testing. Testing software-intensive systems has traditionally focused on verifying and validating compliance and conformance to specifications, as well as some general nonfunctional requirements. Data-intensive systems require a different approach for testing and analysis, moving more toward exploring the system, its elements, behavior, and properties from a big data and analytics perspective to help decision-makers respond in realtime (e.g., for cyber-physical systems). The behavior and, therefore, the expected test result can often not be specified precisely but only with a statistical uncertainty. For instance, the movement of a robot to assemble a part of a car can often not sufficiently be specified in an explicit way but based on machine learning algorithms with uncertainty.

As the area of data-intensive (software) systems and their testing is not well investigated so far, this chapter contributes by defining the main terminology and exploring existing approaches and promising future directions.

The chapter is structured as follows. Section 6.2 presents background on software testing and data quality. Section 6.3 defines what a data-intensive (software) system is and what main challenges to its quality assurance exist. Section 6.4 discusses cyber-physical production systems as an example for data-intensive systems. Section 6.5 presents an exploratory literature study on testing data-intensive software systems. Section 6.6 discusses existing approaches to testing data-intensive systems and future directions. Finally, Sect. 6.7 concludes the chapter.

6.2 Background

In this section, we cover background on software testing and data quality.

6.2.1 Software Testing

Software testing (or software-intensive system testing) is a part of the overall engineering process of cyber-physical systems and can be defined as the process consisting of all lifecycle activities, both static and dynamic, concerned with planning, preparation, and evaluation of software-intensive products, systems and services, and related artifacts to determine that they satisfy specified requirements, to demonstrate that they are fit for purpose, and to detect defects (ISTQB 2012). This broad definition of testing comprises not only dynamic testing activities like classic unit or system testing but also static testing activities, where artifacts are not executed, like static analysis or reviews. The tested software-based system is called *system under test* (SUT). Traditional dynamic testing evaluates whether the SUT behaves as expected or not by executing a *test suite*, that is, a *finite* set of test cases suitably *selected* from the usually infinite execution domain (Bourque and Dupuis 2014). In a generic sense, dynamic testing can also be defined as evaluating software by observing its execution (Ammann and Offutt 2016), which may also subsume *runtime monitoring* of live systems.

After executing a test case, the observed and intended behaviors of an SUT are compared with each other, which then results in a *verdict* (Felderer et al. 2016). Verdicts can be either of *pass* (behaviors conform), *fail* (behaviors don't conform), or *inconclusive* (not known whether behaviors conform) (ISO/IEC 1994). A *test oracle* is a mechanism for determining the verdict. The observed behavior may be checked against user or customer needs (commonly referred to as testing for *validation*) or against a specification (testing for *verification*). So the oracle compares an expected output value with the observed output. But this may not always be feasible, especially in the context of data-intensive software systems. For instance, consider data-intensive systems that produce complex output, like complicated numerical simulations or varying output based on a prediction model, where defining the correct output for a given input and then comparing it with the observed output may be nontrivial and error-prone. This problem is referred to as the *oracle problem*, and it is recognized as one of the fundamental challenges of software testing (Weyuker 1982). *Metamorphic testing* (Segura et al. 2016) is a technique conceived to alleviate the oracle problem. It is based on the idea that often it is simpler to reason about relations between outputs of a program than it is to fully understand or formalize its input-output behavior.

A *failure* is an undesired behavior. Failures are typically observed (by resulting in verdict fail) during the execution of the system under test. A *fault* is the cause of a failure. It is a static defect in the software, usually caused by human error in the specification, design, or coding process. During testing, it is the execution of faults in the software that causes failures. Differing from active execution of test cases, passive testing only monitors running systems without interaction.

Refining previous classifications (Utting and Legeard 2007; Felderer et al. 2016), testing can be classified utilizing the three dimensions: test objective, test level, and execution level (see Fig. 6.1).

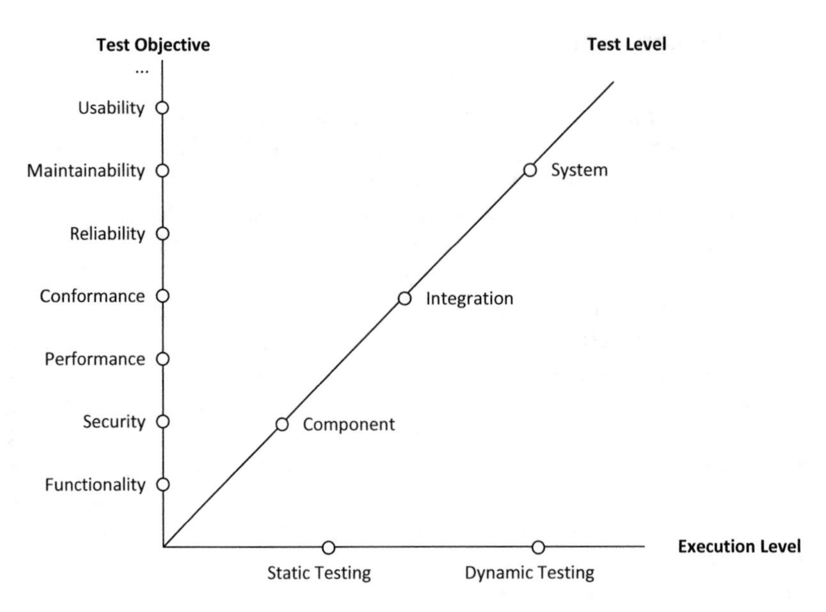

Fig. 6.1 Testing dimensions: test level, test objective, and execution level of traditional software testing

Test objectives are reasons or purposes for designing and executing a test. The reason is to check either the functional behavior of the system or its nonfunctional properties. *Functional testing* is concerned with assessing the functional behavior of an SUT, whereas *nonfunctional testing* aims at assessing nonfunctional requirements with regard to quality characteristics like security, performance, conformance, reliability, maintainability, or usability.

The *test level* addresses the granularity of the SUT and can be classified into component, integration, and system testing. It also determines the *test basis*, that is, the artifacts to derive test cases. *Component testing* (also referred to as *unit testing*) checks the smallest testable component (e.g., a class in an object-oriented implementation or a single electronic control unit) in isolation. *Integration testing* combines components with each other and tests those as a subsystem, that is, not yet a complete system. *System testing* checks the complete system, including all subsystems. A specific type of system testing is *acceptance testing* where it is checked whether a solution works for the users of a system. *Regression testing* is a selective retesting to verify that modifications have not caused side effects and that the SUT still complies with the specified requirements (Radatz et al. 1990).

The *execution level* addresses whether and how the SUT is executed. *Static testing* checks software development artifacts (e.g., requirements, design, or code) without execution of these artifacts. *Dynamic testing* actively executes a system under test and evaluates its results.

The process of testing comprises the core activities test planning, design, automation, execution, and evaluation (ISTQB 2012). According to (ISTQB 2012), *test planning* is the activity of establishing or updating a test plan. A test plan is a document describing the objectives, scope, execution levels, approaches, resources, and schedule of intended test activities. It identifies, among others, concrete objectives, the features to be tested, the test design techniques, and exit criteria to be used and the rationale of their choice. A test objective is a reason, which can be to check either the functional behavior of a system or its nonfunctional or (software) quality properties, for designing and executing a test. Exit criteria are conditions for permitting a process to be officially completed. They are used to report against and to plan when to stop testing. Adequacy criteria like coverage criteria aligned with the tested feature types and the applied test design techniques are typical exit criteria. Once the test plan has been established, test control begins. It is an ongoing activity in which the actual progress is compared against the plan which often results in concrete measures.

During the *test design* phase, the general testing objectives defined in the test plan are transformed into tangible test conditions and abstract test cases. In *criteria-based test design*, one designs test cases that satisfy specific engineering goals such as coverage criteria. In *human-based test design*, one designs test cases based on domain knowledge of the system and human knowledge of testing. *Test automation* comprises tasks to make the abstract test cases executable. This includes tasks like preparing test harnesses and test data, providing logging support, or writing test scripts which are necessary to enable the automated execution of test cases. In the *test execution* phase, the test cases are then executed, and all relevant details of the execution are logged and monitored. Finally, in the *test evaluation* phase, the exit criteria are evaluated, and the logged test results are summarized in a test report.

6.2.2 Data Quality

Quality, in general, has been defined as the totality of characteristics of a product that bear on its ability to satisfy stated or implied needs (ISO 1994). This generic definition can be instantiated to software and data quality, as capability of a software and data product, respectively, to satisfy stated and implied needs when used under specified conditions. For software systems, according to ISO/IEC 25010 (ISO/IEC 2011), these quality characteristics are functional suitability, performance efficiency, compatibility, usability, reliability, security, maintainability, and portability for product quality as well as effectiveness, efficiency, freedom from risk, and context coverage for quality in use. According to ISO/IEC 25012 (ISO/IEC 2008) data quality characteristics in the context of software development can be classified into inherent and system-dependent data characteristics.

Inherent data quality refers to the degree to which data quality characteristics have the intrinsic potential to satisfy stated and implied needs when data is used under specified conditions. From the inherent point of view, data quality refers to data itself, in particular to data domain values and possible restrictions (e.g., business rules governing the quality required for the characteristic in a given application), relationships of data values (e.g., consistency), and meta-data.

System-dependent data quality refers to the degree to which data quality is reached and preserved within a system when data is used under specified conditions. From this point of view, data quality depends on the technological domain in which data are used, and it is achieved by the capabilities of systems' components such as hardware devices or sensors (e.g., to make data available or to obtain the required precision) as well as system and other software (e.g., backup software to achieve recoverability or data processing software).

According to ISO/IEC 25012 (ISO/IEC 2008), inherent data quality characteristics are:

- *Accuracy*, that is, the degree to which data has attributes that correctly represent the true value of the intended attribute of a concept or event in a specific context of use.
- *Completeness*, that is, the degree to which data has attributes that are free from contradiction and are coherent with other data in a specific context of use. It can be either or both among data regarding one entity and across similar data for comparable entities.
- *Credibility*, that is, the degree to which data has attributes that are regarded as true and believable by users in a specific context of use. Credibility includes the concept of authenticity (the truthfulness of origins, attributions, commitments).
- *Currentness*, that is, the degree to which data has attributes that are of the right age in a specific context of use.

According to ISO/IEC 25012 (ISO/IEC 2008), inherent and system-dependent characteristics are :

- *Accessibility*, that is, the degree to which data can be accessed in a specific context of use, particularly by people who need supporting technology or special configuration because of some disability
- *Compliance*, that is, the degree to which data has attributes that adhere to standards, conventions, or regulations in force and similar rules relating to data quality in a specific context of use
- *Confidentiality*, that is, the degree to which data has attributes that ensure that it is only accessible and interpretable by authorized users in a specific context of use
- *Efficiency*, that is, the degree to which data has attributes that can be processed and provide the expected levels of performance by using the appropriate amounts and types of resources in a specific context of use
- *Precision*, that is, the degree to which data has attributes that are exact or that provide discrimination in a specific context of use
- *Traceability*, that is, the degree to which data has attributes that provide an audit trail of access to the data and of any changes made to the data in a specific context of use

- *Understandability*, that is, the degree to which data has attributes that enable it to be read and interpreted by users and are expressed in appropriate languages, symbols, and units in a specific context of use

According to ISO/IEC 25012 (ISO/IEC 2008), system-dependent characteristics are :

- *Availability*, that is, the degree to which data has attributes that enable it to be retrieved by authorized users and/or applications in a specific context of use
- *Portability*, that is, the degree to which data has attributes that enable it to be installed, replaced, or moved from one system to another, preserving the existing quality in a specific context of use
- *Recoverability*, that is, the degree to which data has attributes that enable it to maintain and preserve a specified level of operations and quality, even in the event of failure, in a specific context of use

Big data is a term used to refer to data sets that are too large or complex for traditional data processing software to adequately process them. It is high-volume, high-velocity, and high-variety information assets that demand cost-effective, innovative forms of information processing for enhanced insight and decision-making. Big data can be described by the following characteristics:

- *Volume* refers to the quantity of generated and stored data that determined the value and potential insight.
- *Velocity* refers to the speed at which the data is generated and processed to meet the demands and challenges.
- *Variety* refers to the different types of data that have to be processed, that is, text, images, audio, and video.
- *Veracity* refers to the data quality and the data value that can be measured based on the quality criteria highlighted before.

6.3 Data-Intensive Software Systems

Systems that process large volumes of data have commonly been referred to as "data-intensive." However, the amount of data that is considered as large is relative and changes over time. Thus, information systems that stored and retrieved large volumes of data at their time may be no longer considered as "data-intensive" today. This made the concept of "data-intensive" systems relative to time and recent developments in storage capacities and processing capabilities.

Today, data-intensive systems do not only process large volumes of data but differ from other systems with respect to the following aspects:

- The data is stored and retrieved but depending on the context also collected, generated, manipulated, and redistributed (i.e., processed).

- The data is "big" data and fulfills the characteristics described in the previous section (i.e., volume, velocity, variety, and veracity).
- The data does influence in addition to the operation phase also the analysis, design, implementation, testing, and maintenance phase of the system lifecycle.

These data-intensive systems are more than large data stores that allow the storage and retrieval of large volumes of data. Data-intensive systems additionally comprise the collection of data from different sources, the generation of new data, the manipulation of existing data, and the redistribution of data to data consumers.

The data itself fulfills the characteristics of big data. The different data sources that are used by data-intensive systems provide large volumes of data (volume) in varying formats (variety). Furthermore, the data is processed timely to meet the demands of its users (velocity). Finally, the high dependency on data requires the maintenance of high data quality (veracity) throughout the system.

With respect to the central role that data has in all phases of the system lifecycle (with the consequence that the system behavior is determined by the data), data-intensive systems can be opposed to specification-driven systems where the behavior is explicitly specified and coded in (deterministic) algorithms.

To summarize, data-intensive systems are defined as follows:

> Data-intensive systems are systems that may handle, generate, and process a large volume of data of different nature and from different sources over time orchestrated by means of different technologies and with the goal to extract value from data for different types of businesses. They pose specific challenges in all phases of the system lifecycle, and over time, they might evolve to very large, complex systems as data, technologies, and value evolve.

In the following sections, the architecture and quality assurance of data-intensive systems are discussed.

6.3.1 Architecture of Data-Intensive Systems

Recent years have seen the rapid growth of interest on developing data-intensive systems. These systems include technologies like Hadoop/MapReduce, NoSQL, or systems for stream processing (Casale et al. 2015). Specific technologies like databases, caches, search indexes, stream processing, or batch processing enable the development of data-intensive systems. They provide services like decision support and system behavior monitoring or reporting applicable in different business domains. Thus, a data-intensive system is typically built from standard building blocks that provide commonly needed functionality. For example, many systems need databases, search indexes, or batch processing. As there are various options for database systems, or approaches to caching, the most appropriate technologies for the respective system have to be chosen among the numerous possibilities. Key quality attributes for these decisions are, according to Kleppmann (2017), reliability, scalability, and maintainability.

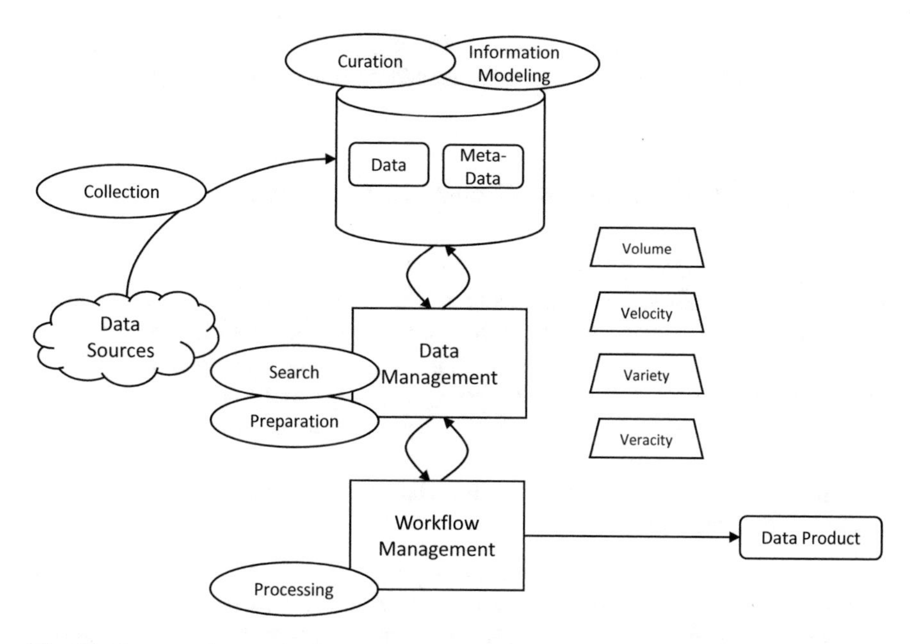

Fig. 6.2 Conceptual structure of a data-intensive system

The building blocks that assemble a data-intensive system are visualized in the conceptual structure of a data-intensive system given in Fig. 6.2, which is based on the system lifecycle of data-intensive systems presented in (Mattmann et al. 2011). Following the path of the data through the system, first, the data is collected from the different data sources with possible varying protocols (e.g., ftp, http) and data formats like domain-specific exchange formats ("Data Sources"). Second, in the staging area, it is ensured that the data conforms to a unified interpretation, sharing, and preservation (Curation). This includes describing the data with meta-data to ease further processing. The meta-data are created, organized, and classified according to the information modeling of the data-intensive system. Thereafter, the data and its meta-data are stored into an archive system ("Data Management"). This part of the system supports searching on the data and further prepares the data for its later usage in the workflow management component. Following the data management, the workflow management is responsible for processing the data. It includes tasks to process the data (e.g., calculations) and workflows that align tasks into a processing pipeline that can make data-based decisions on which concrete tasks to execute. In the last step, the data leaves the data-intensive system as part of a data product. Examples therefore are reports, prediction models, analyses, and recommendations or generated data. It represents the data output of the data-intensive system that may be further processed within a system. The four big data characteristics (see Sect. 6.2.2) that are next to the system components highlight that volume, velocity, variety, and veracity are inherent in every part of the system. Each component and activity (e.g., collection, curation, or processing) has to take these characteristics into account. This reflects the intensive dependency of these systems on data.

6.3.2 *Quality Assurance of Data-Intensive System*

Data-intensive systems process large volumes of data and provide through the processing of it high value to the business. Thus, failures and other quality issues in data-intensive systems are incredibly costly (e.g., production failures), because of their effect on the business. As a result, new types of quality assurance activities and concerns arise. For example, data debugging, that is, discovering system failures caused by well-formed but incorrect data, is a primary issue in the quality assurance of data-intensive systems.

Hummel et al. (2018) identified eight challenges of quality assurance in the context of data-intensive systems.

- *Challenging Visualization and Explainability of Results*. Data needs to be visualized with the right balance between data dimensions and resolution, in order to support the user to understand and assess the validity of the data. Furthermore, the processing that leads to a particular result is difficult to explain (e.g., in deep learning). Thus, trustworthiness and understandability are important challenges for data-intensive systems.
- *Nonintuitive Notion of Consistency*. The large volume of data that is processed by data-intensive systems requires to weaken the notion of data consistency for performance reasons. These inconsistencies are confusing for users not used to softened consistency notions.
- *Complex Data Processing and Different Notions of Correctness*. The numerous processing steps and interconnections between data make the processing of data in data-intensive systems complex. The notion of correctness becomes difficult to define, which complicates the testability of such systems.
- *High Hardware Requirements for Testing*. Data-intensive systems are expected to process large volumes of data. Thus, testing requires to consider issues like performance or scalability that are related to processing of big data. However, this requires hardware similar to its later application environment, which may be not possible because of involved operating costs of such environments.
- *Difficult Generation of Adequate, High-Quality Data*. The data provided for testing should represent realistic and application-specific data in order to meaningfully test a data-intensive system. Thus, all big data characteristics described in Sect. 6.2.2 should be covered, which makes test data generation and management a challenging task, especially when also taking data quality aspects into account.
- *Lack of Debugging, Logging, and Error-Tracing Methods*. The architecture of data-intensive systems results into a distributed system that complicates debugging. Furthermore, logging is scattered over the different components of a

system. As a result, tracing errors back to their origin or understanding system behavior based on the logs become inherently difficult.
- *State Explosion in Verification.* The processing of large volumes of data that requires distributed computing to process its requests results into an exponential state explosion that makes verification approaches complex to apply.
- *Ensuring Data Quality.* The data quality of the large volumes of data that are collected, processed, and aggregated by data-intensive systems is difficult to assess computationally and semantically.

6.4 Cyber-Physical Production Systems as Data-Intensive Systems

Industrial production systems like robots, steel mills, or manufacturing plants are large and long-running cyber-physical systems that generate and process large volumes of data from various sources (see Fig. 6.3) and types (e.g., order data, personnel data, or machine data). Data sources of typical production systems are among others technical production data (like machine or process data), organizational production data (like order or personnel data), and messages from the ERP system. Thus, cyber-physical production systems represent data-intensive systems that are of high value to the business and require testing. The production is tightly coupled to the data-intensive systems, which puts high requirements on the standard of quality assurance of the system and the data it operates on.

The architecture of a large cyber-physical production system as used in manufacturing plant is conceptually consistent with the architecture described in Sect. 6.3.1. Data of various sources (e.g., sensors, orders, processes) is collected, unified, and organized according to the later information needs of the production system (e.g., order status). Employees or intelligent systems (as envisioned by

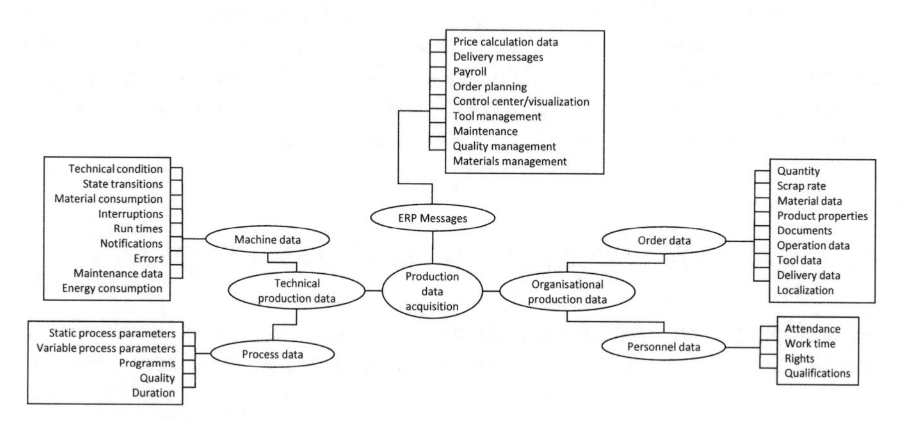

Fig. 6.3 Overview of data sources in the context of production systems

Industry 4.0 (Wang et al. 2016; Foidl and Felderer 2015)) can search within the data and extract relevant information for their specific data needs. Finally, every production system has some kind of workflow management that processes the data in order to generate data products like aggregated reports or prediction models (e.g., for predictive maintenance) to support decision-making.

Production failures or unintended production stops are possible results of software faults in the cyber-physical production system and consequence of inappropriate testing. A cyber-physical production system in a manufacturing plant, which represents a complex and large production system that processes large volumes of data, is a critical business asset that requires careful testing to mitigate production-related risks. Furthermore, the software and data quality of the system has a strong effect on the production itself. Higher-quality standards can lead to optimizations and improvements of the overall production, whereas quality problems can lead to a decline in production. Challenges to the quality assurance (see Sect. 6.3.2) like high hardware requirements for testing (e.g., related to robots or production line) or the difficulty to ensure the quality of the data from various sources (e.g., machines, orders, processes) are specifically present in production systems. Moreover, the trend to smart factories, which, for example, encourage the intelligent communication between robots, further increases the requirements on quality assurance. As a result, the research on testing of data-intensive software systems (see Sects. 6.5 and 6.6) is of great importance for large cyber-physical production systems in general and their quality assurance in particular.

6.5 Testing Data-Intensive Software Systems

Research on testing data-intensive systems is becoming of high academic and practical importance. Only few recent papers (e.g., Hummel et al. (2018) as presented in Sect. 6.3.2) discussed challenges and derived open questions to start guiding research in that field. However, primary studies contributing approaches on testing data-intensive systems are still rare. In this section, we report the result of our literature study on existing research on testing data-intensive systems. The study is not intended to be systematic but rather exploratory to get an initial insight on the aspects of the research that are currently under investigation.

6.5.1 Literature Study on Testing Data-Intensive Systems

We performed an exploratory and preliminary literature study to get initial insights into the research on testing data-intensive systems. Papers have been collected from Google Scholar[1] and the IEEE Xplore Digital Library[2] by searching with keywords "Data-Intensive System" (or "Data-Intensive Systems") and "Software Testing." Included papers are written in English and are journal articles, conference or workshop papers, book chapters, technical reports, or thesis works that cover at least one testing activity as well as big data or data processing aspects. The search initially returned 133 papers. After reading the title and abstract and eventually the body of the work, papers were removed if they:

- Misused the search terms (e.g., "Data-Oriented Systems" as "Data-Intensive System")
- Only used the search term "Data-Intensive System" or "Software Testing" in their citations
- Only referred to data-intensive systems as one of the systems under analysis with no specific discussion or exploitation of the characteristics of data-intensive systems
- Mention "Data-Intensive System" only in the related work section or as an example context in which an approach can eventually work

Taking these criteria into account, we could finally include 16 relevant papers. This indicates that software testing of data-intensive systems is not yet an established and clearly defined term. The 16 papers were further classified according to three major areas (i.e., Testing, System, Data), related categories (e.g., Level of Testing), and factors (e.g., Component, Integration, and System for Level of Testing) based on the following criteria:

- Testing
  - Level of Testing (Component, Integration, System)
  - Test Activity (Test Management, Test-Case Design (Criteria-Based), Test-Case Design (Human Knowledge-Based), Test Automation, Test Execution, Static Analysis, Test Evaluation (Oracle))
- System
  - System Quality (Security, Performance, Conformance, Usability, etc.)
  - System Artifact (SQL Query, System Code, User Form, etc.)
- Data
  - Big Data Aspect (Volume, Velocity, Variety, Veracity, Value)
  - Data Processing Aspect (Collection, Generation, Manipulation, Redistribution)
  - Data Quality Aspect (Integrity, etc.)
  - Data Technology (Database, Cache, etc.)

[1] http://scholar.google.at.

[2] https://ieeexplore.ieee.org.

The applied classification scheme was derived from established classifications as presented in Sects. 6.2 and 6.3, experience of the authors, and an analysis of relevant papers. Open coding (to identify the categories) and axial coding (to relate categories) of grounded theory (Corbin and Strauss 1990) on titles and abstracts of such papers have been used to verify and refine the initial categorization.

6.5.2 Classification of Papers

The 16 relevant papers were classified according to the categories and factors defined in Sect. 6.5.1. The classification has been performed by reading the body of each paper. Papers may be classified into multiple factors of the same category or multiple categories of the same area.

Figure 6.4 illustrates the resulting classification of the included papers by comparing categories and factors characterizing software testing with the ones of data-intensive system (i.e., categories classified as *System* and *Data* in Sect. 6.5.1). One may immediately recognize that not all factors are included in the figure. For instance, Velocity for the Big Data aspects or Cache for the Data Technology aspects are never considered when it comes to software testing in the investigated set of papers. For each factor, the maximum count is four (i.e., a quarter of the investigated papers), which is attained for (1) Database in the Data Technology category at Component and System Level of Testing; (2) Test-Case Design (Human Knowledge-Based) and Test Automation in the Test Activity category for what concerns SQL Query in System Artifacts and at the System Level Testing and Security in System Quality, respectively; and (3) Volume in Big Data aspects concerning System Level of Testing.

6.6 Discussion

In this section, we discuss the current state of testing data-intensive software systems and present directions of research that have the potential to significantly contribute to solving key challenges of testing data-intensive software systems.

The initial literature survey presented in the last section revealed that the automatic search in digital libraries returns a large number of false positives (i.e., 117 out of 133 papers) where the terms of the search are used only marginally, often to mention only future application or related work of a proposed approach. However, we also need to acknowledge that our approach is just exploratory and may have missed some relevant papers. In particular, we decided not to perform snowballing from the papers retrieved from the digital libraries as research on data-intensive system is rather new and backward snowballing seemed not to be promising after a first check of the references of the retrieved papers. With the classification of the relevant papers, we found that some aspects that commonly refer to testing or data-intensive

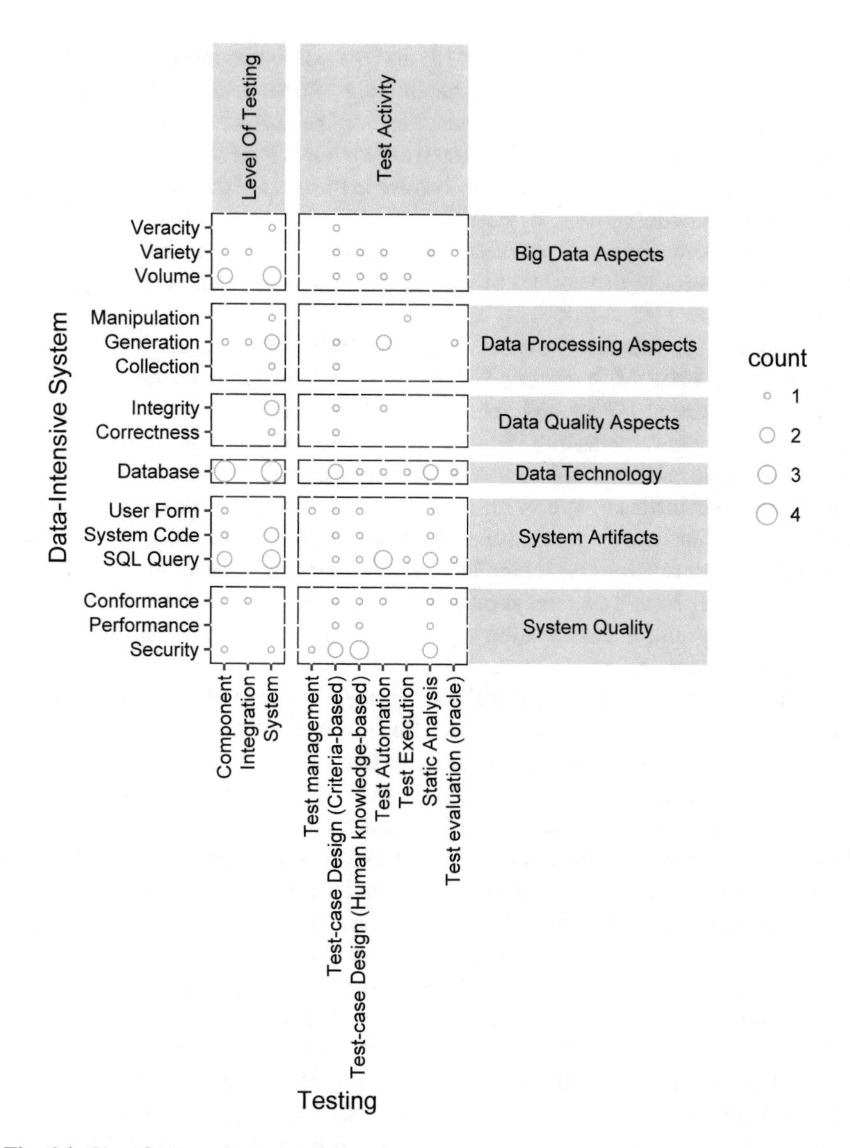

Fig. 6.4 Classification of relevant papers

software system are not yet explored when it comes to testing of data-intensive systems. For instance, for velocity as a big data aspect, suitable testing approaches for data-intensive systems are not available. Velocity is of paramount importance in modern data-intensive systems, especially in the context of real-time analytics, where results are delivered in a continuous fashion and results are only useful if the delay is very short. However, before focusing on testing for velocity, the research community is still struggling with the uncertainty of the output of test cases and the definition of suitable oracles (de Bayser et al. 2015) and the volume and variety of data that data-intensive systems can potentially handle (e.g., Russo et al. 2015; Gadler et al. 2017). A certain number of papers concern testing the correctness and the integrity of the data stored in databases by data-intensive systems. For instance, static analysis has been used to test SQL statements embedded in the code of data-intensive systems to ensure database integrity (Nagy 2013).

However, there are several promising directions of research and new frontiers that are highly relevant in the context of data-intensive systems that have not yet been fully exploited.

Testing of data-intensive software systems is problematic because these systems belong to the class of "non-testable" software, where typically no test oracles exist for verification (Otero and Peter 2015). A proven approach for testing such systems is metamorphic testing (Segura et al. 2016), which requires the discovery of metamorphic relations that can be used to test the validity of machine learning algorithms. For instance, Xie et al. (2011) provide an approach to metamorphic testing of machine learning-based classification algorithms. The authors enumerate the metamorphic relations that classifiers would be expected to demonstrate and then determine for a given implementation whether each relation is a necessary property for the corresponding classification algorithm. If this is the case, then failure to exhibit the relation indicates a fault.

Testing of machine learning algorithms, especially for sophisticated algorithms like deep learning systems, requires specific testing approaches. To that end, recently, several adequacy criteria for deep learning systems like neuron coverage (Pei et al. 2017), surprise adequacy (Kim et al. 2019) or criteria derived from modeling deep learning systems as abstract state transition systems (Du et al. 2018) were defined. However, testing algorithms is not sufficient as the integration of algorithms into systems can be complex, leading to problems and defects being injected along the way. Specific concerns mentioned in the literature include untrustworthy data, bad assumptions, and incorrect mathematics among others (Shull 2013). Tian et al. (2018) propose an approach based on metamorphic testing to test deep learning systems in the context of autonomous driving. As especially the environment is very complex, also simulation plays an important role for testing data-intensive systems as used in the context of autonomous driving, where large, complex, and independent system interact and form the so-called Systems-of-Systems (Dersin 2014).

One step further, testing can even be shifted to the running system. This novel approach to testing of data-intensive systems performs monitoring of the runtime environment to assess system behavior and changes. System changes under test are exposed to a limited number of users. The runtime data generated through their interaction of such users with the changes are compared to the ones generated by users that were not exposed to such changes. The analysis of the collected experimentation data reveals the influence of the chance on predefined characteristics like performance or usability. System change assessment follows also a continuous experimentation approach (Fagerholm et al. 2014) and is promising especially in the context of testing data-intensive systems (Auer and Felderer 2018).

Observing unexpected execution behavior is also used to build self-assessment oracles, a new type of oracles introduced in the anticipatory testing approach (Tonella 2019). Such an approach aims at detecting failures before they even occur. Its oracles can be crucial for data-intensive systems where the volume of the input and the uncertainty of the output are key challenges in designing test cases (Hummel et al. 2018).

Taking the discussion into account, the categories on the testing dimensions for traditional systems shown in Fig. 6.1 have to be extended for testing data-intensive software systems. The testing dimensions and their values for data-intensive software systems are shown in Fig. 6.5. The added categories are shown in italic.

The dimension test level is extended by the values *Algorithm* and *System-of-Systems*. On the one hand, the testing of algorithms in isolation becomes essential (e.g., for classification algorithms or deep learning algorithms). On the other hand, also *Systems-of-Systems*, for instance, in the context of autonomous driving, require new testing approaches where especially simulation of the environment plays an important role. The dimension execution level is extended by *Runtime Monitoring*, where a running system is passively observed to test changes or unspecified system behavior, in contrast to active penetration in case of dynamic testing. Finally, the dimension test objective is extended by *Data Quality*, which comprises testing of data and big data quality properties like correctness or velocity.

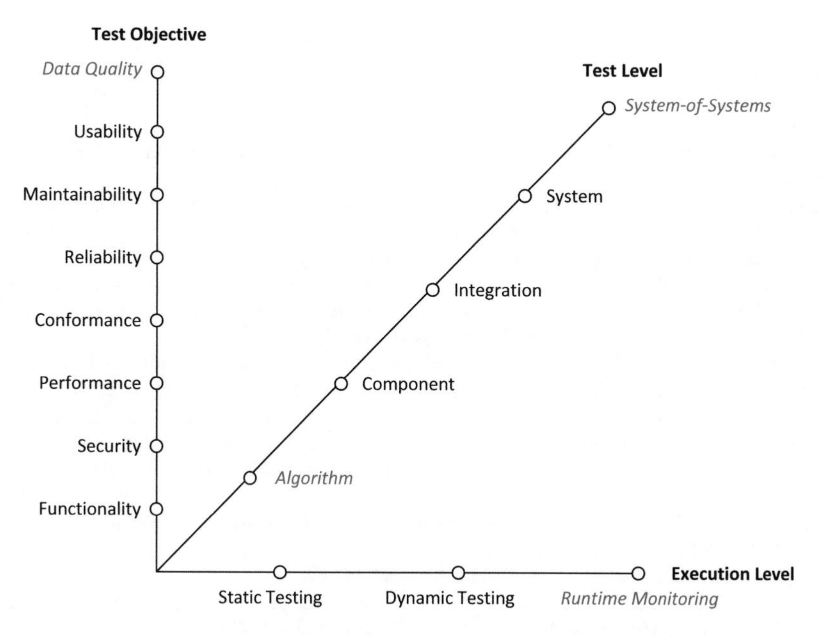

Fig. 6.5 Testing dimensions test level, test objective, and execution level of testing data-intensive software systems

6.7 Conclusion

Software engineering research is needed to fully understand the role of testing in data-intensive software systems (like large cyber-physical production systems) and to provide approaches and frameworks to properly address this challenging and critical issue. The goal of this chapter is to contribute into this direction by providing basic terminology, an overview of the literature, a discussion, and future directions of research.

In this chapter, we first presented the background as well as the basic terminology, and then we discuss the current state and new frontiers on testing data-intensive software systems that are also relevant for testing cyber-physical production systems. To this end, we first provided a definition of data-intensive system (Sect. 6.3) that clearly distinguishes them from other types of software systems and characterized cyber-physical production systems as data-intensive systems. The definition also helped us to derive the major factors characterizing such systems and drive the investigation on the existing research (Sect. 6.5.2). We further discussed the current state and new frontiers on testing data-intensive software systems (Sect. 6.6). The frontiers include metamorphic testing, algorithm testing, testing systems-of-systems, data quality aspects, runtime monitoring, and anticipatory testing. Finally, we extended the testing dimensions for data-intensive software systems accordingly.

References

Ammann, P., & Offutt, J. (2016). *Introduction to software testing*. Cambridge: Cambridge University Press.

Auer, F., & Felderer, M. (2018). Current state of research on continuous experimentation: A systematic mapping study. In *2018 44th Euromicro Conference on Software Engineering and Advanced Applications (SEAA)* (pp. 335–344). Piscataway, NJ: IEEE.

de Bayser, M., Azevedo, L. G., & Cerqueira, R. (2015). Researchops: The case for devops in scientific applications. *In 2015 IFIP/IEEE International Symposium on Integrated Network Management (IM)* (pp. 1398–1404). Piscataway, NJ: IEEE

Bourque, P., & Dupuis, R. (Eds.) (2014). *Guide to the software engineering body of knowledge version 3.0 SWEBOK*. Piscataway, NJ: IEEE. http://www.computer.org/web/swebok

Casale, G., Ardagna, D., Artac, M., Barbier, F., Nitto, E. D., Henry, et al.: (2015). Dice: quality-driven development of data-intensive cloud applications. In *Proceedings of the Seventh International Workshop on Modeling in Software Engineering* (pp. 78–83). Piscataway, NJ: IEEE Press

Corbin, J. M., & Strauss, A. (1990). *Grounded theory research: Procedures, canons, and evaluative criteria* (pp. 3–21). Berlin: Springer.

Dersin, P. (2014). *Systems of systems*. Piscataway, NJ: IEEE.

Du, X., Xie, X., Li, Y., Ma, L., Zhao, J., & Liu, Y. (2018). Deepcruiser: Automated guided testing for stateful deep learning systems. arXiv preprint arXiv:1812.05339.

Fagerholm, F., Guinea, A.S., Mäenpää, H., Münch, J. (2014). Building blocks for continuous experimentation. In *Proceedings of the 1st International Workshop on Rapid Continuous Software Engineering* (pp. 26–35). New York, NY: ACM.

Felderer, M., Büchler, M., Johns, M., Brucker, A. D., Breu, R., & Pretschner, A. (2016). Security testing: A survey. In: *Advances in Computers* (Vol. 101, pp. 1–51). Amsterdam: Elsevier.

Foidl, H., & Felderer, M. (2015). Research challenges of industry 4.0 for quality management. In *International Conference on Enterprise Resource Planning Systems* (pp. 121–137). Berlin: Springer.

Gadler, D., Mairegger, M., Janes, A., & Russo, B. (2017). Mining logs to model the use of a system. In *2017 ACM/IEEE International Symposium on Empirical Software Engineering and Measurement, ESEM 2017* (pp. 334–343). Piscataway, NJ: IEEE.

Hummel, O., Eichelberger, H., Giloj, A., Werle, D., & Schmid, K. (2018). A collection of software engineering challenges for big data system development. In: *2018 44th Euromicro Conference on Software Engineering and Advanced Applications (SEAA)* (pp. 362–369). Piscataway, NJ: IEEE.

ISO. (1994). *ISO 8402:1994 Quality Management and Quality Assurance—Vocabulary*. Tech. rep., ISO.

ISO/IEC. (1994). *Information Technology – Open Systems Interconnection – Conformance Esting Methodology and Framework*. International ISO/IEC multi–part standard No. 9646.

ISO/IEC. (2008). *ISO/IEC 25012:2008 Software Engineering – Software Product Quality Requirements and Evaluation (Square) – Data Quality Model*. Tech. Rep., ISO.

ISO/IEC. (2011). *ISO/IEC 25010:2011 Systems and Software Engineering – Systems and Software Quality Requirements and Evaluation (Square) – System and Software Quality Models*. Tech. Rep., ISO.

ISTQB. (2012). *Standard Glossary of Terms Used in Software Testing. Version 2.2*. Tech. Rep., ISTQB.

Kim, J., Feldt, R., & Yoo, S. (2019). Guiding deep learning system testing using surprise adequacy. In: *Proceedings of the 41st International Conference on Software Engineering* (pp. 1039–1049). Piscataway, NJ: IEEE Press

Kleppmann, M. (2017). *Designing data-intensive applications: The big ideas behind reliable, scalable, and maintainable systems*. Sebastopol, CA: O'Reilly.

Mattmann, C. A., Crichton, D. J., Hart, A. F., Goodale, C., Hughes, J.S., Kelly, S., et al. (2011). Architecting data-intensive software systems. In: *Handbook of data intensive computing* (pp. 25–57). Berlin: Springer.

Nagy, C. (2013). Static analysis of data-intensive applications. In *2013 17th European Conference on Software Maintenance and Reengineering* (pp. 435–438). Piscataway, NJ: IEEE.

Otero, C. E., & Peter, A. (2015). *Research directions for engineering big data analytics software* (pp. 13–19). Piscataway, NJ: IEEE.

Pei, K., Cao, Y., Yang, J., & Jana, S. (2017). Deepxplore: Automated whitebox testing of deep learning systems. In *Proceedings of the 26th Symposium on Operating Systems Principles* (pp. 1–18). New York, NY: ACM.

Radatz, J., Geraci, A., & Katki, F. (1990). *IEEE Standard Glossary of Software Engineering Terminology*. Tech. Rep., Piscataway, NJ: IEEE.

Russo, B., Succi, G., & Pedrycz, W. (2015). *Mining system logs to learn error predictors: a case study of a telemetry system* (pp. 879–927). Berlin: Springer.

Segura, S., Fraser, G., Sanchez, A. B., & Ruiz-Cortés, A. (2016). *A survey on metamorphic testing* (pp. 805–824). Piscataway, NJ: IEEE.

Shull, F. (2013). *Getting an intuition for big data* (pp. 3–6). Piscataway, NJ: IEEE.

Tian, Y., Pei, K., Jana, S., & Ray, B. (2018). Deeptest: Automated testing of deep-neural-network-driven autonomous cars. In: *Proceedings of the 40th International Conference on Software Engineering* (pp. 303–314). New York, NY: ACM.

Tonella, P. (2019). *2019–2023 ERC project: Precrime: Self-assessment Oracles for Anticipatory Testing* Retrieved 2018, January 7th, 2019, from www.pre-crime.eu

Utting, M., & Legeard, B. (2007). *Practical model-based testing: A tools approach*. San Francisco, CA: Morgan Kaufmann Publishers.

Wang, S., Wan, J., Li, D., & Zhang, C. (2016). *Implementing smart factory of industrie 4.0: An outlook* (p. 3159805). London, UK: Sage Publications.

Weyuker, E. J. (1982). *On testing non-testable programs* (pp. 465–470). London, UK: The British Computer Society.
Xie, X., Ho, J. W., Murphy, C., Kaiser, G., Xu, B., & Chen, T. Y. (2011). *Testing and validating machine learning classifiers by metamorphic testing* (pp. 544–558). Amsterdam: Elsevier.

Part II
Engineering Quality Improvement

Chapter 7
Product/ion-Aware Analysis of Collaborative Systems Engineering Processes

Lukas Kathrein, Arndt Lüder, Kristof Meixner, Dietmar Winkler, and Stefan Biffl

Abstract Flexible manufacturing systems, as a vision of Industry 4.0, depend on the collaboration of domain experts coming from different engineering disciplines. These experts often depend on (interdisciplinary) results from previous engineering phases and require an explicit representation of knowledge on relationships between products and production systems. However, production systems engineering organizations, which are set in a multidisciplinary environment, rather than focusing on process analysis and improvement options ranging over multiple disciplines, focus mostly on one particular discipline and neglect collaborations between several workgroups. In this chapter, we investigate requirements for the product/ion (i.e., product and production process)-aware analysis of engineering processes to improve the engineering process across workgroups. We, therefore, consider the following three aspects: (1) engineering process analysis methods; (2) artifact and data modeling approaches, from business informatics and from production systems engineering; and (3) persistent representation of product/ion-aware engineering knowledge and data. We extend existing work on business process analysis methods and BPMN 2.0 to address their limited capabilities for product/ion-aware process analysis. We evaluate the resulting contributions in a case study with domain experts from a large production system engineering company. We conclude that an improved

L. Kathrein (✉) · K. Meixner · D. Winkler
Christian Doppler Laboratory for Security and Quality Improvement in the Production System Lifecycle (CDL-SQI), Institute of Information Systems Engineering, Technische Universität Wien, Vienna, Austria
e-mail: lukas.kathrein@tuwien.ac.at; kristof.meixner@tuwien.ac.at; dietmar.winkler@tuwien.ac.at

A. Lüder
Otto-v.-Guericke University/IAF, Magdeburg, Germany
e-mail: arndt.lueder@ovgu.de

S. Biffl
Institute of Information Systems Engineering, Technische Universität Wien, Vienna, Austria
e-mail: stefan.biffl@tuwien.ac.at

S. Biffl et al. (eds.), *Security and Quality in Cyber-Physical Systems Engineering*,
https://doi.org/10.1007/978-3-030-25312-7_7

product/ion-aware knowledge representation facilitates traceable design decisions as foundation for better quality assurance in the engineering process.

Keywords Production systems engineering · Product-production process-production resource (PPR) relationships · Engineering process analysis · Engineering knowledge representation · PPR knowledge persistence requirements

7.1 Introduction

Production system engineering (PSE) organizations pursue the goal of creating (automated) manufacturing systems satisfying the requirements toward time and cost while meeting quality criteria imposed by customers or standards. In addition, PSE organizations need to tailor their solutions for their customers (Wiesner and Thoben 2017). The insufficient representation of important *relationships between the product, the production process, and production resources* (PPR) (Schleipen et al. 2015) in the PSE process can increase the risk of poor quality and unanticipated costs during the operation phase of an automated manufacturing system. Even though PSE organizations build on experienced domain experts, surprisingly, PPR relationships are not explicitly modeled by default throughout the PSE process.

The relationship of product, production process, and production resource can also be expressed in an *information systems engineering* (ISE) or software engineering (SE) context (Humphrey 1995). The product is equivalent to code produced by developers, which can be anything from a small script to an integrated graphical user interface for an application. In SE, it is considered a best practice to test code early with explicit test setups that closely represent the production environment (Beck 2003). (Staging) environments (Humble and Farley 2010) executing the code can thus be seen as the equivalent of a production process, which executes according to the capabilities of a resource. The concept of a production resource can be expressed for example with web servers or interactive development environments (IDE), which are used by a developer producing/executing code as the product. The risk of miscommunication in PSE translates as follows to the software engineering context: If nonfunctional requirements, such as performance or security, are not communicated to the developers, it may be hard or impossible to add these requirements later on to code or production environments. To address these challenges, the ISE and SE communities have developed methods like SCRUM (Schwaber and Beedle 2002), DevOps (Zhu et al. 2016), rapid prototyping, or test-driven development (Beck 2003).

PSE is conducted in a multidisciplinary environment (Biffl et al. 2017; Jäger et al. 2011), involving, above others, disciplines like mechanical, electrical, and software engineering (Moser et al. 2010; Schafer and Wehrheim 2007). Further, PSE is usually more complex than information systems engineering due to risky hardware, which cannot be rapidly tested and has often much longer feedback cycles than software systems. In addition, it is, most of the time, simply not possible to build a whole (physical) test system that reflects the imagined production system. These factors make it harder to engineer and test the target system. Domain experts tend to deal with these challenges by focusing on their discipline-specific contributions, and may consider product or production process aspects only implicitly throughout the engineering process. This domain-centered view often leads to information silos (Rilling et al. 2008), where workgroups do not optimize their interfaces to other engineering experts for collaboration or coordination. The need to collaborate closely in all stages of the development process in a multidisciplinary engineering environment is critical (Paetzold 2017) for project success. Working in silos increases the risks of miscommunication and loss of access to essential knowledge during the PSE process and the operation phase of a production system.

In this chapter, we build on and extend previous research (Kathrein et al. 2018, 2019). We focus on the capability for the analysis and improvement of multidisciplinary engineering processes that exchange knowledge between workgroups. We are interested in the product/ion (i.e., product and production process)-aware analysis of engineering processes as there is significant potential for improvement in the collaboration and coordination of PSE workgroups by considering and explicitly representing PPR knowledge.

Based on the knowledge hierarchy (Rowley 2007), we define the following terms for further use. An *engineering artifact* is a document, in a digital or non-digital form containing data. These artifacts are potentially hard to process for machines and might contain data. The term *data* refers to all kinds of symbols, ranging from simple text to more complex data, like drawings in proprietary software tools. Data has, however, an underlying data model, which is described using datatypes. An example would be a simple table where each column defines the basic datatype, like integer, for the rows, or a graph, defining which objects are nodes and what the semantics, expressed by edges, are (Sabou et al. 2017). *Engineering information* defines the stakeholder groups that have access to the engineering data, how the underlying data can be processed and gives insights into what, who, where, and when questions. Finally, *knowledge* expresses concepts and provides applications of the underlying data and information models. For this chapter we utilize the PPR concept (see Sect. 7.2) to define PPR knowledge. We further define the term *PPR knowledge* to express (a) success-critical attributes, such as parameters for production processes or configurations for production resource and (b) relationships between products, production processes, and production resources, such as constraint dependencies.

We illustrate the PSE process with the simple *use case: fragile product*, as the use case highlights common challenges in the engineering process and the current situation in many engineering organizations. We assume that a customer requires a production system for producing a fragile product. Therefore, the customer creates plans of the product and its characteristics and hands them over to a PSE company. In the PSE company, a *basic planner* receives the product lifecycle documents provided by the customer and specifies the production process and system according to the product requirements. Throughout the engineering tasks, the basic planner transforms product and process knowledge into resource knowledge, resulting in first sketches of the manufacturing system. A team of *detail planners* then takes over and

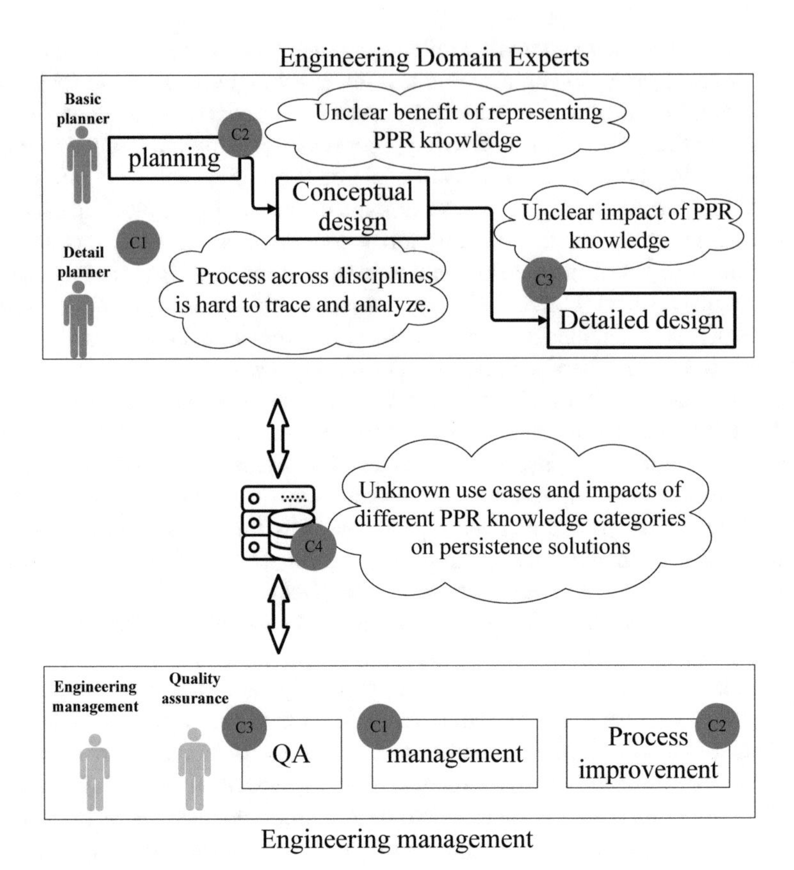

Fig. 7.1 Common challenges in an engineering process

derives discipline-specific detailed plans from the specifications for constructing and operating the production system. This includes a high-throughput transport system, which is required to meet the customer's specifications of parts per minute produced. Unfortunately, during operation of the system, the high acceleration of the transport process damages fragile product parts. This flaw of the production system has many negative effects, such as: extra efforts in rework, uncoordinated communications, and high risk of project failure. These effects all could have been avoided if the missing explicit PPR knowledge on product fragility would have been conveyed in the specifications of the basic planner to the detail planner.

Figure 7.1 illustrates the described engineering process on a high level including the involved stakeholders with their respective challenges. The domain experts for basic and detail planning (orange), represent the operational part of the engineering process, whereas the engineering management with the engineering manager (blue) and quality assurance (green) are more concerned with process planning and improvements.

Figure 7.1 depicts several of the challenges in the use case fragile product, which we describe briefly.

C1. The Engineering Process Between Discipline-Specific Workgroups Is Hard to Trace and Analyze In PSE, workgroups traditionally focus more on intra- than on interprocess improvements. The collaboration of multiple workgroups originates from project needs. Over time, the workgroups may evolve, with new team members joining or team members leaving for another project. Figure 7.1 indicates this through the absence of process/task boundary, which would clearly allow identifying which stakeholder is responsible for which task. There is also no formal process that guides the cooperation or collaboration spanning over multiple disciplines. For the domain experts, this lack of a formal process description makes it hard to trace design decisions throughout the engineering process.

C2. Unclear Benefit of Representing PPR Knowledge Domain experts, who hold a lot of information like the basic planner, are unaware of who would benefit from sharing PPR knowledge. In the described use case this would be the case with the knowledge about the fragility of the product. This knowledge is available to the basic planner through the specifications from the customer. However, the basic planner does not convey this information to the detail planner. In Fig. 7.1, there is no outgoing knowledge from planning into conceptual design. The engineering management again lacks knowledge about the existing knowledge and how it is represented, conveyed, and transformed through the engineering process. This lack of representation makes it also impossible for a quality assurance stakeholder to track or improve engineering artifacts or identify possible reuse scenarios, leading to an improved engineering process.

C3. Unclear Impact of PPR Knowledge Because domain experts do not know what benefit explicit PPR knowledge has (challenge 2), domain experts also do not externalize or document design choices based on product requirements or product design decisions. The product engineer responsible for these decisions simply does not know what impact his decisions might have in the later phases of the engineering of the production system or the operation. In Fig. 7.1, we illustrate this by the two separate "silos" for domain experts and engineering management. The engineering management is not able support the domain experts with this knowledge because they are not aware of project-specific outcomes with possible positive or negative impacts. Explicitly representing PPR knowledge would help both, domain experts and engineering management, to facilitate the analyses of such impacts and highlight dependencies between workgroups that have interfaces for coordination and collaboration. Quality assurance stakeholders have no means on how to improve an engineering process, because they do not know positive or negative impacts that possible new solution approaches might have.

C4. Unclear Use Cases with PPR Knowledge Categories That Require Persistence For software developers, who design and adapt engineering tools for engineering process, it is not clear which are the primary use cases that define requirements for persisting PPR knowledge. Furthermore, it is not clear which categories of PPR data and knowledge exist that may have an impact on the design of data persistence solutions. Addressing the challenges C1–C3 with PPR knowledge representation is not sufficient as the PPR knowledge is not necessarily efficient to search or reuse. For example, engineering managers would require means to query persisting PPR knowledge on project-related information, such as the overall production rate or the percentage of goods with poor quality of projects that include fragile products.

To address challenges C1–C4, we investigate in this chapter a *product/ion-aware engineering process analysis* (PPR EPA) method, based on and extending Kathrein et al. (2018, 2019), resulting in a graphical visualization of the engineering process, classified engineering artifacts and engineering workgroups as a *product/ion-aware data processing map* (PPR DPM). We also investigate use cases to derive requirements for persisting PPR knowledge. The following research questions address these challenges.

RQ1. What Are Main Elements of a PPR EPA Method? To address this research question, we investigate existing EPA solutions and their elements, from both information systems/business informatics and production systems engineering communities. The outcome of this RQ allows identifying buildings blocks for reuse in a new PPR EPA as well as limitations and gaps that a new approach should fill.

RQ2. What Are Main Elements of a PPR DPM Method and Notation? Through applying a PPR EPA, we derive a visualization of the overall engineering process. Because this newly designed artifact is success-critical for the overall application of the PPR EPA, we investigate the main elements that are common, for example, in business process representations from again business informatics and productions systems engineering. In this chapter, we try to close the gap between standard business process representations and extensions that are custom to the PPR DPM approach.

RQ3: What Are Primary Use Cases That Require the Persistence of Different Categories of PPR Knowledge? To address this research question, we build on the use cases coming from RQ1 and RQ2 to elicit primary use cases that stakeholders face in the engineering workflow related to persisting PPR knowledge. The use cases focus on different categories of PPR knowledge present throughout the engineering process and help to define high-level requirements for PPR knowledge persistence.

Main contribution of the conducted research in this chapter allows both ISE and SE as well as PSE communities to gain insights into the other domain. These insights highlight common ground for further research and possible approaches, applicable in both communities, and motivate future research.

The remainder of the chapter is structured as follows: Sect. 7.2 summarizes related work on process analysis approaches, business process notations, and data storage design options. Section 7.3 motivates the research questions and the research approach. Section 7.4 introduces the main elements for the PPR EPA method and PPR DPM artifact, and the treatment designs. Section 7.5 presents the case study conducted with domain experts in a large PSE company. Section 7.6 evaluates the proposed artifacts from RQ1 and RQ2. Motivated by Sects. 7.5 and 7.6, Sect. 7.7

presents PPR knowledge persistence aspects. Section 7.8 discusses the research findings and their limitations and Sect. 7.9 concludes and outlines future work.

7.2 Related Work

This section summarizes related work on product/ion awareness (PPR), on approaches for engineering process analysis, and on notations for representing the analysis results.

7.2.1 Product/ion Awareness in Multidisciplinary Engineering

Technical systems are often distinguished into products and production systems (Biffl et al. 2017). The reason a company exists is often because of its products, that is, products are created in a value-adding process to make profit by selling them (Stark 2015). A production system, however, focuses on creating the products by combining suitable production factors (ElMaraghy 2009). Materials, work-in-progress parts, and production resources (machines) are the most prominent production factors. The product and production system, therefore, have strong dependencies. Schleipen et al. (2015) coined the product-process-resource (PPR) concept for the relationships between products and production systems based on the production process.

This concept of PPR helps to answer questions about the application of engineering data and information and thus, derived from Rowley (2007), is the main building block for the term PPR knowledge used in this chapter.

Figure 7.2 illustrates the relationships between the PPR aspects. We describe the elements of Fig. 7.2 based on the fragile product use case, introduced in Sect. 7.1. The product the customer commissioned contains fragile parts and requires several processes like, gluing, pressing, and transportation. The product has special requirements regarding the transport process, namely, the acceleration of the conveyor belt. Furthermore, the fragile product is processed on an industrial

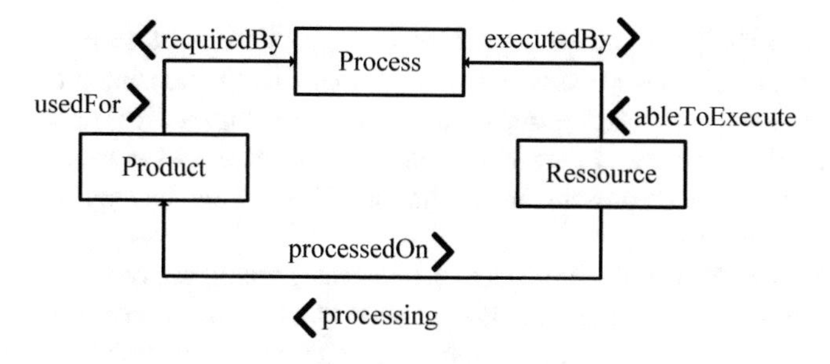

Fig. 7.2 Product-process-resource (PPR) relationships

machine (resource). The link between product and resource has also requirements. For example, the pressing force applied after gluing the fragile parts must range between one and two kilo newton. The resource provides the capabilities that a process needs to be executed with, closing the triangle of Fig. 7.2.

All three concepts can be composed of inner elements, meaning, for example, that a product consists of multiple product parts that are assembled together and make up the final product. This nesting of elements can be described with a pen consisting of the outer shell, the refill, the spring mechanism, and so on. Furthermore, all three concepts of product, process, and resource are interlinked, meaning that they form a graph-like structure, where nodes represent the individual PPR elements and the edges represent links between the individual concepts or hierarchies.

The VDI 3682 standard (VDI 2005) introduces this concept of recursive composition of individual concepts, like the pen example. The standard is further the only visual representation form that has three distinct elements to express, product (parts), processes, and resources.

Other concepts like the ISA95 standard (International Electrotechnical Commission 2003) indirectly allow representing the PPR concept, but are more concerned with describing the interfaces between enterprise resource systems (ERP) and manufacturing execution systems (MES). The goal of the ISA95 standard is to better describe and transfer production order relevant information into the manufacturing system. Furthermore, the standard originates more from batch processing and not so much from discrete manufacturing, which is the primary focus of this chapter. Thus, we do not further consider this option for a solution in this research.

AutomationML (AML) was developed as *glue for seamless automation engineering* (Drath 2009) and uses XML concepts to represent topologies, geometries, as well as behavioral and logical data for production resources. AML became standardized in the open source IEC 62714 standard (International Electrotechnical Commission 2013) and enables representing PPR knowledge as hierarchies with various concepts. Furthermore, AML concepts can be used to model PPR knowledge as a hierarchy of internal elements and linking between the different concepts.

7.2.2 Engineering Process Analysis Methods

To be able to analyze engineering processes and follow the task execution across several workgroups, it is necessary to analyze existing engineering processes on (a) an overview-level of the workgroups and their relationships and (b) detailed analyses of exchanged artifacts and data that identify dependencies between workgroups. These two viewpoints represent the foundation of improving the engineering process between workgroups.

Rosenberger et al. (2018) presents a business process analysis (BPA) method, which determines and defines activities in need of a business context. The presented approach executes a context elicitation, defining contextual functionalities, which in traditional project-based development models is often not done, or simply too much effort. The identified different contexts for various workgroups do not have any implications on other contexts, which makes it hard to use in an engineering process analysis.

To balance exploration and exploitation thinking in a BPA method, Santos and Alves (2017) proposed a three phase BPA, methodologically built on literature surveys, expert opinions, and a case study, all in accordance with the design science cycle from Wieringa (2014). Through the detailed analysis, the results from Santos and Alves allow to identify detailed execution steps, exchanged documents, and a big picture structure of the business process. However, the result does not investigate interfaces between workgroups, as they are predefined and already part of the case study.

Vergidis et al. (2008), who classified several existing business process analysis methods and techniques, highlighted that only a handful of them allows further detailed analysis, or process improvements, which go beyond generic stakeholder, tasks, or input/output artifact identification.

BPA methods allow to easily represent a big picture of a business or engineering process; however, many methods do not consider individual disciplines, interfaces between workgroups, or how the overall collaboration could be improved. The analysis of engineering processes spanning over multiple workgroups requires not only the analysis of the overview on relationships and coexistences of workgroups, but also a more detailed, fine-grained analysis of individual engineering disciplines with specific exchanged artifacts.

On the side of production systems engineering, Jäger et al. (2011) identify the need to "systematically model the engineering workflow, which would allow a deeper knowledge of different engineering aspects and to improve the views of each discipline on the engineering objects." The approach chosen by the authors starts by identifying engineering artifacts and backtracking these artifacts to stakeholders that they belong to. This approach allows the consideration of cause-and-effect analysis in engineering processes, but does not identify interfaces between workgroups and how these could be improved by investigating the engineering artifacts. The process is also driven mainly by engineering documents and not the processes executed by domain experts.

The VDI 3695 standard (VDI 2010) defines the concept of an engineering organization, which conducts its business on a project basis. The engineering organization is further characterized by carrying out the following consecutive engineering activities, depicted in Fig. 7.3: acquisition, planning, realization, commissioning. Such a high-level segmentation of an engineering process does not depict stakeholders, their activities, or artifacts involved. Due to the lack of detail,

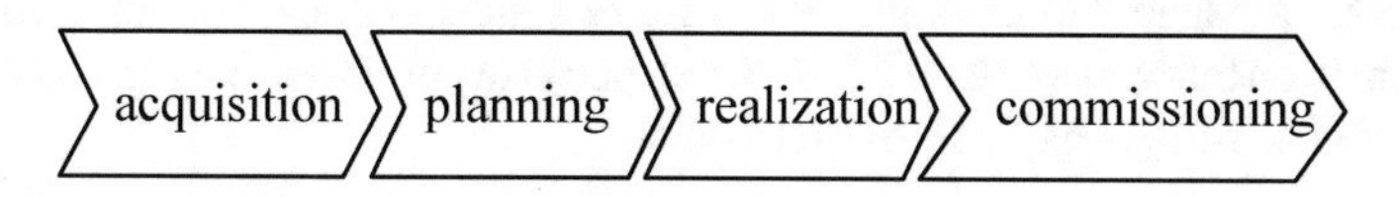

Fig. 7.3 Project-related phases identified by the VDI 3695 guideline (VDI 2010)

it is not possible to identify any interfaces that might exist between workgroups and could be the basis for further analyses. The guideline does also not consider how to improve an engineering process but rather gives rough directions that could be taken to improve the overall engineering process.

Lüder et al. (2012) build upon the presented VDI 3695 standard. The outcome of Lüder et al. (2012) is a more detailed engineering process analysis, which focuses on individual workgroups, their tasks, and a description of engineering artifacts, but with no special focus on PPR knowledge representation. In this approach, it is also not considered how multiple workgroups could better work together for an improved coordination and collaboration in the engineering process. Further, Lüder et al. (2018) investigated common challenges regarding the multidisciplinary aspect of a data exchange process across several workgroups. The authors highlight the importance of an engineering process analysis method that allows the investigation of engineering processes with engineering artifacts and possible dependencies.

The analyzed literature reveals similarities in how the analysis methods of business or engineering processes are conducted, but differ in their focus and results. While BPA methods tend to focus more on the big picture, EPA methods focus more on intra workgroup analyses. A gap that can be identified in both disciplines concerns analysis regarding engineering knowledge exchange between workgroups. Exchanges between workgroups are often the source of missing PPR knowledge, a risk already in traditional production systems engineering, much more for considering flexible manufacturing according to the Industry4.0 vision.

7.2.3 PPR Knowledge Representation in Process Analysis

The previously presented BPA and EPA methods gather a lot of data that needs to be processed in some form. Both communities have different approaches to (graphically) represent the knowledge which is present in an engineering process. This knowledge often contains PPR knowledge aspects and thus, the following existing approaches will be investigated according to their possibilities to represent PPR knowledge and classify data and processes.

IDEF0 (Force 1981; Presley and Liles 1995), for example, is widely used in the engineering domain (Zhang et al. 2010) and provides an overview on processes, their inputs/outputs, controls, and stakeholders. The system analysis standard has only very few distinct elements, namely, arrows and boxes. This limited number of different concepts makes it easy for nonexperts to pick up the modeling approach, but makes it hard to express more complex situations, which would require a richer expression language. For example, is it hard to follow one specific input to output transformation through a large IDEF0 model, because possible other input and output arrows are indistinguishable from each other.

Lüder et al. (2012) introduced a more detailed but not so visual approach, by representing gathered engineering knowledge in tables. This approach allows for a very detailed classification and division of knowledge, but does however become cumbersome to work with when the number of different tables, referencing each other, increases.

Event-driven process chains (EPCs) (Scheer 1998), BPMN 2.0 (Allweyer 2016), or the UML standard (Fowler et al. 2004) are all well-known options to model business processes. Merunka (2017) pointed out that the UML standard has no means to represent product and process knowledge in either one or several combined diagrams. EPCs, extended with data, resources, time, and probabilities are called extended EPCs (eEPC) (Scheer 1998). Both eEPC and BPMN 2.0 are widely used for modeling business processes and have incorporated many similar concepts. Extended EPCs require a more explicit annotation of organizational units for each engineering task, while BPMN 2.0 uses swim lanes for a more compact visualization.

Khabbazi et al. (2013), Huang et al. (2017), and Merunka (2017) proposed the combination of multiple modeling concepts, which should allow overcoming limitations that individual notations have. Even though such a combination allows for a more flexible and detailed notion of processes, the complexity of models also increases for stakeholders, who would like to analyze the underlying models. None of the mentioned authors named the concept of explicitly modeling data and process flows; we use in this chapter the term *data processing map* to express the combined representation of processes with documents.

Unfortunately, PPR knowledge, its flow through an engineering process, or dependencies between tasks and artifacts are not directly expressible in any of the languages discussed in this subsection. The languages do however build a good foundation for closing this gap, by using f. e. BPMN 2.0 and then build custom extensions to express PPR knowledge.

7.2.4 PPR Knowledge Persistence

In this chapter, we use the term PPR knowledge for success-critical attributions, like parameter settings of production resources, of each of the concepts as well as the interrelationships between the individual parts of PPR based on Schleipen et al. (2015). These attributions for product, processes, and resources in combination with the relationships formed between the three concepts need to be represented to allow persistence and retrieval.

We further use the term persistence not as strictly defined as it is in the database community, but we express with it the application of persistence solutions to store PPR knowledge. This can include several different underlying technologies. A designer of persistent PPR knowledge storage should consider established persistence approaches, such as relational databases, NoSQL databases, and AutomationML files, as these fit well to general characteristics of PPR, which essentially are graphs consisting of linked trees in the individual PPR aspects as described in Sect. 7.2.1.

Relational databases have been successfully applied to for persisting business data since the 1970s and gained considerable production experience (Nance et al. 2013). The approach that centers on tables, columns, and rows has been a clear choice for many data-intensive storage and retrieval applications (Vicknair et al. 2010). Relational databases are in general very efficient unless the data is strongly interlinked with many relationships leading to a large number of joins (Vicknair et al. 2010) that reduce access efficiency. A key success factor for relational databases is the fixed structure of each table, which allows for indexing and for using the goal-oriented query language SQL (Date and Darwen 1997). Unfortunately, engineering artifacts often do not follow a predefined fixed structure and may vary from project to project, or depend on customer-specific practices.

NoSQL technologies address this limitation using flexible data models to store schema-less models (Siddiqa et al. 2017). PPR knowledge accumulates in an engineering process and expresses product, process, and resource information as well as the interrelationships in a high number of many-to-many relationships and is to some extent hierarchically structured, which fits NoSQL characteristics presented by Vicknair et al. (2010). Therefore, the available knowledge may also vary depending on project or customer, and thus requires a flexible schema, which is easily changeable, adaptable, and maintainable. NoSQL is not a single solution, but has four major design differentiations to consider for designing an application. These options are key-value, column-oriented, document, or graph databases (Siddiqa et al. 2017). PPR knowledge with its attributions and relationships fits could fit well to a graph-based approach (Vicknair et al. 2010).

Fowler and Sadalage (2013) coined the term polyglot persistence, for using several data storage languages and technologies, each for the use cases it fits best. Nance et al. (2013) pointed out that it is not necessary to make a choice between relational or NoSQL databases but to use both as is seen appropriate. A polyglot data storage approach could help to overcome the requirements of engineering artifact storage by following a "best-of-breed" approach. The solution of polyglot storage requires expertise in several languages and technologies, making the design more complex to understand, implement, test, and operate. Therefore, a key question is what requirements can be derived from use cases and how a sufficiently powerful yet simple design for PPR knowledge persistence might look like.

AutomationML (AML) does not only provide means to express PPR concepts, but also allows representing production systems in XML-like formats. Furthermore, is it possible to represent PPR knowledge for data exchange and logistics storage in AML for small production systems. However, AML files can rapidly grow in size, which may be hard to process efficiently even for medium-sized production systems. Production systems with 5000–10,000 signals may take up 20–50 MB of AML text for its representation, depending on the set of discipline-specific views in the data model.

7.3 Research Questions

By following the design science cycle presented from Wieringa (2014), we address the challenges introduced in Sect. 7.1 by deriving the following research questions for improving the product/ion (i.e., product and production process)-aware analysis of engineering processes.

RQ1. What Are Main Elements of a PPR EPA Method? To address this research question, we build on Kathrein et al. (2018, 2019) and consider the strengths and limitations of approaches from business process analysis and from engineering process analysis to identify promising candidate methods for adaptation and extension. We extend Kathrein et al. (2019) with valuable lessons learned regarding the PPR EPA. We apply a case study design (Runeson and Höst 2009) to elicit what main elements a PPR EPA method needs. These elements need to focus on the design and elicitation of a product/ion-aware engineering process analysis (PPR EPA) method and thus make it possible to identify and collect data on the engineering process. Through focusing on PPR knowledge expression, the EPA method allows to analyze where relevant PPR knowledge is required, created, or lost. From the main elements identified, we derive requirements for a notation to represent the needs and capabilities to represent PPR knowledge.

RQ2. What Are Main Elements of a PPR DPM Method and Notation? Based on Kathrein et al. (2018, 2019), we describe how a product/ion-aware *data processing map* (PPR DPM) can look like. The extended elements serve as foundation for the analysis of gaps regarding PPR knowledge representation in the engineering process. The result of RQ2 highlights elements that are crucial to be able to express in PPR knowledge in an engineering process with the interaction of tasks and engineering artifacts. We follow the design science cycle (Wieringa 2014) and validate both treatments of RQ1 (PPR EPA) and RQ2 (PPR DPM) artifacts, in the context of a case study.

RQ3: What Are Primary Use Cases That Require the Persistence of Different Categories of PPR Knowledge? We use the case study approach from the work of Runeson and Höst (2009) to also investigate common use cases that occur in the engineering workflow and further expand the stakeholders to include software engineering domain experts. These experts, in combination with interviews from RQ1, help to elicit the primary use cases, allowing to derive requirements and different categories of PPR knowledge. The outcome of this RQ allows a three-tier layering of (1) use cases, (2) functions like reuse and search, and (3) persistence technologies like databases. From such a layered outcome, future research and possible new stakeholders can focus on representing PPR knowledge more permanently and make it query-able.

7.4 Product/ion-Aware Analysis of Engineering Processes

This section addresses the limitations of both business process analysis (BPA) methods, such as *context-aware process analysis* and *A2BP* (Rosenberger et al. 2018; Santos and Alves 2017) and engineering process analysis (EPA) methods, such as *mechatronic engineering EPA* and *technical dependency mining* (Lüder et al. 2012; Jäger et al. 2011). We introduce the main elements of a multidisciplinary PPR EPA method (RQ1) as well as the main notation elements of a PPR DPM (RQ2). The goal of the PPR EPA is to focus on product/ion-awareness and have a repeatable process resulting in a PPR DPM. Paetzold (2017) identified the need for a clear and standardized design process, which is connected to the development process and allows efficient and effective work execution. We present in Sect. 7.4.1 requirements for an artifact evaluation, in Sect. 7.4.2 the design of the treatment PPR EPA method, and in Sect. 7.4.3 the design of the treatment PPR DPM artifact proposing an extension of BPMN 2.0 with PPR knowledge elements.

7.4.1 Requirements for PPR Engineering Process Analysis

Following Wieringa (2014) through the design science cycle, this section presents contribution arguments for the PPR engineering process analysis (PPR EPA) and for the PPR data processing map (PPR DPM). A contribution argument is: "an argument, that an artifact, that satisfies the requirements, would contribute to a stakeholder goal in the problem context" (Wieringa 2014). In our case we present the following two sets of requirements, based on Kathrein et al. (2018, 2019), that have been derived from use cases with the involved stakeholders in the case study. The first set of requirements addresses RQ1, the PPR EPA, while the second set focuses on RQ2, the PPR DPM. The requirements are strongly driven by the goal of representing PPR knowledge and are suitable for multidisciplinary PSE organizations and follow the PSE phases basic planning, detail planning, and operation.

RQ1: Main Elements of a PPR EPA To identify the main elements needed for a good solution of a PPR EPA, we present requirements for capabilities of the product/ion-aware PPR engineering process analysis (PPR EPA).

Identification of PPR Knowledge The product/ion-aware PPR engineering process analysis should allow identifying PPR engineering knowledge, for example, product knowledge in initial product drawings coming from the customer, process knowledge conveyed through specifications regarding the transport system.

Process Analysis with PPR Knowledge The PPR EPA method should analyze and focus on: the creation of PPR knowledge in an engineering process, the flow of PPR knowledge through the engineering process, and an indication where relevant PPR knowledge may not be carried on. One example path could look like this: First, production process sequences are created based on process knowledge. Second, a layout for the production system is created with the help of resource knowledge. The process knowledge is not carried on from the first to the second step. Lastly, in step three an offer is submitted to the customer, only conveying resource knowledge.

Identification of PPR Knowledge in Interdisciplinary Interactions The PPR EPA method should allow identifying where engineering disciplines interact with each other, for example, handover phases of project responsibility including artifacts, such as the change from basic to detailed planning where all artifacts are handed over to a new team.

RQ2 Main Elements of a PPR DPM The following set of requirements is motivated by how to represent PPR knowledge in an engineering process after the PPR EPA has been conducted, and what main elements of a PPR DPM visual representation is needed.

PPR-Specific Visual Elements The PPR DPM should provide specific elements for the concepts used in the PPR EPA, including visual elements for roles, tasks, the priority a task has regarding PPR knowledge, artifacts, and the PPR knowledge aspects they contain.

Iterative Refinement It should be possible with the PPR DPM to start with small initial models, only representing the most vital engineering process tasks per discipline, and gradually and iteratively expand the models. With each iteration, the context for collecting more detailed workflows can be expanded and refinements of PPR knowledge classifications of the process steps with stakeholders can be executed.

Process Overview The PPR DPM should provide an overview of the engineering process, including: the involved disciplines with their respective process executions, engineering artifacts and their flow throughout the process, interfaces between workgroups and the sequence that engineering tasks are executed in.

7.4.2 A Product/ion-Aware Engineering Process Analysis Method

To address RQ1, and the limitations of existing business process analysis (BPA) and engineering process analysis (EPA) methods, we identify in this subsection the main elements for a multidisciplinary engineering analysis (PPR EPA). Our approach represents a repeatable two-step process (see Fig. 7.4), resulting in a visual product/ion-aware data processing map (PPR DPM).

Figure 7.4 provides an overview on the steps and tasks of the PPR EPA method. We revisited the proposed PPR EPA from Kathrein et al. (2019) and now present a more detailed description regarding the PPR EPA including some lessons learned. The involved stakeholders are *engineering domain experts* (orange), *engineering management* (blue), *quality assurance* (green), and the new role *EPA facilitator* (red).

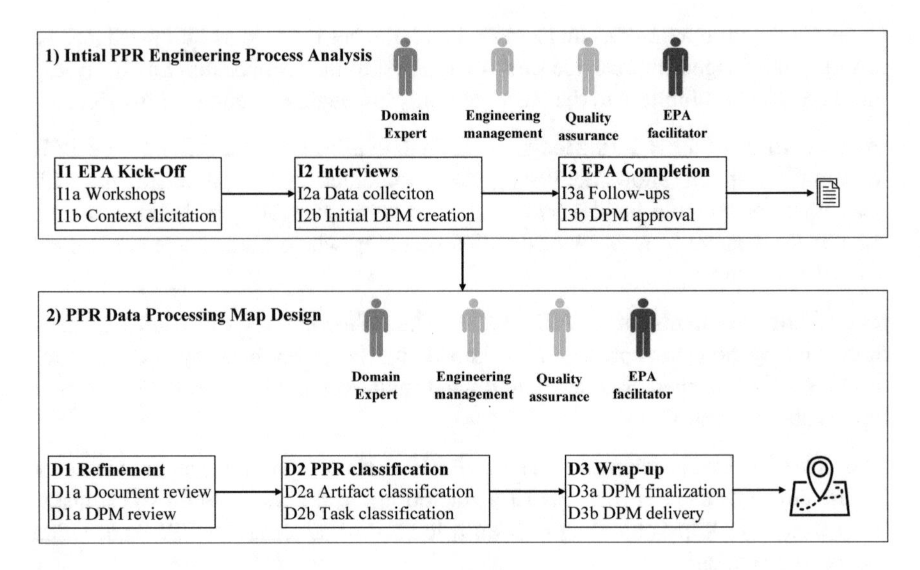

Fig. 7.4 PPR EPA method elements/phases/tasks, based on Kathrein et al. (2018)

The newly introduced role of the EPA facilitator conducts interviews with domain experts and stakeholders, creates initial models for a possible PPR DPM for grading with the domain experts, and holds workshops. This role is similar to the model integrator presented in Fay et al. (2018). All other stakeholders provide insights into their work and are driven to improve the engineering process and optimize existing potential like manual reworks of engineering artifacts due to proprietary engineering tool data formats. The individual tasks of the two phases will be described presently. All tasks prefixed with an "I," represent tasks from the initial PPR EPA phase, and tasks with the "D" prefix correspond to design tasks of the PPR EPA focusing on the PPR DPM.

Phase 1. Initial PPR Engineering Process Analysis starts with initial knowledge about the project under investigation. Outcomes of this phase are interview documentation as notes and audio recording, exemplary files for engineering artifacts, and an initial data processing map depicting a first high-level engineering process.

EPA1 EPA Kick-Off

I1a Workshops. All stakeholders take parts in one or several workshops, stating their role and position that they will play in the PPR EPA.

I1b Context Elicitation. During workshops stakeholders and researchers outline the context of the engineering process under investigation.

Outcome of I1 are documents describing the context, goals, requirements regarding the PPR EPA and PPR DPM and first (hand-drawn) sketches of a DPM.

EPAI2 Interviews
I2a Data Collection. Holding interviews with domain experts allows collecting representative data that is used in a typical engineering project. All captured data should be relevant and put in context to which domain expert and specific task they belong.

I2b Initial DPM Creation. Researchers acting as EPA facilitators elicit PPR knowledge from the domain experts and use this knowledge for an initial PPR classification of engineering artifacts, which results in a first initial DPM.

Outcome of I2 are detailed interview notes and recordings, as well as the initial DPM as basis for further detailing. In regard to Kathrein et al. (2019), we revised the interview task to also contain the initial DPM creation, which allows for a more timely early draft version of a DPM; it is important to not let too much time go by between data collection and initial DPM creation.

EPAI3 EPA Completion
I3a Follow-Ups. The initial DPMN is reassessed, and possible open questions can be discussed with the domain experts. This step is especially important, because it is not guaranteed that the same domain experts will be available in later phases.

I3b DPM Approval. By revisiting domain experts, the modeled initial DPM is either approved or modified to express the engineering process. We propose this additional step as a lesson learned from Kathrein et al. (2019). An early initial approval with the domain experts makes clear that the ground truth for any further work is set and will not be changed.

Outcome of this step is the final basic version of the DPM, representing the basis for further refinements.

Phase 2. PPR Data Processing Map Design is concerned with refining the existing data processing map, and classifying all gathered input data according to PPR and detailing the engineering process model.

DPM1 Refinement
D1a Document Review. All internal data objects (like interview notes) and external data (like engineering artifacts) are investigated more closely and described for following PPR classifications.

D1b DPM Review. The existing basic model is reviewed, potential gaps, notation mistakes and to coarse or detailed tasks are identified and then modeled to represent the as-is engineering process, with references to documents, as closely as possible.

Outcome is a more detailed DPM, identifying engineering artifacts and a data catalogue for easier lookup of exemplary artifacts and data.

DPM2 PPR Classification
D2a Artifact Classification. With the input from F1 Refinement, all engineering artifacts are classified according to product, process, or resource (PPR) knowledge.

D2b Task Classification. All tasks that are present in the PPR DPM are classified regarding their need for PPR knowledge. If so, it further classifies how important PPR knowledge is for a successful execution of the task, including an indication which aspect of PPR is currently available and what additional information would improve the engineering task.

The outcome of this step is the classified DPM, according to PPR.

DPM3 Wrap-Up

D3a DPM Finalization. The PPR DPM is reviewed and all EPA facilitators have a last chance to make small changes to the artifact.

D3b DPM Delivery. The final version is presented to the stakeholders and domain experts and delivered to them for further use.

Outcome is the PPR DPM and all documentation that was accumulated over the course of the PPR EPA.

7.4.3 A Product/ion-Aware Data Processing Map Notation

To address RQ2, and be able to express the knowledge gathered from Sect. 7.4.2, the PPR EPA, we explored business and engineering process analysis notations like UML, BPMN 2.0, or eEPC. In Kathrein et al. (2019), we presented an extension to the BPMN 2.0 standard, which we apply in Fig. 7.5. BPMN 2.0 was chosen because it has already many elements needed to represent business or engineering processes, like events, tasks, documents, gateways. BPMN 2.0 is also a bit "cleaner" than EPCs as it does not require annotating each task with an organizational unit but provides swim lanes to express workgroups.

Our extensions allow to label document content regarding product (P), process (P′), or resource (R) knowledge, as well as to indicate the importance a task has

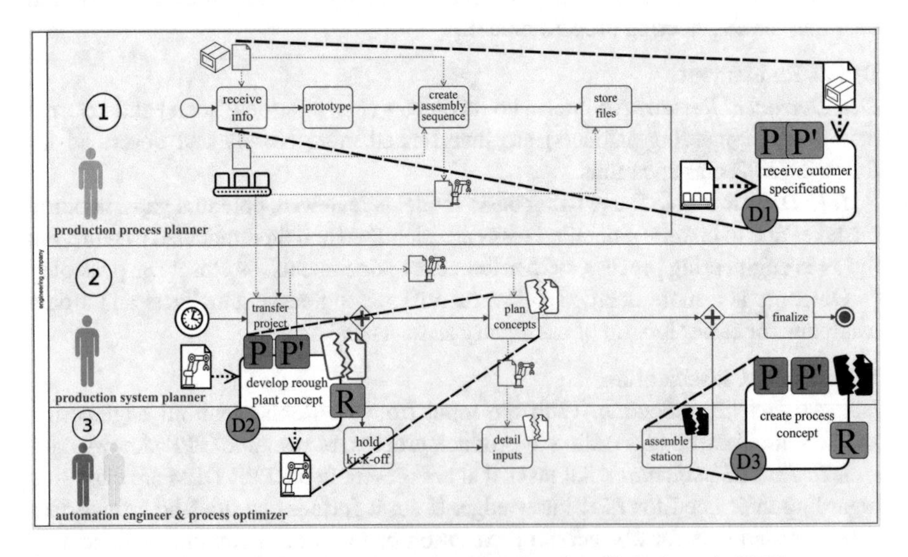

Fig. 7.5 Product/ion-aware PPR Data Processing Map, based on Kathrein et al. (2018)

regarding PPR knowledge. In Fig. 7.5, tags D1 and D2 highlight the use of such a classification. In D1, artifacts containing product (coming from the top) and process (coming from left) information are depicted. In D2, a resource-centered artifact serves as input and another resource containing artifact is created. The individual documents are also graphically distinguishable through annotations in the middle: a package for a product (tag D1), conveyor belt for a process (tag D1), and a robot arm for a resource (D2). This addition to the BPMN 2.0 standard builds the foundation for describing and analyzing a PPR knowledge flow through the engineering process. From this extension possible analyses can be derived, such as where PPR knowledge is created, transformed, or lost.

Further, we provide *PPR knowledge requirements*. These requirements are expressed by (a) annotations of P, P′, and R surrounding the task outline (see Fig. 7.5 tags D1, D2, and D3), and (b) white/black broken documents, if the task misses at least one of the PPR aspects (see Fig. 7.5 tags D2 and D3). The annotations of P, P′, and R indicate what information the task currently receives (colored in green) and what information would additionally be needed but is missing (colored in red). In Fig. 7.5, tag D1, for example, requires and receives product and process information; in tags D2 and D3, the same information and resource information is needed, but only resource information is received. This leads to the red coloring of the product and process annotations. The white broken document highlighted in Tag D2 indicates that for a task execution it is important to receive PPR knowledge; however, the execution is not hindered if this knowledge is not present. This annotation allows indicating which tasks could be executed more efficiently or with better quality if additional PPR aspects, like parameter settings, were present. However, the knowledge can be derived, even if this is not time-efficient. Black broken documents, such as in tag D3, indicate that the role cannot execute this task properly if PPR knowledge is absent. It is absolutely crucial for the task execution to have PPR knowledge present or otherwise will run into efficiency, quality, or cost issues. In a situation where PPR knowledge is crucial, it is not possible for the domain expert to derive this knowledge, make assumptions about settings, or start a communication process.

We evaluate the proposed extensions for the PPR DPM notation, with a case study conducting the proposed PPR EPA (see Sect. 7.5).

7.5 Case Study

We conducted a case study following Runeson and Höst (2009) to evaluate the proposed approaches PPR EPA (RQ1) and the PPR DPM (RQ2). Researchers took the role of the EPA facilitator, which is described in Sect. 7.4.2. The EPA facilitator followed the proposed PPR EPA, executing each task with domain experts. We collected data on the existing engineering process as well as representations of PPR knowledge in the current setting. All domain experts voiced their needs regarding the PPR EPA and how the PPR DPM should look like to better support their work packages.

Study Subject The case study on the proposed *engineering process analysis* (EPA) method was conducted with domain experts at a large production system engineering and manufacturing company. The company focuses on discrete manufacturing systems and can be seen as representative for systems engineering enterprises that conduct their business on a project basis. The company did not consider PPR knowledge at the point of the case study. The case study for collecting data on the PPR EPA method and on the PPR DPM notation spanned over nearly 2months from the initial kick-off to the final version of the data processing map and the final feedback from the involved stakeholders. In the case study, six domain experts, five stakeholders for the engineering process, and three software engineering stakeholders were interviewed. This allowed us to execute the PPR EPA and model the PPR DPM, as well as gather input for data storage requirements, which will be presented in Sect. 7.6. In the context of this case study, one project, focusing on one manufacturing system, was investigated. This means, that the production system and all engineering processes focused on one product, with a set of processes and adequate selected resources for the execution.

Study Execution We followed the PPR EPA approach presented in Sect. 7.4.2 by starting with a project kick-off, consisting of workshops that helped elicit the context. This first step allowed the company stakeholders to introduce their work field context, context, and current problems to the three researchers, who took on the role of the EPA facilitator.

Following the kick-off, each domain expert and stakeholder was interviewed separately for 2 h. The interviews followed a funnel approach (Runeson and Höst 2009), meaning that the question started broad, for example, regarding context and general responsibilities, and became more detailed later, concerning individual work aspects.

Breaks after the interviews allowed creating the initial DPM (Step I2b in the PPR EPA), and collecting feedback from the domain experts. On a separate day, the team completed the EPA with follow-ups, a small presentation of the DPM model, and a check if all needed exemplary documents were given to the researchers for phase 2, the design of the PPR DPM.

All gathered information was reexamined, reviewed, and ordered for easier retrieval. The gathered artifacts were carefully classified regarding the information on the product, process, or resource; an example can be seen in Table 7.1.

The classification builds on a mapping proposed by Hundt and Lüder (2012), who map between different engineering phases and engineering artifacts, such as electrical or mechanical plans, which are present in the detailed engineering phase. In addition, we reexamined the identified engineering tasks and expressed their requirements for PPR knowledge as *no need*, *important need*, or *crucial need*. Figure 7.5 illustrates a representative part of the final version of the PPR DPM.

Table 7.1 Classification of engineering artifacts and PPR knowledge, based on Kathrein et al. (2018)

PPR EPA concept	Collected data
Stakeholder	Domain expert engineering
Process step number	1
Process step name	Receive customer product life cycle management documents
Input artifact name	Product variations
Description	The artifact provides a mapping of which individual parts are used in which product families and created on which part of the production resource. The knowledge is usually stored in an excel document
Product relevant knowledge:	Individual parts used in the product Mapping from part to product family Product name given by the customer Identification numbers from the customer for the individual parts
Relevant process knowledge	None
Relevant resource knowledge	The mapping between which part is created, or processed on which resource part
Output artifact name:	No output artifact is created

The *production process planner* (light orange and swim lane number one), starts each individual project. He receives product and process information from the customer, presented in detail tag *D1*. From the product and process information he is the one to create first new resource knowledge and convey this to the next role. The problem here is that the product and process information is not transported alongside the resource knowledge.

The second stakeholder, the *production system planner* (purple, swim lane number two), receives the resource knowledge and holds an internal kick-off meeting for all other involved workgroups (indicated by the clock symbol). Tag *D2* depicts that for the development of rough plant concepts the production process planner needs PPR knowledge, but only receives the R part.

In swim lane number three, the *automation engineer* (dark orange) and the *production process optimizer* (yellow), work in parallel. Each domain expert delivers a more detailed view regarding the system under construction. For the creation of process concepts, tag *D3*, the workgroups are in need of PPR knowledge but again only receive the R part. For the domain experts it is crucial to receive all possible knowledge and through manual uncoordinated communication with other domain experts, the automation engineer and production process optimizer try to get hold of additional information. The execution of this task is thus highly risky, due to missing PP knowledge, and can lead to unsupported decision making and in later phases to bad quality.

7.6 Evaluation of PPR EPA Visualizations

This section reports on a comparison between the outcomes of different data processing map notations in an initial feasibility case study (Runeson and Höst 2009) with domain experts at a large multidisciplinary systems engineering company.

We evaluate in this section (a) the visualization of engineering processes currently used at the company, discipline-specific EPC workflows, (b) a standard BPMN 2.0 model, and (c) in Sect. 7.4.3, the proposed PPR extensions to the BPMN 2.0 standard.

The evaluation was conducted in an engineering company that creates custom, project-based, automation systems. We conducted interviews with the engineering manager as well as involved domain experts that gave feedback for the parts that were relevant for them. All interviewees could rate the approaches regarding usability, usefulness, and effort based on a 3-point Likert scale (+, 0, −). "+" indicates fulfillment of the criterion, "0" represents neutral fulfillment of the criterion and "−" indicates disagreement that the approach fulfills the criterion (Table 7.2).

The current approach at the company, using EPC workflow diagrams in selected workgroups, is not very usable due to a high level of detail, and changes always imply high rework efforts. The approach is only useful to a limited number of people conducting intra process optimizations.

A standard BPMN 2.0 model was rated usable because it is easy to understand and has concepts like tasks, swim lanes, and documents. The overall creation and adaptation effort was rated adequate as well. However, the standard BPMN 2.0 model is not useful for any PPR-related analyses, due to missing classifications regarding engineering artifacts.

The last approach, the product/ion-aware BPMN 2.0 model, was rated overall very positive. It is as useful as the standard version of BPMN 2.0, but has a much higher usefulness due to the classification of PPR knowledge in engineering artifacts. This classification has a minor drawback and needs a bit more effort to work with than for example the standard BPMN 2.0 model.

The case study results reveal that our proposed approach of extending a well-known standard, in this case BPMN 2.0, allows breaking out of the existing "information silos" that exist in the engineering company. Also, it is much simpler and more useful to classify engineering artifacts regarding PPR knowledge and use these insights. We also learned from the case study and the evaluation that it is a

Table 7.2 Evaluation results, based on Kathrein et al. (2018)

Approaches—>criteria	Current DPM approach: Discipline-specific EPC workflows	Standard BPMN 2.0 model	Product/ion-aware BMPN 2.0 model
Usability	−	+	+
Usefulness	0	−	+
Effort	−	+	0
Overall DPM quality	−	0	+

good first step to represent PPR knowledge explicitly in the form of a PPR DPM, but that it is also vital to investigate possible PPR knowledge persistence solutions. For the involved domain experts, it is not enough to exchange PPR artifacts but they have the need to query and reuse PPR knowledge currently represented in the artifacts. This need is based on use cases that occur in the engineering process and are drivers for further research. In Sect. 7.7, we introduce primary use cases that are relevant for PPR knowledge persistence.

7.7 PPR Knowledge Persistence Use Cases and Data Categories

To address RQ3, we built on the case study presented in Sect. 7.5 to gain insights into the current persistent representation of engineering knowledge. We interviewed three team leaders of software engineering projects responsible for the development of engineering tools, for production machine programming, and for data mining.

PPR Persistence Use Cases The following use cases describe and motivate requirements of software systems that use the PPR knowledge persistence system as foundation for deriving technology requirements.

UC1 Product/ion-Aware Engineering Tool Support
Advanced engineering tool functions based on PPR knowledge, such as checking whether the characteristics of a production process fit the characteristics of the product to be produced, require a *programmable interface* to PPR knowledge. The stakeholders in the engineering process phases have both common and different needs.

UC1a. Basic Engineering. For designing the production process, the basic engineer requires the definition and access to mapping of product parts to process steps characteristics, which are currently stored in excel tables providing only poor possibilities to execute this task. For identifying a set of useful resources for a specified product feature, the basic engineer requires the access to mapping of product features to production resource characteristics. For finding and comparing promising production process variants, the basic engineer requires the capability to discern between the desired process (customer requirements or product manager of a family of similar systems) and the possible process variants (a) derived from a product specification or (b) derived from the set of resource components and their combinations. For reusing PPR knowledge in a family of products or production systems, the basic engineer requires the capability for variant management in a PPR context.

UC1b. Detail Engineering. For designing a production system from an early rough sketch to a detailed construction plan, the detail engineer requires the capability to define and enhance the design of a resource from the viewpoint of one discipline and describe design dependencies across disciplines, for example, for machine configurations, which could be stored again in excel files or relational databases. For designing a production system part from reusable components, the detail engineer requires the capability to discern between information on a specific product and on a library of products and resources with detailed information on product and resource types, for example, a tree of motors, electrical motors, and specific motor types and instances. In a PPR context, this resource-specific view shall be linked to product/ion-relevant characteristics. For validating his design decisions, the detail engineer requires traceability of design decisions back to basic engineering by mapping the configuration of the production system parts back to parameters of the product to be produced and the planned production process.

UC2 PPR-Based Run-Time Data Analysis

UC2a Run-Time Process Data Analysis. For comparing the intended (specified) production process to the actual operation process, the production process optimizer requires capabilities for defining and comparing planned and actual production processes. To do this, operational data logs of the resource are needed as well as test data and if possible simulation results.

UC2b Run-Time Data Mining. For better understanding the impact of engineering and operation factors on the production process results, the production process optimizer requires capabilities for data integration and aggregation of production operation data with engineering data. This requirement is based on improvements for (a) the production process and (b) the capabilities of the production system family. For data integration, the production process optimizer requires capabilities for linking operation data to engineering data, for example, mapping of identifiers in data sets coming from a variety of sources like configuration files, operational data, and planned layouts from basic engineering.

PPR Data Category Characteristics The current technology landscape of the company consists of several in-house development tools used in the engineering process and of applications for configuring and analyzing the operation of manufacturing systems. These tools are only focused on expressing resource knowledge, neglecting the potential that a full PPR knowledge base could have. PPR knowledge could be used for expressing (a) success-critical attributes, such as parameters for production processes or configurations for production resource and (b) relationships, such as constraint dependencies, between products, production processes, and production resources. The three major groups identified with the domain experts currently in use areas follows:

1. *Engineering data* is all data that is created during the engineering process, for example, for designing a robot work cell, ranging from engineering artifacts, such as CAD drawings, to data tables, such as Excel files, hierarchically structured product parts, and PPR knowledge, such mappings between processes and resources in the robot work cell. Engineering data structures may differ from project to project and consist most of the time of complex engineering artifacts, objects with attributes, or graphs.

2. *Configuration data* includes data that describes the resource (machinery), such as relationships between production components or configurations or parameter settings for machines and devices. This data can be described and stored in classical table structures, consisting of many primitive values, like integers and strings. Configuration data schemas are rather stable; challenges come from keeping track of the semantics of changes in versions that may differ only in numerical/textual changes and linking these configuration values to outcomes in run-time data files.
3. *Run-time data* consists of all data accumulated during the operation of the manufacturing system. Analyses, logs, quality measurements, and so forth are all representatives of run-time data as foundation for data mining. Run-time data can be characterized as time series data, which is written once and read many times. The underlying schema may change with every new quality metric or sensor added, making it challenging to keep track of the semantics of the collected data.

Although these data categories have very different characteristics, they are often stored in a large relational database, which introduces challenges regarding technical debt, understandability, performance, and maintainability of data definition and access. Through mapping the different characteristics of these data categories into one shared schema many PPR knowledge aspects, like relationships between the individual concepts, might be lost, for example, if there is only a focus on configuration data for resources, there might be no concept for storing process or product-relevant data.

PPR Persistence Requirements From the discussion of these use cases with the software domain experts, we derive the following major requirements for PPR persistence design.

Data Representation for the Different PPR Knowledge Groups UC1 and UC2 target different phases of an engineering process. UC1 focuses on the early engineering phases where the planning and creation of PPR knowledge is the main objective. In these phases, a lot of the configuration data is initially created to be then detailed in later phases. UC2 aims at the run-time perspective of an engineering system, where large amounts of quality data in different forms are accumulated. Due to these different foci of the use cases, it is a requirement for a PPR persistent solution to be able to handle different data groups and their characteristics like fixed schema tables, graphs expressing relationships between PPR concepts, and time series consisting of quality metrics measured by the production system.

Programmable Interface A PPR persistence solution consisting of many different data aspects and data groups has a high potential for reuse, spanning over different disciplines and engineering phases. To avoid the accumulation of technical debt, a PPR persistence solution requires a programmable interface, an *API to the PPR knowledge base*. This API should represent the only entry point for accessing PPR knowledge and possible metadata representations, like for example who or what tool changed which part of the PPR knowledge representation. This requirement is based on the different existing tools present in an engineering company, which all support their individual specialized use case like in UC1, basic versus detail planning, resulting in different engineering artifacts.

Flexibility Derived from the two previous use cases and the different requirements of the data groups is flexibility, also a requirement for a PPR persistence solution. For example, UC1 provides two different views regarding PPR knowledge. In basic planning, stakeholders plan a production process and design the resources. Following this phase, detail planning is interested in the actual and more detailed process and the concrete realization of the design. These two use cases might have different requirements for a PPR knowledge persistence solution, requiring *flexibility* and easy to maintain data model implementations. UC2 also motivates this requirement, because the use case is interested in how the production system performs and how possible optimizations might look like, requiring adaptations to existing solutions and their persistence.

Usability and Usefulness A possible new solution should provide *usability* for the developers that need to work with the new technology and should also be *useful* and provide reusability in similar but different projects. As already identified, the mapping of different data groups into one technical solution may lead to high technical debt; also, this approach does impose many restrictions onto the developers that are responsible for the development of engineering tools. These restrictions can be seen currently in high development cycles and nearly unusable solutions, where even custom-made software leads to a vendor lock-in, making it virtually impossible to adapt a solution. Also these solutions do not provide any reusability in different projects. A new solution thus should focus beyond the PPR knowledge representation on providing useable and useful concepts for domain experts responsible for the technical implementation and maintenance.

Performance The presented use cases derived from UC2 focus on data mining and process data analysis. These use cases impose with increasing data sizes requirements regarding the performance. Performance can be expressed in the time period needed from measuring the quality/run-time data until it is analyzed and ready to provide again insights into the engineering of current or future systems.

Reusability of PPR Knowledge Engineering companies often have similar but not the same requirements regarding production systems and their design. For each new contract the two use cases UC1a and UC1b are executed, requiring the involved domain experts often to start from scratch or reuse, through many years of experience, existing solutions. Even though many products or systems could be classified and aggregated into families of products and production systems, this is not done, resulting in high rework efforts. A new PPR knowledge persistence solution should provide means of *reusability* for the engineering domain experts, providing libraries for reusing already existing PPR knowledge, mappings of (a) product to processes and (b) process to resources. Especially, these mappings often are based on reoccurring requirements from customers or imposed limitations from production resources.

Overall, the use cases revealed important requirements for PPR persistence that are hard to meet with the typical traditional persistence technology mix of (proprietary) engineering artifacts, Excel tables, XML configuration files, and relational databases.

7.8 Discussion

This section reports on a discussion of the overall process execution, observations, and lessons learned and extend Kathrein et al. (2019). It discusses results regarding the research questions introduced in Sect. 7.1 and in detail in Sect. 7.3.

RQ1. What Are Main Elements of a PPR EPA Method? Both business process analysis (BPA) and engineering process analysis (EPA) methods are concerned with investigating an existing process, involved stakeholders, and exchanged artifacts. Whereas BPA approaches like Santos and Alves (2017) and Rosenberger et al. (2018) focus more on the big picture of an engineering process, and do not allow for very sophisticated and detailed analysis (Vergidis et al. 2008), EPA approaches like Lüder et al. (2012), Jäger et al. (2011), and VDI (2010) tend to represent more individual workgroups and their procedures. Our presented approach in Sect. 7.2.2 combines the existing solutions and identifies the main elements in a repeatable two-phase process, resulting in a visual product/ion-aware representation, namely, the PPR data processing map (DPM). The proposed main elements: kick-off, interviews, refinement, and PPR artifact classification were evaluated in a holistic case study (Runeson and Höst 2009).

To support the proposed PPR EPA and execute its tasks, we introduced the role of the *EPA facilitator*. This role mediates the interests of all involved stakeholders and is responsible for choosing the right level of detail of the EPA as well as for choosing an adequate visual representation. In the conducted case study, three researchers took on this role. The execution and enactment of the proposed PPR EPA with its steps provide a first outline of how multidisciplinary engineering processes can be investigated. However, possible open issues that may surface in practice are still open for investigation and should be addressed.

The PPR EPA method allows collecting data, which is passed through the engineering process and records the current engineering process with links to engineering artifacts. A special focus lies on identifying tasks that create, require, or lose PPR knowledge and prioritizing the need of PPR knowledge for certain tasks and stakeholders. All involved stakeholders found the PPR EPA method suitable and useful. The PPR EPA further gave the stakeholders insights into not only their own line of work but also beyond and into other workgroups.

Both, independent investigations of workgroups and a high-level analysis for improvement potential for cooperating and collaborating stakeholders is possible with the proposed PPR EPA and further brings the benefit of explicit PPR knowledge identification.

The proposed process concept can also be used for the identification of technical depth and the identification of necessary security measures. Within the planning phase, information flow and therefore necessary user access privileges for the project can be derived. Furthermore, responsibilities of certain workgroups for certain components can be defined and non-repudiation can be ensured. This can be done either on a system level, or by applying cryptographic measures, which has the benefit of being independent of file- and operating system. A key challenge thereby lies in the nonintrusive support of employees in their daily work, which allows them to execute their work tasks as efficiently as before. One possible solution could allow for "weak" access rights, where users can execute tasks they are not responsible for, based on the engineering process description. Such an overstepping of a security boundary could be allowed, which should, however, be monitored, logged, and traced in a comprehensible way for project members and managers.

RQ2. What Are Main Elements of a PPR DPM Method and Notation? Section 7.2 briefly gave an overview of existing visualization notations for process analysis. In Sect. 7.2.3, we introduced the PPR DPM notation based on the BPMN 2.0 standard. The result is a PPR DPM, allowing a stakeholder to classify engineering artifacts regarding product, process, or resource knowledge and how these artifacts interact with certain engineering tasks.

The main elements from the standard BPMN 2.0 notations are: tasks, gateways, documents, and events. The newly introduced product/ion-aware notation elements are: annotations for documents regarding product, process, or resource knowledge classifications. We extend the task concept by annotating which of the PPR concepts is currently available, as well as further information that would be needed for an ideal task execution. A second extension to the task notation is an importance level, distinguishing important or crucial PPR knowledge dependencies, depicted as white/black broken documents.

By using a well-known and easy-to-use notation, the number of different concepts was minimized, which kept the level of complexity lower than in other approaches like Khabbazi et al. (2013), Huang et al. (2017) and Merunka (2017).

For the application of the new PPR notation, the stakeholders required a little bit of training but evaluated the PPR DPM as usable, useful, and a little bit less effort than the existing eEPC modeling approach.

RQ3. What Are Primary Use Cases That Require the Persistence of Different Categories of PPR Knowledge? From the case study for evaluating the PPR EPA and PPR DPM, we collected use cases on *Product/ion-Aware Engineering Tool Support* (UC1) and on *PPR-Based Run-Time Data Analysis* (UC2) to gain insights into the current technical landscape at the engineering company. These use cases build the first layer of a possible PPR knowledge persistence solution. Combining the insights from the use cases with interviews lead to the identifying of the characteristics of PPR knowledge categories and requirements on how to store and access PPR knowledge.

While the engineering tools currently focus on functions that use production system engineering data, advanced engineering tool functions require capabilities for defining and accessing PPR data and knowledge. The PPR knowledge categories of engineering data, configuration data, and run-time data indicate conflicting requirements for the persistence of mainly engineering artifacts, tables, graphs, and time series data. The requirements for PPR persistence were found hard to meet with the traditional persistence technology, such as repositories for engineering artifacts, structured text, and relational tables and databases. Also these requirements, combined with the PPR knowledge categories, provide functional requirements for the second layer of the PPR knowledge persistence solution. The third layer of the solution can in parts be addressed with the combination of use cases, requirements, and the knowledge gathered from the current situation at the company, but which requires further research.

While relational databases are a good choice for table-based data persistence (Vicknair et al. 2010), repurposing table-based data storage technologies for applications that require rapid changes of schemas or an altogether schema-less data model accumulates technical debt. Siddiqa et al. (2017) argued for the advantage of NoSQL data storage technologies for further flexibility of data definition and analysis in the development and operation phases.

As comparable persistence challenges can be found in business informatics, Sadalage and Fowler (2013) and Nance et al. (2013) pointed out that a combination of relational and NoSQL database technologies could be used for persistence design. However, this means redesigning the existing solution with new concepts and a clean data model leading to risks from data migration and from introducing a persistence design that uses considerably more complex technologies beyond the expertise of the domain experts, who often have an engineering background, but not from engineering large and heterogeneous software systems of systems. Therefore, we see future research work in exploring PPR knowledge persistence designs that allow addressing the use cases elicited in this chapter regarding their strengths and limitations in theory and in empirical studies with typical domain experts.

Limitations As all empirical studies, the presented research has some limitations that require further investigation.

Feasibility Study To evaluate the PPR EPA and the PPR DPM, we focused on specific use cases, which were chosen in cooperation with domain experts from an engineering company. The company is representative in size and domain for systems engineering enterprises, conducting business on a project basis. The focus of the engineering company lies on the manufacturing of production systems, without PPR knowledge management. All of our evaluation results are based on a limited sample of engineering projects, involved stakeholders, as well as different data models. The approach thus did not investigate situations where multiple products or variants are created and how this might affect the overall engineering process. We plan to overcome these limitations by expanding the case study in other domains and application contexts and further investigate possible issues of the PPR EPA that might arise.

Expressiveness of the PPR DPM Notation The notation of the PPR DPM enabled the involved stakeholders of the feasibility study to better express which PPR knowledge concerns are present in engineering documents. The proposed notation is not yet formalized or described and only presents a first visual aspect of ongoing research. There are also more advanced applications and analyses in prospects like constraint modeling or variation modeling. Constraint modeling would require extending the current PPR DPM notation to have an even higher expressiveness at hand, possibly exploiting concepts of ISA 95 (International Electrotechnical Commission 2003) or formal process specification given in VDI Guideline 3682 (VDI 2005). The involved stakeholders have also expressed the desire to model basic variations of products or product families, ranging from simple color adaptations to more complex process and system variations, which would affect the whole manufacturing system.

PPR Knowledge Flow and Artifact Exchange Investigation As mentioned previously, the PPR DPM notation is solely a visual extension to the BPMN 2.0. Even though it is possible to investigate an engineering process regarding the flow of knowledge, our proposed the notation come short regarding concrete dependencies between stakeholders and content of engineering artifacts. It was discussed that domain experts depend on intermediary results of one another; however, in some cases there might be only a partial dependency between a stakeholder and an artifact or one concrete value out of this artifact. The proposed PPR DPM only classifies the artifacts regarding PPR knowledge and does not detail the artifacts very much. This is however addressed in Chap. 8, with several approaches and methods to investigate such data logistics dependencies across several domains.

PPR Knowledge Persistence Use Cases and Requirements We collected and analyzed the use cases and requirements with domain experts at a single company. While we expect these use cases and requirements to be relevant for a wider application context, the focus on one company introduces bias that should be addressed by extending and validating the use cases and requirements with researchers and domain experts from a wider and representative set of data sources.

7.9 Conclusion and Future Work

The work environment of domain experts in systems engineering organizations is characterized by many different, collaborating disciplines and, from project to project changing of personnel. In such a multidisciplinary environment, many workgroups focus solely on improving their own local processes, tools, and methods. Little to no thought is given on how improvements of engineering interfaces for better collaboration and coordination could look like. This mindset leads to information silos, where only the bare minimum effort is fulfilled to have a working project collaboration.

The domain experts of systems engineering organizations also tend to focus more on the technical aspects of a system and product or process aspects are often neglected. This one-sided view on the PPR concept bears the risk of not communicating crucial parameter settings and endangering the project success and operation phase, as was described in Sect. 7.1 with the use case fragile product.

In this paper, we investigated a product/ion-aware method for an engineering process analysis (PPR EPA) method, as well as a notation for product/ion-aware data processing map (PPR DPM). Both contributions were based on elicited use cases from the systems engineering domain and should help domain experts, including the newly introduced role of an EPA facilitator, with a systematic repeatable approach to represent PPR knowledge in an engineering process. The introduced PPR EPA method is capable of tracing PPR knowledge throughout an engineering process. A special focus and capability is that tasks can be investigated regarding PPR knowledge requirements. The investigation of engineering artifacts further builds a main building block for analyzing PPR knowledge gaps that are present in an engineering organization and its process. Such analyses are a first step toward closing this knowledge gap.

The PPR EPA method provides the foundations for addressing the characteristics of Responsible Information Systems, such as flexibility, trustworthiness, and security. In respect to security, it allows the EPA to investigate possible security measures based on involved domain experts and their security clearance as well as classified engineering artifacts. Such an investigation finally can be the basis for planning necessary countermeasures and secure the intellectual property of an engineering organization. The EPA specifically addresses the major challenges introduced in Sect. 7.1.

C1. The Engineering Process Between Discipline-Specific Workgroups Is Hard to Trace and Analyze The outcome of the proposed PPR EPA approach visualizes a multidisciplinary engineering process. The visualization allows identifying PPR knowledge flows throughout the engineering process, highlighting tasks that create, transform, or lose PPR knowledge, as well classifying engineering artifacts regarding PPR knowledge aspects. This makes it possible to trace process executions and engineering artifacts through the engineering process. The PPR EPA also identifies interfaces between different disciplines and creating descriptions of which tasks are executed under which responsibility.

C2. Unclear Benefit of Representing PPR Knowledge Through visualizing the different involved disciplines of the engineering process, and further focusing on expressing the importance a task has regarding PPR knowledge, it is possible to analyze the whole engineering process and explicitly express PPR knowledge gaps. This product/ion-aware processing map (PPR DPM) can be analyzed regarding high-risk tasks and estimating the cost and effort it takes to explicitly represent PPR knowledge in engineering artifacts. Through this approach, domain experts see what information is available in which engineering phase and can match this to the actual PPR knowledge they receive and demand to close possible gaps or losses of knowledge along the engineering process.

C3. Unclear Impact of PPR Knowledge The PPR EPA and PPR DPM are able to assess the impact of PPR specific knowledge aspects, leading to considerations as to which PPR knowledge should be explicitly modeled. This is based on expressions regarding engineering tasks that need PPR knowledge for their execution. The PPR DPM addresses this challenge by indicating the priority an engineering task has regarding PPR knowledge. This allows all involved domain experts to identify especially critical tasks and address possible high-risk issues. The PPR DPM also refines the awareness and impact of early design decisions by domain experts.

C4. Unclear Use Cases with PPR Knowledge Categories That Require Persistence To address this challenge, we elicited primary use cases on *Product/ion-Aware Engineering Tool Support* (UC1) and on *PPR-based Run-time Data Analysis* (UC2) and the main PPR knowledge categories: engineering data, configuration data, and run-time. These use cases revealed a range of requirements for PPR knowledge persistence to guide software engineers who design and adapt engineering tools. Unfortunately, these requirements are conflicting and hard to address with traditional relation-based methods and technologies. Therefore, the initial research results on requirements suggest exploring a combination of persistence technologies regarding their technical capabilities to support advanced product/ion-aware use cases and regarding their usability and usefulness in typical application contexts.

Future Work Future work will include further applications and evaluations of the PPR EPA method and the PPR DPM notation in other engineering domains and application areas. Possible evaluations include the execution of the PPR EPA in a second engineering company, to cross-evaluate how the PPR EPA performs and if the found strengths and limitations are comparable between these two case studies, or if there is a bias based on the engineering organizations and their domains that should be investigated. The following aspects are of special interest for future research.

Advanced PPR Knowledge and Artifact Flow Investigation As discussed in the previous section, the presented approach does not provide means to investigate data logistics issues. Chapter 8 presents, however, several options on how dependencies between single values of engineering artifacts and dependencies on a more granular level can be addressed. Thus, in future work the proposed PPR EPA and also PPR DPM should be investigated with regard to how they can be combined with a possible data logistics approach.

Advanced PPR Knowledge Representation To be able to annotate PPR knowledge aspects directly onto engineering artifacts, there is the requirement and need to represent PPR knowledge explicitly in an engineering process. In future, these annotations should not only be visualized but also stored for further processing, analyses, and knowledge queries. The actual representation and storage of PPR knowledge could allow domain experts and stakeholders to move from general artifact representations to specific PPR knowledge aspects, which is also part of the Industry 4.0 vision.

Traceable Design Decisions Through expressing PPR knowledge explicitly, the relationships between the concepts and inherently made design decisions build the foundation for analyzing rationales and give insights into the early phases of an engineering process. Especially, the systems engineer gains understanding on how certain values for operational system parameters were chosen.

Generation of System Design Aspects From explicitly modeling PPR aspects and having traceable design decisions, it could be possible to derive design parameters from product/ion design decisions and engineering design patterns. Through efficiently deriving system designs and reusing these systems for whole production system families, an engineering company can achieve a considerable business advantage against its competitors.

Exploration of PPR Knowledge Persistence Requirements and Design Options We plan to explore PPR knowledge persistence designs that address the use cases and requirements elicited in this chapter. Possible designs need to be investigated regarding their strengths and limitations in theory and in empirical studies with typical domain experts.

IT Security Considerations The PPR EPA presents a detailed set of documentation regarding the engineering processes currently implemented in an engineering organization. This knowledge allows analysis of data flows across workgroups and could thus be interesting to a potential IT security attacker. Such threats to the integrity of the collected PPR knowledge and further even industrial espionage have to be researched in future work.

Apart from that, an interesting advancement can be the (semi-/fully) automatic detection of intentional/unintentional wrongdoing: the system should recognize if a certain step may result in bad engineering quality. The main challenge here is to recognize such possible results. One approach could be to let people define quality within the context of the project in an early project phase.

In terms of security, a next step for PPR can be to integrate secure software lifecycle processes, such as NIST SP 800-64 (Kissel et al. 2008) or ISO/IEC 27034-3:2018 (ISO 2018) (and future versions).

Acknowledgments The financial support by the Christian Doppler Research Association, the Austrian Federal Ministry for Digital and Economic Affairs, and the National Foundation for Research, Technology and Development is gratefully acknowledged.

References

Allweyer, T. (2016). *BPMN 2.0: Introduction to the standard for business process modeling*. Norderstedt: BoD–Books on Demand.

Beck, K. (2003). *Test-driven development: By example*. Boston, MA: Addison-Wesley.

Biffl, S., Gerhard, D., & Lüder, A. (2017). Introduction to the multi-disciplinary engineering for cyber-physical production systems. In *Multi-disciplinary engineering for cyber-physical production systems* (pp. 1–24). Cham: Springer.

Date, C. J., & Darwen, H. (1997). *A guide to Sql standard* (Vol. 3). Reading, MA: Addison-Wesley.

Drath, R. (Ed.). (2009). *Datenaustausch in der Anlagenplanung mit AutomationML: Integration von CAEX, PLCopen XML und COLLADA*. Berlin: Springer.

ElMaraghy, H. A. (2009). Changing and evolving products and systems–models and enablers. In *Changeable and reconfigurable manufacturing systems* (pp. 25–45). London: Springer.

Fay, A., Löwen, U., Schertl, A., Runde, S., Schleipen, M., & El Sakka, F. (2018). Zusätzliche Wertschöpfung mit digitalem Modell. *atp magazin, 60*(6–7), 58–69.

Force, U. A. (1981). *Integrated computer aided manufacturing (ICAM) architecture part II*. Volume IV-functional modeling manual (IDEF0), Air Force Materials Laboratory, Wright-Patterson AFB, Ohio.

Fowler, M., Kobryn, C., & Scott, K. (2004). *UML distilled: A brief guide to the standard object modeling language*. Boston, MA: Addison-Wesley.

Huang, Y., Huang, J., Wu, B., & Chen, J. (2017). Modeling and analysis of data dependencies in business process for data-intensive services. *Communications, 14*(10), 151–163.

Humble, J., & Farley, D. (2010). *Continuous delivery: Reliable software releases through build, test, and deployment automation*. London: Pearson.

Humphrey, W. S. (1995). *A discipline for software engineering*. Boston, MA: Addison-Wesley.

Hundt, L., & Lüder, A. (2012, September). Development of a method for the implementation of interoperable tool chains applying mechatronical thinking—use case engineering of logic control. In *Proceedings of 2012 IEEE 17th international conference on Emerging Technologies & Factory Automation (ETFA 2012)* (pp. 1–8). IEEE.

International Electrotechnical Commission. (2003). *IEC 62264-1 Enterprise-control system integration–Part 1: Models and terminology*. Geneva: IEC.

International Electrotechnical Commission. (2013). *Engineering data exchange format for use in industrial systems engineering – Automation Markup Language AML*. Retrieved from http://www.automationml.org/

ISO/IEC 27034-3. (2018). *Information technology – Application security – Part 3: Application security management process*. Retrieved March 6, 2019, from https://www.iso.org/standard/55583.html

Jäger, T., Fay, A., Wagner, T., & Lowen, U. (2011). Mining technical dependencies throughout engineering process knowledge. *Emerging Technologies & Factory Automation (ETFA), 2011 IEEE 16th Conference* (pp. 1–7).

Kathrein, L., Lüder, A., Meixner, K., Winkler, D., & Biffl, S. (2018). *Process analysis for communicating systems engineering workgroups; Technical report CDL-SQI-2018-11*, TU Wien. http://qse.ifs.tuwien.ac.at/wp-content/uploads/CDL-SQI-2018-11.pdf

Kathrein, L., Lüder, A., Meixner, K., Winkler, D., & Biffl, S. (2019). Product/ion-aware analysis of multi-disciplinary systems engineering processes. In *Proceedings of 21st International Conference on Enterprise Information Systems, ICEIS* (Vol. 2, pp. 48–60). Setúbal: SciTePress. ISBN 978-989-758-372-8. https://doi.org/10.5220/0007618000480060

Khabbazi, M. R., Hasan, M. K., Sulaiman, R., & Shapi'i, A. (2013). Business process modeling in production logistics: Complementary use of BPMN and UML. *Middle East Journal of Scientific Research, 15*(4), 516–529.

Kissel, R. L., Stine, K. M., Scholl, M. A., Rossman, H., Fahlsing, J., & Gulick, J. (2008). *Security considerations in the system development life cycle* (No. Special Publication (NIST SP)-800-64 Rev 2). National Institute of Standards and Technology.

Lüder, A., Foehr, M., Köhlein, A., & Böhm, B. (2012). Application of engineering processes analysis to evaluate benefits of mechatronic engineering. In *Emerging technologies & factory automation (ETFA), 2012 IEEE 17th conference* (pp. 1–4). IEEE.

Lüder, A., Pauly, J., Kirchheim, K., Rinker, F., & Biffl, S. (2018). *Migration to AutomationML based tool chains –Incrementally overcoming engineering network challenges*. Retrieved January 2, 2019, from https://www.automationml.org/o.red/up-loads/dateien/1548668540-17_Lueder_Migration-ToolChains_Paper.pdf

Merunka, V. (2017). Symmetries of modellingconcepts and relationships in UML -Advances and opportunities. *Lecture Notes in Business Information Processing, 298*, 100–110.

Moser, T., Biffl, S., Sunindyo, W. D., & Winkler, D. (2010, February). Integrating production automation expert knowledge across engineering stakeholder domains. In *Complex, intelligent and software intensive systems (CISIS), 2010 international conference* (pp. 352–359). IEEE.

Nance, C., Losser, T., Iype, R., & Harmon, G. (2013). *Nosql vs rdbms-why there is room for both. SAIS 2013 proceedings* (p. 27).
Paetzold, K. (2017). Product and systems engineering/CA$_*$ tool chains. In *Multi-disciplinary engineering for cyber-physical production systems* (pp. 27–62). Cham: Springer.
Presley, A., & Liles, D. H. (1995). The use of IDEF0 for the design and specification of methodologies. In *Proceedings of the 4th industrial engineering research conference*. Citeseer.
Rilling, J., Witte, R., Schuegerl, P., & Charland, P. (2008). Beyond information silos—An omnipresent approach to software evolution. *International Journal of Semantic Computing, 2*(04), 431–468.
Rosenberger, P., Gerhard, D., & Rosenberger, P. (2018). Context-aware system analysis: Introduction of a process model for industrial applications. In *ICEIS* (Vol. 2, pp. 368–375). Setúbal: SciTePress.
Rowley, J. (2007). The wisdom hierarchy: Representations of the DIKW hierarchy. *Journal of Information Science, 33*(2), 163–180.
Runeson, P., & Höst, M. (2009). Guidelines for conducting and reporting case study research in software engineering. *Empirical Software Engineering, 14*(2), 131–164.
Sabou, M., Ekaputra, F. J., & Biffl, S. (2017). Semantic web technologies for data integration in multi-disciplinary engineering. In *Multi-disciplinary engineering for cyber-physical production systems* (pp. 301–329). Cham: Springer.
Sadalage, P. J., & Fowler, M. (2013). *NoSQL distilled: A brief guide to the emerging world of polyglot persistence*. London: Pearson.
Santos, H., & Alves, C. (2017). Exploring the ambidextrous analysis of business processes: A design science research. In *International conference on enterprise information systems* (pp. 543–566). Cham: Springer.
Schafer, W., & Wehrheim, H. (2007, May). The challenges of building advanced mechatronic systems. In *Future of software engineering, 2007 (FOSE'07)* (pp. 72–84). IEEE.
Scheer, A.-W. (1998). *ARIS: Vom Geschäftsprozeß zum Anwendungssystem/August-Wilhelm Scheer*. Berlin: Springer.
Schleipen, M., Lüder, A., Sauer, O., Flatt, H., & Jasperneite, J. (2015). Requirements and concept for plug-and-work. *at-Automatisierungstechnik, 63*(10), 801–820.
Schwaber, K., & Beedle, M. (2002). *Agile software development with scrum* (Vol. 1). Upper Saddle River: Prentice Hall.
Siddiqa, A., Karim, A., & Gani, A. (2017). Big data storage technologies: A survey. *Frontiers of Information Technology & Electronic Engineering, 18*(8), 1040–1070.
Stark, J. (2015). Product lifecycle management. In *Product lifecycle management* (Vol. 1, pp. 1–29). Cham: Springer.
VDI 3682. (2005). *Formalised process descriptions*. Berlin: BeuthVerlag.
VDI 3695. (2010). *Engineering of industrial plants, evaluation and optimization, Part 1*. Berlin: BeuthVerlag.
Vergidis, K., Tiwari, A., & Majeed, B. (2008). Business process analysis and optimization: Beyond reengineering. *IEEE Transactions on Systems, Man, and Cybernetics, Part C, 38*(1), 69–82.
Vicknair, C., Macias, M., Zhao, Z., Nan, X., Chen, Y., & Wilkins, D. (2010, April). A comparison of a graph database and a relational database: A data provenance perspective. In *Proceedings of the 48th annual southeast regional conference* (p. 42). ACM.
Wieringa, R. (2014). *Design science methodology for information systems and software engineering*. Berlin: Springer.
Wiesner, S., & Thoben, K. D. (2017). Cyber-physical product-service systems. In *Multi-disciplinary engineering for cyber-physical production systems* (pp. 63–88). Cham: Springer.
Zhang, C., Chen, X., Feng, Y., & Luo, R. (2010, June). Modeling and functional design of logistic park using IDEFO method. In *2010 7th international conference on service systems and service management* (pp. 1–5). IEEE.
Zhu, L., Bass, L., & Champlin-Scharff, G. (2016). Devops and its practices. *IEEE Software, 33*(3), 32–34.

Chapter 8
Engineering Data Logistics for Agile Automation Systems Engineering

Requirements and Solution Concepts with AutomationML

Stefan Biffl, Arndt Lüder, Felix Rinker, Laura Waltersdorfer, and Dietmar Winkler

Abstract In the parallel engineering of large and long-running automation systems, such as *Production Systems Engineering* (PSE) projects, engineering teams with different backgrounds work in a so-called *Round-Trip Engineering* (RTE) process to iteratively enrich and refine their engineering artifacts, and need to exchange data efficiently to prevent the divergence of local engineering models. Unfortunately, the heterogeneity of local engineering artifacts and data, coming from several engineering disciplines, makes it hard to integrate the discipline-specific views on the data for efficient synchronization.

In this chapter, we introduce the approach of *Engineering Data Logistics* (EDaL) to support RTE requirements and enable the efficient integration and systematic exchange of engineering data in a PSE project. We propose the concept of EDaL, which analyzes efficient *Engineering Data Exchange* (EDEx) flows from data providers to a consumer derived from data exchange use cases. Requirements for EDEx flows are presented, for example, the definition and semantic mapping of

S. Biffl (✉)
Institute of Information Systems Engineering, Technische Universität Wien, Vienna, Austria
e-mail: stefan.biffl@tuwien.ac.at

A. Lüder
Otto-v.-Guericke University/IAF, Magdeburg, Germany
e-mail: arndt.lueder@ovgu.de

F. Rinker · L. Waltersdorfer · D. Winkler
Christian Doppler Laboratory for Security and Quality Improvement in the Production System Lifecycle (CDL-SQI), Institute of Information Systems Engineering, Technische Universität Wien, Vienna, Austria
e-mail: felix.rinker@tuwien.ac.at; dietmar.winkler@tuwien.ac.at

S. Biffl et al. (eds.), *Security and Quality in Cyber-Physical Systems Engineering*,
https://doi.org/10.1007/978-3-030-25312-7_8

engineering data elements for exchange. We discuss main requirements for and design elements of an EDaL information system for automating EDaL process capabilities. We evaluate the benefit and cost of the EDEx process and concepts in a feasibility case study with requirements and data from real-world use cases at a large PSE company in comparison to a traditional manual point-to-point engineering data exchange. Results from the feasibility study indicate that the EDEx process flows may be more effective than the traditional point-to-point engineering artifact exchange and a good foundation to EDaL for more agile engineering.

Keywords Multidisciplinary engineering · Production systems engineering · Cyber-physical production systems · Engineering process · Process design · Data exchange · Data integration

8.1 Introduction

Engineering industrial production systems, so-called cyber-physical production systems, for example, long-running and safety-critical systems for assembling automotive parts or for producing metal, is the business of multidisciplinary *production system engineering* (PSE) companies (Biffl et al. 2017; Vogel-Heuser et al. 2017).

Planning such systems involves parallel engineering: due to increased production cycles and multiple disciplines, such as mechanical, electrical, and simulation engineering, developing their engineering and artifacts, such as plans, models, software code, or machine configurations, independently. Nevertheless, they have to consider dependencies between the engineering disciplines in order to build a common system. To shorten the duration of PSE projects, parallel engineering takes place, the so-called *Round-Trip Engineering* (RTE) process to iteratively enrich and refine engineering artifacts (Drath 2009). Similar to RTE in software engineering (Medvidovic et al. 1999), its main goal is the synchronization of two or more engineering artifacts to ensure consistency.

Unfortunately, the heterogeneity of local engineering artifacts and data coming from several engineering disciplines complicate the integration of discipline-specific views on the data for efficient synchronization and lead to late communication of changes and ineffective version management of engineering data.

Therefore, a traditional *engineering data exchange* (EDEx) flow, in other words the flows of single engineering artifacts that contain essential information, such as parameters for configuration or formulae, is a cumbersome, manual process whose effort lies in the hand of the data consumer.

To overcome these disadvantages and be able to perform agile RTE for consistent distributed data management, the following requirements have to be met: (a) a process for negotiating data elements requested by data consumers and matching to data elements coming from data providers and (b) an efficient engineering data exchange method and mechanism for conducting the agreed data exchanges between data sources and sinks of domain experts.

A key success factor for the *round-trip engineering* (RTE) process pattern (Biffl et al. 2019a) to enrich and improve engineering data iteratively along the engineering process and with information coming back from testing, simulation, and operation is the capability for *Engineering Data Logistics* (EDaL). In this chapter, we define EDaL to be *the planned management of multiple engineering data exchanges from a data provider to a data consumer according to the consumer's requirements in a consistent and valid manner.* In other words, EDaL provides the capability to exchange selected data in the engineering artifacts with related domain experts efficiently and in a timely manner in order to reduce rework due to inconsistencies in diverging local data views. Therefore, EDaL extends EDEx by more complex capabilities such as data versioning, consistency checks, and automation of work processes reduce work effort, delivery delays, and enhance the overall data quality. The enhanced data quality for production systems simulations can enhance the security of the resulting system due to a minimized rate of errors and faults. Moreover, EDaL builds the foundation for agile PSE that can adapt to changes communicated from backflows in the PSE process due to design changes or errors and builds the foundation for cyber-physical systems and more complex data analysis.

We illustrate the EDaL approach and *Engineering Data Exchange* (EDEx) (Biffl et al. 2019b, c) processes with use cases from simulation in PSE. Simulation is a major activity for engineering data to assess the safety and business risks of a production system before system construction. The simulation engineer combines engineering data from several data providers such as rotation speed, torque, control signals, or power consumption of a motor as foundation for calculating and analyzing the movement of work pieces and robots over time. The purpose of these calculations is to explore dynamic properties of the designed production system, such as throughput or the physical feasibility of production steps. These simulation models can also be used to provide feedback on issues with the engineering data and on design issues that need to be addressed by the domain experts, for example, changing the placement of a robot to improve the material throughput in a work cell.

In the *traditional EDEx process* (Biffl et al. 2014a, 2018), domain experts communicate their engineering artifacts point-to-point (P2P), typically in the form of spreadsheet tables, pdf, or XML files. Unfortunately, in the traditional EDEx process, Lüder et al. (2018a, b) identified the following major challenges that also make EdaL, which builds on EDEx, more difficult (see Fig. 8.1).

C1. Data Exchange Requirements Are Not Clear or Are Conflicting While the domain experts know their partners in the engineering process, there is surprisingly little concern for the data exchange requirements of data consumers and the impact of ineffective or inefficient EDaL on the project team performance. As EDaL is not a formal engineering activity but a necessary cost factor, comparable to the transport activity between production tasks, every engineer tries to minimize locally, overall at the expense of the cost to the engineering team. In many cases, data consumers bear the cost and risk of EDaL due to missing awareness for and support by an EDaL process and infrastructure. Potential data providers often are not aware of who needs

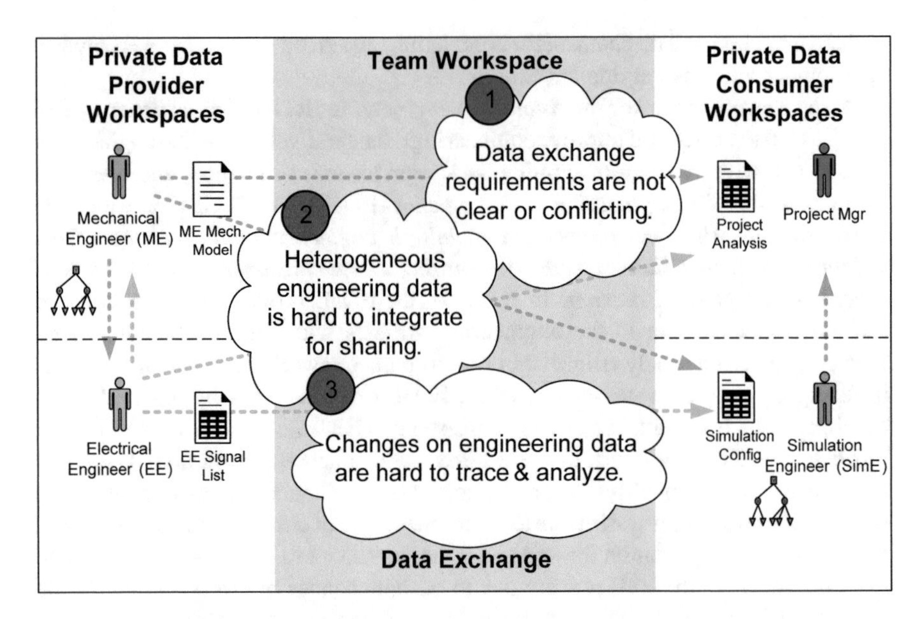

Fig. 8.1 Challenges in data exchange in parallel systems engineering, based on Biffl et al. (2019b)

what kind of data at what point in time in the project. General dependencies between stakeholders might be known; however, the specific relations between engineering artifacts and their content within an engineering project can change during the project execution.

C2. Heterogeneous Engineering Data Is Hard to Integrate for Sharing Engineering tools are tailored to the specific needs of their specific discipline due to historical and practical reasons and therefore unfit for the use of exchanging engineering data. The different disciplines may share some common concepts (Sabou et al. 2017; Winkler et al. 2017), such as the concept of a machine, a device, or a signal. However, these concepts are not consistently modeled across disciplines, making data integration for sharing error-prone and hard to automate. Therefore, extracting relevant information from engineering artifacts exchanged by providers requires extensive effort and knowledge on the side of consuming domain experts. This manual, unstructured process hinders comprehensive automated processing of the engineering data buried within the engineering artifacts. While the reusable representation of explicit semantic relationships between similar concepts of data providers and consumers may be costly for an informal point-to-point data exchange, EDaL support for the EDEx in an engineering team can build on the explicit representation of common concepts as in semantic links between heterogeneous engineering data sets to enable automation of EDEx and analyses.

C3. Changes on Engineering Data from Round-Trip Engineering Are Hard to Trace and Analyze A data consumer in the RTE process has to be keep track of the changes in the data versions he or she receives to enable analyses of the received data and metadata, for example, for identifying missing or inconsistent data. Unfortunately, using point-to-point (P2P) data exchange makes it very hard for a consumer to trace and analyze the set of data versions exchanged that may come from several providers as there is no EDaL support to keep track of EDEx flows, including roles and rules for process conduct.

In this chapter, we introduce a process for efficient *Engineering Data Logistics* (EDaL) to address these challenges and to automate data logistics in order to improve the value and reduce the risks of EDEx. This chapter extends EDEx process and information system considerations, introduced in Biffl et al. (2019c), in the EDaL context. We investigate the following *research questions* (RQs) based on *Design Science* research methodology (Wieringa 2014) and the use cases in Biffl et al. (2018, 2019a).

RQ1: *What are main information system mechanisms that enable engineering data logistics for Multidisciplinary System Engineering?*

RQ2: *What are main elements of an effective and efficient engineering data exchange (EDEx) process in Multidisciplinary System Engineering?*

RQ3: *What are main information system mechanisms that enable engineering data logistics for Multidisciplinary System Engineering?*

From the research we expect the following contributions for the *information systems engineering* (ISE) community. The use cases and EDaL approach give ISE researchers insight into the PSE domain, the foundation for Industry 4.0 applications. The EDEx process contributes capabilities for designing and investigating agile processes and information systems in PSE, a foundation for conducting engineering projects for cyber-physical production systems economically.

The remainder of this chapter is structured as follows. Section 8.2 introduces use cases, collected in workshops with stakeholders at a large PSE company, to illustrate RTE requirements for an *Engineering Data Logistics* (EDaL) approach and information system that enable the efficient integration and systematic exchange of engineering data in a PSE project. Section 8.3 summarizes related work on approaches for data logistics in multidisciplinary *production systems engineering* (PSE), information systems and software engineering. Section 8.4 motivates the research questions and the research approach. Section 8.5 introduces steps for an *EDaL approach* to address the requirements identified in Sect. 8.2. Section 8.6 discusses main design elements for effective and efficient *EDaL information system* (EDaLIS) mechanisms to address the requirements identified in Sect. 8.2. Section 8.7 reports on an evaluation of the effectiveness and effort of the proposed EDaL process with EDaLIS mechanisms in a feasibility case study with requirements and data from real-world use cases with domain experts at a large PSE company. Section 8.8 discusses the research findings and limitations. Section 8.9 concludes and proposes future research work.

8.2 Engineering Data Logistics Use Cases

Section 8.2 introduces use cases, collected in workshops with stakeholders at a large PSE company (Biffl et al. 2018), to illustrate RTE requirements for an *Engineering Data Logistics* (EDaL) approach and information system that enable the efficient integration and systematic exchange of engineering data in a PSE project. This section introduces a use case with illustrative data for the data exchange of the data consumer simulation with several data providers.

8.2.1 Engineering Data Exchange Use Cases for Requirements Analysis

We consider the following use cases, starting from the traditional basic collection/provision of engineering artifacts in a point-to-point (P2P) network of domain experts (UC1), progressing to stepwise enrichment of engineering artifacts (U2), to the parallel iterative enrichment of engineering artifacts (U3), and finally to consider backflows in the engineering network (U4) as foundation for true round-trip engineering. Strictly speaking, UC1 *artifact provision* and UC2a *simple sequential enrichment* only represent EDEx (Engineering Data Exchange) use cases.

UC2b *sequential enrichment with updates*, UC3 *parallel enrichment with updates*, and UC4 *backflows of artifacts* incorporate more complex requirements. Thus, they represent typically use cases that require EDaL (Engineering Data Logistic) capabilities, such as version management or consistency checks to truly implement an agile round-trip engineering process.

In the use cases, we assume a team of domain experts involved in designing a work cell as part of a larger production system.

UC1 Artifact Provision Figure 8.2a shows a set of domain experts in an engineering project, the plant planner (PP), the machine engineer (ME), the electrical engineer (EE), and the control programmer (CP) as providers of engineering artifacts, and the simulation engineer (SimE) and the project manager (PM) as consumers of engineering data. The orange arrows in Fig. 8.2a illustrate the provision of engineering artifacts by the PP, ME, EE, and CP to the SimE, who has to extract the data from the engineering artifacts, to integrate the data from heterogeneous sources, and to clarify issues with each data provider. We describe the artifact provision as data exchange in the notation (PP, ME, EE, CP)—>SimE, that is, the SimE requires a data set from the other four domain experts.

UC2 Sequential Enrichment of Artifacts In a typical sequential engineering process, the PP starts with providing the structure of the production system, then the ME selects and designs the mechanical parts, then the EE designs the electrical mechanisms for providing the system with energy and information connections, and then the control programmer designs the software and configurations to automate

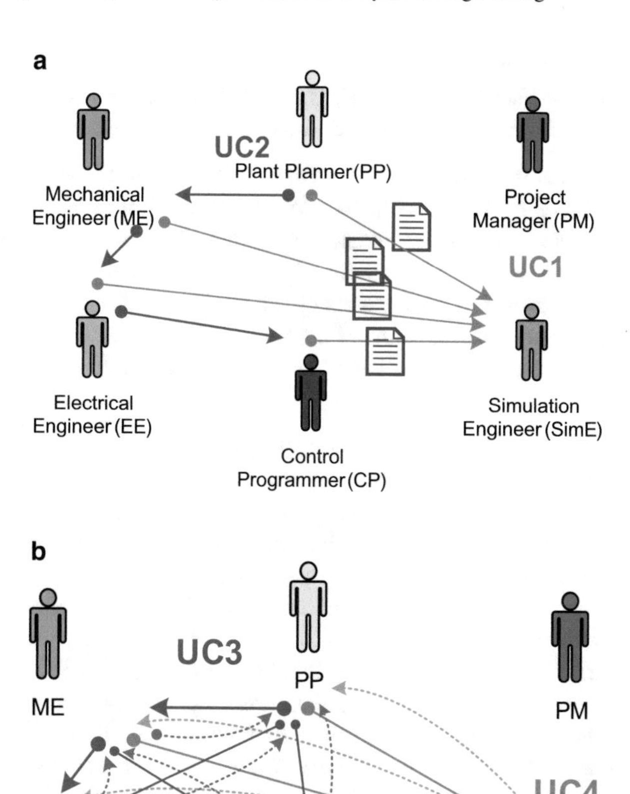

Fig. 8.2 (**a** and **b**) *Engineering Data Logistics* (EDaL) use cases in the *round-trip engineering* (RTE) process, based on Biffl et al. (2019b)

system parts. The violet arrows in Fig. 8.2a illustrate the same sequence of engineering artifact exchanges between the domain experts.

UC2a Simple Sequential Enrichment In the simplest case, the domain experts conduct one sequence of engineering artifact exchanges for enrichment and then deliver to the SimE: PP—>ME, ME—>EE, EE—>CP; (PP, ME, EE, CP)—>SimE.

UC2b Sequential Enrichment with Updates In an advanced case, each domain expert may improve his/her engineering artifacts and propagate the updated engineering artifact version along the engineering chain, resulting in a sequence of follow-up updates. In this case, the domain engineers require a mechanism to efficiently identify changes between artifact versions. PP (PP.A)—>ME, PP (PP.A′)—>ME denotes that the PP sends the ME first his/her artifact A and then an updated version A′.

UC3 Parallel Enrichment of Engineering Artifacts In a parallel engineering process, the domain experts start in parallel with a rough design; they refine their designs in parallel and exchange updates as needed. The violet arrows in Fig. 8.2b illustrate the multitude of point-to-point exchanges in parallel engineering, making it hard to keep an overview on the artifact versions and their dependencies, as the sequence of updates and their time of communication in the team is not known. SimE—>(PP, ME, EE) denotes a backflow from simulation to earlier engineering activities.

UC4 Backflows of Artifacts Changes to engineering artifacts may come from backflows in the engineering process due to changed requirements, errors found in the engineering design, or feedback from tests, simulations, and operation. The dashed arrows in Fig. 8.2b illustrate the multitude of potential backflows in the engineering team. Unfortunately, there is, in general, no effective and efficient process for systematic backflows, making it uncertain to what extent backflow information is considered or lost.

The data exchange requirements, specified by these use cases in the data logistics network, result in a set of engineering data exchange (EDEx) flows, for example, PP—>ME. In the following, we focus on an individual EDEx flow to identify EDEx requirements and solution options. The *round-trip engineering* (RTE) process (Biffl et al. 2019a) provides the foundation for consistent distributed data management and requires (a) a process for negotiating data elements requested by data consumers and matching the data elements coming from data providers and (b) a data exchange method and mechanism for executing the agreed upon data exchanges between domain experts and their data sources and sinks.

Figure 8.3 illustrates the EDEx data flow for UC1 *Artifact provision*. The data flows from a data provider to a data consumer. In the traditional engineering artifact communication, the provider sends an artifact that contains the data relevant for the customer. However, the consumer has to bear the effort for extracting and validating the data from the artifact. In addition, there is no systematic traceability and quality assurance in the process. Therefore, we propose introducing an *engineering data logistics information system* (EDaLIS) as foundation for traceable, quality-assured, and efficient data flows in an engineering team.

8.2.2 Engineering Data Exchange Use Cases for Evaluation

Based on observations and discussions with our industry partners conducted during the case study (Runeson and Höst 2009), we identified two illustrative *use cases* (UCs) that show the benefit of improved EDEx (see also Biffl et al. 2019c) and the implementation of EDaL: engineering data collection for *production system simulation* and for *production system engineering project monitoring*. The

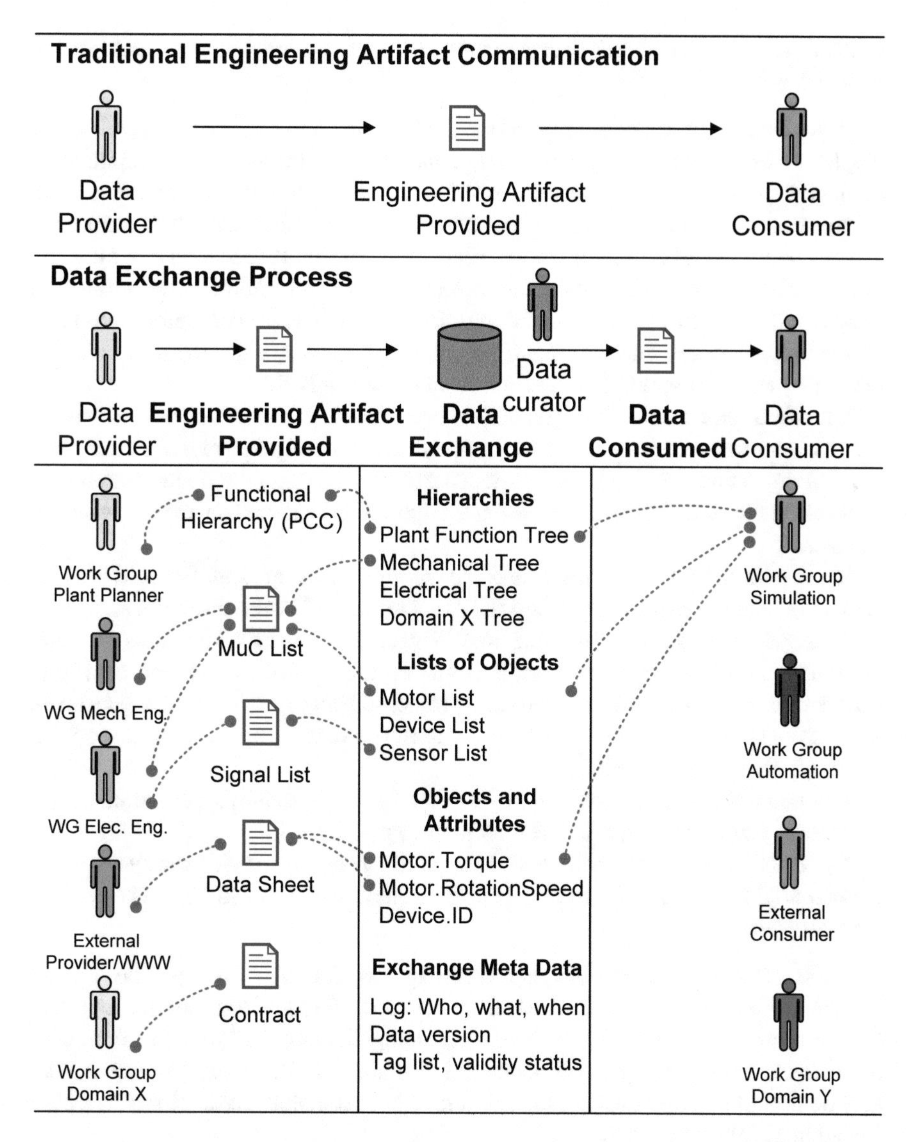

Fig. 8.3 Simple data processing map illustrating one data flow from several data providers to one simulation expert, based on Biffl et al. (2019b)

engineering of a typical industrial *production system* (PS), such as automotive assembly, requires at least the collaboration of—and EDEx—between the *plant planner* (PP), who plans the layout of the PS, *mechanical engineer* (ME), *electrical engineer* (EE), and *control programmer* (CP). Each domain expert designs and updates complex and heterogeneous local models that are hard to understand by other domain experts. A more detailed description of newly introduced roles, engineering artifacts, and process phases derived from our EDaL approach will be presented in Sect. 8.5.4.

UC Sim. Data Exchange for Production System Simulation In a typical advanced engineering environment, a *simulation engineer* (*SimE*) designs and runs simulation models to check the engineering results and to optimize production system parameters, such as safety risks, production throughput, and energy consumption. These designs of the simulation models depend on the input of several other domain experts, such as the configuration parameters of motors and conveyers in a transport system and requirements of production processes, such as process duration (s) and production resources, such as length (m), size (m^2 or m^3), mass (t), heat radiation (kW), power consumption (kW), or maximal noise level (db).

The *SimE* requires this input from data providers to calculate characteristics, for example, power consumption or movement dynamics, of a system part, for example, a drive chain, to find out whether the system part will behave as intended and to provide feedback to the contributing engineering disciplines on risks and on necessary design changes.

If the simulation identifies infeasible system plans or significant risks, the engineers have to cooperate to adapt the plans in the individual disciplines.

A single change in a discipline may trigger a chain of adaptations in other disciplines and lead to unclear implications on the overall system and avoidable rework in later project phases. Therefore, project stakeholders would like to evaluate defined constraints as early as possible for each relevant change of a local model, a capability that is linked to EDaL.

The manual synchronization of these data typically requires additional effort, tends to be error-prone, and induces avoidable project risks.

The early detection of errors and faults in the design of production systems has the potential to increase the overall security of the system and to minimize costs in the total operation.

UC PM. Production System Engineering Project Monitoring The project manager (PM) wants to use the input from data providers to the simulation engineer to assess project progress by analyzing the completeness and quality of data with respect to the project phase and planned deliverables. Missing or inconsistent data may be fine in an early design phase, but may pose a major risk in closer to a later design milestone and require action by the PM.

8.3 Related Work

This section summarizes related work on approaches for data logistics in multidisciplinary *production systems engineering* (PSE), information systems, and software engineering.

8.3.1 Data Logistics in Multidisciplinary Systems Engineering

Analogue to the real world, *Engineering Data Logistics* describes the flow of data elements from a data provider to a data consumer according to the customer's requirements. As in traditional logistics, the change of even single parts of the data transport and exchange network may affect the characteristics of the data logistics system, such as duration, quality, or cost of logistics, important aspects for the cost and risk of the engineering system using the data logistics, as well as the business advantages of enabling frequent and cheap data updates between work groups that work in parallel (Hell 2018; Andersen et al. 2018).

In *Production Systems Engineering* (PSE) (Biffl et al. 2017, 2019a), the content of the exchanged artifacts is important as these artifacts contain only part of the local models of the domain experts. Due to the inherent dependencies between these local models, such as dependencies between mechanical engineering defining cable routes, electrical engineering defining the applied wires and their location on cable routes, and communication system engineering defining used communication lines all effecting the possible impact of electrical fields on communication system quality, domain knowledge is required on both the customer and the provider data models to interpret the content of the exchanged data. Therefore, it is necessary to move from delivering engineering artifacts to *engineering data exchange* (EDEx), as a stepping stone to implementing agile engineering data logistics (EDaL). Agility in the industrial contexts is defined as the ability to cooperate within an organization (data providers and data consumers) and coping with change and uncertainty for example by implementing flexible planning and management processes (Jackson and Johansson 2003; Gould 1997).

Business *process analysis* (OMG 2011) has proven to illustrate relevant stakeholder groups, activities, and exchanged engineering artifacts to improve the overall process quality in an organization. However, additional data modeling is required to represent the knowledge required for EDEx. Thus, the workflow analysis shall cover the aspects engineering decisions (engineering activities made), applied engineering tools, created and required artifacts covering engineering information, and involved humans with skills and competences (Schäffler et al. 2013).

While EDEx is already important and difficult for traditional PSE, the migration toward cyber-physical systems is a complex task that requires an extensive solution, covering technical, operational, and human dimensions (Calà et al. 2017). Due to this multidimensional complexity, traditional information systems have not yet adequately addressed the challenges imposed by collaboration in multidisciplinary engineering systems: heterogeneous tools and data formats, diverging views on artifacts and their versioning are the most pressing ones (Drath et al. 2011). Optimizing and enriching the currently available engineering data and data exchange is a feasible strategy that can be achieved by integrating EDEx (Sabou et al. 2017) based on the machine understandable representation of knowledge on how exchanged data elements fit the local data models of the data providers and consumers.

While there are engineering tool suites that integrate several engineering functions in one set of tools with a common data model that greatly simplifies EDEx, most engineering projects use many tools with heterogeneous data models that are challenging to integrate (Biffl et al. 2017). The traditional EDEx process (Biffl et al. 2014a) is a *point-to-point exchange* of engineering artifacts between domain experts via email, repository, or USB stick, typically in the form of spreadsheet tables, pdf, or XML files.

Lüder et al. (2018a, b) introduce an architecture for engineering data logistics, based on *AutomationML* (IEC 62714 2018; Vogel-Heuser et al. 2017), an open, XML-based format for the exchange of engineering data. The proposed architecture allows exchanging data between discipline-specific data models with varying hierarchical key systems. While this approach is useful in an *AutomationML* environment, the approach does not consider how to negotiate the EDEx between many data consumers and providers. Often the data providers tend to provide all kinds of data that someone might find useful in the future, leading to a pile of data that is expensive to provide and hardly used.

8.3.2 Multi-model Dashboard for Data Integration and Monitoring

The *Multi-model Dashboard* (MMD) approach (Biffl et al. 2014a, b) extends the *Decision Board* approach (Holl et al. 2012) by adding the concept of constraints that use shared model parameters, and by automating the data extraction and integration of parameter values from heterogeneous data sources (Biffl et al. 2014a). The tool-supported MMD process guides the systematic definition, design, monitoring, and evaluation of MMD parameters and constraints, visualized on the MMD (see also Fig. 8.7). A dashboard provides the semantically integrated values of parameters and of constraints to the domain experts, as parameter values in various local models change during the project. The MMD provides promising capabilities for data extraction from engineering artifacts, often engineering models.

The MMD concepts of private workspaces and common team workspace in a heterogeneous System-of-System environment fit well to typical parallel systems engineering environments. The roles in the MMD approach, *data subscriber* and *publisher*, can be mapped well to the *data consumer* and *provider* in the context of this chapter. While we can build on the MMD strengths as foundation for the EDEx research in this chapter, the following limitations of the MMD approach require adaptation for data exchange in a system engineering project. The MMD does not consider the provision of data to consumers but focuses on the evaluation of engineering parameters and constraints. In practice, the MMD assumption of well-defined common concepts may not hold, if several disciplines may cooperate without one discipline clearly leading. The MMD software architecture based on an *AML Hub*[1] (Biffl et al. 2014a) is a limitation for a more general EDEx software architecture. The research questions in Biffl et al. (2014a) focused on identifying common concepts, software design options for change monitoring and for awareness design in a heterogeneous System-of-System environment, while the focus of this chapter is engineering EDaL based on EDEx definition and operation.

8.3.3 Data Exchange in Information Systems and Software Engineering

Business process management approaches, such as UML class diagrams (Brambilla et al. 2017) or BPMN (OMG 2011), can be a good foundation for EDEx definition by characterizing involved stakeholders, systems and, to some extent, data types and their relationships. Workflow management systems allow the automated setup, performance, and monitoring of previously defined processes; a common tool for industry use cases is the *Aris*[2] tool set. However, these generic methods need to be adapted to heterogeneous engineering data integration (Rosemann and vom Brocke 2015). In specialized domains, such as medicine, science, and engineering, new approaches may be needed to optimize data exchange according to domain-specific requirements (Jimenez-Ramirez et al. 2018; Putze et al. 2018).

Semantic Web technologies facilitate data exchange across applications and organizations and have proposed engineering data integration approaches following *the interchange standardization approach* (Sabou et al. 2017). However, the manifold types of dependencies in PSE data models differ from typical Semantic Web requirements (Kovalenko and Euzenat 2016) and the Semantic Web technology stack is therefore currently seldom used in engineering environments.

Model-driven software engineering (Brambilla et al. 2017) is a well-established software methodology, in which the abstraction of the problem domain is utilized to facilitate automated code generation, testing, and verification. *Seamless Model-Based Development* (Broy et al. 2010) is a desirable strategic goal that is hard to achieve in the current heterogeneous engineering reality with less-than-willing tool vendors who may prefer vendor lock-in to open standards. However, as domain-specific languages such as *AutomationML* gain acceptance in PSE, a foundation for model-based approaches is likely to become stronger.

Design patterns (Hohpe and Woolf 2003) encapsulate best practices of software system design for commonly occurring problems, in our case data and tool integration. In the context of this work, we build on design patterns, such as *message passing* and *publish-subscribe*, to support the loose coupling of work groups and tools.

---

[1] http://www.amlhub.at/

[2] https://www.softwareag.com/ch/products/aris_alfabet/bpa/default.html

8.3.4 Technical Data Exchange Formats

To facilitate data exchange, technical data exchange formats have to be able to cover, possibly all but at least most of the information required and/or produced within the PSE process by *data consumers* and *providers*. For these data exchange formats, there are a set of (sometimes contradicting) requirements to be fulfilled (Lüder and Schmidt 2017):

- The data format shall be adaptable to different application cases and flexible with respect to extensions and changes.
- The data representation shall be efficient.
- The data representation shall be human readable.
- The data representation shall be based on international standards.

These requirements lead to an XML-based data format, which makes engineering tools standardized data exchange formats like STEP (Xu 2012) and AutomationML (Drath 2009) preferable as they represent a tree structure similar to the topologies common in engineering, such as functional, mechanical, or electrical hierarchies.

Following Diedrich et al. (2011), the data exchange between engineering tools requires two levels of standardization, the syntax level and the semantic level. The syntax level defines the correct technical representation of the data objects in the data exchange format, including the vocabulary of the data exchange. In contrast, the semantic level defines the interpretation of data objects, that is, the conceptual meaning of objects in the engineering tool chain. With respect to the intended EDEx approach, both levels are relevant, but the semantic level is more important as it enables the identification of common information exchanged between the *data provider* and *consumer.*

Technical data exchange formats can be defined in two ways, either they define syntax and semantics together, as in the STEP approach or the approach defined in VDI Guideline 3690 (VDI 3690 2012–2017), or they define syntax and semantics separately, as in the AutomationML or the XMI approach (Grose and Doney 2002). Since the separate definition of semantics enables better flexibility and adaptability of a data exchange format to application cases, this approach seems to be preferable.

8.4 Research Questions and Approach

This section motivates the research questions and the research approach. In this chapter, we introduce a process for efficient Engineering Data Logistics (EDaL) to address these challenges and to automate data logistics in order to improve the value and reduce the risks of EDEx. We investigate the following research questions (RQs) based on *Design Science* research methodology (Wieringa 2014).

RQ1. What Are Main Elements of an Engineering Data Logistics (EDaL) Approach in Round-Trip System Engineering? To address this research question, we define in Sect. 8.5.1 the term *Engineering Data Logistics* (EDaL) and analyze key requirements for effective and efficient EDaL, such as support for clarifying data consumer and provider win conditions that may conflict or patterns for EDaL for the enrichment and backflow of engineering data in a *round-trip engineering* process. Section 8.5.2 discusses EDaL design considerations to address the EDaL requirements. A key EDaL capability is the effective organization and management of exchanging engineering data. Therefore, we derive in Sect. 8.5.3 requirements for defining and negotiating the required individual data flows between data providers and consumers.

RQ2. What Are Main Elements of an Effective and Efficient Engineering Data Exchange (EDEx) Process in Multidisciplinary System Engineering? To address this research question, Sect. 8.5.3 builds on contributions from Biffl et al. (2019c) to discuss requirements for the EDEx process collected in workshops with stakeholders at a large PSE company, and proposes steps from an artifact-based exchange toward a consumer-driven *EDEx process* that address these requirements by defining, prioritizing, and designing EDEx data flows in a project team. A key capability of EDEx is to support the data integration of heterogeneous engineering data by representing the implicit relationships between engineering data coming from different domains as foundation for a common view on and efficient sharing of data.

For designing the EDEx process, Sect. 8.5.4 builds on contributions from Biffl et al. (2019c) to adapt the *Multi-model Dashboard* approach (Biffl et al. 2014a) from constraint evaluation to EDEx and replace the design requirement of an initial common concept model, which may not be available, with direct links between consumer and provider data elements. As the manual conduct of EDEx is inefficient, Sect. 8.5.6 builds on contributions from Biffl et al. (2019c) to derive requirements for an EDEx information system that automates functions in EDEx process steps.

RQ3. What Are Main Information System Mechanisms That Enable Engineering Data Logistics for Multidisciplinary System Engineering? To address this research question, Sect. 8.5.6 derives requirements for effective and efficient *EDaL information system* (EDaLIS) mechanisms: capabilities for data set specification and for the representation of dependency relationships as foundation for data integration and transformation. We discuss in Sect. 8.6 the design of an EDaLIS that supports efficient tracing of data flows in an engineering team as foundation for analyzing the exchanged data and the EDEx process. Section 8.7 reports on an evaluation of the effectiveness and effort of the proposed EDExprocess with EDaLIS mechanisms in a feasibility case study with requirements and data from real-world use cases with domain experts at a large PSE company.

8.5 Engineering Data Logistics Process

In this work, *Engineering Data Logistics* (EDaL) is the controlled, traceable exchange of engineering data between data providers and data consumers in an engineering process, extending the concept of *Engineering Data Exchange* (EDEx) that focuses on a single data flow. EDaL describes a sequence of exchange steps that may follow a process pattern, such as *Round-Trip Engineering* (RTE). Agile EDaL workflows depend on the efficient implementation of the EDEx concepts.

Following the *design science cycle* in Wieringa (2014), we set up an initial problem investigation with workshops (Biffl et al. 2018), outlining the context and problem space of research, and deriving the following requirements for EDEx capabilities that allow addressing the challenges introduced in Sect. 8.1: *C1. Data exchange requirements are not clear or conflicting* and *C2. Heterogeneous engineering data is hard to integrate for sharing*.

Section 8.5 derives requirements for EDaL from the use cases described in Sect. 8.2 and introduces steps for an *EDaL process* to address these requirements by the EDEx process for defining single data flows in an engineering team. We propose the concept of an EDaL process that can handle different data formats and discipline-specific views and derive requirements for EDEx capabilities, such as guidance for the definition and semantic mapping of engineering data elements for exchange.

8.5.1 Requirements for an Engineering Data Logistics

From a workshop with domain experts and subsequent discussion of use cases, we derived the following requirements for an EDaL process.

Capa EDaL1. EDaL Scope Analysis The EDaL approach should guide the collection and analysis of data consumers, providers, the engineering artifacts and data they want to exchange in the scope of an EDaL use case as foundation for clarifying win conditions for both data consumer and provider, such as compensation for extra effort coming from conducting EDaL tasks.

Capa EDaL2. EDaL Use Case Analysis The EDaL approach should guide the design of an EDaL by identifying and configuring EDaL patterns to derive a sequence of individual EDEx data flows specified in an EDaL language. EDaL designs should allow addressing the use cases described in Sect. 8.2 on (UC1) engineering data provision, (UC2) sequential enrichment of engineering data, (UC3) parallel enrichment of engineering data, and (UC4) backflows of engineering data.

Capa EDaL3. EDEx Specification The EDaL approach should guide the design of an individual EDEx data flow in an EDEx language.

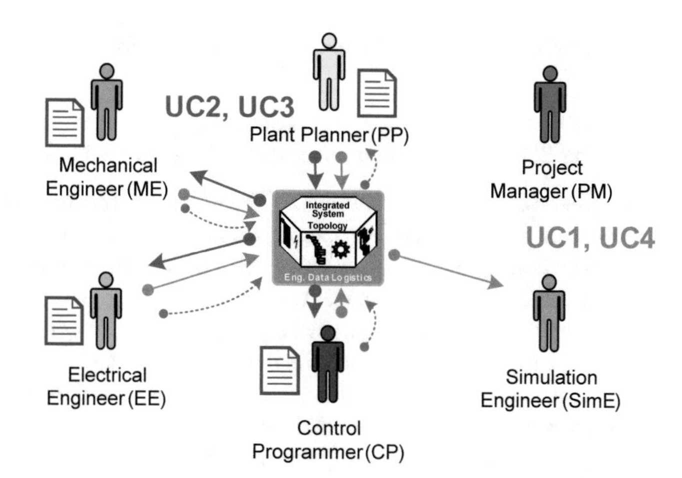

Fig. 8.4 *Engineering Data Logistics* (EDaL) process concepts to address *round-trip engineering* (RTE) requirements, based on Biffl et al. (2019b)

8.5.2 Engineering Data Logistics Design

To address the EDaL requirements in Sect. 8.5.1, we derived the following EDaL process for designing an EDaL solution.

EDaL Step 1. EDaL Requirements Analysis The role *EDaL Data Curator* conducts an analysis with candidate data consumers and providers on their data exchange requirements, the engineering artifacts and data they want to exchange. Result is an EDaL requirements document and a data processing map. This data processing map shows a network of data providers and consumers represented as nodes in the network and a set of data flows between a provider and consumer depicted as arrows (see Fig. 8.4).

EDaL Step 2. EDaL Use Case Design The *EDaL Data Curator* designs with candidate data consumers and providers use cases that address their requirements using EDaL design patterns, such as the RTE pattern and the engineering backflow pattern. Result is a use case description including an initial set of EDEx data flows, for example, data provider (artifact: data set)—>data consumer. Figure 8.4 illustrates a solution design based on a central EDaLIS that mediates the data flow between providers and consumers. Providers send their engineering artifacts into the EDaLIS that extracts the data relevant for consumers for distribution to the consumers.

EDaL Step 3. List of EDEx Flows The *EDaL Data Curator* derives a list of EDEx specification candidates from EDaL patterns. Result is a refined set of EDEx data flow specifications, for example, data provider (artifact: data set)—>data consumer, with a detailed description of the data set as a set of data elements specified in a domain-specific language, such as AutomationML.

8.5.3 Requirements for an Engineering Data Exchange Process

From a workshop with domain experts and subsequent discussion of use cases, we derived the following EDEx process requirements (see also Biffl et al. 2019c).

Capa EDEx1. Engineering Data Representation The EDEx approach should allow representing typical engineering data structures, such as tree hierarchies of the functions of a production system, (e.g., a work cell consists of devices), lists of objects (e.g., list of motors), and objects and their attributes (e.g., motor torque or rotation speed) and relationships forming networks (e.g., a work cell with an electric motor requires an electric power supply), both for data consumers and data providers. In addition, the technical data representation needs to be considered by identifying data storing and data exchange technologies that can be applied on the data consumer and data provider sides for encoding the engineering information to be exchanged.

Capa EDEx2. Semantic Link Knowledge Representation This capability concerns the representation of semantic links as a means for data integration between selected consumer and provider data elements. Representing the knowledge explicitly shall enable the automated data integration and transformation, as well as reasoning of the data.

Capa EDEx3. Process Data Representation This capability concerns the representation of metadata on the EDEx process, for example, data providers, timestamps, versioning, data quality, and validity (e.g., unclear/checked valid, invalid data).

Capa EDEx4. Consumer- and Benefit-Driven EDEx Planning The EDEx approach should be consumer-driven (with EDEx curator) and consider the likely cost–benefit of setting up and conducting a specific EDEx for prioritization in planning (not a value-neutral approach focusing on technology without considering economic benefits). The EDEx approach should help to identify what data to exchange, how to structure and integrate the data for exchanging.

8.5.4 Engineering Data Logistics Process Design

To address the required capabilities defined in Sect. 8.5.3, and the use cases described in Sect. 8.2, we introduce the main elements of an *engineering data exchange* (EDEx) process, as described in Biffl et al. (2019b), a treatment design according to Wieringa (2014), based on the knowledge gathered in workshops with domain experts. The EDEx process adapts and extends the *Multi-model Dashboard* process (Biffl et al. 2014a) in the research scope of cooperating multidisciplinary engineering work groups in a production system engineering project. The EDEx process is independent of a concrete implementation technology. This process description builds on and extends the EDEx process description in Biffl et al. (2019c).

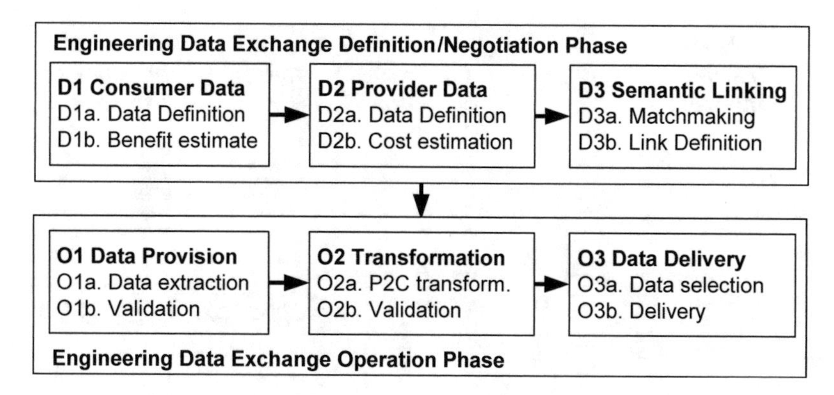

Fig. 8.5 Engineering data exchange process, based on Biffl et al. (2019b, c), © Springer 2019 (reprinted with permission)

Figure 8.5 gives an overview on the EDEx and operation phases. The EDEx operation phase assumes an agreement between data consumers and data providers on the data model and concepts for EDEx. Therefore, a negotiation of the data requested by consumers and the data published by providers is required, similar to a marketplace of well-defined data products. In this section, we introduce the roles and processes for a data negotiation marketplace as foundation for the data extraction and exchange between data providers and consumers. Key roles are the data consumer, the data provider, and the data curator. The *data consumer* requests data according to their local consumer data model from providers to conduct business processes more effectively or more efficiently. The *data provider* has artifacts that contain data that is relevant to a data consumer and knows how to extract from the artifacts this data following the local provider data model. The *data curator* has background knowledge on the business and relevant data models of all domain experts to mediate between data consumers and data providers using their local data models. The data curator has the capability to link the local data models of consumers and providers with appropriate linking formulae.

Data Exchange Negotiation Phase/Process (see illustrative example in Figs. 8.6 and 8.7). The EDEx process consists of three main steps to identify feasible and beneficial data exchange instances.

D1. Consumer Data Definition and Prioritization *D1a. Consumer Data Definition.* Project stakeholders, who want to receive data from providers, have to define their data requests. In general, domain experts in PSE have to find out where to collect the data they need for conducting their engineering processes. Therefore, these data consumers know what data is available from which data providers. Outcome of this step is a data model of the local consumer data view, for example, in UML, SysML, or AutomationML, or a sufficiently precise description in natural language based on the modeling concepts and vocabulary of the data consumer.

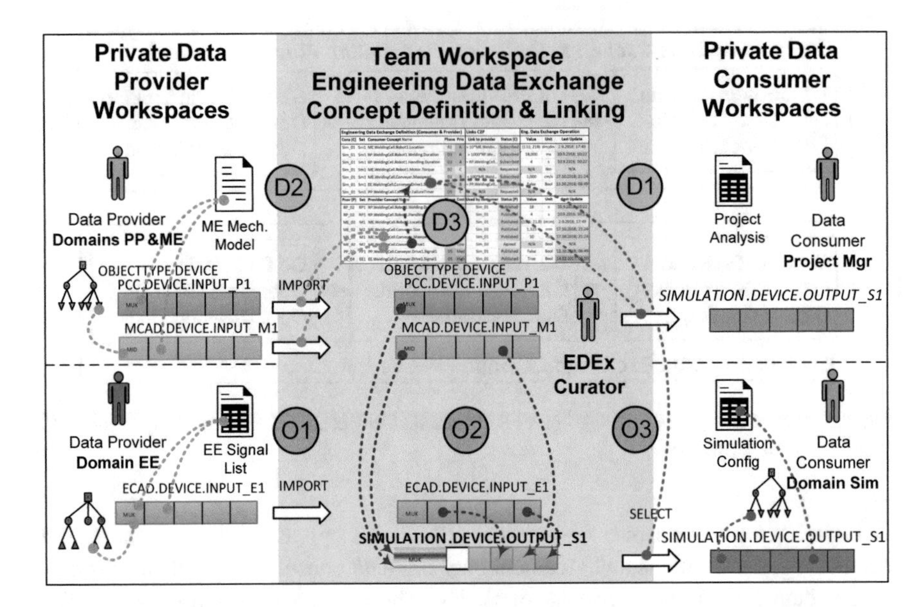

Fig. 8.6 Engineering data exchange definition/operation for one customer data set, based on Biffl et al. (2019a, b, c), © Springer 2019 (reprinted with permission); tags in green circles refer to EDEx steps in Fig. 8.5

Data Exchange Definition/Negotiation Phase | **Operation Phase**

Engineering Data Exchange Definition (Consumer & Provider)					Links C2P		Eng. Data Exchange Operation		
Cons (C)	Set	Consumer Concept Name	Phase	Prio	Link to provider	Status (C)	Value	Unit	Last Update
Sim_01	Sm1	ME.WeldingCell.Robot1.Location	R1	A	= 10*ME.Weldin...	Subscribed	(133; 218)	dm;dm	2.9.2018; 17:49
Sim_01	Sm1	RP.WeldingCell.Robot1.Welding.Duration	D3	A	= 1000*RP.We...	Subscribed	18,000	ms	10.9.2018; 10:22
Sim_01	Sm1	RP.WeldingCell.Robot1.Handling.Duration	D3	A	= RP.WeldingC	cribed	4	s	10.9.2018; 10:22
Sim_01	Sm1	ME.WeldingCell.Robot1.Motor.Tor	D2	C	N/A	uested	N/A	N	N/A
Sim_01	Sm1	ME.WeldingCell.Conveyer.Maxspee	D2	B	= 100*ME.Weld.	bed	1,000	cm	0.2018; 21:24
Sim_01	Sm1	EE.WeldingCell.Conveyer.Drive1.Signal1	D5	B	= PP.WeldingCell...	ribed	False	Bool	13.10.2018; 06:49
Sim_01	Sm1	PP.WeldingCell.Conveyer.FailureTimer	D5	C	N/A	Requested	N/A	s	N/A
Prov (P)	Set	Provider Concept Name	Phase	Cost	Used by consumer	Status (P)	Value	Unit	Last Update
RP_02	RP1	RP.WeldingCell.Robot1.Welding.Duration	D3	Low	Sim_01	Published	18	s	10.9.2018; 10:22
RP_02	RP1	RP.WeldingCell.Robot1.Handling.Duration	D3	Low	Sim_01	Published	4	s	10.9.2018; 10:22
ME_03	M1	ME.WeldingCell.Robot1.Location	R1	Med	Sim_01	Published	(13.3; 21.8)	(m	.2018; 17:49
ME_02	M2	ME.WeldingCell.Conveyer.Size	R2	Low	Sim_03	Published	1,325	m	0.2018; 21:24
ME_02	M1	ME.WeldingCell.Conveyer.Maxspeed	D2	Low	Sim_01	Published	10	m/s	17.10.2018; 21:24
ME_02	M2	ME.WeldingCell.Conveyer.Drive1	D1	Low	Sim_03	Agreed	N/A	Bool	N/A
PP_03	PP1	PP.WeldingCell.Conveyer.Drive1.Signal1	D5	Med	Sim_01	Published	False	Bool	13.10.2018; 06:49
EE_04	EE1	EE.WeldingCell.Conveyer.Drive1.Signal1	D5	High	Sim_01	Published	True	Bool	14.12.2017; 06:50

Fig. 8.7 EDEx definition/negotiation and operation overview table, based on Biffl et al. (2019a, b, c), © Springer 2019 (reprinted with permission); tags in green circles refer to EDEx process steps in Fig. 8.5

D1b. Cost–Benefit Estimate and Prioritization. The EDEx curator validates with the consumer the definition of the requested data and estimates the likely benefit and cost of providing the data in order to focus on the most relevant EDEx instances first. Outcome of this step is a set of data model elements in the local consumer data view, with a semantic description that is understandable both to the EDEx curator and prospective data providers based on the modeling concepts and vocabulary of the EDEx curator (see Fig. 8.7 for examples). Note that this step can be repeated, if data consumers need to define additional data elements later in the project. A required mechanism for this step is a team workspace (see Fig. 8.6) that allows sharing the data requests on project level with prospective data providers. The EDEx overview (see Fig. 8.7, tag D1) shows the status of the data elements agreed for provision.

D2. Provider Data Definition and Cost Estimation *D2a. Provider Data Definition as Source for Consumer Data.* An EDEx provider can react to consumer data requests by *agreeing* to publish data that is semantically equivalent to (parts/aspects of) the requested consumer data. In general, providing the data will involve extracting the data elements from suitable engineering artifacts, often export results from a specific engineering tool, for example, the mechanical structure of a work cell. Outcome of this step is a set of data model elements in the local provider data view, with a semantic description that is understandable both to the EDEx curator and prospective data providers based on the modeling concepts and vocabulary of the data provider (see Fig. 8.7 for examples).

D2b. Data Provision Cost Estimation. Extracting data from engineering artifacts can take significant effort and cost, even to an expert. Therefore, the data curator has to validate that the provided data is equivalent to (relevant parts/aspects of) requested data items and elicit the likely cost for data extraction and transformation in a format that is suitable for EDEx, such as AutomationML. Outcome of this step is feedback to the provider whether the data is of sufficient quality and cost to continue setting up the EDEx. The EDEx overview (see Fig. 8.7, tag D2) shows the status of the data elements *agreed* for provision.

D3. Consumer-Provider Mediation and Semantic Link Definition *D3a. Economic Matchmaking Between Data Consumers and Providers.* For each promising consumer data request, the EDEx curator tries to find sets of data providers that would allow providing the requested data. These candidate providers should cover both the required and available data and the technical capability (for example suitable data exchange formats) applicable to exchange data. In the simplest case, one provider can provide the requested data in exactly the required data format. However, in typical cases, the data elements will come from several data providers in a variety of data formats (see the example in Fig. 8.6). Outcome of this step is a set of EDEx providers that could, together, provide the input data for transformation into the requested data elements. If there are several solutions, the solution options could be ranked by data quality and cost considerations.

D3b. Semantic Linking Between Consumer and Provider Data Models. For a suitable set of data providers that would allow providing the requested data, the EDEx data curator tries to establish for each requested data item a formal semantic link, that is, a formula that specifies how to calculate the consumer data item value from one or more *published* provider data item instances using the modeling concepts and vocabulary of the EDEx curator. A semantic link formula can describe in a simple case, semantic identity between provider and consumer data elements. More advanced semantic relationships (Kovalenko and Euzenat 2016) include basic string operations, mathematical calculations, and parameterized function calls to semantic transformation algorithms (see Fig. 8.9). Outcome of this step is a set of customer data, semantically linked to a set of provider data as foundation for designing the EDEx operation. The EDEx overview table (see Figs. 8.6 and 8.7, tag D3) shows the status of the linked data elements. The EDEx process provides the foundation for conducting the *EDEx Operation* process.

EDEx Operation Phase/Process (see illustrative example in Figs. 8.6 and 8.7). EDaLIS data structure of consumers subscribing to provider data enables flexible data exchange in engineering.

O1. Data Provision and Validation *O1a. Data Extraction and Transformation.* The data provider extracts the data elements as agreed in the EDEx Negotiation process from their local engineering models and/or engineering tool outputs. Then the data provider transforms the extracted data into a data model and format that the EDEx IS can import (see Fig. 8.6, tag O1). Outcome of this step is a data set for import into the EDEx IS.

O1b. Traceable Validation of Data Provision to Data Logistics. The data provider and the EDEx curator agree on a procedure to validate the data from extraction to input to the EDExIS to ensure that only correctly transformed data is imported. The data curator imports valid data into the EDEx IS. Outcome of this step is the import of valid data into the EDEx IS and feedback to the data provider on the validity of the provided data. The EDEx overview (see Figs. 8.6 and 8.7) shows the status of the imported data elements.

O2. Data Transformation and Validation *O2a. Semantic Transformation of Provider to Consumer Data Model.* The EDEx IS propagates the provided data along the semantic links to fill in or update consumer data sets (see Fig. 8.6, tag O2).

Outcome of this step are updated consumer data sets.

O2b. Validation of Semantic Transformation. The data curator can follow the propagation of the provided data along the semantic links to consumer data sets to check the correctness of the transformation. Outcome of this step is feedback on the validity of the semantic transformation of the most recently imported provider data set.

O3. Data Selection and Delivery *O3a. Data Selection by Consumer.* The data consumer selects consumer data instances by providing the EDaLIS with the type of requested data and information to select the desired data instances, such as data identifiers or selection conditions, similar to an SQL query to a database. Outcome of this step is a set of selected data in the EDaLIS for delivery to the data consumer.

O3b. Data Delivery from Data Logistics to Consumer. Finally, the EDExIS delivers the result data to the data consumer (see Fig. 8.6, tag O3). Outcome of this step is the data set at the consumer in the agreed upon data format.

Illustrating Use Cases Figure 8.6 introduces the roles, engineering artifacts, and exchanged data for the EDEx definition/negotiation and operation processes (see Fig. 8.5) for one consumer data set, in this case device parameters collected for the *SimE*. The data providers and data consumers, such as the PP, ME, EE, and *SimE*, operate in *private workspaces*. The *team workspace* contains shared data views as foundation for preparing and operating the EDEx processes (Biffl et al. 2019c).

Parameter Exchange for Production System Simulation The *SimE* needs a set of parameters from data providers to set up the simulation model for a device (see Fig. 8.6, lower right-hand part, red bar), such as a robot or conveyer. The *SimE* requests the set of parameters from providers, such as the PP, ME, EE, and CP, who in return can agree to publish their local engineering data fitting the consumer request (see Fig. 8.7, left-hand part). The *EDEx curator's* task is to link the set of requested parameters requested with the published parameters from the providers PP, ME, EE, and CP (see Fig. 8.7, middle part for the ME and EE data).

During the EDEx operation phase, the *team workspace* receives updates of provider data instances in engineering artifacts from the *private workspaces* of the PP, ME, EE, and CP (see Fig. 8.6, left-hand side for the ME and EE) and transforms this input data according to the semantic links into output data for delivery to the *SimE* (see Fig. 8.6, right-hand side, and example output data in Fig. 8.7, right-hand upper part). Notifications of new or changed data are possible, so the *SimE* can consider when to retrieve which part of the currently available data.

Production System Engineering Project Monitoring The advantage of the PM in this use case consists of simple or advanced analyses. A simple analysis could be to subscribe to the same data sets as the *SimE* and analyze at specific points in the project for which data elements the engineering data is expected but missing.

Figure 8.7 illustrates the EDEx overview table during operation: provided data instances have been processed according to the linking formulae to fill in data instances for consumers (tags O1, O2, O3). For consumers, the EDEx overview (tag D1) shows the status of the data elements as *requested*, *agreed* for provision, or *subscribed* for delivery. The EDEx overview table (tag D3) shows the status of linked data elements. For a requested data element, there may be several providers; therefore, the EDEx overview table (see Fig. 8.7) indicates the cost of providing a data element and the engineering process phase, in which the data will be available with sufficient precision, to support making an informed choice on the best provider. For example, *EE*... *Signal1* could be obtained from *PP*... *Signal* at lower cost.

The concepts illustrated in Figs. 8.6 and 8.7 are the foundation for prototype designs as input to the evaluation with domain experts in Sect. 8.7.

8.5.5 Data Modeling Across (AML-1) and Within (AML-2) Engineering Disciplines

One necessary foundation of the EDaL is the appropriate modeling of the engineering data within the different involved disciplines and within the data logistics to bridge gaps between disciplines. Lüder et al. (2018b) presented a suitable view-based approach. Obviously, within the EDaLIS only the engineering artifacts of data providers and consumers and the central data storage are relevant to be considered. The data model of the central data storage has to be the union of the engineering data models of the individual tools.

Thus, in the case of the intended EDaL, we can postulate two types of data models, Type 1 models and Type 2 models that can be represented by AutomationML. Type 2 models (identified as AML-2 data models) correspond to the engineering artifact data models of the involved engineering tools, typically local to an engineering discipline. AML-2 data model scan be modeled by a tool-related set of role classes and interface classes to cover the relevant conceptual objects and system unit classes to represent their hierarchical structuring. Type 1 models (identified as AML-1 data models) correspond to the data model of the central data storage. They represent the union of all sets of AutomationML *role classes* and interface classes of the involved Type 2 models and Type 1 special AutomationML *system unit classes* to represent all possible hierarchical structures.

8.5.6 Requirements for an Engineering Data Logistics Information System

From a workshop with domain experts and subsequent discussion of use cases, we derived the following requirements for EDExIS (Biffl et al. 2019c) and EDaLIS mechanisms.

Capa EDExIS1. EDEx Management and Overview The EDExIS has to provide the capabilities of the *EDEx overview table* illustrated in Fig. 8.7, including *EDEx definition functions* to request, agree on providing, publishing, and subscribing to data elements (see EDEx process steps D1–D3), as well as setting relevant attributes of and searching the table for understanding the status of the EDEx definition in the project team.

Capa EDExIS2. EDEx Data Definition Languages The EDExIS has to process the languages for the specification of consumer and provider data sets using the modeling concepts and their vocabulary, and the language for semantic link definition specifying (a) the dependencies between consumer and provider data sets and (b) the transformation of imported provider data into consumer data best based on the modeling concepts and vocabulary of the EDEx curator.

Capa EDExIS3. EDEx Operation Capabilities The EDExIS has to be able (a) to import and validate provider data, (b) to store imported data versions including their metadata for processing, (c) to analyze the data and semantic links in order to correctly propagate the provider data to consumer data structures, and (d) to select and export consumer data.

Capa EDaLIS1. Validation and Versioning of Exchanged Engineering Data The EDaLIS should provide the capability to define validity conditions for exchanged data elements as foundation for checking the validity of data elements along the EDaLIS process. The EDaLIS should provide the capability to compare engineering data versions as foundation for the detection and analysis of changes.

Capa EDaLIS2. Consistency Checking and Change Propagation The EDaLIS should provide the capability to define consistency checks between semantically related provider data, knowing that these relationships and checks may differ according to the engineering phase, for example, inconsistencies or large differences between disciplines may be okay in an early engineering phase but not in a later engineering phase. The EDaLIS should provide the capability to check the consistency between semantically related provider data and report the results as foundation for a systematic conflict detection and resolution process. The EDaLIS should provide the capability to define rules for change propagation between semantically related provider data as foundation for a semi-automated change propagation process.

Capa EDaLIS3. Provider and Consumer Notification The EDaLIS should provide the capability to define notifications to providers and consumers on changes that are relevant to them as foundation for awareness in the engineering team on changes to relevant data and for analyses that support the effective and efficient resolution of missing, invalid, or inconsistent data while preventing unwanted notifications.

Together, the capabilities for an EDExIS and an EDaLIS provide the foundation for considering information system design options.

8.6 Data Logistics Information System Design

This section discusses main design elements for effective and efficient *EDaL information system* (EDaLIS) mechanisms to address the requirements identified in Sect. 8.5.6 on capabilities for data set specification and for the representation of dependency relationships as foundation for data integration and transformation. We discuss main design elements of an EDaL information system to provide these engineering data exchange capabilities for automating the EDaL process. As there is no suitable out-of-the-box technology to link discipline-specific views on data, we introduce a software architecture with a data model based on *AutomationML* data models that address these challenges.

The EDaLIS provides capabilities for the EDEx operation phase (Biffl et al. 2019a). We assume that the EDaLIS can handle AutomationML (AML) files in the so-called AML-2 and AML-1 formats (Biffl et al. 2019a). The data curator models AML-2 and AML-1 templates in the *EDEx Definition Phase*.

The AML-1 data model defines the central/core model of the EDExIS to transform data between several providers and consumers. Therefore, the AML-1 data model needs to represent several discipline-specific hierarchies that share some common concepts, such as machines or devices. The AML-1 data model consists of a set of AutomationML *RoleClasses*, *InterfaceClasses*, and *SystemUnitClasses*. *RoleClasses*, and *InterfaceClasses* define the data types and elements for discipline-specific hierarchies that the data curator can build on to define EDEx data flows.

An AML-2 data model defines a discipline-specific view on a provided engineering artifact. AML-2 uses a subsect of the AML *RoleClasses*, *InterfaceClasses*, and *SystemUnitClasses* defined in the AML-1 core model to model the structure and content of an engineering artifact. The AML-2 data model is the foundation to configure a transformer that transforms an input engineering artifact into an AML-2 data structure. Therefore, there is a specific AML-2 data model for each engineering artifact.

8.6.1 EDaLIS System Components

Figure 8.8 gives an overview on the conceptual system design of an EDaLIS. The EDaLIS consists of two main components: The service-oriented *backend* exposes the system capabilities, and the *web application* is the entry point for data consumers and data providers (and the data curator).

The *web application* represents the *EDEx team workspace* consisting of several pages for data consumers and providers, such as the *Data ImportPage*, the *Project Browser*, and the *Data Export Page*. The web application communicates via software interfaces and the EDaLIS service API with the *backend*, which consists of the *Data Import Service*, the *CoreModel Service*, the *Merge Service*, the *Transformation Service*, the *Validation Service*, the *Merge Service*, the *Data Repository*, a *Workflow Engine*, a *Rule Engine*, and the *Data Export Service*. The *EDEx team workspace* facilitates the import of provider data by communicating via the EDaLIS service API, as well as the export of required data to data consumers. The *Project Browser* allows to display an overview on the AML-1 data as well as data analyses, for example changes to single data instances.

In the backend, the *CoreModel Service* orchestrates the communication with the different services: AML-2 input data is validated by the *Validation Service* and *Rule Engine*; changes to data in the AML-1 core model are compared via the *Compare Service* and merged by the *Merge Service* to achieve a consistent new AML-1 data version for storing in the repository.

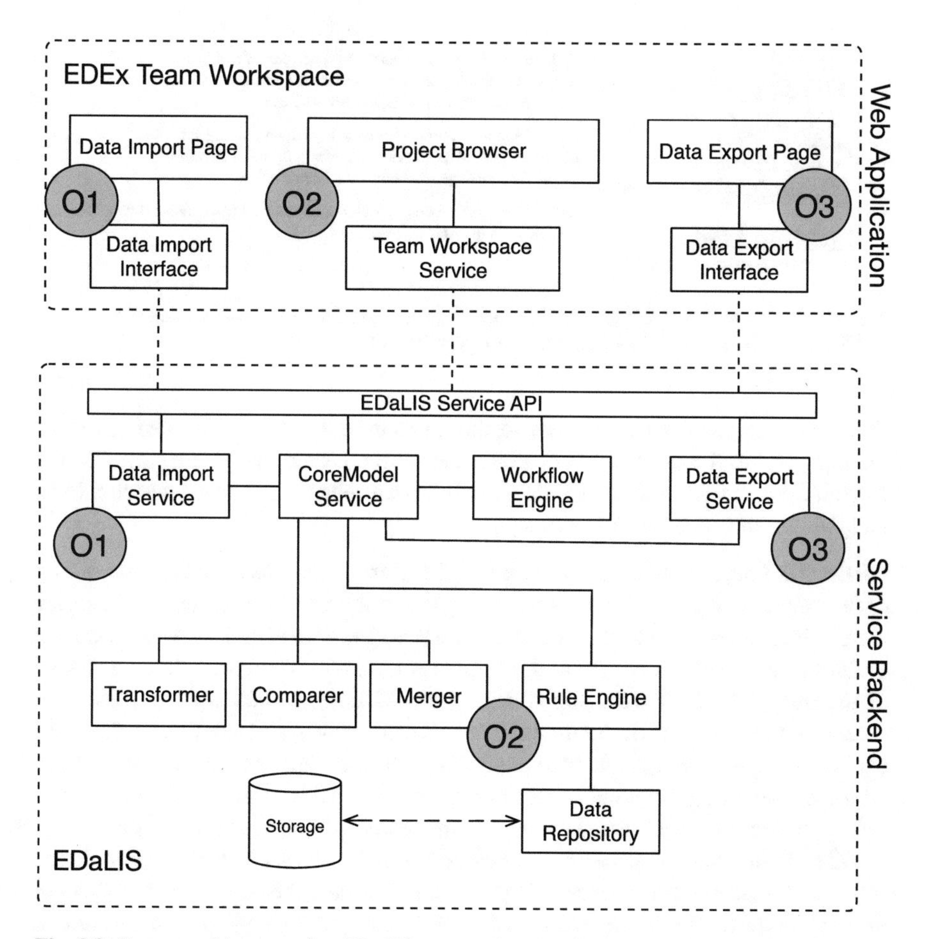

Fig. 8.8 System architecture of an EDaLIS system, based on Biffl et al. (2019b)

8.6.2 EDaLIS Contributions to the EDEx Operation Phase

In this subsection, we assume the *EDEx Definition Phase* (Sect. 8.5.4, steps D1–D3) to be completed by the involved data providers and data curator. According to the EDEx operation phase, we discuss for each step the EDaLIS contributions.

EDEx O1. Data Provision and Validation *O1a. Data Extraction and Transformation.* The data provider prepares an engineering artifact for import into the EDaLIS web application via the *Data Import Page*. In the future, the EDaLIS can be integrated with engineering tools to automate this process step by automatically transforming the engineering tool data into AML-2 and importing the AML-2 data into the system.

O1b. Traceable Validation of Data Provision to Data Logistics. The data provider uploads the engineering data via the web application, from where it is transported to the backend for transformation and *validation*. If the data is valid, the *Compare*

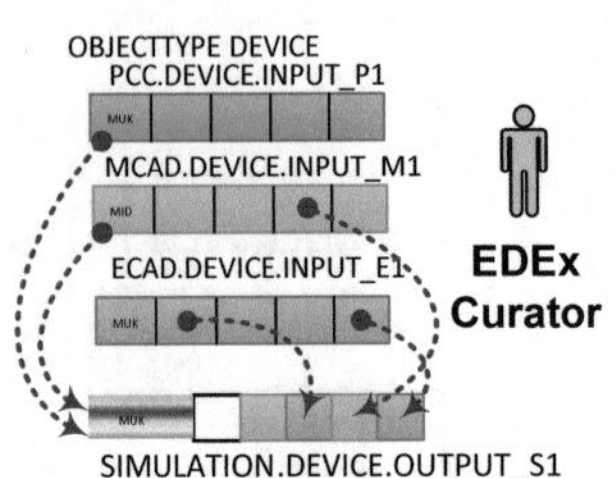

DL1. <consumer data element> := <provider data element>
Sim_01.EE.WeldingCell.Conveyer.Drive1.Signal1 :=
EE_04.EE.WeldingCell.Conveyer.Drive1.Signal1

DL2. <consumer data element> := a*<provider data element> + b
Sim_01.RP.WeldingCell.Robot1.Welding.Duration := 1000*
RP_02.RP.WeldingCell.Robot1.Welding.Duration

DL3. <consumer data element> := function f(set of <provider data element>)
Function call to calculate result

Fig. 8.9 Semantic link definition between consumer and provider data models, based on Biffl et al. (2019b, c), © Springer 2019 (reprinted with permission)

Service compares the new dataset to the current core model. The result is a list of changes that the provided data would cause to the core model, displayed in the *Project Browser*, from which the data provider can select the changes that should be merged into the AML-1 core data repository.

EDEx O2. Data Transformation and Validation *O2a. Semantic Transformation of Provider to Consumer Data Model*. The selected changes from the provider data are merged into the AML-1 core model data with the *Merge Service*. The *Transformation Service* links the provider data to the consumer data sets. Therefore, the role classes of the provider data were mapped during the *EDEx Definition Phase* to the role classes in the core model. Figure 8.9 displays examples of semantic link definitions between consumer and provider data models. In simple cases, the transformation just requires converting the input values to appropriate scales of units, more advanced links may require defining and evaluating the results of complex algorithms.

O2b. Validation of Semantic Transformation. In this step, the validity of the semantic transformation is checked by the *Rule Engine*, for example, that all links in the core model are set correctly. The data curator can check the validity of the content in the AML-1 repository against the original input data in the provided engineering artifact.

O3. Data Selection and Delivery *O3a. Data Selection by Consumer*. The consumer can request certain data via the project browser. This can be realized by SQL[3]-like queries, or similar to XPATH[4] to specify the required data.

O3b. Data Delivery from Data Logistics to Consumer. The requested data can finally be exported to a consumer AML-2 representation and downloaded via the EDaLIS *Data Export Page* to the private workspace of the consumer.

[3]https://standards.iso.org/ittf/PubliclyAvailableStandards/c053681_ISO_IEC_9075-1_2011.zip
[4]https://www.w3.org/TR/xpath-31/

8.6.3 EDaLIS System Support for EDaLIS Mechanisms

According to the requirements in Sect. 8.5.6, we investigate the capabilities of the EDaLIS mechanisms.

Capa EDExIS1. EDEx Management and Overview *Private and Team Workspace.* The EDaLIS serves as an interface between the private and the team workspaces. All involved parties (data curator, provider, and consumer) have one single point of entry for the required data, the *EDEx team workspace.* The workspace can display both discipline-specific views as well as the common model. This can be achieved by implementing a web application using the *Spring Boot*[5] framework for resource and service orchestration as well as providing the REST-interfaces[6] (*Representational State Transfer*). The system can also manage requests and subscriptions of consumers, and publishing data by providers, e.g., based on the *publish-subscribe* design pattern.

Capa EDExIS2. EDEx Data Definition Languages The discipline-specific views share common concepts, such as machines, devices, and signals, which link the views across disciplines; however, the discipline-specific views differ in their hierarchies, such as the hierarchy of the mechanical structure for the ME, the hierarchy of electrical circuit areas for the EE, or the hierarchy of software functions for the CP. Furthermore, links, interfaces, and roles need to be displayed. An adequate EDEx data definition language needs to facilitate the appropriate representation of such data structures. File formats such as CAEX or AutomationML (AML) were developed for such industry use cases (Lüder et al. 2018b).

Capa EDExIS3. EDEx Operation Capabilities *Data Import.* The EDaLIS needs to be able to import and transform provided according to the agreed upon data formats and common concepts. Data input formats could be CSV spreadsheets or XML files, or AML-2 models (Biffl et al. 2019a). Combined with the data export this allows addressing *UC1 Artifact provision* introduced in Sect. 8.2.1.

Storing of Input Data. Another essential capability is to store the input data in order to process it. The transfer can be handled by REST or a similar data transfer protocol to an XML database to store it in AML-1 data structure, a graph consisting of linked discipline-specific trees.

Analysis of Data and Semantic Links. The logic of the EDaLIS must be capable to analyze the input data and make the semantic links to transform the provider data to the given consumer data structures to represent them in the AML-1 core model. This transformation requires semantically similar attributes and identifiers (common concepts) and can be specified for processing by *XPath* for accessing XML-based files.

[5]http://spring.io/projects/spring-boot

[6]https://www.w3.org/TR/2004/NOTE-ws-arch-20040211/#relwwwrest

Selection and Export of Data. The enrichment of data in the context of RTE is an essential feature of an EDaL process, to enable the backflow of information. Therefore, the EDaLIS also needs to provide an export function, to download specific and general views of the core model in valid file formats such as AML-1 or CSV/XML-files.

Capa EDaLIS1. Validation and Versioning of Exchanged Engineering Data *Validator of Input Data.* The core model service validates the provider data by checking the engineering artifacts with the given core model. Due to the parallelism of the multiple disciplines this validation is essential to verify, for example, attributes to be compliant with the current core model. If this is not the case the merge process cannot be completed.

Versioning of Input Data. Versioning of engineering data is an important aspect of the EDaLIS. The repository allows storing and versioning engineering data (including metadata) as commonly known in software engineering by version control systems to compare different engineering data versions. Commonly used version control technologies such as *Git*[7] or SVN[8] can be used. Clear benefits are the traceability of changes and possibility to roll back to older versions if inadequate or incomplete data has been imported by a data provider, which enables *UC2b sequential enrichment with updates* in Sect. 8.2.

Capa EDaLIS2. Consistency Checking and Change Propagation A language such as the *object constraint language*[9] (OCL) can be used in the system design to check automatically whether the engineering data complies with previously defined constraints, for example, that all semantic links have to be connected in the AML-1, or to identify and resolve dead links. The system should be able to check whether specific attributes or elements have valid attributes, for example, rotation speed cannot be negative, which could also be used to model dependencies between components. Change propagation is executed by the display of the change attributes. Together, these mechanisms enable *UC3 parallel enrichment of engineering artifacts* in Sect. 8.2.1.

Capa EDaLIS3. Provider and Consumer Notification For this use case, a workflow engine such as *Camunda*[10] or *Activiti*[11] can be integrated into the EDaLIS to further automate the EDaL process, for example, by automatically notifying providers that data is requested from them or consumers as soon as the data provider has imported the required data into the system.

All together, the EDaLIS supports the EDaL use cases, implementing all mechanisms required for addressing *UC4 backflows of artifacts* in Sect. 8.2.1.

[7]https://git-scm.com

[8]https://subversion.apache.org

[9]https://www.omg.org/spec/OCL/

[10]https://camunda.com/de/

[11]https://www.activiti.org

8.7 Evaluation

This section derives a conceptual evaluation for the EDaL, and reports on the evaluation of the *engineering data exchange* (EDEx) process and requirements (a) in an initial feasibility study with domain experts at a large *production systems engineering* (PSE) company, a systems integrator for metallurgic production systems, and (b) in a cost–benefit comparison of the EDEx definition and operation processes to the traditional process of point-to-point exchange of engineering artifacts between domain experts, closing an iteration of the design cycle (Wieringa 2014) and providing knowledge for guiding future research (Biffl et al. 2019b, c).

8.7.1 Conceptual Evaluation of the EDaL Process Study

Goal of the conceptual evaluation is to discuss to what extent the EDaL process, introduced in Sect. 8.5.2, allows addressing the EDaL requirements, defined in Sect. 8.5.1, regarding the use cases, introduced in Sect. 8.2.

EDaL Step 1, EDaL requirements analysis, addresses the capability *EDaL scope analysis* by systematically collecting the candidates for data consumers and providers as well as an initial set of data that could be exchanged between consumers and providers. Chapter 7 in this book describes a method for deriving a data processing map that helps identify engineering artifacts and the relevant engineering data they contain.

EDaL Step 2, EDaL use case design, addresses the capability *EDaL use case analysis* by systematically considering process design patterns, such as the RTE pattern and the engineering backflow pattern, to identify a complete set of EDEx data flows for a use case context. The simple EDaL language describing an EDEx flow as *data provider (artifact: data set)—>data consumer* allows defining the main elements for a data flow as foundation for a more detailed analysis in the EDEx process definition phase. This approach goes beyond the EDEx approach to ensure that all relevant EDEx data flows are considered to address an EDaL use case.

EDaL Step 3, List of EDEx flows, details the data flows as input to the EDEx definition phase as foundation for the consumer-driven design and implementation of the data flows. The EDaL list of data flows can act as a checklist to ensure the EDEx process to finally result in a network of EDEx data flows that allow fulfilling the EDaL design pattern required for addressing the required use cases.

Therefore, the EDaL process elements based on the underlying EDEx process, allows addressing the general use cases introduced in Sect. 8.2 *UC1 Artifact provision*, *UC2/3 Sequential/Parallel enrichment of artifacts*, and *UC4 Backflows of artifacts* as well as the specific use cases *UC Sim. Data exchange for production system simulation* and *UC PM. Production system engineering project monitoring*.

8.7.2 Feasibility Study of EDEx Process

We evaluated the basic concept of the EDEx process with domain experts by following the steps of the EDEx process description (see Sect. 8.5.4, Fig. 8.5, and Biffl et al. (2019c)). Based on the use cases in Sect. 8.2.2, we designed prototypes of selected user interface elements, such as the overview table, data specification, linking, and retrieval as electronic mock up artifacts with data from domain experts. We collected data on the usability and usefulness of the EDEx process based on the *Technology Acceptance Model* questionnaire (Davis 1985; Biffl et al. 2019a).

Further, we developed technology prototypes of the EDaLIS capabilities to explore the feasibility of designing the EDaLIS concepts with available technologies, including *AutomationML* for data specification (Lüder et al. 2018a), an *Excel* dialect for the specification of dependency links, *Java* code for transformations, and *BaseX* as data storage. We conducted and discussed the EDEx steps in a workshop with domain experts representing the roles *data provider* (PP, ME, EE, CP in the use cases), *data consumer* (*SimE*, PM), and *EDEx curator*.

Overall, the domain experts found the EDEx process feasible, useful, and usable for basic cases that make up most of the data exchange use cases in their typical project context, assuming that the EDaLIS provides effective tool support to automate the data transformation, storage, and selection tasks. The domain experts provided improvement suggestions for the user interfaces, and for describing the data transformation and linking formulae in their context. Further, the domain experts noted that more complex cases may take considerable effort to design and automate; therefore, cost-benefit estimates in the EDEx process are important to guide planning the EDEx implementation. Nevertheless, they indicated that more advanced cases, such as the EDaL use cases, described in Sect. 8.2.1, will also enable more advanced engineering data usage by exploiting trusted and quality-ensured data that enables the automation of engineering steps leading to significant cost reductions within the overall engineering process.

8.7.3 Cost–Benefit Considerations

Costs and benefits of the EDEx process via a *team workspace* in comparison to the traditional manual process of point-to-point e-mail-based EDEx are evaluated (see also Biffl et al. (2019c)). Needs and estimates from domain experts are used, who are responsible for engineering and project management of large-scale metallurgic production system projects.

The results are presented in Table 8.1 for the EDEx process steps of the use case *parameter exchange for production system simulation* by comparing the benefits, that is, correct and useful results for a task, and the cost, that is, the effort in person hours for processing a set of typical inputs, of a stakeholder conducting a task. We applied a 5-point Likert-Scale (++, +, 0, −, −−), where "++" indicates very

Table 8.1 Comparison of the benefit, cost, and risk of traditional manual and EDEx processes, based on Biffl et al. (2019b, c), © Springer 2019 (reprinted with permission)

	Benefit		Cost	
EDEx process step	Manual	EDEx	Manual	EDEx
D1. Consumer data definition and prioritization	0	+	+	0
D2. Provider data definition and cost estimation	−−	+	+	−
D3. Consumer-provider semantic link definition	−−	++	n/a	−
O1. Data provision and validation	−	0	+	0
O2. Data transformation and validation	−−	+	−−	+
O3. Data selection and delivery	−	++	−−	++
Risks from EDEx to engineering project				
Risk of unplanned rework due to defects in EDEx	−−	++	−−	++
Risk of defects in engineering from low-quality EDEx	−	+	−	+

Legend: ++ very good, + good, or average, − weak, −− very weak

positive effects, and "−−" very negative effects. Positive effects refer to high benefit of the investigated approaches, to low cost for implementation and application, and to low risk from EDEx to the engineering project.

Regarding benefit, the EDEx process was found effective to very effective by the interviewed stakeholders, both providers and consumers. The reason being they were able to exchange data elements in a traceable and validated way. In the traditional approach, the data consumers had to define, procure, transform, and validate the required data with significant cost and prone to errors. The downside is, however, that the application of the EDEx process incurs extra cost, especially during the EDEx definition (D2) and linking (D3), in particular for providers and for the new role of the *EDEx curator*.

On the other hand, the linking step (D3) significantly improves the representation of shared knowledge in the engineering team over the previously implicit dependencies between the engineering roles. Domain experts and the PM can always get a current overview on the status of data deliveries and can identify missing engineering data and unfulfilled requests by consumers. In addition, the EDaLIS can provide the benefit of immediate feedback on changed engineering data elements efficiently, without additional cost to the domain experts.

Regarding risk from EDEx to the engineering project, the participants know that traditional EDEx often leads to unplanned rework due to defects in EDEx and expect major benefits from a lower rate of defects introduced from EDEx and cost reductions from less effort for unplanned rework. Further, the participants know that traditional EDEx may lead to defects in engineering artifacts, for example, due to divergent views in the engineering disciplines from infrequent or incomplete EDEx. The participants expect from low-cost EDEx a faster synchronization between the disciplines, which lowers the risk of defects and facilitates more agile PSE as changes can be propagated both correctly and faster.

8.8 Discussion

This section discusses the evaluation results regarding the research questions introduced in Sect. 8.4, and extending Biffl et al. (2019c).

RQ1: *What are main elements of an Engineering Data Logistics (EDaL) approach in round-trip System Engineering?*

Section 8.5.2 introduced the EDaL elements in the process for EDaL, an EDaL *curator* identifying data providers and consumers, their candidate engineering artifacts, and data to exchange according to EDaL design patterns, described by an EDaL specification language. In an initial conceptual evaluation, we found the EDaL approach adequate to address the core use cases for *round-trip System Engineering* introduced in Sect. 8.2.1, assuming an effective underlying EDEx process for the data flow between a consumer and her data provider(s).

As a next step in the design science approach, the initial conceptual evaluation will be the foundation for an empirical study to investigate what methods and mechanisms typical domain experts will require to apply the EDaL approach effectively in their engineering context.

RQ2: *What are main elements of an effective and efficient engineering data exchange (EDEx) process in Multidisciplinary System Engineering?*

Section 8.5.4 introduced as main EDEx process elements EDEx roles, process steps, and data structures. The new role of the *EDEx curator* mediates between data consumers and providers. In the feasibility study, a domain expert filling this role informally was identified. The EDEx data structures represent the necessary knowledge on engineering data, semantic links between consumer and provider data, and the status on the EDEx process as foundation for effective EDEx for the use cases introduced in Sect. 8.2.2 and according to the required capabilities for EDEx in multidisciplinary engineering, discussed in Sect. 8.5.3. Further, the EDEx process facilitates efficient EDEx (a) by considering the benefits of EDEx for consumers and the cost for providers to focus first on the data sets with the best cost–benefit balance and (b) by automating the EDEx operation with support by the EDaLIS.

As potential drawback of the EDEx process, the domain experts noted the need to convince data providers to take over the task and extra effort of extracting requested data from their engineering artifacts. For this task, specific tool support will be required according to the project context as well as appropriate compensation for the extra effort. A company internal cost balancing scheme shall be investigated, enabling the transfer of cost reductions at consumer side to the data provider side that can be organized by the EDEx curator (see also Biffl et al. 2019c).

From a data model point of view, the local data models in a discipline-specific view of a provider or consumer are, in general, trees. The common model in EDaLIS links these trees to a graph via semantically equivalent concepts, such as system part, device, or signal. However, the effective and efficient identification of relevant semantically equivalent concepts may take considerable effort and requires research on methods for supporting the EDEx data curator.

RQ3: *What are main information system mechanisms that enable engineering data logistics for Multidisciplinary System Engineering?*

The EDaLIS mechanisms for management and overview, data definition languages, and operation capabilities addressed the requirements for EDEx capabilities in Sect. 8.5.6 on a conceptual level. Together, the EDaLIS mechanisms facilitate efficient round-trip engineering among domain experts, that is, the enrichment of common engineering concepts in iterations from several disciplines (use cases 2 and 3 in Sect. 8.2.2), as the domain experts may act both as consumers and providers. The design of an operational EDaLIS will have considerable impact on the efficiency of the EDEx process in the application context and requires further investigation regarding the interfaces to domain experts and their tools, regarding the languages to specify EDaL and EDEx aspects, and regarding data structures to process and store the data required for addressing the EDEx and EDaL use cases.

Limitations As all empirical studies the presented research has some limitations that require further investigation (see also Biffl et al. 2019c).

Conceptual Evaluation of EDaL We evaluated the EDaL concepts with typical use cases in the context of a large PSE company. However, these use cases may be specific to the company and not representative for typical PSE companies. Therefore, we plan to evaluate the EDaL concepts in a wider set of representative PSE companies.

Feasibility Study We evaluated the EDEx process approach with focus on specific use cases in cooperation with domain experts in a typical large company in PSE of batch production systems that can be seen as representative for systems engineering enterprises with project business using a heterogeneous tool and technology landscape. The evaluation results are based on observations from a limited sample of projects, stakeholder roles, and data models. To overcome these limitations, we plan a more detailed investigation in a wider variety of domains and application contexts.

The *expressiveness of data specification and linking languages*, used in the evaluated prototype, can be considered as a limitation. The prototype is able to address an initial set of simple data types, while industrial scenarios showed that value ranges and aggregated ranges have to be expressible in the desired data and link languages for specification and validation. While the evaluation worked well with data provided in tables, the evaluation of advanced data structures such as trees or graphs remains open.

8.9 Summary and Outlook

This section summarizes the findings of the book chapter and proposes future research work. Digitalization in *production system engineering* (PSE) (Vogel-Heuser et al. 2017) aims at enabling flexible production toward the Industry 4.0 vision and at shortening the engineering phase of production systems. This results in an increase of parallel PSE, where the involved disciplines have to exchange updates of engineering information for synchronization due to dependency constraints between the engineering disciplines.

In this chapter, we introduced and investigated PSE use cases for *engineering data logistics* (EDaL) and, based on Biffl et al. (2019c), for the *engineering data exchange* (EDEx) process to provide domain experts in parallel PSE with a systematic approach to define and efficiently exchange agreed upon sets of data elements between heterogeneous local engineering models as foundation for agile, traceable, and secure PSE. EDaL and the EDEx process provide the foundations for addressing the major challenges introduced in Sect. 8.1.

C1. Data Exchange Requirements Are Not Clear or Conflicting The EDEx definition phase results in an EDaL network of stakeholders linked via data representing engineering information they exchange as foundation for EDaL patterns, such as RTE.

This EDaL network improves the cooperation within the organization and the data providers' insights into needs of other project participants: Data providers now know specific data requirements of project participants at any point of time. The EDaL network can grow iteratively, going beyond the insight of a one-time process analysis, as the specific relations between engineering artifacts and their content within an engineering project can change during the project execution. The data in the EDaL network enables the analysis of stakeholder priorities and relationships in an engineering project to provide the knowledge on which stakeholders require what data by when in the PSE process.

C2. Heterogeneous Engineering Data Is Hard to Integrate for Sharing While the data provided by engineering tools is typically specific for a discipline and not designed for use with other disciplines or with the project they contribute to, semantic linking allowed the integration of heterogeneous data in the evaluated EDEx use cases. The semantic linking enables seamless traceability in the EDEx process that, for the first time, gives all stakeholders the opportunity to know and analyze which role provided or received which kind of engineering data, which addresses a major awareness shortcoming in the traditional EDEx process. EDaL support for the EDEx in an engineering team can build on the explicit representation of common concepts, as semantic links between heterogeneous engineering data sets enable automation of EDEx and analyses. Furthermore, the EDEx semantic linking improves the representation of shared knowledge in the engineering team in a way that is understandable for machines, a prerequisite for introducing Industry 4.0 applications by supporting knowledge preservation in an aging engineering society.

C3. RTE Changes on Engineering Data Are Hard to Trace and Analyze The EDaL approach provides a data processing map, a network of stakeholders and the engineering artifacts and data they exchange during the engineering process, as foundation for automating analysis of changes to the content of exchanged data. Therefore, a data consumer in the RTE process can efficiently track back changes in the data versions he/she receives from several data sources to enable analyses of the received data and metadata, for example, for identifying missing or inconsistent data. The EDaL support in the EDaLIS keeps track of EDEx flows, including roles and rules for process conduct. Therefore, the EDaLIS facilitates frequent synchronization between work groups to reduce the risk of divergent local designs, rework, and project delays and to enhance the overall efficiency, agility, and security of the system design. The price to pay is the introduction of a new stakeholder role, the EDEx curator, having the knowledge and responsibility to coordinate the EDEx definition phase and to supervise the EDEx operation process.

Future Work We foresee the following avenues of future research work to investigate applications of the EDaL and EDEx capabilities and to address limitations of the research in this work.

Case Study on EDaL Concepts To explore the EDaL approach, we will conduct an empirical study to investigate what methods and mechanisms typical domain experts require to apply the EDaL approach effectively in their engineering context. Based on the single EDEx data flow discussed, we will explore defining a network of EDEx flows, supported by an EDaLIS prototype.

Advanced Analyses on the Exchanged Data and Associated Metadata The EDEx data will enable consumers and researchers to conduct advanced analyses, such as on expected but missing values, data validity and consistency, and symptoms for security risks. The EDEx metadata allows analyses of PSE process characteristics.

Semantic Linking Between Consumer and Provider Data Models During the use of EDEx, the complexity of links may grow considerably with the number of data elements, consumers, and providers, which will require research on the scalability of EDEx. While the EDEx process identifies direct links between consumer and provider data sets, it may be more efficient on a larger scale to identify common concepts (Sabou et al. 2017) in the engineering data model and link the consumer and provider data via these common concepts.

IT Security Considerations Centralizing knowledge in the EDaLIS requires research on threats to the integrity of collected knowledge and of industrial espionage.

EDEx and EDaLIS Application Future work will include the application and evaluation of the EDEx process and an operational EDaLIS in various engineering domains and application areas.

Acknowledgments The financial support by the Christian Doppler Research Association, the Austrian Federal Ministry for Digital and Economic Affairs, and the National Foundation for Research, Technology and Development is gratefully acknowledged.

References

Andersen, A. L., ElMaraghy, H., ElMaraghy, W., Brunoe, T. D., & Nielsen, K. (2018). A participatory systems design methodology for changeable manufacturing systems. *International Journal of Production Research, 56*(8), 2769–2787.

Biffl, S., Winkler, D., Mordinyi, R., Scheiber, S., & Holl, G. (2014a, September). Efficient monitoring of multi-disciplinary engineering constraints with semantic data integration in the multi-model dashboard process. In *Emerging technology and factory automation (ETFA)*. IEEE.

Biffl, S., Lüder, A., Schmidt, N., & Winkler, D. (2014b, October). Early and efficient quality assurance of risky technical parameters in a mechatronic design process. In *Industrial electronics society, IECON 2014-40th annual conference of the IEEE* (pp. 2544–2550). IEEE.

Biffl, S., Gerhard, D., & Lüder, A. (2017). Introduction to the multi-disciplinary engineering for cyber-physical production systems. In *Multi-disciplinary engineering for cyber-physical production systems* (pp. 1–24). Cham: Springer.

Biffl, S., Eckhart, M., Lüder, A., Müller, T., Rinker, F., & Winkler, D. (2018). *Data interface for coil car simulation (case study) part I*, Technical Report, CDL-SQI-M2-TR02, TU Wien.

Biffl, S., Eckhart, M., Lüder, A., Müller, T., Rinker, F., & Winkler, D. (2019a). *Data interface for coil car simulation (case study) part II - Detailed data and process models*, Technical Report, CDL-SQI-M2-TR03, TU Wien.

Biffl, S., Lüder, A., Rinker, F., Waltersdorfer, L., & Winkler, D. (2019b). *Introducing engineering data logistics for production systems engineering*, Technical Report, CDL-SQI-2018-10, TU Wien. October, 2018. http://qse.ifs.tuwien.ac.at/wp-content/uploads/CDL-SQI-2018-10.pdf.

Biffl, S., Lüder, A., Rinker, F., Waltersdorfer, L., & Winkler, D. (2019c). Efficient engineering data exchange in multi-disciplinary systems engineering. In *Proceeding of international conference on advanced information systems engineering* (pp. 17–31). Cham: Springer.

Brambilla, M., Cabot, J., & Wimmer, M. (2017). Model-driven software engineering in practice. In *Synthesis lectures on software engineering* (Vol. 3, 2nd ed., pp. 1–207). San Rafael, CA: Morgan & Claypool.

Broy, M., Feilkas, M., Herrmannsdoerfer, M., Merenda, S., & Ratiu, D. (2010). Seamless model-based development: From isolated tools to integrated model engineering environments. *Proceedings of the IEEE, 98*(4), 526–545.

Calà, A., Lüder, A., Cachada, A., Pires, F., Barbosa, J., Leitão, P., & Gepp, M. (2017, July). Migration from traditional towards cyber-physical production systems. In *Industrial informatics (INDIN), 2017 IEEE 15th international conference* (pp. 1147–1152). IEEE.

Davis, F. D. (1985). *A technology acceptance model for empirically testing new end-user information systems: Theory and results*. Doctoral dissertation, MIT.

Diedrich, C., Lüder, A., & Hundt, L. (2011). Bedeutung der Interoperabilität bei Entwurf und Nutzung von automatisierten Produktionssystemen. *at-Automatisierungstechnik Methoden und Anwendungen der Steuerungs-, Regelungs- und Informationstechnik, 59*(7), 426–438.

Drath, R. (Ed.). (2009). *Datenaustausch in der Anlagenplanung mit AutomationML: Integration von CAEX, PLCopen XML und COLLADA*. Berlin: Springer.

Drath, R., Fay, A., & Barth, M. (2011). Interoperabilität von Engineering-Werkzeugen. *at–Automatisierungstechnik, 59*(7), 451–460.

Gould, P. (1997). What is agility? *Manufacturing Engineer, 76*(1), 28–31.

Grose, T. J., & Doney, G. C. (2002). PhD. Stephan A. Brodsky. *Mastering XMI. Java programming with XMI, XML, and UML*. Wiley.

Hell, K. (2018). *Methoden der projektübergreifenden Wiederverwendung im Anlagenentwurf: Konzeptionierung und Realisierung in der Automobilindustrie*, Dissertation, Fakultät Maschinenbau, Otto-v.-Guericke Universität Magdeburg.

Hohpe, G., & Woolf, B. (2003). *Enterprise integration patterns: Designing, building, and deploying messaging solutions*. Boston, MA: Addison-Wesley.

Holl, G., Thaller, D., Grünbacher, P., & Elsner, C. (2012). Managing emerging configuration dependencies in multi product lines. In Proceedings of the Sixth International Workshop on Variability Modeling of Software-Intensive Systems (pp. 3–10). ACM.

IEC 62714. (2018). *Engineering data exchange format for use in industrial automation systems engineering - automation markup language*, 4 Parts 1, IEC, 2014–2018.

Jackson, M., & Johansson, C. (2003). An agility analysis from a production system perspective. *Integrated Manufacturing Systems, 14*(6), 482–488.

Jimenez-Ramirez, A., Barba, I., Reichert, M., Weber, B., & Del Valle, C. (2018, June). Clinical processes-the killer application for constraint-based process interactions? In *International conference on advanced information systems engineering* (pp. 374–390). Berlin: Springer.

Kovalenko, O., & Euzenat, J. (2016). Semantic matching of engineering data structures. In S. Biffl & M. Sabou (Eds.), *Semantic web for intelligent engineering applications*. Berlin: Springer.

Lüder, A., & Schmidt, N. (2017). AutomationML in a nutshell. In *Handbuch Industrie 4.0 Bd. 2* (pp. 213–258). Berlin: Springer.

Lüder, A., Pauly, J.-L., Kirchheim, K., Rinker, F., & Biffl, S. (2018a). Migration to *AutomationML* based tool chains – Incrementally overcoming engineering network challenges. In *Proceeding of 5th AutomationML user conference*, Göteborg, 24/25.10.2018; AML Association. https://www.automationml.org/o.red/uploads/dateien/1548668540-17_Lueder_Migration-ToolChains_Paper.pdf.

Lüder, A., Pauly, J.-L., Rosendahl, R., Biffl, S., & Rinker, F. (2018b). Support for engineering chain migration towards multi-disciplinary engineering chains. In *2018 14th IEEE international conference on automation science and engineering (CASE 2018)* (pp. 671–674). Germany, August 2018, IEEE.

Medvidovic, N., Egyed, A., & Rosenblum, D. S. (1999, September). Round-trip software engineering using UML: From architecture to design and back. In *Proceeding of the second international workshop on object-oriented reengineering (WOOR'99)* (pp. 1–8). Cham: Springer.

OMG. (2011). Business process model notation (BPMN) version 2.0. *OMG specification* (pp. 22–31). Object Management Group.

Putze, S., Porzel, R., Savino, G. L., & Malaka, R. (2018, June). A manageable model for experimental research data: An empirical study in the materials sciences. In *International conference on advanced information systems engineering* (pp. 424–439). Cham: Springer.

Rosemann, M., & vom Brocke, J. (2015). The six core elements of business process management. In *Handbook on business process management* (Vol. 1, pp. 105–122). Berlin: Springer.

Runeson, P., & Höst, M. (2009). Guidelines for conducting and reporting case study research in software engineering. *Empirical Software Engineering, 14*(2), 131–164.

Sabou, M., Ekaputra, F. J., & Biffl, S. (2017). Semantic web technologies for data integration in multi-disciplinary engineering. In S. Biffl, A. Lüder, & D. Gerhard (Eds.), *Multi-disciplinary engineering of cyber-physical production systems*. Cham: Springer.

Schäffler, T., Foehr, M., Lüder, A., & Supke, K. (2013, May). Engineering process evaluation: Evaluation of the impact of internationalisation decisions on the efficiency and quality of engineering processes. In 2013 IEEE International Symposium on Industrial Electronics (pp. 1–6). IEEE.

VDI. (2012–2017). *VDI-Richtlinie: VDI/VDE 3690 XML in der automation*. Berlin: Beuth.

Vogel-Heuser, B., Bauernhansl, T., & ten Hompel, M. (Ed.). (2017). *Handbuch Industrie 4.0*, Bände 1-4, VDI Springer.

Wieringa, R. (2014). *Design science methodology for information systems and software engineering*. Berlin: Springer.

Winkler, D., Sabou, M., & Biffl, S. (2017). Improving quality assurance in multi-disciplinary engineering environments with semantic technologies. In L. D. Kounis (Ed.), *Quality control and assurance – An ancient Greek term ReMastered*. Book Chapter 8 (pp. 177–200). London: INTEC Publishing.

Xu, X. (2012). From cloud computing to cloud manufacturing. *Robotics and Computer-Integrated Manufacturing, 28*(1), 75–86.

Chapter 9
Efficient and Flexible Test Automation in Production Systems Engineering

Dietmar Winkler, Kristof Meixner, and Petr Novak

Abstract *Context and background*: In Production Systems Engineering (PSE), software and systems testing are success-critical along the production automation life cycle to identify defects early and efficiently. Although test automation concepts enable continuous integration and tests during engineering and maintenance, tool chains are often hardwired, less flexible, and inefficient. Thus, there is a need for more flexible tool chains to support verification and validation of control code variants. *Objective*: In this book chapter, we (a) describe a *flexible Test Automation Framework (TAF)* to enable continuous integration and tests and (b) provide an adapted maintenance process to enable efficient verification and validation of control code variants. *Method*: We build on best practices from Software Engineering and Software Testing to establish a flexible TAF based on *Behavior-Driven Testing*. We use the *Abstract Syntax Tree (AST)* as foundation for human-based verification and validation. We developed an initial prototype derived from industry partners and used an *Industry 4.0 Testbed* for evaluation. *Results and conclusion*: First results of the prototype implementation with selected testing tools showed the capability of the TAF concept for supporting flexible configurations of testing tool chains. The AST concept can support the human-based verification and validation of control code variants.

Keywords Production systems engineering · Test automation · Behavior-driven test · Model quality assurance · Verification and validation

D. Winkler (✉) · K. Meixner
Christian Doppler Laboratory for Security and Quality Improvement in the Production System Lifecycle (CDL-SQI), Institute of Information Systems Engineering, Technische Universität Wien, Vienna, Austria
e-mail: dietmar.winkler@tuwien.ac.at; kristof.meixner@tuwien.ac.at

P. Novak
Czech Technical University, Prague, Czech Republic
e-mail: petr.novak@cvut.cz

S. Biffl et al. (eds.), *Security and Quality in Cyber-Physical Systems Engineering*,
https://doi.org/10.1007/978-3-030-25312-7_9

9.1 Introduction

In *Production Systems Engineering (PSE)* projects, the growing share of software components and the need for flexibility make engineering projects more complex and risky (Broy 2006). Software and system testing approaches represent key activities along the production systems engineering life cycle to identify defect early and efficiently (Vogel-Heuser et al. 2015). A key question is how to implement software and system testing within the PSE life cycle that can address new challenges, related to flexibility in context of *Industry 4.0* (Biffl et al. 2016). In the industrial practice of the production system life cycle, we observed a set of sequential process steps, such as system design, system construction, implementation, test and commissioning, and operation (Winkler et al. 2017c). In engineering phases, engineers, coming from the electrical, mechanical, and the software domain, typically work in parallel in small iterations, comparable to agile software development approaches (Lindstrom and Jeffries 2003). These small iterations include frequent system changes that need to be tested accordingly. In system maintenance projects, operators and system maintenance engineers should/could reuse these test cases to verify and validate the system behavior during/after executing maintenance tasks. In this context, test automation can help to automate frequent test runs in context of continuous integration and test during system development, and system maintenance (Duvall et al. 2007). Although test automation enables continuous integration and tests during engineering and maintenance, these tool chains are often hard-wired, less flexible, and inefficient (Winkler et al. 2018a). Adapting and/or exchanging tools in existing testing tool chain often require considerable human effort to reconfigure these tool chains during system development. System maintenance tasks are often executed by dedicated organizations such as third-party organizations that are not necessarily similar to the development organizations. Thus, adaptations in the testing tool chain might be required as well. Thus, we see the need to provide mechanisms for a flexible test automation approach that enables effective and efficient reconfiguration of testing tool chains and/or closing gaps in the test automation tool chain.

Beyond these tool chain reconfiguration issues, caused by the availability of different testing tools during development and/or maintenance, another critical issue can come up during system maintenance activities in PSE: there could be the need for exchanging selected and already implemented devices in the production system, such as robots. We observed different programming languages for different device types, coming from vendor-specific characteristics or even caused by evolved programming languages used in newer device versions. Note that we observed different programming languages used by different vendors, but also different programming languages used even by the same vendor. Thus, there is the need during maintenance tasks to (a) transfer and/or rewrite device control code from one device type to the other and (b) verify and validate expected system behavior of both, device code implementations and control code models. In this chapter, we focus on test automation and verification and validation of device control code (model) variants.

The *overall key question*, addressed in this book chapter, *is how to support flexibility in test automation in PSE*. Results can support engineers (a) to reconfigure testing tool chains efficiently and effectively and (b) to verify and validate control code model variants.

In context of the overall research questions introduced in Chap. 1, this chapter addresses the following research questions: In Chap. 1, RQ1a and RQ1b focus on typical characteristics and requirements of engineering processes for long-running software-intensive technical systems. In this chapter, we describe a test automation approach that is applicable along the product lifecycle, that is, during engineering and maintenance phases. Characteristics include the need for flexibility of the test automation framework and the capability for human based verification and validation of changes. In Chap. 1, RQ2 focuses on how successful business informatics approaches can be adapted for the engineering of large cyber-physical systems. To address this research question, we build on best practices from Software Engineering (Schatten et al. 2010) and Software Testing (Spillner et al. 2014) to establish a flexible TAF based on *Continuous Integration and Test* (Duvall et al. 2007) concepts and *Behavior-Driven Testing (BDD)* (Soeken et al. 2012).

The basic idea of BDD is to use simple (standardized) natural language constructs to define test cases. Typically, domain experts can formulate these test cases according to the expected behavior of the system without knowing (software-related) characteristics of test case implementation. Our observations at industry partners confirmed that often these type of tests, for example, on integration or acceptance test level, require the expertise of domain experts *and* software engineer *and* software test experts (Winkler et al. 2018a). The flexible TAF should also separate responsibilities and knowledge of testing roles and provide capabilities for effective and efficient reconfiguration of the testing tool chain. The verification and validation of (device) control code variants will support engineers in evaluating different versions of implemented and similar functional behavior of devices (i.e., control code of two robot variants), which should be exchanged during system maintenance. We use *Abstract Syntax Trees (AST)* (Jones 2003) as foundation for the verification and validation of control code variants. An AST represents a hierarchical, logical representation of a software artifact structure, like the source or test code of a program, omitting code characteristics that are specific to the applied used programming language. For example, code characteristics include indention characters and parentheses. This abstraction allows identifying structural elements such as branches and loops, their conditions, relevant fields and variables, or method calls. Therefore, an AST represents a model of the software code that can be analyzed within a programming language family such as Java and C#, independent of different language formalism or programming styles. Furthermore, methods exist that allow to extract and store such models in an exchangeable data format. Based on implemented control code versions, AST models are generated/derived using a model named *Abstract Syntax Tree Model* (ASRM) published by the Object Management Group.[1]

[1]OMG ASTM: www.omg.org/spec/ASTM

Thus, Quality Assurance (QA) experts can verify and validate different derived AST versions by using model quality assurance approaches, such as reviews or inspections (Zhu 2016).

For evaluation purposes, we apply two prototype implementations, derived from industry partners and the *Industry 4.0 Testbed*,[2] located at Czech Technical University in Prague. First results of the prototype implementation with a selected tool chain showed that the flexible TAF concept is capable of supporting configuration options of testing tool chains in a continuous integration and test environment more effectively and efficiently. However, the individual characteristics of different testing tools and tool suites in PSE still require the consideration of human effort to support more effective and efficient tool chain configurations. In context of verification and validation of control code variants, we showed the feasibility of the AST model approach. However, due to the complexity of derived AST models and the structure of these models, tool support is required to enable engineers in efficiently verifying and validating control code variants with AST models. This tool support can include concepts from model quality assurance with human computation to evaluate different models in PSE more effectively and efficiently. Although the AST is promising, there is the need for human effort to bridge gaps in the testing tool chains that could not be fully automated.

The contributions of this chapter address the scientific communities of *Information Science* in *Automation and Industrial Manufacturing* and *Cyber-Physical Production Systems*. We are especially aiming at improvements of software and testing quality in the engineering and maintenance process. This is done by providing novel approaches for a flexible exchange of testing tools based on layered framework and a flexible, model-based approach to validate and verify control code for industrial resources such as robot arms. Furthermore, we address *Process Improvement* and *Software Quality Assurance* communities by adapting and extending approaches from *Software Engineering* to the *Automation Systems* domain. *Practitioners* can benefit from the reported results in terms of applying and extending the flexible test automation framework in individual contexts to make software and system testing processes more flexible, effective, and efficient. Finally, process improvement can help to implement quality assurance mechanisms in engineering and maintenance processes.

The remainder of this book chapter is structured as follows: Sect. 9.2 summarizes background and related work on Production Systems Engineering (PSE), Software and System Testing, Continuous Integration Processes and Tools, and model quality assurance with human computation. We focus on the research issues in Sect. 9.3 and describe the illustrative use case in Sect. 9.4. Section 9.5 summarizes the flexible test automation framework and Sect. 9.6 focuses on the verification and validation approach of control code variants. Finally, we discuss findings, limitations, and future work in Sect. 9.7.

[2]Industry 4.0 Testbed: www.ciirc.cvut.cz/testbed/

9.2 Related Work

This section summarizes related work on Production Systems Engineering (PSE), Software and System Testing, Continuous Integration Processes and Tools, and Model Quality Assurance with Human Computation.

9.2.1 Production Systems Engineering

Production systems typically consist of mechatronic objects that incorporate components of multiple engineering disciplines (Moser et al. 2012), such as electronics, mechanics, and control software. Engineers of different disciplines use various tools, programming languages, terminology, and models for problem description and solving. From the physical perspective, mechatronic components are tightly coupled. However, from the system engineering point of view, mechatronic components consist of a set of heterogeneous data models constructed and maintained by domain experts using a variety of different engineering tools. These data models and tools should be well connected with a bunch of interfaces to guarantee correct operation of the final mechatronic objects and to overcome technical and semantic heterogeneities of tools and data models (Winkler et al. 2017c). Common data exchange approaches, such as common concepts or a data description language (such as *AutomationML*[3]) for data exchange represent the foundation for effective and efficient data exchange (Winkler et al. 2018a).

Automation and control of production systems is typically realized by PLCs (i.e., programmable logic controllers). PLCs can be programmed with a family of PLC programming languages that is standardized as IEC 61131. Although there exist vendor-neutral and standardized programming languages, individual vendors include numerous extensions and variations of these basic languages to support specific characteristics of their devices (e.g., for optimization of the defined behavior). Consequently, code reuse and re-deployment still pose issues, even though a standardization effort in the frame of a standard PLCopen[4] tries to bridge these gaps between vendor-specific implementations. Besides PLC programming, on a higher level of the automation architecture (Vogel-Heuser et al. 2009), different programming languages such as C, C#, C++, Python, or Java are used for implementing functional behavior. This heterogeneity also hinders reuse and re-deployment of software components within a maintenance project. For example, industrial robotics typically utilizes proprietary programming languages or vendor specific programming languages, for example, the robot vendor KUKA[5] uses the

[3]AutomationML: www.automationml.org

[4]PLCopen: www.plcopen.org

[5]KUKA: www.kuka.com

language called KRL for the traditional industrial robots representing the most significant part of KUKA portfolio, whereas new and advanced types of cooperative robots use the Java language. Therefore, if one robot type needs to be replaced by another robot type, for example, during a system maintenance project, there is the need to (a) rewrite the robot control code and (b) verify and validate the implemented code versions (similar to an engineering project). Due to the use of a large set of different programming languages in PSE, the verification and validation of correct operations and correct systems behavior is challenging. Note that in context of this book chapter, verification refers to compliance of the implementation and the related specification while validation refers to compliance of the implementation and customer expectations, requirements (i.e., expected functional behavior).

Verification and validation are also issues during systems maintenance and evolution because even minor updates or changes of the system during production system life cycle imply high testing effort and costs. Therefore, there is the need for supporting engineers in verifying and validating robot control code in PSE to ensure that corresponding code snippets in different languages are equivalent from the behavior point of view. While reviews and inspections or (software) tests can be used to evaluate these language constructs, there is still need for a more effective and efficient evaluation approach, for example, based on control code models.

9.2.2 Software and Systems Testing with Behavior-Driven Tests

Modern approaches in software engineering that follow agile principles (Van Bennekum et al. 2001), such as Scrum (Schwaber and Beedle 2002) and XP (Lindstrom and Jeffries 2003), are based on short development iterations to continuously evolve and improve the software under development. A prerequisite for these approaches is quick feedback for software developers early in the development process, which determines the correctness of the source code.

Test-Driven Development (TDD) is a common software engineering best practice that also supports quality assurance in software engineering (Beck 2003), which requires developers to write tests before implementing the functionality of the software. Using this approach, the executed tests fail initially until a developer implements the functionality correctly. As a consequence, the test cases themselves act as acceptance criteria for the developed methods. Furthermore, writing the tests before implementing the software tends to increase the productivity of the developers as well as the number of tests (Erdogmus et al. 2005). However, such tests often represent the view of the software developer of the system and its state, rather than the system's behavior (Solis and Wang 2011) as desired by the product owner for example.

Behavior-Driven Development (BDD) aims at overcoming these limitations by enabling stakeholders, such as the product owner or application designers, to formulate their acceptance criteria as fine-grained scenarios, containing a sequence

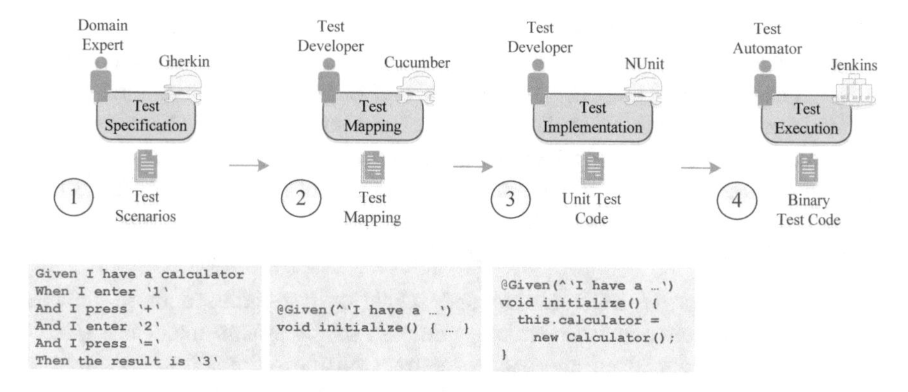

Fig. 9.1 Behavior-driven development process with examples

of steps that act as specifications for the behavior of the system under test. On the one hand, a primary goal of BDD is to utilize *Keyword-Driven Testing (KDT)* (ISO/IEC/IEEE 2016), which aims at providing a common vocabulary that reflects the relevant domain-specific concepts as building blocks of acceptance test cases in natural language (Soeken et al. 2012; Solis and Wang 2011). Furthermore, BDD aims at making the resulting BDD scenarios executable (Solis and Wang 2011) for easier automation of the test process. Today, a wide range of BDD frameworks, such as *Cucumber*,[6] *Radish*,[7] or *Specflow*,[8] exist for various programming languages like Java, Python, and .Net. A widely used *Domain-Specific Language (DSL)* to describe BDD scenarios is the *Gherkin* notation (Micallef and Colombo 2015).

Figure 9.1 depicts a representative simple example of the BDD tool chain and an exemplary test process in combination with typically used technologies.

1. *Test scenario definition*: First, a domain expert specifies the acceptance criteria in several test scenarios, written as scenario steps in a quasi-natural language with related parameters (e.g., by using the *Gherkin* language). We need to mention that domain experts and test developers need to negotiate the wording of the scenario steps and their parameters to create shared meaning. Figure 9.1 shows such a simple test scenario in *Gherkin* notation on the bottom left. The initial execution of these defined scenarios with a test execution tool or a build server fail will fail, as an implementation is missing at that point.
2. *Test mapping*: A Test Developer maps the scenario steps in a "Test Mapping" process step to test stubs using frameworks such as *Cucumber*. For example, when using *Cucumber* and *Java*, the test developer annotates the Java test stubs with the relevant *Gherkin* keyword and the language construct from the scenario

[6]Cucumber BDD Framework for Java: http://cucumber.io

[7]Radish BDD Framework for Python: http://radish-bdd.io

[8]Specflow BDD Framework for .Net: http://specflow.org

step that contains placeholders for the parameter. This parameterization allows the reuse of the scenario steps with different test data and thus the combination of scenario steps to a variety of scenarios, that is, test case specifications.
3. *Unit test case development*: The Test Developer implements the test code as, for example, Unit Test Code in a particular testing framework like *JUnit*.
4. Finally, in the *Test Execution activity*, the Binary Test Code of the unit tests is executed using an integrated development environment, such as *Eclipse*, or, in a more advanced setting, on a build server like, for example, *Jenkins*.

To further improve BDD testing and scenarios it requires a more suitable vocabulary that better fits the particular domain of the system under test (Häser et al. 2016) introduced an approach to extract business domain concepts from the application domain's body of knowledge. They observed, through a controlled experiment, that the use of such a domain-specific vocabulary allows significantly faster creation of BDD scenarios. The authors further claim that the use of such vocabularies could improve existing BDD toolkits and support various domains such as software development in PSE. However, to achieve the goal of improving the vocabulary for BDD needs a better integration of systems and testing activities such as collective glossaries (Musil et al. 2015).

9.2.3 Continuous Integration Processes and Tools

As already introduced, software and system testing represent key activities along the production systems engineering life cycle to identify defect early and efficiently (Vogel-Heuser et al. 2015). However, assuring the quality and validity of software is not only a necessary task in today's software and systems engineering best practices but—due to the complexity of modern software systems—also allows a better collaboration between engineers (Whitehead 2007). Proper software testing is gaining even more attention with the recent DevOps movement, which aims at moving development and operations closer together to deliver better software faster and in shorter iterations (Roche 2013).

Continuous Integration (CI) is a standard method in the field of Software Engineering (SE) to test and improve the quality as well as shorten release cycles, by enabling developers to submit their code frequently, automatically build and test new features, and integrate new developed code into a shared code base (Duvall et al. 2007). Figure 9.2 shows a simplified version of a typical CI process with activities and stakeholders, but also some common, exemplary tools or technologies used for these activities.

1. *Source and test code*: In the first step of the CI process, developers commit the source code implemented in the development activities, shown in the light gray box on the left of Fig. 9.2, to a shared Code Repository. This source code can either be the software source code of the system under test from the Software

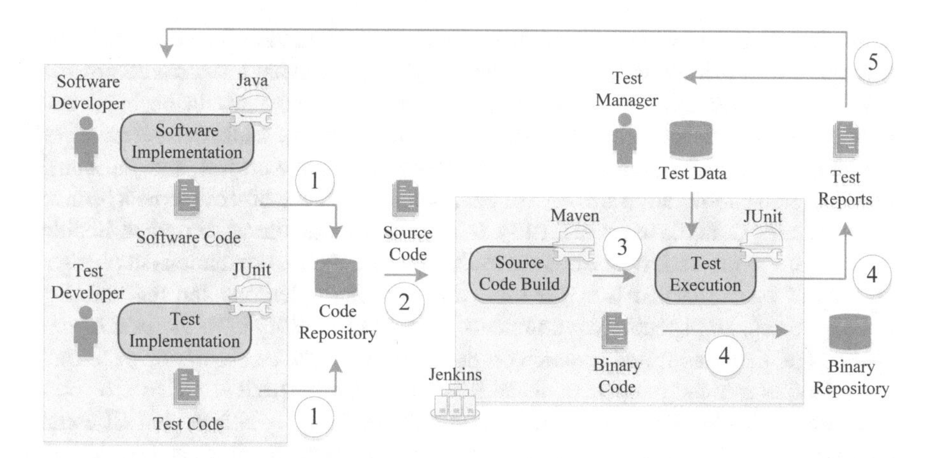

Fig. 9.2 Continuous integration and test process based on (Sunindyo et al. 2010) and (Winkler et al. 2018a)

Implementation task, which is developed, for example, in Java, or test source from the Test Implementation that represents the executable test cases and is, for example, developed in JUnit.

2. *Source code build process step*: The build server, for example, Jenkins, which is shown in the light-gray box on the right of Fig. 9.2, either is triggered by the code repository or periodically checks the code repository for updates. If an update occurs, the build server, in a second step, checks out the Source Code and translates it to Binary Code using a build tool such as Maven employing a predefined ruleset.
3. *Test execution process step*: If the software and test source code are built without problems, as a next step, the test cases are executed in a Test Execution activity that uses Test Data that is provided by a Test Manager.
4. *Binary and test reports*: In the final step, the build server commits the Binary Code to a Binary Repository and generates the Test Reports.
5. *Developer feedback*: Finally, the test reports are sent as feedback to the test manager and the developers either as verification and validation document, or to address bugs and improve the quality of the code. Although some authors (e.g., Laukkanen et al. 2017) argue that Acceptance Testing goes beyond CI and belongs to the broader context of Continuous Delivery, which also takes the deployment to production or a staging area where the code is released from, into account, we count acceptance testing and committing binaries to a repository to the CI process.

Although CI can mean faster integration of code in better quality, researchers found that there are several issues in the adoption of CI in practice. Ståhl et al. reported that instead of a general CI approach, companies need to develop their very own variant for a CI process, which is not always easy in practice (Ståhl and

Bosch 2014). In a case study, Mårtensson et al. found that their interviewees found it frustrating that tools form the CI process instead of having a structured process that uses the best tools to make CI fast and simple, which would be important improvement for developers (Mårtensson et al. 2017, 2018). Laukkanen et al. argue that inflexible builds in the CI system, changing roles in the project, and additional need for team coordination are often impediments for a successful implementation of CI in practice (Laukkanen et al. 2017). In their systematic literature review, Shahin et al. illustrate the variety of different tools used and their combinations in practice and report on the lack of suitable tools due to their limitations and the inability to adapt to custom-specific environments (Shahin et al. 2017). To the best of our knowledge, there is limited research on the issues of cross-configuration of testing tools. However, we can report on these issues from own experience of over 6 years of CI practice with industry partners. In context of PSE, we believe that CI could gain significant benefits during system development, maintenance, and evolution. Furthermore, we argue that due to the additional complexity of the multidisciplinary context the problems of CI in PSE are at least comparable.

9.2.4 Model-Quality Assurance with Human Computation

In *Production System Engineering (PSE)*, engineering models play an important role along the production system life cycle, for example, for project planning, execution, evaluation, operation, and maintenance (Winkler et al. 2017a). Models can focus on the system structure and/or system behavior of the planned systems. Model examples include *AutomationML*, CAEX (e.g., plant topology), COLLADA (geometry and kinematics), PLCopen (logical behavior), programming language data constructs (e.g., C, C+, or Python), models for software and system testing, UML models for designing the structure (e.g., class diagrams) or behavior (e.g., state charts or activity diagrams), or data base modeling (e.g., EER). Nevertheless, the heterogeneity of data models in various disciplines in a multidisciplinary engineering environment requires mechanisms for model construction and model evaluation. Although there exist methods and tools for model construction and basic mechanisms for model evaluation (within one model type), quality assurance across different model types is often limited. Thus, human activities are needed for verification and validation of models and model variants. Because models typically represent the foundation for later engineering phases, for example, for detailed planning of certain PSE artifacts based on underlying basic planning models, mechanisms are required for identifying and removing defects early and efficiently in the production system life cycle (Brambilla et al. 2012).

In Software Engineering, static quality assurance approaches, such as *Software Inspections*, established quality assurance approaches (Aurum et al. 2002) aim at identifying defects early and efficiently in engineering artifacts without the need for executable artifacts such as software code. In contrast to static quality assurance, dynamic quality assurance methods, such as software testing (Spillner et al. 2014) aim at identifying defects based on test cases and the execution of (existing) software control code. In this chapter, we focus on early human-based quality assurance approaches of engineering artifacts with software reviews and inspection.

Reviews and Inspections (Zhu 2016) are formal verification and validation approaches in teams that enable defect detection already early in the (software) engineering life cycle. Traditional inspections typically focus on checking an inspection artifact (such as a generated/constructed model) with respect to a reference document (such as requirements or a specification document). While the reference document is considered to be correct, defects should be identified in inspection artifacts, such as a model that is based on requirements, specifications, or existing models from previous engineering steps. Typical review and inspection processes are based on human experts and include several phases such as preparation, individual defect detection (inspection), defect collection and aggregation (team meetings or nominal team defect aggregation), defect correction, and follow-up activities with focus on quality assessment and decision making (Laitenberger and DeBaud 2000). Typical inspection processes include a set of human experts such as a moderator, who is responsible for managing and controlling the overall inspection process, 5–6 individual inspectors (experts) for defect detection and team meetings, and domain experts for inspection support (Gilb et al. 1993). The involvement of (expensive) experts includes risks for inspection planning and inspection execution, for example, based on the availability of experts and time constraints as typical inspections are scheduled for about 2 hours. This time constraint also limits the size of inspection artifacts. Therefore, inspections planning, embedded within a project plan is challenging for project managers. In context of PSE and traditional human-based inspection processes, we observed a set of limitations: (a) *Cost and availability of experts*. Expert inspectors are expensive and it is challenging to organize inspection sessions including all relevant stakeholders; (b) *limited inspection duration and scope*. The upper time duration of about 2 hours for an inspection session limits scope, complexity, and the size of inspection artifacts and reference documents. Often, multiple inspection sessions are required to cover all inspection artifacts in sufficient detail; and (c) *supporting material*. Depending on the domain and inspection artifacts, several supporting material, such as checklists, guidelines for inspection (e.g., reading techniques) need to be configured according to the inspection context. However, as software inspection does not require executable code (compared to testing), this quality assurance approach is applicable to different types of artifacts and models.

To overcome limitations of traditional inspection and to gain benefits of inspection in defect detection of PSE artifacts, adapted approaches are needed to enable high-quality defect detection processes. In Winkler et al., we introduced an improved model inspection process with human computation and crowdsourcing (Winkler et al. 2017b) and presented evaluation results with focus on defect detection performance and evolution steps of the defect detection tasks in controlled experiments (Winkler et al. 2018b; Sabou et al. 2018). The key idea is to split up inspection tasks into smaller pieces of manageable work to enable distributed and fine-grained defect detection. A group of experts (i.e., an expert crowd) can participate in the inspection

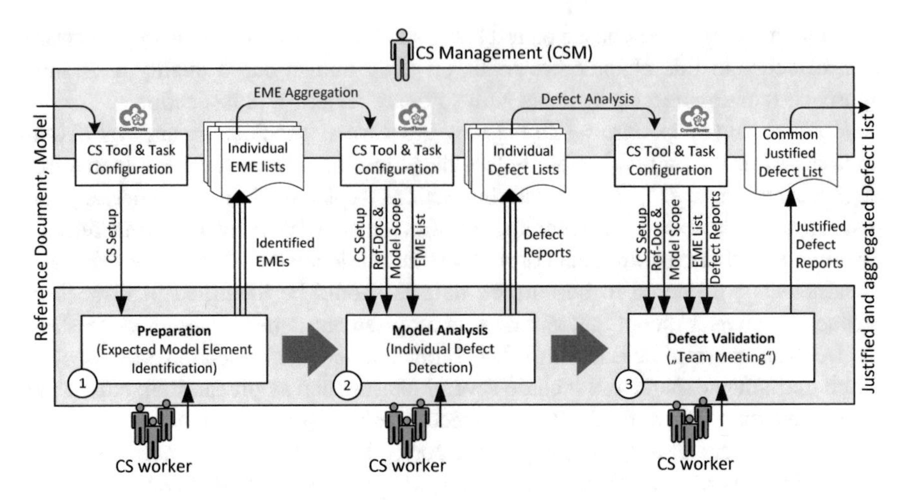

Fig. 9.3 Basic model inspection process with human computation and crowdsourcing

process steps via a crowdsourcing platform, such as *FigureEight*.[9] However, there is the need for a *CS-Management (CSM)*, who is responsible for setting up the crowdsourcing platform. Figure 9.3 presents the basic steps of the model inspection process with human computation and crowdsourcing with related input and output: (1) *Preparation*, (2) *Model Analysis*, and (3) *Defect Validation*. The *Crowdsourcing Management (CSM)* role prepares and controls the inspection process and supports all process steps.

In the *Preparation* phase (1), the so-called *Expected Model Elements (EMEs)* are identified by a crowd of experts. EMEs represent building blocks of the model, such as *Entities*, *Relationships* between entities, and *Attributes* and are extracted from the reference documents by crowd workers (CS workers). This approach is comparable to reading a document and marking/annotating relevant pieces of information. Input of this step is a set of tasks, represented by sentences of the reference document, prepared by CSM. Output includes a set of EMEs that need to be aggregated accordingly (i.e., removing duplicates, handling synonyms or typos). Note that the CSM role is responsible for this task, supported by natural language processing approaches or executed manually.

The *Model Analysis* phase (2) takes as input individual EMEs, identified in the previous step, the scope of the model, and the reference documents. Again, CSM is responsible for setting up tasks in the crowdsourcing platform that provides CS workers with relevant information for defect detection. CS workers report candidate defects or confirm the correctness of the model fragment. Note that this step refers to the individual inspection approach in traditional review and inspection processes. In contrast to traditional inspections where defects are reported, this approach allows

[9]Crowdflower/FigureEight: www.figure-eight.com

to measure coverage of reference documents and model parts because CS workers also have to report the correctness of the model elements. Output of this step is a list of reported defects by several CS workers within the related scope. However, these defect reports need to be aggregated in the defect validation process phase.

In the *Defect Validation* phase (3) several candidate defect lists are available. Similar to the Model Analysis Phase, CS workers receive individual tasks to justify identified candidate defects, that is, to check whether or not a defect is a real defect. CSM is responsible for the task design and for enabling the validation of individual candidate defects. Output of this phase is an aggregated list of defects with related justifications. If there is an agreement on a candidate defect it will go into the common defect list (i.e., classified as defect) or into the false positive list (i.e., classified as non-defect). If there is a disagreement, there is the need for discussion among a small group of experts to come to a consensus. Note that this process step is comparable to a (distributed) team meeting. This *Model Quality Assurance* (MQA) process has been evaluated in a Software Engineering context with *Extended Entity Relationship* (EER) diagrams and specification documents, for example, in Winkler et al. (2017b, 2018b) and Sabou et al. (2018).

However, the MQA process approach can be applied to engineering models in the PSE context and can support the verification and validation of engineering models as described in Sect. 9.2.1. In this chapter, we build on software inspection and model quality assurance concepts that aim at bridging the gap in the maintenance process, as it could not be executed automatically but need human experts to verify and validate software control code variants.

9.3 Research Questions

Based on the need for *more flexible test automation* in the PSE context, we focus on (a) how to support the reconfiguration of testing tool chains and (b) to verify and validate control code model variants with human inspection. Therefore, we derive two related research questions.

RQ.1: How Can We Support Flexibility in Test Automation to Allow Reconfiguration of Testing Tool Chains in PSE Contexts? In PSE industry, test automation of software and systems is still not widely established because of several reasons, identified in industrial environments: (a) Large and complex systems require a manual execution of defined tasks such as preparing and resetting physical components. (b) Testing processes and tasks are often bound to a specific person due to specific knowledge of underlying technologies, such as software test aspects and techniques, and tools, like particular connecting interfaces and simulations for robot arms. Thus, especially the roles of domain experts and test experts are intertwined in current solutions. (c) Testing tool chains are often hard-wired with limited flexibility that require considerable effort for reconfiguration, for example, if new testing tools are required in project consortia or during maintenance projects, where originally applied tool chains are not available anymore. To increase flexibility, it requires new approaches that enable flexible reconfigurations of testing tool chains to facilitate an increased degree of test automation and the separation of testing roles. A flexible *Test Automation Framework (TAF)* aims at overcoming these limitations by providing suitable mechanisms and methods.

To address RQ.1, we (a) identify key stakeholders and challenges for test automation in PSE and derive requirements that should be covered by a solution, (b) develop a concept for a flexible test automation framework based on software engineering and testing best practices, (c) develop a prototype implementation, and (d) conceptually evaluate the prototype with requirements and challenges in a real-world application use case, derived from our industry partners.

In the context of system maintenance, typical tasks include exchanging defined components, such as robots. Because of the different programming languages used by different device types (or robot types), there is the need for supporting engineers in (a) transferring control and test code from one device type to the other and (b) for validating the expected behavior in two or more robot versions. In this chapter, we focus on the validation part, that is, evaluating the behavior of two or more implemented control code versions with a human inspection. Therefore, we derived the second research question.

RQ.2: How Can We Evaluate Control Code Versions Based on the Expected System Behavior Effectively and Efficiently? This research question focuses on the validation of robot control code in the context of the software testing terminology. We build on the concept of *Abstract Syntax Trees (ASTs)* that focus on the abstract description of the control code structure. Based on this control code structure we see two main benefits for engineers: (a) The *AST*, which results from the original control code implementation, can be used as a blueprint for writing control code for the new (target) device type, as the basic structure is already available. (b) Given two *AST* versions (i.e., source and target control code) as the base for a human-based inspection to evaluate (i.e., validate) the structure of both *AST*s, respectively control code versions, can be compared.

To answer RQ.2 we (a) revisit and extend key stakeholders in the context of system maintenance, identify new challenges and derive related requirements in context of system maintenance; (b) adapt the basic maintenance process with human inspection to enable efficient verification and validation of control code variants based on Model-Driven Engineering and Inspection best practices; and (c) conceptually evaluate the process prototype with requirements and challenges in a real-world engineering context, that is, an *Industry 4.0 Testbed*.

9.4 Sample Production System and Illustrative Use Case

For illustration purposes, in this section, we introduce a sample production system that is capable of addressing both research questions, (a) with focus on the need for flexible testing tool chains to evaluate a (part of a) PSE system (RQ.1) and (b)

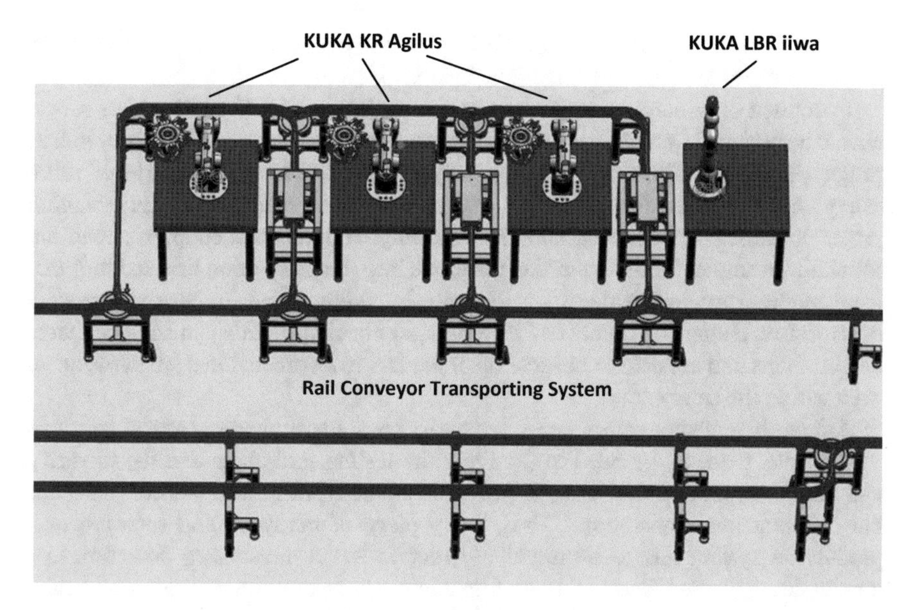

Fig. 9.4 Schematic overview of the *Industry 4.0 Testbed* at Czech Technical University in Prague, with three KUKA KR Agilus robot arms, one KUKA LBR iiwa robot arm and the rail conveyor transporting system

with focus on the evaluation of various (abstract) control code variants in the context of systems maintenance (RQ.2). For evaluation purposes, we use the *Industry 4.0 Testbed*, located at the Czech Technical University in Prague, CIIRC. This system, mainly intended for training and education, represents a simple production system with transportation and handling units. The PSE is equipped with four industrial robots, five shuttles, a set of rail conveyors, and six workstations. Figure 9.4 presents a schematic overview of this production system. Main working tasks focus on final assembling activities, that is, robots are equipped with grippers as tools for manipulating and handling material, semi-products, and final products. Note that this type of robotic handling system is called *pick and place* and these types of robotic workstations are *pick and place units*.

Industry 4.0 Testbed is equipped with robots of two types, that is, KUKA KR Agilus and KUKA LBR iiwa. Although both types are from the same vendor, they are programmed in significantly different languages and programming styles. The KUKA KR Agilus robot type is a fast industry robot that is frequently used in high-performance environments such as in the automotive industry sector for car assembling. For safety and security reasons, physical barriers are needed to prevent damages or human physical injuries. KUKA KR Agilus uses the vendor-specific KUKA WorkVisual IDE and the programming language KRL. The second robot type is KUKA LBR iiwa, a cooperative robot that should not make any injury to manufacturing stuff. There is no need for a safety zone (like fences, cages, or optical barriers) as this robot type can cooperate with humans. For programming, this robot type uses KUKA Sunrise Workbench IDE that is built on top of the Eclipse Framework and Java as a programming language.

In context of a maintenance and evolution project, all KUKA KR Agilus robots should be replaced by modern KUKA LBR iiwa robots. Main reason for this evolution project is to minimize safety and security concerns, that is, to get rid of robot safety cages required for the KUKA KR Agilus robot. Benefits of this evolution project include: (a) *Improved material handling*. Humans can cooperate with the robot, for example, they can put raw materials into the production line and they can handover final products without considering robot safety zones. (b) *Improved process observation*. Human operators can continuously check the quality of semi-products and products and are able to observe the manufacturing process in a small distance, even within the safety zone.

Although replacing robot types seems to be a small change from a software perspective, there is the need to (a) adapt the testing tool chain and (b) to verify and validate robot control code as involved robot types require similar behavior but different implementations. Thus, every piece of hardware and software in a production system has to be tested in order to avoid unexpected downtimes or unintended behavior. Each robot and its control code are thus tested individually, but also in the integrated environment including interactions with other devices in a production system. Furthermore, when changing the robot types, it is often not possible to reuse previously implemented tests to validate and verify the expected robot behavior but tests need to be adapted due to different programming languages, different programming styles, different integrated development environments, and relationships to other components.

9.5 Flexible Test Automation Framework

Software and systems testing is a success-critical activity in the PSE life cycle (Vogel-Heuser et al. 2015) to ensure sufficient quality of the production automation system. Therefore, there is the need for suitable testing tool chains. Changes in the engineering and/or maintenance environment raise the need for more flexibility to address changes in the testing environment and/or changes in PSE. In industry settings, current solutions of testing tool chains are often inefficient and lack in flexibility that is needed to use such tool chains in different project contexts.

For better understanding of the issues in context of the PSE life cycle, we first describe basic requirements for and challenges of test automation in the production systems engineering domain. In a second step, we present the underlying architecture for a flexible *Test Automation Framework (TAF)* that aims at addressing these requirements and challenges. Finally, we present a prototype implementation to show the feasibility of the introduced TAF in context of the *Industry 4.0 Testbed*.

9.5.1 Challenges and Requirements for Test Automation in PSE

Identifying and describing challenges and requirements of a flexible test automation approach in Production Systems Engineering (PSE) are the basis for designing and evaluating a test automation framework. In context of testing production automation systems a set of stakeholders are involved (see Figs. 9.1 and 9.2 for selected stakeholders in test automation). For clarity and completeness, we summarize these stakeholders and their responsibilities aligned with characteristics of the stakeholders derived from testing roles defined in the literature (Spillner et al. 2014; ISO/IEC/IEEE 2013):

- *Test Managers* are responsible for test planning, such as selecting test strategies, resource allocation, like assigning the needed employees, and managing the overall test process based on test reports. Furthermore, test managers are responsible for organizing sources for test data and test uses.
- *Domain Experts* are responsible for deriving logical test cases from the System under Test (SuT) as they are familiar with technical aspects and the functional behavior of the production system, for example, particular movement of robot arms or actions a tool should execute.
- *Software Developers* implement (control) software of the SuT and adapt the software based on the test results and reports, for example, bug fixing or refactoring.
- *Test Developers*: In addition to software developers, *Test Developers* implement logical test cases based on concrete test scenarios and test cases received from domain experts. These logical test cases are based on a set of appropriate technologies, such as unit tests. Furthermore, in cooperation with domain experts, test developers provide related test data for test execution.
- *Test Automation Engineers* create and maintain the *continuous integration and test (CI&T)* pipeline by providing required scripts and the infrastructure to execute implemented tests automatically.
- *Test Environment Engineers* are responsible for *setting up*, *maintaining*, and *tearing down* the test environment. In contrast to traditional software engineering, in PSE, these activities often require manual intervention and bear the threat of safety issues if, for example, the engineer does not reset robot arms correctly.

Stakeholder and roles in the test automation process of PSE include a set of challenges that need to be addressed by a flexible TAF.

TAF-C1. Separation of Roles In practice, a single expert usually does not have one distinct role but represents a combination of the described roles. Thus, responsibilities, tasks, and activities are often intertwined, for example, incorporated by testing tools that support different tasks, testing techniques, and required knowledge. This leads to a lack of separation of roles and is challenging for project management, as different activities are required in different phases of the PSE project. For example, although domain experts for robot arms might not be the best developers for test code, they might need to implement tests because of their specific domain knowledge of robot arm processes and procedures and the lack of tools supporting their domain-specific language.

TAF-C2. Human in the Loop The testing process includes a set of steps such as test planning and control, analysis and design, implementation and execution, evaluation, and reporting, and closing and follow-up (Spillner et al. 2014). Specific testing tools can support software and systems testing in one or more project phases. However, in PSE, there is often a gap in the tool chain that is related to the testing process. Thus, if testing tool capabilities are missing, human experts often replace these capabilities by either executing tasks manually or by developing temporary solutions, such as auxiliary scripts, that often remain in use longer than expected. Thus, a challenge refers to closing gaps in the testing tool chain or supporting human experts in overcoming these limitations.

TAF-C3. Hard-Wired Tool Chains Existing solutions often use hard-wired tool chains developed in a defined context, caused by compatibility issues of tools and related artifacts, missing alternatives, or tools that offer several required capabilities. However, these hardwired tool chains are often tailored initially for a specific project purpose and, thus, include limitations regarding flexibility. Thus, when project requirements change or a set of projects have different needs, these hardwired tool chains hinder the test automation process and impede an efficient and effective testing practice.

TAF-C4. Artifact Management Engineers regularly use various methods for the application of different testing techniques, like test strategy selection, test data generation, or test case generation. These methods typically produce artifacts that should be included into the testing tool chain to make them available for other applications in a structured way. The application of source code or binary artifact management solutions can support artifact management in test automation.

Based on the challenges, that is, TAF-C1 to TAF-C4, and discussions with our industry partners and experts, we derived a set of requirements (TAF-R1 to TAF-R6) that need to be addressed by an underlying architecture for a flexible Test Automation Framework (TAF). Table 9.1 summarizes identified challenges and requirements and indicates whether or not a requirement is related to a challenge.

TAF-R1. Clear Role Definition A separation of roles in the testing process aims at a clarification of responsibilities, which distributes the work load of individuals, improves the quality of artifacts due to specialization and experience, and eases resource management, for example, for test managers. A clear separation of roles makes it easier to define who is responsible for what particular artifacts and their integration to the tool chain and testing process. Requirement *TAF-R1* addresses challenges *TAF-C1* and *TAF-C4*. Challenge *TAF-C1* concerns the entanglement of the roles in test automation of production systems engineering, which results in overlapping responsibilities of, for example, the tasks of deriving logical test cases and actually implementing the code for these test cases. Challenge *TAF-C4* concerns the inclusion of artifacts for testing into the testing tool chain.

Table 9.1 Challenges and requirements toward a flexible test automation framework based on Winkler et al. (2018a)

		Challenges			
		TAF-C1	TAF-C2	TAF-C3	TAF-C4
Requirements		Separation of roles	Human in the loop	Hardwired tool chains	Artifact management
TAF-R1	Clear role definition	✓			✓
TAF-R2	Configurable tool chains		✓	✓	✓
TAF-R3	Extensibility of tool chains		✓	✓	
TAF-R4	System under test management			✓	
TAF-R5	Method and tool support		✓	✓	✓
TAF-R6	Test code management	✓			✓

TAF-R2. Configurable Tool Chains Configurable tool chains aim at more modular integration of the regularly isolated and strictly defined testing tools across all testing steps. The gained flexibility allows the exchange of tools or a combination of their capabilities based on the particular requirements of various projects and, therefore, enabling better maintainability. Requirement *TAF-R2* addresses challenge *TAF-C2* by means of decreasing the recurring manual work of engineers by automating such tasks as far as possible and making them configurable for easier reuse. Challenge *TAF-C3* is addressed to omit rigid and hardwired tool chains that are hard to change on new requirements. *TAF-C4* is addressed by enabling a better integration and management of various test artifacts that are generated during testing and support the different involved roles.

TAF-R3. Extensibility of Tool Chains The extensibility of testing tool chains targets at using open and extensible interfaces and artifacts models to support an easy integration of various tools and technologies. The requirement tries to tackle challenge *TAF-C2* by bridging the gaps in the testing tool chain that are managed by humans often with considerable effort. For example, engineers need to substitute missing test automation servers by manual test runs. Challenge *TAF-C3* is addressed by supporting a flexible exchange of tools to use the best available for the purpose and allowing to model a testing pipeline without the limiting factors of proprietary tool interfaces and artifacts.

TAF-R4. System Under Test Management In automation and production systems, engineering tests are still often solely run on the physical system. However, recent developments, like advanced simulations and Digital Twins (Rosen et al. 2015), allow running of tests in an artificial environment before actually testing them with physical, often sensitive, equipment. Allowing managing the System under Test and addressing the capabilities of both system types, for example, during commissioning aims at tackling challenge *TAF-C3*.

TAF-R5. Efficient Tool and Method Support For testing, a wide range of different methods and testing tools, both vendor-specific and open source, are available to support different phases of the testing process. Examples are *Polarion* for test management or *Cucumber* for Behavior-Driven Testing. However, often it relays to the project manager to select approaches that best fit the project requirements and the project context and the testing team (i.e., a team incorporating different testing roles) to implement selected approaches. Thus, the requirement for efficient tool and method support aims at addressing challenges *TAF-C2*, by supporting advanced automation, *TAF-C3*, by enabling more flexibility, and *TAF-C4*, by allowing efficient and effective artifact management to support multiple roles.

TAF-R6. Efficient Test Code Management This requirement focuses on providing an efficient test code management throughout the testing process and environment. Such a management solution is needed because in industry practice, even if test cases are formally available, the software test code and data to be executed are still not always available project-wide or even better company-wide. Challenge *TAF-C1* should be addressed by decoupling engineering roles. Nowadays, these roles are often responsible for several tasks that compete with each other or require the knowledge of multiple domains.

Based on relevant test automation stakeholders and the six identified requirements (TAF-R1 to TAF-R6), we propose a test automation framework that is capable of addressing identified challenges (TAF-C1 to TAF-C4).

9.5.2 Flexible Test Automation Framework

The proposed *Test Automation Framework (TAF)* (see Fig. 9.5) includes four basic layers, that is, (L1) the *Test Management Layer*, (L2) the *Test Specification Layer*, (L3) the *Test Execution Layer*, and (L4) the *Test Environment Layer* (Winkler et al. 2018a).

Each layer can contain several specialized testing tools or methods, for example, *qTest* on the Test Management Layer, *Cucumber* on the Test Specification Layer, or *Jenkins* on the Test Execution Layer. Simulations or physical devices are located on the Test Environment Layer. Note that many tools also vertically span over some of the layers and provide, for example, a mechanism to manage tests and run their implementation done in *JUnit* on an internal build server. Different layers or capabilities of defined testing tools are connected by dedicated interfaces that allow efficient and automated data exchange. An exemplary testing tool chain that illustrates a path through the layers is shown in Fig. 9.5 in the gray boxes on the right-hand side. The selection of the tools is based on discussions with industry partners on real-world use cases, that is, the *Industry 4.0 Testbed* and from our experience in software development and software testing.

The flexible configuration of the set of selected testing tools between the layers, for example, *qTest* and *Cucumber*, is based on connections that are made by defined and shared interfaces and models that represent relevant information of the source layer and the destination layer. Figure 9.5 also describes the most relevant testing

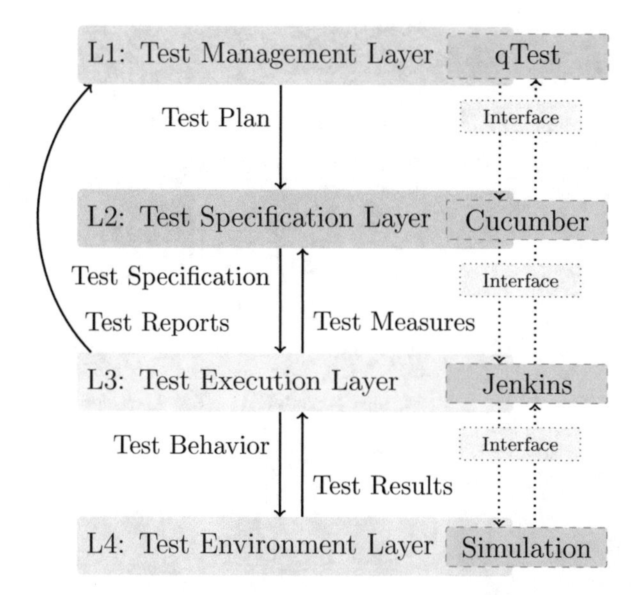

Fig. 9.5 Architecture for a flexible test automation framework, based on Winkler et al. (2018a)

artifacts exchanged between the different layers of the framework according to ISO 29119 (ISO/IEC/IEEE 2013). The interfaces and models provided by the *TAF* and located between the tools are shown as light-gray boxes in the figure. These interfaces and models are based on common concepts (Winkler et al. 2017c) that allow a quick and straightforward method to connect tools and artifacts of different origin. The negotiation of common concepts between the roles that are responsible for the layers and the implementation of shared interfaces require expertise from several disciplines such as software and knowledge engineering. Furthermore, the implementation of the common concepts and the shared interfaces as well as the initial configuration of the *TAF* require effort from engineers of different domains (Biffl et al. 2016), especially by the test manager and test automation engineer.

9.5.3 Test Automation Framework Prototype and Evaluation

In context of the *Industry 4.0 Testbed* and the *TAF* architecture, we developed a prototype implementation to show the feasibility of the proposed *TAF* approach.

Figure 9.6 illustrates the testing process of the *TAF* prototype application for robot arm testing and involved tools.

1. *Test Management*: In the prototype, we applied *Behavior-Driven Development*, that is, *Gherkin* scenarios, as technique for testing the functionality of a robot arm from a robot-engineering point of view. In this step, *Domain Experts* and *Test Developers* negotiate the most relevant test steps by using natural-language

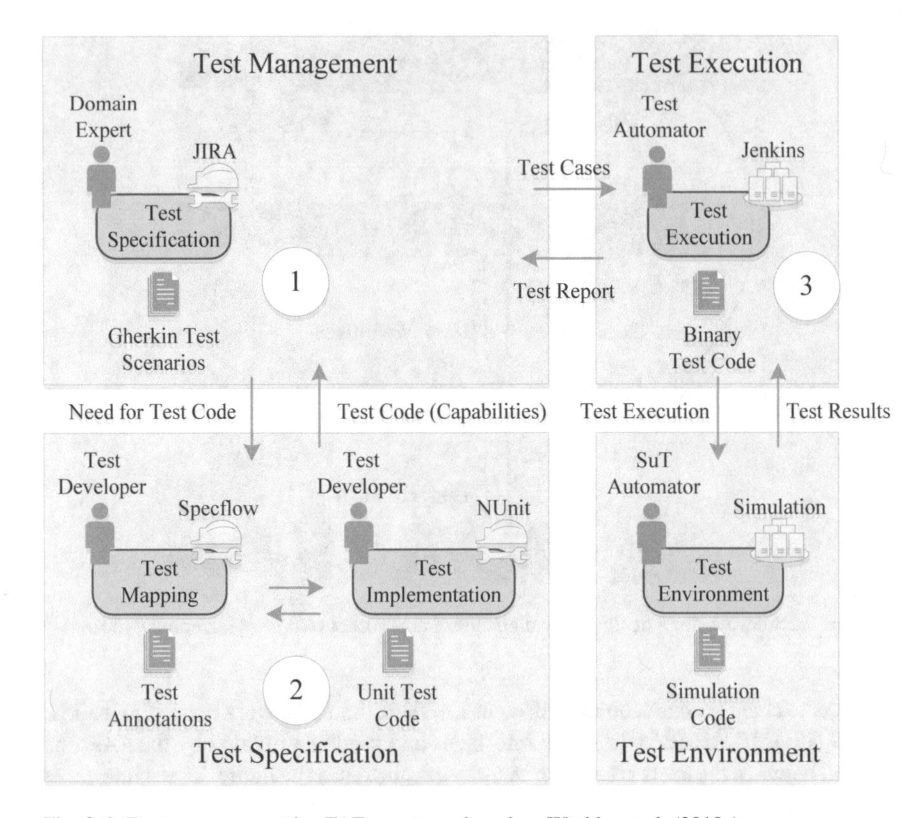

Fig. 9.6 Testing process with a TAF prototype, based on Winkler et al. (2018a)

constructs. These constructs represent the *Need for Test Code*. This task also includes a clarification of the needed test parameters. As soon as the negotiations are completed, the domain experts can start writing test scenarios for acceptance or integration tests of the robot arm based on test step fragments. We selected *Jira*,[10] a well-established issue tracking system in software engineering, as test management tool and implemented a plugin for test scenario configuration.

2. *Test Specification*: In the second step, the test developers annotate stubs of test code with the negotiated test step fragments and thus create a *Test Mapping* between the scenario steps and the test code, as illustrated in Fig. 9.1 in the code snippet in the middle. During the compilation of these stubs, tools like *Specflow* link the (*Gherkin*) test step fragments to the (*JUnit*) test source code to allow relations between test scenarios and unit tests during test execution. For convenience, already existing test fragments are provided to the *Jira* plugin to allow the auto-completion of test steps for domain experts. Therefore, a small application analyzes and reads existing test step templates from a Git code repository such as

[10] Atlassian JIRA: www.atlassian.com/software/jira

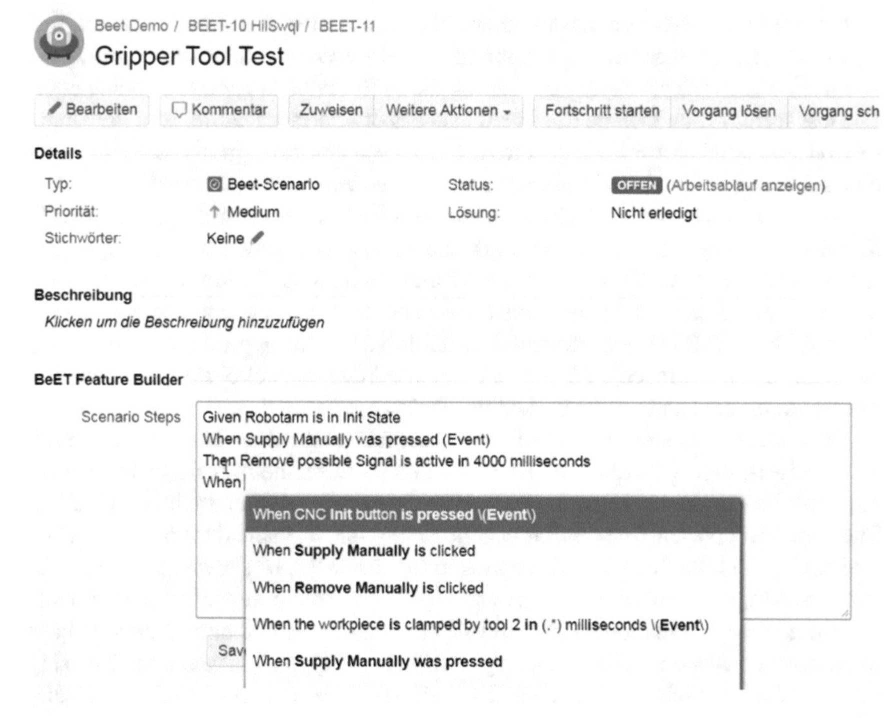

Fig. 9.7 Creating behavior-driven development scenarios in the Jira plugin with test steps from a shared code repository

Github[11] and provides them via *REST* interface to the plugin. During this step, the *Test Developers* also implement the tests that are subsequently committed to a shared repository so that they can be used for test execution.

3. *Test Execution and Test Environment*: In the third step, *Domain Experts* trigger test runs on the build server, that is, *Jenkins*, via the *Jira* plugin. Therefore, *Jenkins* pulls the *Test Cases* from Jira and executes them on the *System under Test (SuT)*, which is in this case a *Simulation* controlled by an *OPC UA server*[12] to abstract the robot arm. After the test execution, *Jenkins* collects the *Test Results* and processes them for better understandability. These results are then sent back to *Jira* as fast feedback to enable the domain experts to inspect the results and draw their conclusions. When the domain experts use test fragments that are not yet supported by the corresponding test code, the test report can be used to directly communicate the need for the particular fragment to the test developer.

Figure 9.7 presents screenshots derived from the prototype implementation and shows the specification of a test case, that is, test *BDD* scenario, in the prototype

[11]Github: www.github.org

[12]OPC UA: https://opcfoundation.org

of the Jira plugin. The screenshot shows the perspective of a domain expert, that is, a robot arm engineer that is supported by test step recommendations, which are extracted from a *Git* repository containing the *BDD* annotated unit test source code. The test scenario can then be transferred via a user interface feature from the plugin to a *Jenkins* build server that executes the test steps described in the scenario either on a simulation or on an isolated robot arm embedded in the *Industry 4.0 Testbed.*

For evaluation purposes, we built on the identified requirements (TAF-R1 to TAF-R6) and the prototype implementation and conceptually evaluated our approach for a flexible *TAF* to show its strengths and limitations. Note that we included additional requirements for implementation and maintenance, and test automation (TAF-R7 to TAF-R11). We compared a traditional manual approach (without using any automation approach), a less flexible and hardwired tool chain (as a common practice approach) and the novel flexible *TAF*.

Table 9.2 presents the results of the conceptual evaluation. The rows represent requirements, that is, TAF-R1 to TAF-R11 that focus on industry needs in the PSE domain who want to implement a test automation process within their organization. The columns represent three different approaches, that is, manual testing, hardwired tool chains, and the flexible TAF approach: (a) *Manual Tool Combination* refers to isolated testing tools that are applied manually in a PSE environment by experts. Note that there is no automation supported tool chain available and human experts have to handle exchanged artifacts manually. Artifacts are typically transferred by using communication mechanisms such as E-Mail; (b) *Hardwired Tool Chains* typically utilize a set of selected tools for a particular project that are configured in such a way that the tests run semiautomatically based on a particular rule set, such as commit triggers. Note that there is limited flexibility in case of required changes in the test environment, test approach, or the SuT; and (c) *Flexible Tool Chain* based on our

Table 9.2 Support of requirements in different test automation variants based on Winkler et al. (2018a)

Requirements		Test automation variants		
		Manual tool combination	Hardwired tool chains	Flexible tool chains
TAF-R1	Clear role definition	–	O	+
TAF-R2	Configurable tool chains	O	–	+
TAF-R3	Extensibility of tool chains	O	–	+
TAF-R4	System under test management	O	–	+
TAF-R5	Method and tool support	O	O	O
TAF-R6	Test code management	O	+	+
TAF-R7	Initial implementation effort	O	–	–
TAF-R8	Maintaining the test approach	O	–	+
TAF-R9	Integrated application	O	O	+
TAF-R10	Frequent test runs	–	+	+
TAF-R11	Test and result traceability	–	O	+

"–" weak/no support; "O" neutral support; "+" well supported proposed *TAF* including common interfaces for data exchange and maintainability capabilities in case of changes.

For the *Manual Tool Combination* approach of testing tools, the flexibility of the approach depends on the capabilities of individual engineers and their knowledge. Therefore, most of the values are set to neutral. However, the manual approach often needs experts that unify several roles to execute the tests and interpret them. This is often a bottleneck for testing and, furthermore, results in less frequent test runs, which lowers the quality of the software. Furthermore, the traceability of these manual runs is weak. *Hardwired Tool Chain* supports a separation of the roles to a certain extent, but still often one engineer has to have the knowledge of several roles or needs to perform tasks of other roles. The hardwired approach also does not allow an easy extension and configuration of the testing tool set due to missing interfaces and configuration options. Compared to the manual approach, the initial implementation effort to set up a hardwired tool chain is considerable higher. Nevertheless, such tool chains allow more frequent test runs, for example, on code commit basis, that are more traceable for other stakeholders of the testing process. Finally, the *flexible TAF approach* supports most of the requirements well compared to the manual and the hardwired approach. We, especially, laid our attention on the extension and the configuration of the tool chains, to support a variety of testing tools that best support the particular requirements of certain projects. Another advantage is that the resulting tool chain can be managed from a single interface to avoid the clutter of different technologies. However, the initial effort for the design and implementation of such a flexible *TAF* should not be underestimated.

In context of maintenance projects, a single robot arm within the *Industry 4.0 Testbed*, the *flexible TAF*, can support experts in rerunning predefined test cases based on natural-language constructs and implemented test code. However, in context of our sample production system and the illustrative use case, introduced in Sect. 9.4, where a robot type should be exchanged by a newer version, that is, exchanging KUKA KR Agilus with KUKA LBR iiwa, there is still the issue of different underlying programming languages, required by different robot types. Therefore, test cases might fail because of incompatible software (control) code and test code constructs despite similar test scenarios and expected behavior. Thus, there is the need for some support for transferring robot control code and test control code.

9.6 Validation of Different Software Control Code Variants

The process of *Production Systems Engineering (PSE)* contains several phases from the requirements engineering, to system design, implementation, and testing, as well as the operation and maintenance phase (Broy 2006). An important phase in *PSE* projects is the maintenance phase, where often parts of the system need to be reconfigured, updated, or exchanged with newer or more efficient soft- and hardware components.

A common problem in the maintenance phase is that individual devices in the production system often use vendor-specific programming languages, which might even diverge between different device types from the same vendor. Therefore, engineers have to execute three main activities: (a) *derive requirements* directly from specification documents (which might have changed over time), running systems (as they behave like expected), or control code implementation; (b) *develop software and test code* from scratch in the "new" programming language; and (c) *verify and validate software and test code* according to the expected behavior. Because of these time-consuming activities, new approaches are required to support engineers in deriving requirements, developing new software control and test code, and validating new code components. Note that test code components can be easily reused in the code based of the *Test Automation Framework (TAF)*, as described in Sect. 9.5.

In this section, we focus on analyzing the structure of software code as foundation for (a) using the structure as a blueprint for the human-based development of software code in the new programming environment and (b) for validating the structure of different code variants (i.e., old software code vs. new software code) for compliance and quality assessment. Given a manual implementation of new software control code and test cases and with respect to the 4-eyes principle, where the construction and validation of software code should be based on different sources (such as software code and a model), we focus on the validation of different code variants based on the code structure.

Therefore, we identify challenges and requirements in the context of a maintenance and evolution project in *Production Systems Engineering (PSE)* and propose and conceptually evaluate an approach that utilizes an *Abstract Syntax Tree (AST)* model to improve the validation and verification of the control code of the robot arms from the use case.

9.6.1 Challenges and Requirements of PSE Maintenance

In Sect. 9.5.1, we identified stakeholders that are mainly relevant for the testing phase in *PSE* projects. However, in context of a maintenance phase, a similar set of stakeholders is relevant. Furthermore, operators and system maintenance experts have to join the group of experts. In this section, we summarize the most important stakeholders, especially in context of test automation (please refer to Sect. 9.5.1 for more details):

- *Maintenance Engineers* are typically responsible for a maintenance and/or evolution project, often as some project management role. Thus, the expertise typically includes a comprehensive view on related disciplines.
- *Operators* who are responsible for keeping the system running can bring in experiences from the runtime perspective. In addition, improvement suggestions could bring relevant input for the maintenance project.

In context of testing and test automation, similar aspects, compared to an engineering project, are relevant because changed system parts (such as exchanging a robot type) need to be tested accordingly. Referring to software control code that needs to be rewritten and validated, testing becomes a critical issue that could be handled *by the Test Automation Framework (TAF)*, described in Sect. 9.5. However, support for engineers is needed for the creation and verification/validation of control code components. Selected important stakeholders include:

- *Domain Experts* who have specific expertise and insight to the technical details of the *System under Test (SuT)*, in this case the robot arms and their desired behavior in the context of the production system. They are responsible for the derivation of the logical test cases and the correct description of what the robots should do.
- *Software Developers* who implement the desired behavior of the system, that is, the robot arms, based on the specifications of the domain experts in the programming language specific to the particular device. Furthermore, they are responsible for refactoring and rewriting software, for example, in the context of a maintenance project.
- *Test Developers* who are responsible for the implementation of the concrete test cases in combination with the relevant test data that results from the logical test cases of the domain experts.

A common industry goal in context of maintenance projects is to keep system downtimes as short as possible, even if critical changes are needed. Typically, these changes can require high human effort for software control code construction and for verification and validation. Thus, engineers need support in (a) deriving/writing new software code (in a different programming language) and (b) to effectively verify and validate newly constructed software control and test code components. Main challenges include the following.

AST-C1. Requirements Elicitation In the engineering phases, specification documents typically hold explicitly defined requirements and or design decisions. During operation, maintenance engineers often implement changes without updating engineering documents accordingly. Thus, it is often hard to elicit (implemented) requirements in a maintenance project as software control and test code hold these implemented requirements implicitly. Therefore, it is challenging to elicit the functional behavior based on the implementation manually. Thus, there is the need for supporting the requirements elicitation process.

AST-C2. Code Construction Because of different programming languages used in the evaluation use case, that is, two robot types with different programming languages, there is the need for transferring software control and test code to the new development environment. Due to different programming paradigms, this software code has to be rewritten manually. However, engineers can benefit from the structure of the old code to support the development of the new code. Thus, this challenge addresses limitations in context of deriving the structure of code components, as these code components have to be analyzed manually in a time-consuming and error-prone way.

AST-C3. Engineering Knowledge In *PSE*, engineers often have a very specific engineering knowledge that helps them to perform their tasks. However, often engineers have to gain additional knowledge that they do not need for their everyday tasks to be able to maintain a system. This knowledge is then often superficial knowledge as it is not properly applied and refreshed. Thus, it is necessary to support engineers in gaining this knowledge based on the existing implementation.

AST-C3. Verification and Validation In a maintenance project, where a device type (such as a robot) has to be replaced by another device type, the verification and validation of code variants is success critical, as the behavior should be similar. In industry practice, this verification and validation task is time consuming as engineers have (a) to inspect code components manually or (b) have to evaluate the expected system behavior based on test cases and by executing the new software components. Furthermore, programming language characteristics need to be considered accordingly. Thus, there is the need for supporting engineers in the verification and validation tasks by lifting implementation specific characteristics to a more abstract level, for example, by using an abstract representation.

Based on these challenges, that is, AST-C1 to AST-C4, and discussions with our industry partners and experts, we derived a set of requirements (AST-R1 to AST-R5) that need to be addressed by a modified system maintenance process with human-based verification and validation based on abstract representation of different control code variants. Table 9.3 summarizes identified challenges and requirements and indicates their relationships.

AST-R1. Clear Role Definition The separation of role definitions aims at distributing the workload of experts. This separation addressed all challenges (AST-C1 to AST-C4) in that engineers have clearly defined responsibilities and capabilities and

Table 9.3 Challenges and requirements toward human-based verification and validation of software control code variants

Challenges		AST-C1	AST-C2	AST-C3	AST-C4
Requirements	Support for...	Requirements elicitation	Engineering knowledge	Software code construction	Verification and validation
AST-R1	Clear role definition	✓	✓	✓	✓
AST-R2	Abstract code representation	✓		✓	✓
AST-R3	Deriving the code structure	✓	✓		
AST-R4	Extensibility of code representations		✓		✓
AST-R5	Extensibility of language elements		✓	✓	✓

that they do not need to share and maintain specific knowledge, for example, for requirements elicitation based on code, for code construction, or for verification and validation based on models such as an *Abstract Syntax Tree (AST)* model.

AST-R2. Abstract Code Representation An abstract representation of the code addresses challenge AST-C1, AST-C3, and AST-C4 as it abstracts certain parts of the control code that are not needed and lowers the verification effort as the concept can be faster grasped and evaluated. Based on the abstract model, requirements can be derived based on the structure of code components (AST-C1). Developers are able to reuse abstract representations for code construction (without knowing details of the existing code written in the origin language) (AST-C3). Finally, quality assurance experts can use two abstract representations for human-based inspection based on the structure of the code components, without knowing details on the detailed implementations.

AST-R3. Deriving the Code Structure Similar to AST-R2, deriving the software code structure is important for retrieving (a) requirements and (b) engineering knowledge and design decisions based on existing software control and test code. These abstract representations and knowledge represent the foundation for developers to construct/create software control code and test code in the new environment and as "blueprint" of the functional behavior of the system. Furthermore, derived abstract representations from existing and new software code represent the foundation for human-based inspection for quality assurance experts.

AST-R4. Extensibility of Code Representation An extensibility of the code representation aims at supporting different representations of the *AST*. An *AST* is a tree-like structure that can get quite large and complex, which makes them hard to inspect. When different *AST* representations of the source code exist, for example, partial or navigable trees, the required engineering knowledge does not need to be as proficient (AST-C2) and the verification and validation task can be eased through better tool support (AST-C4), like visual highlighting of elements.

AST-R5. Extensibility of Language Elements The extensibility of language elements addresses challenges C6–C8 as it allows further integration of programming languages and constructs, for example, device-specific languages or standardized languages such as IEC 61131-3[13] language families. Therefore, more engineering disciplines can be supported by abstracting engineering knowledge needed for the validation (AST-C2). An extension of the language elements also supports engineers during code construction, as they are able to inspect old code snippets in an abstract representation (AST-C3). Furthermore, the verification and validation tasks are obviously made easier when ASTs for several artifacts exist then just for a few.

[13]IEC 61131-3: https://webstore.iec.ch/publication/4552

Identified challenges (AST-C1 to AST-C4) and requirements (AST-R1 to AST R5) represent the foundation for an adapted maintenance process with human-based verification and validation based on the structure of source and test code components.

9.6.2 Adapted PSE Maintenance Process with Human-Based Verification and Validation

Figure 9.8 presents the adapted *PSE Maintenance Process with human-based verification and validation* including (a) maintenance planning, (b) analysis and implementation, (c) verification and validation (V&V), (d) deployment, and (e) operation. In the *planning phase* (Fig. 9.8A), the required maintenance tasks are planned according to customer needs. In the evaluation example, the maintenance task includes exchanging a robot arm of type KUKA KR Agilus with a KUKA LBR iiwa. Because of the different languages used to program these robot types, control and test code needs to be transferred to a new engineering environment. Thus, the *Analysis and Implementation phase* (Fig. 9.8B), includes two steps: (B1) *Derivation of Requirements* based on existing (but maybe outdated) specification documents and existing control code implementations as well as (B2) *Code Implementation* of robot control code in the new engineering environment. The *Verification and Validation (V&V)* (Fig. 9.8C) process phase is required to ensure similar behavior of the new variant of the software control code. This V&V includes three subprocess steps. First, the *Validation Planning* (C1) step focuses on selecting the validation approach, for example, traditional reviews, trial-and-error and testing, or human-inspection. The *Human Inspection* step (C2) includes an expert-based comparison of two abstract software code models, that is, *ASTs*, in order to evaluate both models (similar behavior) and to identify defects. *Model-Quality Assurance (MQA)* techniques, introduced in Sect. 9.2.4, can be used to distribute the workload and to use an expert crowd to identify defects effectively and efficiently. However, for

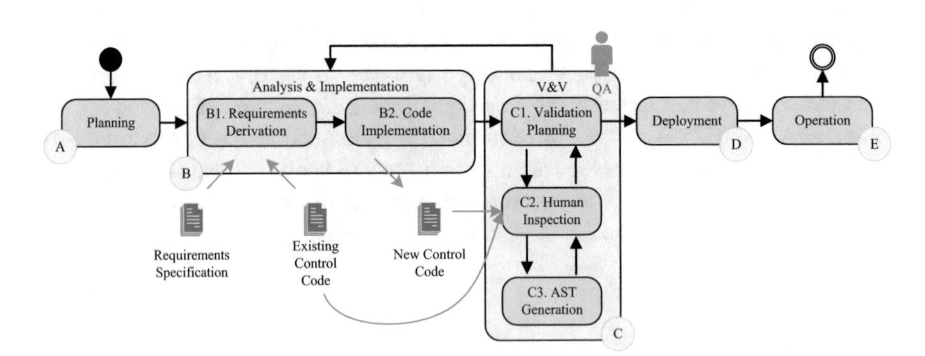

Fig. 9.8 PSE maintenance process with human-based verification and validation (V&V) of software code variants

applying *MQA*, models and/or specification documents need to be available. Thus, in the third process step, that is, *AST Generation* (C3), AST models need to be derived. This means they are either generated or created manually from the existing software control code (i.e., as a reference) and the new software control code (that should be inspected for correctness). Note that there is a feedback loop to the previous process phase in case of deviations, that is, defects. While, as a next step, the *Deployment Phase* (Fig. 9.8D) focuses on deploying (and testing) the new software control code in the target environment, the Operation Phase (Fig. 9.8E) focuses on continuing the operation of the production system in the daily business.

The *Abstract Syntax Tree* (AST) can represent the foundation for (a) requirements elicitation, (b) as a blueprint for control code development, and (c) for human-based verification and validation. Note that, in the context of this chapter, we focus on the human-based V&V approach. Figure 9.9 illustrates the subprocess of deriving an abstract representation of the software control code. The figure shows a snippet of the robot control code on the bottom-left side. In the first step, indicated with 1, a parser that can interpret the particular programming language reads this source code and creates an internal model of the code. In a second step, marked with 2 in the figure, the internal model of the parser is transformed to a more generic model, that is, the *ASTM* model of the *OMG*, shown on the bottom right. In our case, the Java code of the Kuka LBR iiwa robot arm is read with the standard Eclipse Java parser that translates it to the internal JDT source model. Then this JDT source model is transformed into a Java model that we created from the specification of the *ASTM* EMF model. The figure shows, for example, how the `switch` or the `getProgramNumber` call elements from the code on the left-hand side are represented in the *AST* on the right-hand side. Having a single model like the *ASTM* that maps to different but similar programming languages has the advantage that a single model can be used

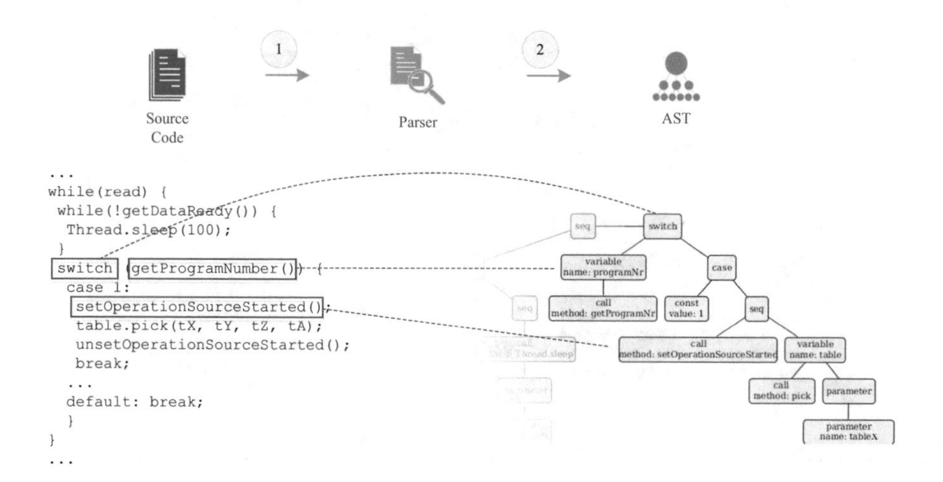

Fig. 9.9 Concept for transferring control code into an abstract syntax tree (AST) representation

for different types of visualization, for example, using GraphML and related tools or further processing of the *AST*.

9.6.3 Prototype and Evaluation of Human-Based Verification and Validation of Software Control Code Variants

In the context of the evaluation use case (see Sect. 9.4), we have implemented a prototype to show the feasibility of the proposed approach. Figure 9.10 presents the application of the adapted *PSE* maintenance process with human inspection based on an abstract code representation of two control code variants. The existing robot control code for the KUKA KR Agilus, which uses the KUKA KRL programming language, including the derived AST model, is shown in Fig. 9.10 on the left-hand side. The right-hand side of Fig. 9.10 presents the new software control code and the derived *AST* model for the KUKA LBR iiwa, using *Java* as the programming language.

In the context of the adapted maintenance process introduced in Sect. 9.6.2 and Fig. 9.8, we focus on the V&V process step (C), where different AST model

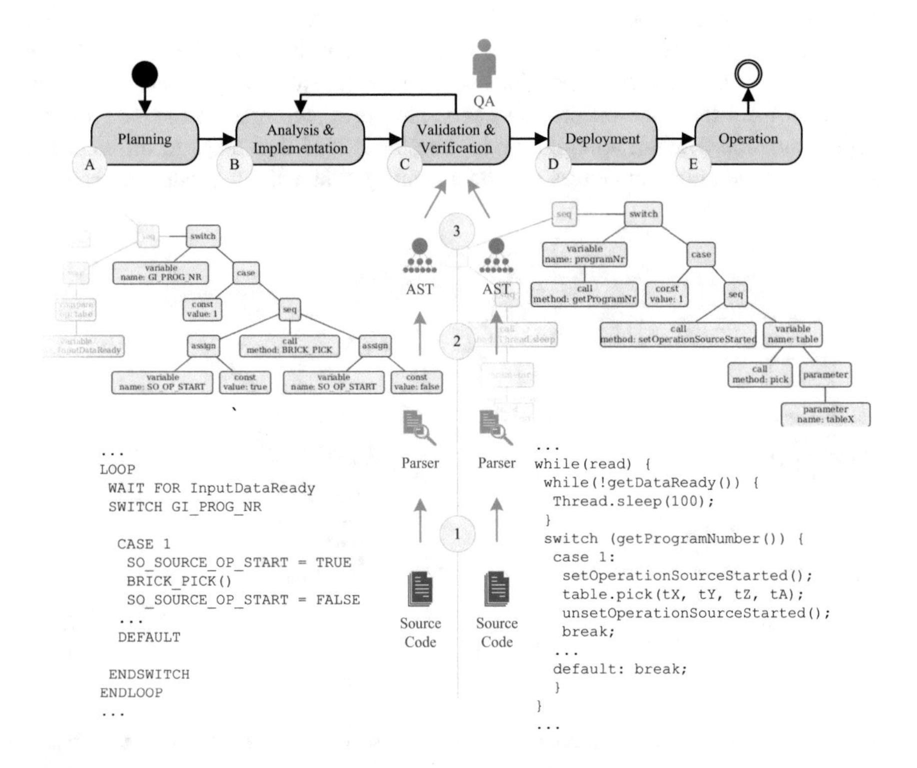

Fig. 9.10 Validation process for control code variants based on AST

representations (a) from existing software control code and (b) the new software control code are available. We follow the approach to derive an abstract representation of source code in an abstract syntax tree (see process steps 1–3 in Fig. 9.10). For deriving *AST* representations, we developed, utilizing existing parsers related to each programming language. These parsers extract relevant code constructs from the source code components and build up two *AST* versions.

Quality assurance engineers take both *AST* models and check, whether or not these *AST* representations implement similar systems behavior. Therefore, we followed the *MQA* process approach as described in Sect. 9.2.4 for human-based inspection. Note that in the evaluation part we did not apply any crowdsourcing approach but focused on a paper-based approach for inspection including three main steps: (a) *Preparation*, (b) *Model Analysis*, and (c) *Defect Validation*. In the preparation phase, inspectors identified *Expected Model Elements (EMEs)*, comparable to language constructs in existing software code components. In the model analysis part, inspectors take these language constructs and check whether or not these constructs are represented correctly in the second AST model, derived from the new software code. Inspectors report defects by taking notes on the identified deviation in a paper-based way. Note that three experts execute the inspection tasks, which results in three candidate defect lists. In a team meeting, the experts discuss their findings and derive a common list of defects. In this initial evaluation, we focused on a paper-based inspection approach, following the *MQA* process. However, in future work, this paper-based process approach will be replaced by a crowdsourcing based *MQA* process approach to improve defect detection capabilities, reduce inspection effort, and distribute the workload of paper-based inspection.

To evaluate the adapted maintenance approach with human-based inspection and *AST* models, we compared different aspects (i.e., manual maintenance process, AST-Based Development, and AST-Based V&V) of the maintenance process concerning identified requirements.

Table 9.4 presents the results of the conceptual evaluation. The rows represent requirements as discussed in Sect. 9.6.1 that we derived from discussions with industry partners. Note that we added two requirements, relevant for setup and maintenance of AST generation. The columns represent different aspects of a PSE maintenance project. (a) *Manual Maintenance Project*, where engineers typically have existing code components and need to derive requirements based on existing code components, develop new software code, and verify and validate these new code components manually. (b) *AST-Based Development* focuses on using an AST model as a blueprint (representative for systems requirements) for developing new software control code. (c) The *AST-Based Verification and Validation* (V&V) approach takes two AST models, derived from existing and new software control code as input for evaluation. With a focus on AST-R1 (*Clear Role Definitions*) there is no support in the manual process but there exist defined roles for the AST-Based approaches. *AbstractCode Representations* (AST-R2) and *Code Structures* (AST-R3)are not available in the manual approach (but have to be created manually, if needed) and are generated for the AST approaches by using parsers. AST-R4 (*Extensibility of Code Representations*) and AST-R5 (*Extensibility of Language Elements*) are not

Table 9.4 Support of requirements in different system maintenance processes

Requirements		System maintenance process		
		Manual maintenance process	AST-based development	AST-based V&V
AST-R1	Clear role definition	O	+	+
AST-R2	Abstract code representation	–	+	+
AST-R3	Deriving the code structure	–	+	+
AST-R4	Extensibility of code representations	O	+	n/a
AST-R5	Extensibility of language elements	O	+	n/a
AST-R6	AST generation	n/a	+	+
AST-R7	Setup of AST parsers	n/a	–	–

"–" weak/no support; "O" neutral support; "+" well supported; "n/a" not applicable

supported by the manual approach but help in the *AST-based development*. Note that for *AST-based V&V*, these requirements are not relevant as no extensions are required for inspection. *AST Generation* (AST-R6) is not relevant for the manual process but necessary for *AST-based development* and *AST-based V&V*. If available, AST models can be generated frequently on an automated basis if needed. However, there is a need for setting up *AST parsers* (AST-R7), which could take some effort. However, these parsers can be easily reused for similar problems, as they build on language-specific code constructs.

9.7 Discussion, Limitations, and Future Work

In this chapter, we focused on the key question to enable *flexibility in test automation* throughout the *Production System Engineering (PSE)* life cycle. Based on discussions with industry partners and domain experts, we identified the *maintenance phase* as a critical life cycle phase as changes in this phase could have a major impact on system downtime and engineering/maintenance effort. Based on the need for more flexible test automation in *PSE*, we focus on (a) how to enable an efficient and effective reconfiguration of testing tool chains and (b) to verify and validate control code model variants with human computation. We base the underlying use case on the *Industry 4.0 Testbed*, located at the Czech Technical University in Prague, where several robot arms and a rail conveyor form a production system.

9.7.1 *RQ.1: Flexible Test Automation*

During development, but more critical during system maintenance, testing tools may change in the testing tool chain. Such changes hinder automated testing of modified system parts, such as robots. Thus, we observed the need for a flexible test automation framework that is capable of handling system environment changes. Based on this requirement, we identified the following research question.

RQ.1: How Can We Support Flexibility in Test Automation to Allow Reconfiguration of Testing Tool Chains in PSE Contexts? In Sect. 9.5, we described related stakeholders, challenges, and requirements for a flexible test automation tool chain in *PSE* and presented an architectural solution approach, the *Test Automation Framework*. Furthermore, we showed the feasibility of the proposed approach based on a prototype and an illustrative use case. In comparison to manual tool combinations, and hardwired tool chains, we identified strong benefits of the proposed *Test Automation Framework (TAF)* regarding clear role definitions of testing stakeholders, support for extensibility and adaptability because of introduced interfaces between tool layers, and test code management that enable reuse and frequent test runs. Although the prototype shows the feasibility of the *TAF* approach, additional effort is required to setup interfaces for individual tools related to test automation layers.

Limitations In the context of this chapter, we focused on a selected set of software testing tools (assigned to individual layers) for the prototype implementation. Nevertheless, additional testing tools need to be integrated and might require adaptations of the interface architecture. Furthermore, we identified testing tools that cover more than one layer of the testing framework, for example, test management and test case design. These kinds of tools include implicit (hardwired) interfaces that cover two or more layers that need to be considered in the proposed *TAF*. We evaluated the flexible *TAF* in the context of a component in the *Industry 4.0 Testbed*, that is, a robot arm located within an isolated workstation. Although the TAF concept seems to be capable of handling larger systems, there is a need for more detailed evaluation in a broader context.

Future Work Based on the limitations of the prototype implementation, we see the need for future investigation in three areas. These areas are: (a) extending the set of selected tools and tool suites to improve and extend the TAF toward industry applications, (b) in-depth evaluations of the proposed prototype implementation to gain additional insights in the performance of the TAF concept, and (c) industry evaluations with more extensive systems to investigate scalability issues of the proposed approach.

9.7.2 RQ.2: Evaluation Support for Control Code Variants

The support of *Behavior-Driven Tests* and the *Gherkin* language for test case design still requires software test code that is needed to execute test cases. In industry settings (and also in our illustrative use case) we identified the need for software control code transfer from one development environment to another. A typical task in PSE maintenance and evolution projects is to exchange device types, which could use different programming languages, such as KUKA KR Agilus and KUKA LBR iiwa. Therefore, we see the need for supporting engineers in transferring software control and test code. Based on this need we identified the second research question.

RQ.2: How Can We Evaluate Control Code Versions Based on the Expected System Behavior Effectively and Efficiently? In Sect. 9.6, we identified stakeholders, challenges, and requirements, proposed an adapted systems maintenance process with the focus on verification and validation, and presented a prototype implementation in the context of the *Industry 4.0 Testbed.* Changing the programming language requires additional engineering steps such as requirements elicitation (from specification, running systems, or software code), developing new software components based on derived requirements, and verification and validation according to the expected system behavior (as the system behavior may not change). In common practice, these steps are usually executed manually by maintenance experts. We identified the *Abstract Syntax Tree (AST)* as an abstract representation of (existing) source code components that can be used (a) as a blueprint for code construction and (b) for verification and validation of different code variants. The *AST* as models allows to bridge gaps in human inspection as concepts can be better represented for experts from various domains. In the conceptual evaluation, we identified benefits of the AST application as a foundation for developing new software components and as a foundation for human-based inspection based on AST models (i.e., an AST model based on existing software code and an AST model based on the new model). Although the AST generation could be automated, an initial effort is required to derive the code constructs from source code, for example, by developing programming language-specific parsers.

Limitations Although we see the adapted maintenance process approach useful for supporting engineers in systems maintenance activities, the AST model approach has two limitations: First, even simple software code can result in complex AST models that need tool support for inspection (e.g., searching, automated comparison, or filtering). Second, based on the underlying software code structure, a human-based comparison is still challenging because of different derived language constructs. We evaluated the adapted maintenance process in a robot cell of the *Industry 4.0 Testbed.* Although this setting is representative for industry, its size is limited. Therefore, additional investigations are required to focus on larger and more complex system environments. Finally, we used two selected robot types (coming from the same vendor but with different programming languages). Given a wide range of vendors and robot types, there is still the need for investigating the variability of different robot types but also for other devices, relevant in a PSE. A concept that also seems promising is to use behavioral models of program control code that can be compared, for example, using approaches like fUML, which provides a run-time environment that is capable of executing activity diagrams. However, the parsing and translation to a behavioral model would be much more complicated.

Future Work Based on the initial prototype in the *Industry 4.0 testbed*, research effort is needed in the context of different robot types and—more generally—different device types from the application point of view. With a focus on code construction and verification and validation, alternative approaches need to be investigated regarding the capability for requirements elicitation (from source code), code construction (in different languages), and verification and validation support. The MQA process approach has been found useful in the evaluation context. However, research is required for different model types and for tool support that can handle the verification and validation of models that are based on software code.

Based on the overall key question, how to enable flexibility in test automation in PSE, we see the flexible *Test Automation Framework (TAF)* as promising for supporting engineers in PSE in defining and automating their testing processes. Furthermore, automation-supported Maintenance Processes including human-based verification and validation based on software control and test code can represent the foundation for the TAF in context of testing but also for a human-based quality assurance approach based on models.

Acknowledgments The financial support by the Christian Doppler Research Association, the Austrian Federal Ministry for Digital and Economic Affairs, and the National Foundation for Research, Technology and Development is gratefully acknowledged. The research done by Petr Novak has been supported by the DAMiAS project funded by the Technology Agency of the Czech Republic.

References

Aurum, A., Petersson, H., & Wohlin, C. (2002). State-of-the-art: Software inspections after 25 years. *Journal of Software Testing, Verification and Reliability, 12*(3), 133–154.

Beck, K. (2003). *Test-driven development: By example*. Boston, MA: Addison-Wesley.

Biffl, S., Lüder, A., & Winkler, D. (2016). Multi-disciplinary engineering for Industrie 4.0: Semantic challenges, needs, and capabilities. In S. Biffl & M. Sabou (Eds.), *Semantic web for intelligent engineering applications*. New York: Springer.

Brambilla, M., Cabot, J., & Wimmer, M. (2012). *Model-driven software engineering in practice*. San Rafael, CA: Morgan & Claypool.

Broy, M. (2006). The 'grand challenge' in informatics: Engineering software-intensive systems. *IEEE Computer, 39*(10), 72–80.

Duvall, P. M., Matyas, S., & Glover, A. (2007). *Continuous integration: Improving software quality and reducing risk*. London: Pearson.

Erdogmus, H., Morisio, M., & Torchiano, M. (2005). On the effectiveness of the test-first approach to programming. *IEEE Transactions on Software Engineering, 31*(3), 226–237.

Gilb, T., Graham, D., & Finzi, S. (1993). *Software inspection*. Boston, MA: Addison Wesley.

Häser, F., Felderer, M., & Breu, R. (2016). Is business domain language support beneficial for creating test case specifications: A controlled experiment. *Information and Software Technology, 79*, 52–62.

ISO 29119-2. (2013). Software and systems engineering – Software testing – Part 2: Test processes, ISO/IEC/IEEE291192.

ISO 29119-5. (2016). Software and systems engineering – Software testing – Part 5: Keyword-driven testing, ISO/IEC/IEEE 29119-5.

Jones, J. (2003). Abstract syntax tree implementation idioms. In *Proceedings of the 10th Conference on Pattern Languages of Programs (PLOP)* (pp. 1–10). North Carolina: The Hillside Group.
Laitenberger, O., & DeBaud, J. M. (2000). An encompassing life cycle centric survey of software inspection. *Journal of Systems and Software, 50*(1), 5–31.
Laukkanen, E., Itkonen, J., & Lassenius, C. (2017). Problems, causes and solutions when adopting continuous delivery—A systematic literature review. *Information and Software Technology, 82*, 55–79.
Lindstrom, L., & Jeffries, R. (2003). Extreme programming and agile software development methodologies. In *IS management handbook* (8th ed.). Abingdon: Taylor & Francis.
Mårtensson, T., Hammarstrom, P., & Bosch, J. (2017). Continuous integration is not about build systems. In *Proceedings of the 43rd Euromicro Conference on Software Engineering and Advanced Applications (SEAA)*. IEEE.
Mårtensson, T., Ståhl, D., & Bosch, J. (2018). Enable more frequent integration of software in industry projects. *Journal of Systems and Software, 142*, 223–236.
Micallef, M., & Colombo, C. (2015). Lessons learnt from using DSLs for automated software testing. In *Proceedings of 8th International Conference on Software Testing, Verification and Validation Workshops (ICSTW)*. IEEE.
Moser, T., Mordinyi, R., & Winkler, D. (2012). Extending mechatronic objects for automation systems engineering in heterogeneous engineering environments. In *Proceedings of the 17th International Conference on Emerging Technologies & Factory Automation (ETFA)*. IEEE.
Musil, J., Musil, A., Weyns, D., & Biffl, S. (2015). An architecture framework for collective intelligence systems. In *Proceedings of the 12th Working IEEE/IFIP Conference on Software Architecture*. ACM.
Roche, J. (2013). Adopting DevOps practices in quality assurance. *Communications of the ACM, 56*(11), 38–43.
Rosen, R., von Wichert, G., Lo, G., & Bettenhausen, K. D. (2015). About the importance of autonomy and digital twins for the future of manufacturing. *IFAC-PapersOnLine, 48*(3), 567–572.
Sabou, M., Winkler, D., Penzerstadler, P., & Biffl, S. (2018). Verifying conceptual domain models with human computation: A casestudy in software engineering. In *Proceedings of the 6th AAAI conference on human computation and crowdsourcing (HCOMP)*. Zurich: AAAI.
Schatten, A., Biffl, S., Demolsky, M., Gostischa-Fanta, E., Östereicher, T., & Winkler, D. (2010). *Best Practice Software-Engineering: Eine praxiserprobte Zusammenstellung von komponentenorientierten Konzepten, Methoden und Werkzeugen*. Springer.
Schwaber, K., & Beedle, M. (2002). *Agile software development with Scrum*. London: Pearson.
Shahin, M., Babar, M. A., & Zhu, L. (2017). Continuous integration, delivery and deployment: A systematic review on approaches, tools, challenges and practices. *IEEE Access, 5*, 3909–3943.
Soeken, M., Wille, R., & Drechsler, R. (2012). Assisted behavior driven development using natural language processing. In *Objects, models, components, patterns, tools* (pp. 269–287). Berlin: Springer.
Solis, C., & Wang, X. (2011). A study of the characteristics of behaviour driven development. In *Proceedings of the 37th EUROMICRO Conference on Software Engineering and Advanced Applications (SEAA)*. IEEE.
Spillner, A., Linz, T., & Schaefer, H. (2014). *Software testing foundations: A study guide for the certified tester exam*. San Rafael, CA: Rocky Nook.
Ståhl, D., & Bosch, J. (2014). Modeling continuous integration practice differences in industry software development. *Journal of Systems and Software, 87*, 48–59.
Sunindyo, W. D., Moser, T., Winkler, D., & Biffl, S. (2010). Foundations for event-based process analysis in heterogeneous software engineering environments. In *Proceedings of the 36th EUROMICRO conference on software engineering and advanced applications (SEAA)*. IEEE.
Van Bennekum, A., Cockburn, A., Cunningham, W., Fowler, M., Grenning, J., Highsmith, J., Hunt, A., Jeffries, R., Beck, K., & Beedle, M. (2001). *Manifesto for agile software development*. Alicante: Universidad de Alicante.

Vogel-Heuser, B., Kegel, G., Bender, K., & Wucherer, K. (2009). Global information architecture for industrial automation. In *Automatisierungstechnische Praxis (atp)*. Munich: Oldenbourg-Verlag.

Vogel-Heuser, B., Fay, A., Schaefer, I., & Tichy, M. (2015). Evolution of software in automated production systems: Challenges and research directions. *Journal of Systems and Software, 110*, 54–84.

Whitehead, J. (2007). Collaboration in software engineering: A roadmap. In *Future of software engineering (FOSE)*. IEEE.

Winkler, D., Wimmer, M., Berardinelli, L., & Biffl, S. (2017a). Towards model quality assurance for multi-disciplinary engineering.Needs, challenges, and solution concepts in an AutomationMLcontext. In S. Biffl, A. Lüder, & D. Gerhard (Eds.), *Multi-disciplinary engineering for cyber-physical production systems, chapter 16* (pp. 433–457). Cham: Springer.

Winkler, D., Sabou, M., Petrovic, S., Biffl, S., Kalinowski, M., & Carneiro, G. (2017b). Improving model inspection with crowdsourcing. In *Proceedings of the 4th International Workshop on Crowdsourcing in Software Engineering*. Buenos Aires: ACM/IEEE International Conference on Software Engineering (ICSE).

Winkler, D., Sabou, M., & Biffl, S. (2017c). Improving quality assurance in multidisciplinary engineering environments with semantic technologies. In *Quality control and assurance - an ancient Greek term re-mastered*. London: Intec Publishing.

Winkler, D., Meixner, K., & Biffl, S. (2018a). Towards flexible and automated testing in production systems engineering projects. In *Proceedings of the 23rd International Conference on Emerging Technologies and Factory Automation (ETFA)*. IEEE.

Winkler, D., Sabou, M., Petrovic, S., Biffl, S., & Kalinowski, M. (2018b). Investigating a distributed and scalable model review process. *Centro Latinoamericano de Estudios en Informatica (CLEI), 21*(1).

Zhu, Y.-M. (2016). *Software reading techniques: Twenty techniques for more effective software review and inspection*. New York: Apress.

Chapter 10
Reengineering Variants of MATLAB/Simulink Software Systems

Alexander Schlie, Christoph Seidl, and Ina Schaefer

Abstract In a variety of industrial domains, quality and security are paramount factors during software system development. Model-based languages such as MATLAB/Simulink can improve software quality and are used for the development of safety-critical functionality. To comply with changing customer demands, product portfolios oftentimes emerge ad hoc by copying and modifying existing systems in an undocumented fashion. The proliferation of redundant, almost alike assets adversely affects the quality, maintenance, and evolution of the variant portfolio. To reinstate sustainable development, we describe a holistic approach to migrate the portfolio toward managed reuse by collapsing redundant parts and reengineering specific relations between similar, almost alike system parts. We elaborate on a technique to capture course-grained variability by assessing the portfolio as a whole, which identifies and groups together similar and redundant system parts. Such groups are analyzed further using a fine-grained comparison procedure, which captures their variability by means of common and varying system assets. The result is a variability model with redundant parts collapsed and reusable parts identified. By that, the overall size of the product portfolio is reduced, allowing for better quality assurance and improved maintenance. Furthermore, with knowledge about the relations between variants, affected software systems can be identified across variant boundaries, mitigating security concerns for the entire product portfolio.

Keywords MATLAB/Simulink · Legacy systems · Variability · Reverseengineering · Software product lines

A. Schlie (✉) · C. Seidl · I. Schaefer
TU Braunschweig, Braunschweig, Germany
e-mail: a.schlie@tu-braunschweig.de; c.seidl@tu-braunschweig.de; i.schaefer@tu-braunschweig.de

S. Biffl et al. (eds.), *Security and Quality in Cyber-Physical Systems Engineering*, https://doi.org/10.1007/978-3-030-25312-7_10

10.1 Introduction

Various industrial domains, such as avionics, automotive, and automation, rely on embedded software systems (Ryssel et al. 2010a). In such fields where large, complex, and safety-critical systems are developed, their reliability and maintainability are pivotal (Deissenboeck et al. 2008). Rather than implementing functionality using imperative languages, model-driven engineering, a paradigm known to improve reliability, is predominantly used in these domains (Cretu and Dumitriu 2014). Model-based languages such as *MATLAB/Simulink* can support modularity and are considered to enhance maintainability (Pressman 2005; Sullivan et al. 2001) and reliability by facilitating a test-driven development process for all design stages (Ryssel et al. 2010b). Despite allowing for functionality to be implemented on a more intuitive level for engineers, overall development and evolution of embedded systems remains a challenging and time-intensive task.

Changing customer requirements are commonly addressed by copying and subsequently modifying existing assets, typically referred to as *clone-and-own* (Riva and Rosso 2003). Saving both time and workload in the beginning, this approach entails considerable problems in the long run and has severe implications on maintenance and evolution of the emerging product portfolio (Pham et al. 2009). During *clone-and-own*, proper documentation is usually not cherished (Dubinsky et al. 2013). As a result, relations between variants are oftentimes rendered incomprehensible in the aftermath (Lapeña et al. 2016). With this form of unstructured and undocumented reuse, system quality can be negatively affected (Sullivan et al. 2001). For instance, crucial data might accidentally not be updated after modification and faulty parts can be unintentionally transferred to the cloned system. If not precisely detected in the first place, such parts can accumulate and account for inexplicable malfunctions at a later time (Schlie et al. 2017b). With a proliferation of redundant, almost alike assets, engineers then have to manually identify every single affected model variant and version separately and apply patches accordingly (Deissenboeck et al. 2010; Rubin and Chechik 2012). Such identification and rectification induce a vast manual overhead but are unavoidable to redress malfunctions and to reinstate sustainable development and evolution.

With evolution adversely affected and an increased cost to system quality in maintenance, *clone-and-own* can be a source of technical debt (Codabux and Williams 2013; Ernst et al. 2015). Characteristic to *clone-and-own* practices, bad architectural choices and a lack of proper documentation are driving factors of technical debt (Ernst et al. 2015), which accumulates and negatively impacts software systems quality and security (Ramasubbu et al. 2015). Unfortunately, *clone-and-own* prevails, especially with only minor rectification presumed necessary to comply with changing requirements (Rubin and Chechik 2013b). In addition, the specific workload required to rectify a system might be greater than to re-implement existing functionality again from scratch. Evidently, this procedure only amplifies variant diversity (Schlie et al. 2017b) and, thereby, technical debt. By collapsing redundancies and specifying common and varying parts, *Software Product Lines*

(SPLs) (Pohl et al. 2005; van der Linden et al. 2007) can facilitate strategic reuse and promote software maintainability and overall system quality (Kim 2010). However, actively migrating variants toward an SPL or taking other measures to reinstate sustainable development requires the systems' variability information (Grönniger et al. 2014) to be reengineered, which remains an enormous challenge for practitioners (Dubinsky et al. 2013; Kastner et al. 2014; Martinez et al. 2015b).

We address this issue and describe a holistic approach to allow for legacy product variants to be assessed and migrated toward managed reuse. Firstly, we transform models into a different representation that allows for the comparison of large quantities of data and, thus, an entire family of similar variants. All variants are compared with each other, and identical, similar, and completely distinct model parts are identified. Identical parts can be collapsed to reduce overall portfolio size, whereas comparisons of distinct parts can be omitted to allow further analysis to efficiently target systems that are at all similar to each other. Secondly, all remaining systems are then subject to a second, fine-grained procedure that captures the relations between individual system parts and, thus, their variability prior to migration or revamp decisions. To this end, we create a *family model*, which represents common and varying system parts as well as their relations across all variants. The family model thereby allows for a better quality assurance for the entire product portfolio. With relations between variants and their system parts reengineered, faulty parts can be identified for the entire portfolio, thereby mitigating quality and security concerns for the entire product portfolio. In this chapter, we make the following contributions:

- We propose a holistic approach to reengineer *MATLAB/Simulink* model variants. We illustrate a coarse-grained analysis, identifying similar structures between all models, and a fine-grained analysis, comparing respective structures in greater detail. As a result, we create a *family model* to represent implementation-specific variability of the variant portfolio.
- We demonstrate the feasibility of our approach using an industry-inspired example portfolio, comprising four *MATLAB/Simulink* model variants evolved from *clone-and-own*. We show our approach to identify redundancies, to precisely capture implementation-specific variability, and to create a comprehensible *family model*.

We extend upon work on model-based languages (Holthusen et al. 2014; Wille et al. 2016), family model creation (Wille 2014), and work specific to *MATLAB/Simulink* models (Schlie et al. 2017b, 2018) to create a holistic approach. This chapter is structured as follows. We provide background information on *MATLAB/Simulink* models and their variability in Sect. 10.2 and give an overview of our approach to reengineering *MATLAB/Simulink* model variants in Sect. 10.3. We detail our coarse-grained analysis of the entire portfolio in Sect. 10.4 and the subsequent fine-grained analysis of individual model parts in Sect. 10.5 and elaborate on the variability model creation in Sect. 10.6. We assess the feasibility of our approach using an industry-inspired portfolio with four model variants and discuss produced results in Sect. 10.7. Related work is discussed in Sect. 10.8. We detail future work and conclude our chapter in Sect. 10.9.

10.2 Preliminaries

In this section, we state basic information on *MATLAB/Simulink* models and properties utilized for our approach (cf. Sect. 10.2.1) and provide basic terminology for representing implementation-specific variability (cf. Sect. 10.2.2). Complementary material and further details are given online.[1]

10.2.1 MATLAB/Simulink Models

MATLAB/Simulink is a block-based behavioral modeling languages that utilizes *functional blocks* and *signals* to specify certain software system functionality. It is heavily used for the development of safety-critical functions in a variety of industrial domains, such as automated production systems, automotive, avionics, and rail (Deissenboeck et al. 2010; Ryssel et al. 2010a). Such models are the central development artifact and used for simulation, test case creation, and source code generation (Merschen et al. 2011). To capture complex system behavior, models can grow to enormous size and complexity, comprising thousands of blocks (Deissenboeck et al. 2008). Blocks have syntactical and semantical properties, which allow for their identification and comparison. Table 10.1 lists block properties of interest for the remainder of this chapter. For engineers to maintain an overview, logically connected blocks are commonly grouped in a *Subsystem (SM)* block. SMs can be nested as they structure the model vertically, constituting a *model hierarchy* (Ryssel et al. 2010a). Each SM is located on a specific *hierarchical layer* δ_j, which corresponds to its nesting depth. Industrial *MATLAB/Simulink* models can comprise a *hierarchical depth* of ten layers and more (Haber et al. 2013). Figure 10.1 depicts a simple *MATLAB/Simulink* model M_E. Contained SMs, labeled *2*, *3*, and *6*, are highlighted gray. For clarity, all models within this chapter contain unique block labels only.

Highlighted in gray, respectively, M_E's corresponding graph representation includes an artificial SM named *root*. It comprises all blocks residing on the first hierarchical layer δ_0 to show M_E's model hierarchy being a tree structure. We denote the SM labeled *5* on the second hierarchical layer δ_1 as $M_{E:5}$. We refer to $M_{E:5}$ as the *child system* of its respective *parent system* $M_{E:1}$, the SM labeled *1* in Fig. 10.1. Any SM on any layer δ_j is the root of its respective subtree. Models structured with SMs can be depicted as a graph, precisely a *tree*.

[1] Online material: https://www.isf.cs.tu-bs.de/cms/team/schlie/material/SQ2019.

Table 10.1 Properties of *MATLAB/Simulink* functional blocks and signals

Property	Description
• *name*	The textual label of a block
• *function*	Constitutes the semantic meaning of a block, that is, what the block is used for
• *interfaces*	A block's interface consists of:
– *inports*	For incoming data (usually signals). A block contains an arbitrary number of inports, each directly connected to exactly *one* outport
– *outports*	Constitute the interface for outgoing data. A block contains an arbitrary number of outports, each directly associated with *one or more* inports
• *connector*	A directed edge, connecting in- and outports
• *attributes*	Semantical properties, defining the blocks' internal behavior

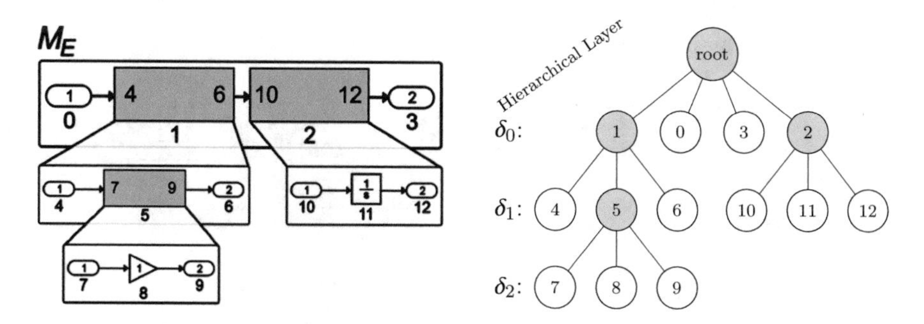

Fig. 10.1 MATLAB/Simulink model and its graph representation

10.2.2 Variability in Models

MATLAB/Simulink models resulting from *clone-and-own* differ. We refer to locations where two similar models differ as *variation points (VPs)*. Even when comprising only few blocks, correct identification of all VPs is crucial. *150%* models combine sharable and variability-containing assets from multiple products in a central development artifact (Schulze et al. 2013). To this end, we represent implementation-specific variability (Grönniger et al. 2014), including all VPs, using a *family model*, which is the final *150%* model representing all analyzed input models. Such *150%* and, thereby, the *family model* are a valuable asset for developing and managing SPLs (Holthusen et al. 2014; Pohl et al. 2005). Figure 10.2 shows the family model for the two models $M_{C:3}$ and $M_{D:3}$ and shows every block annotated with its variability category and specifics such as varying *attributes* (cf. *Switch* blocks labeled 8).

The variability categories from Fig. 10.2 reflect how similar individual blocks are, which we assess with our fine-grained analysis (cf. Sect. 10.5). Highlighted by circles in both *MATLAB/Simulink* models, blocks 6 and *INV* constitute a Variation Point (VP), which is stated as a *VariantSubsystem* in the family model. Furthermore, Fig. 10.2 illustrates the family model to comprise hierarchical layers and, thus, to capture all implementation details of the models. The portfolio

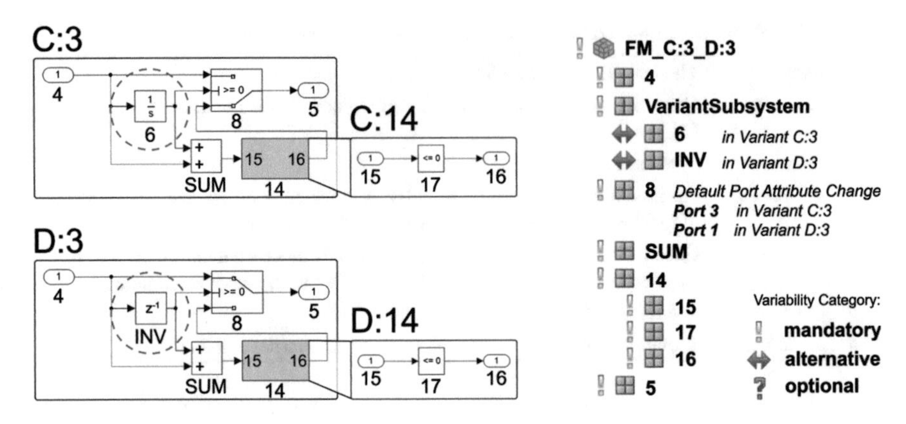

Fig. 10.2 Schematic family model with annotated variability categories

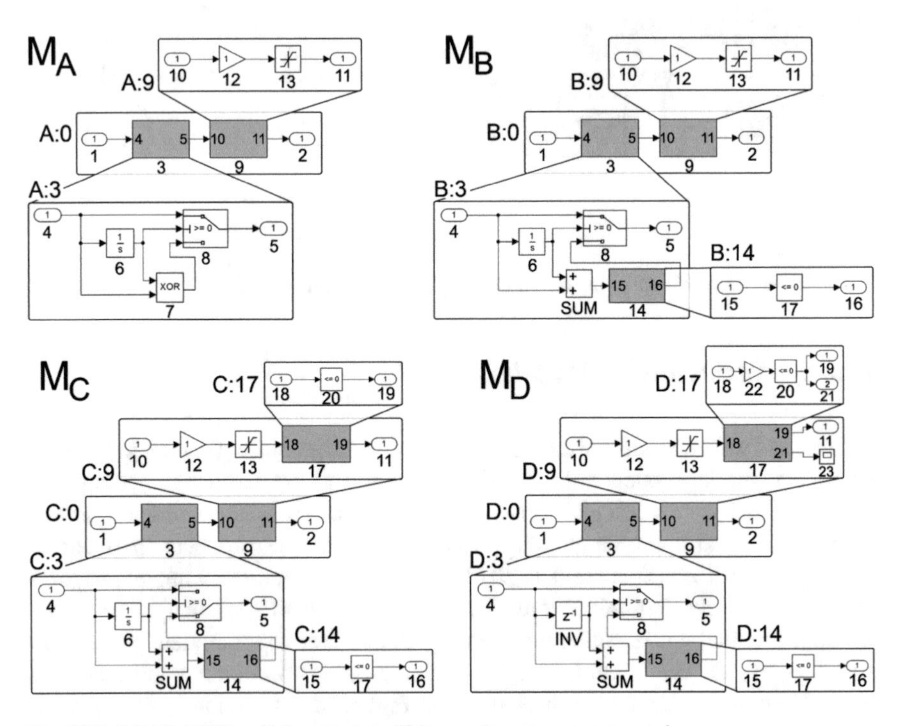

Fig. 10.3 MATLAB/Simulink variant portfolio used as a running example

in Fig. 10.3 is used as a running example for this chapter. It is not of industrial size but depicts variant diversity resulting from *clone-and-own*. The models vary at multiple locations and, thus, exhibit various VPs.

For instance, evaluating the models M_A and M_B reveals the block labeled *7* in M_A to have been replaced by two blocks, *SUM* and *14*, with the latter introducing an additional SM $M_{B:14}$. However, comparing M_A with M_D reveals them to differ at *further* and *other* locations than M_A and M_B (blocks *6* and *INV*, $M_{A:9}$ and $M_{D:9}$). Therefore, variability itself can be highly diverse within the variant family as VPs, and their specific extent vary between model combinations. In addition to structural changes, the *Switch* blocks, labeled *8* in all models, illustrate semantic change. Within M_C, the respective block exhibits a different *internal default port* compared to the remaining model variants.

10.3 Our Approach in a Nutshell

Our approach to reengineering *MATLAB/Simulink* model variants is a two-level process. It contains an initial coarse-grained analysis of the entire portfolio to remove those pairings not warranting further analysis. Secondly, relevant model parts are compared in greater detail using a fine-grained analysis. Direct application of fine-grained analysis procedures to the portfolio as a whole is typically impractical. Scalability remains an evident limitation, and the sheer amount of produced results renders their usability void. Therefore, our holistic approach as illustrated in Fig. 10.4 combines course-grained and fine-grained analyses to reengineer *MATLAB/Simulink* model variants.

First, the entire portfolio is subject to a coarse-grained analysis. All SMs are transformed them into a simpler representation, a *descriptor*, which are compared to identify similar and dissimilar structures between all variants, regardless of their extent and location within the models (cf. Sect. 10.4). We refer to such structures as *trees*, which together constitute a *forest*. Our descriptor is *instance-agnostic*, thereby not capturing instancespecifics such as individual block labels but high-level information only. With our descriptor, we capture distinct or highly dissimilar parts early on, which are paramount, as it allows to precisely target relevant parts only, improving overall scalability.

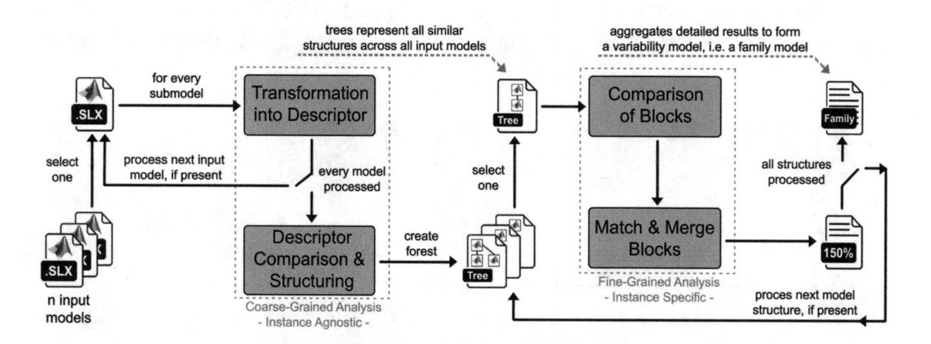

Fig. 10.4 Workflow of our approach to reengineer *MATLAB/Simulink* variants

With unnecessary comparisons omitted, usefulness of the remaining results is improved. Yielding groups of similar model structures, *trees*, we subsequently compare the functional blocks they comprise in greater detail using a fine-grained analysis and metric to capture their implementation-specific variability. With relations between models reengineered, we restructure portfolio assets by removing redundancies and by collapsing similar parts, hence, merging them, while preserving their variability information. As a result, we create a *family model* and a *MATLAB/Simulink 150% model*, which combine sharable and variability-containing assets in a single platform (cf. Sect. 10.5).

10.4 Coarse-Grained Analysis of the Variant Portfolio

In this section, we detail our coarse-grained analysis procedure to compare an entire variant portfolio, the *Static Connectivity Matrix Analysis (SCMA)* (Schlie et al. 2018). Illustrating the overall workflow of SCMA in Fig. 10.5, we first provide information on our *model descriptor*, the *Connectivity Matrix* in Sect. 10.4.1, which abstracts from SMs and allows for multiple models to be compared in their entirety. We outline SCMA in Sect. 10.4.2.1 and explain how it identifies all similar and dissimilar parts between analyzed variants. Finally, we detail how results are used to direct our fine-grained analysis to relevant parts only, thereby omitting workload when unnecessary.

10.4.1 The Connectivity Matrix: A Model Descriptor

For an entire portfolio of *MATLAB/Simulink* models to be assessed, their complexity needs to be reduced first. For our approach, we utilize *descriptors* (Duan 2010), which abstract relevant system information from complex systems, here *MATLAB/Simulink* models, to represent them in a simpler format. Descriptors are applied extensively in various fields, such as network and electrical circuit design and image processing (Zhang and Lu 2003; Jeong et al. 2013; Tan 2012). Moreover, descriptors

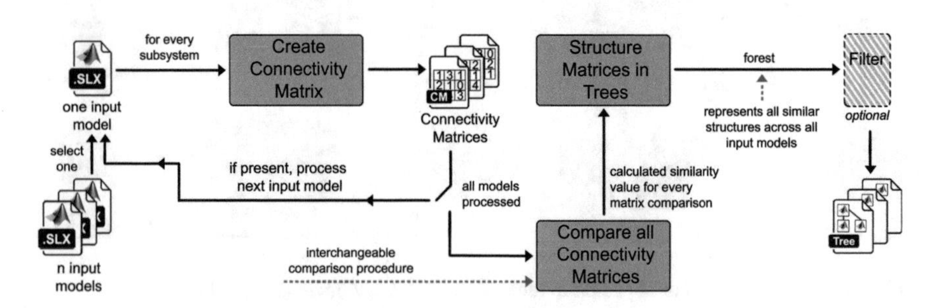

Fig. 10.5 Coarse-grained analysis—the Static Connectivity Matrix Analysis

can be compared efficiently and, thereby, allow even for large quantities of data to be assessed (Benner et al. 2005). Such descriptor should be easy to extract from the original model, and there should be a low probability of two distinct models resulting in the same descriptor (Tuytelaars and Mikolajczyk 2008). *Matrices* are predominantly used as a descriptor format and can be compared efficiently (Olshevsky 2001; Antoulas 2005; Benner et al. 2005). Furthermore, matrices can generally represent graphs, and with their numerical efficiency, they are intrinsically suitable for large-scale graph transformation and analysis procedures (Henley 1973). As depicted in Fig. 10.1, block-based models such as *MATLAB/Simulink* inherently constitute such graph structure. Consequently, we use matrices in our technique to abstract from complex *MATLAB/Simulink* models and to derive a descriptor, the *Connectivity Matrix (CM)*. Such models and SMs, respectively, are the composition of directly connected *functional blocks*. We exploit this inherent characteristic to abstract SMs into matrix form.

Precisely, a CM represents which two block functions (cf. Table 10.1) directly connect and how often they connect within the evaluated SM. For the remainder of this chapter, we refer to a *connection* as the *function* of a specific signals source block and the *function* of that signals target block. All CMs are *static* and, thus, identical in their row and column construction. Therefore, they have the same dimensions, and we ensure this by preprocessing all input models and creating a dictionary of all distinct block functions, which then constitutes both axes of every CM. The order of functions in the dictionary can be chosen arbitrarily but must be fixed for subsequent CM creation. For the models $M_{C:17}$ and $M_{D:17}$ in Fig. 10.6, an excerpt from the running example in Fig. 10.3, we show the corresponding CMs created by SCMA, $CM_{C:17}$ and $CM_{D:17}$, in Fig. 10.6. The illustrated models contain six signals as well as eight blocks with a total of four distinct functions. For each block, its specific function is given in Fig. 10.6 and pointed out to using arrows. The CMs in Fig. 10.6 highlight in gray the single connection present in both abstracted models $M_{C:17}$ and $M_{D:17}$, connecting the functional block types *Zero* and *Outport*. For readability reasons, non-present connections are left blank.

Shown in Fig. 10.6, a connections' *source function* always resides on the x-axis, while the associated *target function* is located on the y-axis. Thus, CMs are $n \times n$ matrices and provide means to store every possible connection present in the entire

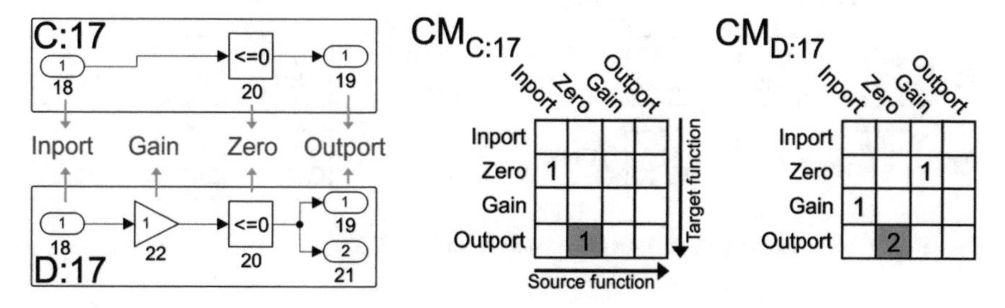

Fig. 10.6 The static connectivity matrix—an instance agnostic system descriptor

portfolio. As depicted in Fig. 10.5, every input model is processed separately, and each SM is transformed into a CM. If a specific connection is present multiple times within an SM, the corresponding CM entry is incremented (cf. *Zero - Outport* in $M_{D:17}$ in Fig. 10.6). CMs preserve the SM's parent-child relation (cf. Fig. 10.1) and retain the model's hierarchy. Thus, CMs also exhibit a *hierarchical depth*, corresponding to its SM. CMs are agnostic to instance-specific changes commonly performed during *clone-and-own*, such as relocations of SMs or renaming of blocks. However, CMs capture changed syntactical compositions of SMs and, thus, changed functionality.

10.4.2 Static Connectivity Matrix Analysis

As illustrated in Fig. 10.5, SCMA processes each variant of the portfolio separately, transforming every SM into its corresponding CM representation. To capture blocks residing on the top hierarchical layer (cf. Fig. 10.1), SCMA creates an additional CM for those elements. Consequently, for a model comprising k SMs, SCMA creates $k+1$ CMs, each CM holding an explicit reference to the SM it represents. For the model portfolio in Fig. 10.3, we schematically depict all created CMs in Fig. 10.7, illustrating the preserved model hierarchy. For the model M_A from Fig. 10.3, containing two SMs, the respective set of CMs CM_A is shown in Fig. 10.7, comprising three CMs. Respectively, the set CM_B shows the CM $CM_{B:14}$ on the third hierarchical layer δ_2, representing the SM $M_{B:14}$ from the model M_B (cf. Fig. 10.3).

The variant portfolio from Fig. 10.3 comprises a total of 12distinct block functions (cf. Table 10.1). The model M_A contains eight different functions, whereas, for instance, M_B additionally introduces *Summation* and *Zero* in its SMs $M_{B:3}$ and $M_{B:14}$ (cf. blocks *SUM & 17* in Fig. 10.3). Ergo, every CM created for the portfolio exhibits a 12×12 dimension as stated in Fig. 10.7. All CMs are compared with each other to identify similar SMs and reuse potential across all variants, regardless of location or hierarchical depth.

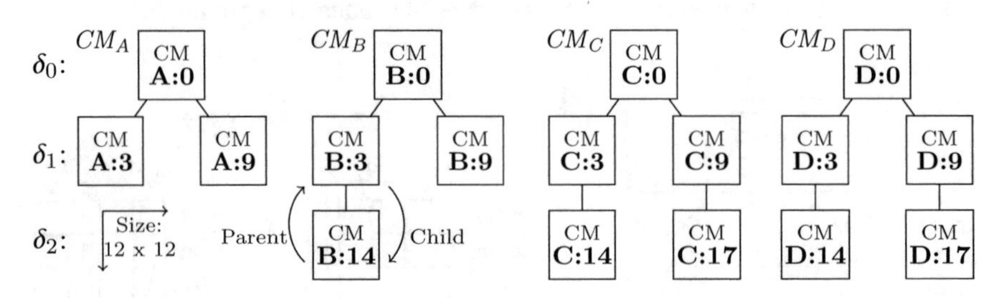

Fig. 10.7 CMs created by SCMA for the models from Fig. 10.3

10.4.2.1 Comparing the Variant Portfolio

For the 17 CMs from Fig. 10.7, representing the variant portfolio in Fig. 10.3, a total of 107 pairwise CM comparisons[2] are performed by SCMA. For every comparison, a similarity value ω is calculated, ranging between zero and one, hence $0 \leq \omega \leq 1$. Precisely, the similarity value is calculated as the normalized distance of all two entries with the same x- and y-coordinates. We only consider entries and, thus, *connections*, which are present in either CM. Including non-present connections in the similarity calculation would adversely increase the similarity value and negatively affect its soundness. Specifically, for two non-empty matrix entries, the minimum is divided by the maximum. When both entries are zero, no division occurs. Otherwise, we can divide as the maximum entry value is greater or equal to one. For two CMs A and B with dimensions n, their similarity value is calculated as follows.

$$\omega(A, B) = \frac{\sum_{x=1}^{n} \sum_{y=1}^{n} \begin{Bmatrix} 0 & A(x,y) = B(x,y) = 0 \\ \frac{min(A(x,y), B(x,y))}{max(A(x,y), B(x,y))} & else \end{Bmatrix}}{\sum_{x=1}^{n} \sum_{y=1}^{n} \begin{Bmatrix} 0 & A(x,y) = B(x,y) = 0 \\ 1 & else \end{Bmatrix}}$$

Consequently, the CM comparison is commutative. If two CMs exhibit a similarity value of $\omega = 0$, this shows complete structural dissimilarity and suggests entirely different functionality implemented by the respective SMs. Contrary, $\omega = 1$ indicates total structural similarity and suggests equal functionality within the corresponding SMs. Looking at the two CMs from Fig. 10.6 in isolation, a total of four non-empty entries exist, thus, *connections* present in either abstracted SM. Hence, the similarity calculation is as follows.

$$\omega(CM_{C:17}, CM_{D:17}) = \frac{1/2 + 0/1 + 0/1 + 0/1}{4} \Rightarrow 0.125$$

For the CMs from Fig. 10.7, we list all comparisons along with their similarity values ω in Table 10.2. For readability reasons, comparisons with zero similarity are left blank, while remaining values are rounded. Gray and dashed entries reflect comparisons that can be omitted, as CMs (a) originate from the same model and (b) are redundant due to the calculation being commutative.

The data given in Table 10.2 gives an overview of similar CMs and, thus, SMs. For instance, $\omega(CM_{A:0}, CM_{B:0})$, representing the top hierarchical layers of M_A and M_B (cf. Figs. 10.3 and 10.7), exhibit a similarity value of $\omega = 1$, which indicates total similarity. The data in Table 10.2 also allows for a preliminary assessment of more than two models. $CM_{A:0}$ is similar to both $CM_{B:0}$ and $CM_{C:0}$, suggesting related SM structures across multiple models (cf. Fig. 10.3). Furthermore, both children of $CM_{A:0}$ and $CM_{C:0}$, thus, $CM_{A:9}$ and $CM_{A:9}$, also exhibit total similarity, showing similar structures across hierarchical layers.

[2]All pairwise combinations, except when two CMs originate from the same model.

Table 10.2 Similarity values for CMs from Fig. 10.7

ω	A:0	A:3	A:9	B:0	B:3	B:9	B:14	C:0	C:3	C:9	C:14	C:17	D:0	D:3	D:9	D:14	D:17
B:0	1.0																
B:3		0.4															
B:9			1.0														
B:14																	
C:0	1.0			1.0													
C:3		0.4			1.0												
C:9	0.2		0.4	0.2		0.4											
C:14							1.0										
C:17							1.0										
D:0	1.0			1.0				1.0									
D:3		0.2			0.5				0.5								
D:9	0.1		0.3	0.1		0.3		0.1		0.8							
D:14							1.0				1.0	1.0					
D:17										0.2	0.1	0.1					

δ_0 δ_1 ... δ_m / δ_1 ⋮ δ_m — CMs for each model comparison

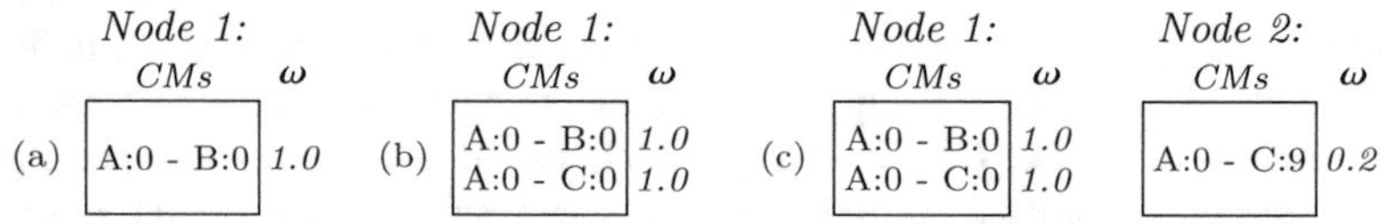

(a) *Node 1:* CMs ω — A:0 - B:0 *1.0*

(b) *Node 1:* CMs ω — A:0 - B:0 *1.0*; A:0 - C:0 *1.0*

(c) *Node 1:* CMs ω — A:0 - B:0 *1.0*; A:0 - C:0 *1.0* — *Node 2:* CMs ω — A:0 - C:9 *0.2*

Fig. 10.8 Schematic illustration of CMs processing and grouping in nodes

10.4.2.2 Identifying Similar Structures in the Variant Portfolio

Within the variant portfolio, multiple similar SM structures may exist, varying in their extent and location. To derive all structures, across model and hierarchy boundaries, we utilize the CM comparisons given in Table 10.2, which shows CMs to be ordered in a descending order based on hierarchical depth. Since CMs preserve the hierarchical depth of the SMs and their *parent-child* relation, we thereby ensure that a *parent* CM is always processed prior to their *child* CMs. Specifically, we start with comparisons for $CM_{A:0}$ from Table 10.2, beginning with $CM_{B:0}$ and ending with $CM_{D:17}$. For the first three comparisons, we illustrate the process to group together similar CMs in Fig. 10.8. Similar CMs, that is, $CM_{A:0}$ and $CM_{B:0}$, are grouped within *nodes* (cf. Fig. 10.8a). Proceeding with $CM_{C:0}$, we evaluate whether that CM can be added to an existing node; here, only *Node 1* exists. $CM_{C:0}$ is similar to both $CM_{A:0}$ and $CM_{B:0}$. Moreover, all CMs reflect *top-level* SMs and, thus, have the same *parent-child* relation. Thus, we add $CM_{C:0}$ to *Node 1* (cf. Fig. 10.8b), which now represents a similarity between *three* models, M_A, M_B, and M_C. Figure 10.8c shows

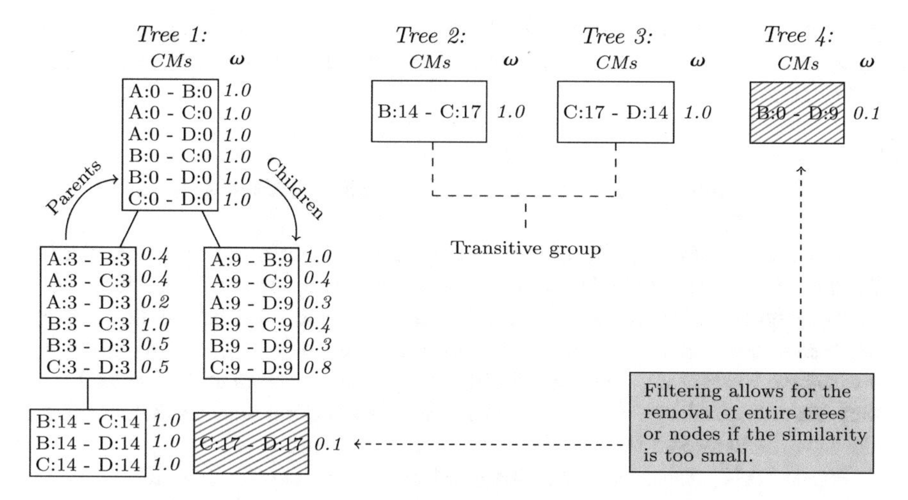

Fig. 10.9 Excerpt from forest created by SCMA for comparisons from Table 10.2

$CM_{C:9}$ to create *Node 2*, rather than enlarging *Node 1* as nodes only comprise CMs and, thus, abstracted SMs from different models.

When processing CM comparisons, we exploit their parent-child relation. If a node exists that contains both *parent* CMs, a new node is created for their *child* CMs and a reference between them set. Hence, nodes form *trees*, each reflecting a similar hierarchical SM structure between multiple models. When assessing a CM comparison, it can either enlarge an existing node, extend an existing tree, or become the root of a new tree. We refer to Schlie et al. 2018 for more details on the structuring of CM comparisons and the forest creation. Processing all CM comparisons from Table 10.2, representing the model portfolio in Fig. 10.3, yields a forest that comprises 11 trees. We depict an excerpt containing four trees in Fig. 10.9. We provide the entire set online.[1]

Illustrating the largest similar structure, comprising two hierarchical layers, *Tree 1* contains CMs from all four input models. It also shows similarities on the third hierarchical layer, which only the models M_B, M_C, and M_D exhibit. Listed on the right for each CM comparison, the similarity values reveal certain SM structures to be more similar between some models than for others. In *Tree 1* from Fig. 10.9, for instance, the systems $CM_{B:3}$ and $CM_{C:3}$ seem to be more similar than $CM_{B:3}$ and $CM_{D:3}$. Additionally, *Trees 2 and 3* reveal identical functionality, which, however, resides at completely different locations within the input models M_B, M_C, and M_D (cf. Fig. 10.3). The corresponding CMs form transitive groups as illustrated in Fig. 10.9. These can indicate reusable SMs, which may be suitable for encapsulation in a library system. Furthermore, CMs comparisons in Table 10.2 show over 70%[3] to exhibit zero similarity, motivating the necessity for a coarse-

[3] 107 pairwise CMs exists for which 76 (≈71%) exhibit zero similarity.

grained analysis. This reduction allows for the fine-grained analysis to precisely target only those model parts, which warrant the reverseengineering of their variability information.

10.4.2.3 Isolating Groups of Model Parts for Fine-Grained Analysis

By collapsing transitive CM groups and by exploiting their parent-child relation, the amount of CM comparisons passed on for a detailed analysis can be reduced further. Figure 10.10 shows both transitive relations and similarity-based relations between all compared CMs from Table 10.2. Illustrated in Fig. 10.10, CMs exhibiting a transitive relation are collapsed if either they exhibit a similarity of $\omega = 1.0$ or they are connected by solid lines. Furthermore, similar but not transitively related CMs identified by SCMA are illustrated with dashed lines.

Figure 10.10, reveals four transitive CM groups and depicts them in Fig. 10.11. Such groups correspond to the hierarchical structure shown in Fig. 10.9 and represent relations between SMs. For instance, the first group in Fig. 10.11 represents top level SMs of the portfolio (cf. Fig. 10.3), grouped together in the top node of *Tree 1* in Fig. 10.9. Contrary to the forest (cf. Fig. 10.9), transitive groups isolated from trees can comprise CMs originating from the same model, and we highlight them gray in Fig. 10.11.

To this end, we identify groups of similar CMs, thereby similar groups of *MATLAB/Simulink* SMs. However, their implementation-specific variability

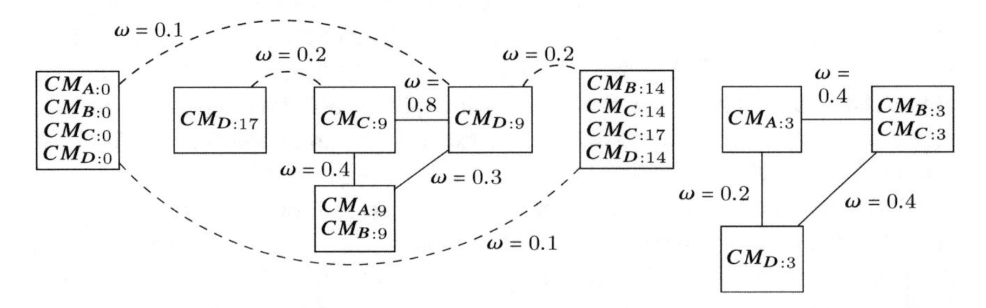

Fig. 10.10 Transitivity relations between CM comparisons from Table 10.2

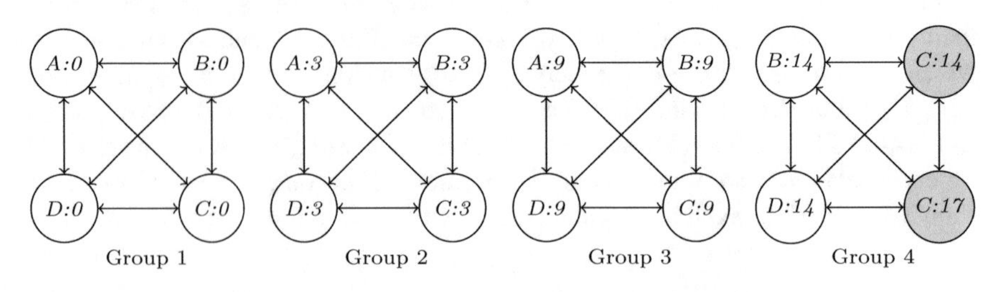

Fig. 10.11 Transitive groups isolated by SCMA for further fine-grained analysis

information such as common and varying assets has not been reengineered. Consequently, every CM group identified by SCMA, respectively, the SMs they abstract, are compared using a fine-grained analysis procedure, assessing the SMs implementation and, thus, the *functional blocks* they comprise.

10.5 Fine-Grained Analysis of Individual Model Parts

In this section, we detail our fine-grained analysis procedure to compare related *MATLAB/Simulink* model parts and to reengineer their variability information. Showing the approaches' workflow in Fig. 10.12, we first elaborate on the metric we utilize to compare *functional blocks* in Sect. 10.5.1 and variability categories used to classify comparison results in Sect. 10.5.2. We explain our fine-grained comparison procedure, an adaptation of *N-Way Model Merge (N-Way)* (Rubin and Chechik 2013a) and detail its application to groups of *MATLAB/Simulink* SMs in Sect. 10.5.3. We show necessary adaptations performed to N-Way to cope with *MATLAB/Simulink* models in Sect. 10.5.4.

10.5.1 Detailed Comparison Metric

Identifying implementation-specific variability in *MATLAB/Simulink* models requires a detailed comparison of *functional blocks* and, thus, an appropriate, instance-specific metric. Extending upon previous work by Wille et al. 2013 and introducing further *properties* by Schlie et al. 2017b, we utilize an extensible, user-adjustable metric and list the block properties p, which the metric assesses in Table 10.3. The weights w indicate the significance of the corresponding property p when comparing blocks. Assigned 75% by default, we argue that a block's *function* is most salient for comparison, while remaining weights distribute equally. However, practitioners can adjust the weights based on their demands. The comparison of two blocks results in a similarity value λ, which is the weighted average of all

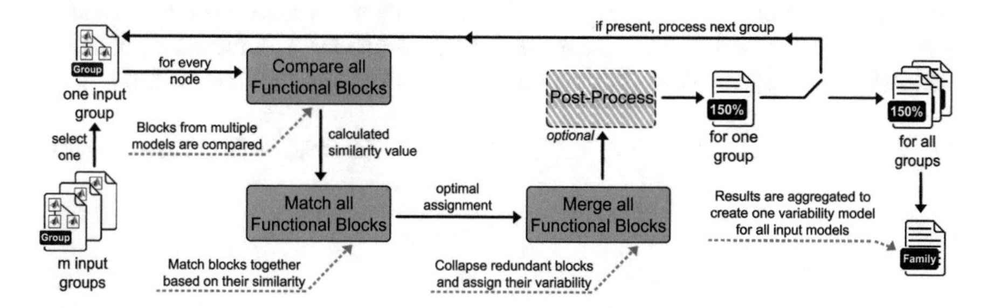

Fig. 10.12 Fine-grained comparison using N-way model compare

Table 10.3 Properties for a detailed comparison of *MATLAB/Simulink* blocks

Property p	Weight w (%)	Computation/description
name	5	LD^a of the blocks' names
function	75	$\begin{pmatrix} 1 & \text{Blockshavethesamefuntion} \\ 0 & \text{else} \end{pmatrix}$
#inports	5	$\sum_{i \in Inports} (i) / \lvert Inports \rvert$
#inport-functions	5	$\sum_{f \in F_{Inport}} (\#f) / \lvert F_{Inport} \rvert^b$
#outports	5	$\sum_{o \in Outports} (o) / \lvert Outports \rvert$
#outport-functions	5	$\sum_{f \in F_{Outport}} (\#f) / \lvert F_{Outport} \rvert^b$

[a]Levenshtein distance (Levenshtein 1966)
[b]Max. number of inport- and outport-functions

properties p. Thus, we evaluate blocks by comparing values for certain properties and, precisely, the extent to which values are similar. For two *MATLAB/Simulink* blocks A and B, we calculate their similarity as follows.

$$\lambda(A, B) = \left\{ \sum_{i=1}^{k} similarity(A, B, p_i) \cdot w_i \quad \mid \quad 0 \leq \lambda \leq 1 \right\}$$

We denote p_i to be a specific metric property, that is, comparing the blocks *functions* (cf. second row in Table 10.3), and w_i to be the assigned weight (here 75%). Every property is evaluated for both blocks, here *similarity*(A, B, p_i) with the result stating their similarity with respect to the property p_i, ranging between zero and one. The metric in Table 10.3 comprises *k=5* properties, and the similarity value λ is the weighted average over all metric properties.

For instance, the two blocks labeled *18* in C:17 and D:17 from Fig. 10.6 both have identical *names,functions* (here *Inport*), and the same number of *inports* and *outports* with $|Inports|$=1 and $|Outports|$=1. However, their respective *outport-functions* differ, being *Zero* in C:17 while being *Gain* in D:17. The distinction between these characteristics of ports allows for a more fine-grained comparison of the blocks' interfaces. By taking both the number and function of ports into consideration, the surrounding structure of the compared blocks is evaluated to an extent that produces a more context-based similarity for the compared blocks. The metric calculates a similarity value of $\lambda = 0.95$ for the two *Inport* blocks from Fig. 10.6 as follows.

$$\lambda(\in {}^{18}_{M_C}, \in {}^{18}_{M_D}) = (1 \cdot 5\%) + (1 \cdot 75\%) + (1 \cdot 5\%) + (1 \cdot 5\%) + (1 \cdot 5\%) + (0 \cdot 5\%)$$

10.5.2 Variability Categorization

Based on the similarity value, we define the *relation* between two implementation artifacts, here *functional blocks* A and B, according to the following, user-adjustable thresholds (Schlie et al. 2017b). Two blocks with a similarity value of $\lambda = 0.95$, hence 95% and more, are classified to be *mandatory*. Respective blocks will be collapsed within the final variability model, and information on minor changes, that is, varying *names* (cf. Table 10.1), will be annotated. The *Switch* blocks compared in Sect. 10.5.1 would therefore be collapsed (cf. Fig. 10.3). Blocks present in only one analyzed variant exhibit zero similarity and are declared *optional*. Remaining comparisons, for which the blocks' similarity value ranges between those thresholds, are classified as *alternative* and either facilitate a new VP or contribute to an existing VP. The metric in Table 10.3 and the variability thresholds can be adjusted.

$$Relation(A, B) = \left\{ \begin{array}{ll} \text{mandatory} & 0.95 <= sim(A, B) \\ \text{alternative} & 0 < sim(A, B) < 0.95 \\ \text{optional} & 0 = sim(A, B) \end{array} \right\}$$

10.5.3 Adapted N-Way Model Merge

Originally proposed by Rubin and Chechik 2013a, N-Way Model Merge introduces a solution to *compare* multiple *Unified Modeling Language (UML) class diagrams* at once and to *merge* them into one variability model. However, *function block diagrams* and, by that, *MATLAB/Simulink* models are fundamentally different from UML class diagrams and render the original approach non-applicable. Specifically, a UML class is uniquely identifiable by its name within its diagram. *MATLAB/Simulink* blocks can be assigned the same name multiple times within a model. Furthermore, UML classes differ in their attributes, thereby their *properties*. For *MATLAB/Simulink* blocks, however, properties are the same, and only their specific *value* differs. Consequently, we re-implemented N-Way and performed significant adaptations to cope with *MATLAB/Simulink* models and SMs instead. In the following sections, we provide a brief overview of N-Way prior to stating specifics on the comparison of SM groups and the identification of their implementation-specific variability. Finally, we elaborate on a post-processing step to assess assigned variability categories prior to the final restructuring.

10.5.3.1 Comparison of Individual Blocks

We use the metric detailed in Sect. 10.5.1 to compare functional blocks with each other. Hence, we calculate their similarity value by evaluating their properties as detailed in Table 10.3 and, precisely, to what extent these are similar. Every group of similar *MATLAB/Simulink* SMs (cf. Fig. 10.11) identified by SCMA is processed

$$A_{\substack{C:3,\\D:3}} = \begin{array}{c|cccccc} \lambda & 4 & 6 & SUM & 8 & 14 & 5 \\ \hline 4 & 0.98 & 0.07 & 0.02 & 0.02 & 0.05 & 0.0 \\ INV & 0.07 & 0.2 & 0.08 & 0.06 & 0.1 & 0.05 \\ SUM & 0.02 & 0.8 & 0.97 & 0.1 & 0.1 & 0.03 \\ 8 & 0.02 & 0.06 & 0.1 & 0.98 & 0.07 & 0.02 \\ 14 & 0.05 & 0.1 & 0.08 & 0.07 & 0.99 & 0.1 \\ 5 & 0.0 & 0.05 & 0.03 & 0.02 & 0.05 & 1.0 \end{array}$$

Total Weight: 5.12

$$A_{\substack{C:14,\\D:14}} = \begin{array}{c|ccc} \lambda & 15 & 17 & 16 \\ \hline 15 & 1.0 & 0.03 & 0.08 \\ 17 & 0.08 & 1.0 & 0.08 \\ 16 & 0.02 & 0.08 & 1.0 \end{array}$$

Total Weight: 3.0

Fig. 10.13 Comparison results for blocks from $M_{C:3}$, $M_{D:3}$ and $M_{C:14}$, $M_{D:14}$

separately by N-Way. First, every SM of the respective group is retrieved and, more precisely, every functional block it comprises. For instance, *Group 2* from Fig. 10.11 contains four SMs, while Fig. 10.3 shows those SMs to comprise a total of 23 functional blocks. Subsequently, such blocks are compared in a pairwise fashion resulting in 253 comparisons. Illustrated in Fig. 10.13, such pairwise comparisons can be stored in matrix A, which lists functional blocks on its axes and their comparison result as entries. Throughout this section, we omit full representation of matrices for the sake of simplicity. For instance, Fig. 10.13 does not picture the 23×23 matrix A required for *Group 2* from Fig. 10.11 but shows an excerpt of two models only. Contrary to SCMA (cf. Table 10.2), matrix A does not contain SMs but individual functional blocks they comprise. Thus, matrix A is always *square*, even if SMs differ in size (cf. $M_{A:3}$ and $M_{D:3}$ in Fig. 10.3).

N-Way as proposed by Rubin and Chechik 2013a considers comparisons of UML classes originating from the same input model as *invalid* and assigns zero similarity automatically. However, such comparisons are necessary to identify reusable artifacts as *Group 4* in Fig. 10.11 highlights two SMs from one input model. Our re-implementation of N-Way automatically checks this and adapts accordingly. Figure 10.13 lists the calculated similarity values for pairwise block comparisons of the systems $M_{C:3}$, $M_{D:3}$ and $M_{C:14}$, $M_{D:14}$ with matched blocks highlighted gray. The matrix shown left in Fig. 10.13 pictures the *Inport* blocks labeled 4 in both SMs $M_{C:3}$ and $M_{D:3}$ to exhibit a similarity value of $\lambda = 0.98$. For these two blocks, Fig. 10.3 reveals one of three *outport-functions* (cf. Table 10.1) to be different, accounting for their similarity to decrease slightly. Compared blocks are stored within a *Comparison Element (CE)*, along with specific results based on the metric (cf. Table 10.3) and their similarity value.

10.5.3.2 Matching of Individual Blocks

The comparison procedure results in a vast number of CEs, most of which exhibit a low similarity (cf. Fig. 10.13). During *matching*, the *best* combinations and, thus, the ones with the highest similarity value are retrieved and chained together to form larger similar groups of related *functional blocks*. The matching is based on the

$8 \in M_{A:3} \rightarrow 8 \in M_{B:3}$
$8 \in M_{B:3} \rightarrow 8 \in M_{C:3}$
$8 \in M_{C:3} \rightarrow 8 \in M_{D:3}$
$8 \in M_{D:3} \rightarrow 8 \in M_{A:3}$

Excerpt from optimal assignment for block 8

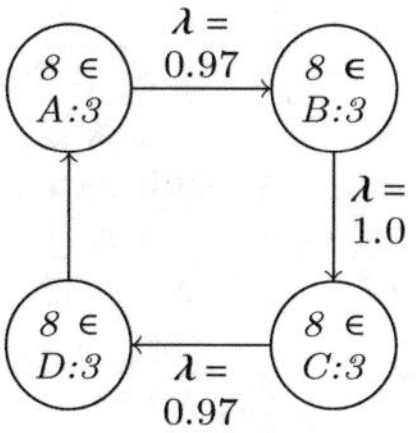

8 ∈ A:3
λ = 0.97
8 ∈ B:3
λ = 1.0
8 ∈ C:3
λ = 0.97
8 ∈ D:3

Fig. 10.14 Excerpt from assignments showing transitivity between matched blocks

Hungarian algorithm (Kuhn 1955), which takes as input a square matrix with numeric values and returns an optimal assignment such that no other combination of x- and y-axis entries exist with a higher cumulative similarity value. This can be reduced to the bipartite graph matching. Figure 10.13 highlights in gray the assignments and for both matrices their respective *total weights*, which describe the assignments overall quality. Hence, there exists no combination of assignments that increases the *total weight*, but we acknowledge that there can be multiple optimal assignments. When comparing more than two SMs and matching their blocks, transitive relations can exist between them. Capturing those is necessary to identify relations between blocks from more than two input models. For instance, when matching all four SMs in *Group 2* from Fig. 10.11, all *Switch* blocks labeled 8 would be matched together as depicted with the assignment in Fig. 10.14.

The Hungarian algorithm can retrieve an optimal assignment, which, however, relates only pairs of blocks, thereby only *two* input models. To cope with multiple input models and, hence, to state the relation between them, CEs must be able to hold more than one pair. Consequently, assigned pairs, thereby CEs, are processed further and *chained together* and, thus, collapsed within a *Chained Comparison Element (CHE)* if appropriate. Specifically, two CEs, CE_x and CE_y, can be collapsed if (a) they have a specific block in common and (b) the recalculated similarity value of the resulting CHE is higher than both similarity values for CE_x and CE_y. As a result, CHE comprise multiple CEs, each containing two functional blocks and their similarity value λ. Their average then determines a CHE's similarity, which calculates as follows.

$$sim(CHE) = \frac{\sum_{i=1}\sum_{j=i+1}^{\#CE}\lambda(i, j)}{\binom{\#CE}{2}}$$

All CHEs are stored in the chain, and CEs are either prepended or appended to an existing CHE within the chain or used to create a new CHE in the chain. For the four assignments in Fig. 10.14, each being a pair of *MATLAB/Simulink* blocks, we process $CE_{\{8 \in M_A, 8 \in M_B\}}$ first. The chain is empty, and, therefore, we add the CE, making it $CHE_{\{8 \in M_A, 8 \in M_B\}}$. Proceeding with $CE_{\{8 \in M_B, 8 \in M_C\}}$, the block $8 \in M_B$ is present in both the CE and created CHE. We can collapse them within $CHE_{\{8 \in M_A, 8 \in M_B, 8 \in M_C\}}$, which now contains blocks from three input models. For *Group 2* from Fig. 10.11,

Chain:

CHE_1 :	CHE_2 :	CHE_3 :	CHE_4 :	CHE_5 :	CHE_6 :
$-5 \in M_{A:3}$	$-4 \in M_{A:3}$	$-6 \in M_{A:3}$	$-8 \in M_{A:3}$	$-7 \in M_{A:3}$	
$-5 \in M_{B:3}$	$-4 \in M_{B:3}$	$-6 \in M_{B:3}$	$-8 \in M_{B:3}$	$-S^* \in M_{B:3}$	$-14 \in M_{B:3}$
$-5 \in M_{C:3}$	$-4 \in M_{C:3}$	$-6 \in M_{C:3}$	$-8 \in M_{C:3}$	$-S^* \in M_{C:3}$	$-14 \in M_{C:3}$
$-5 \in M_{D:3}$	$-4 \in M_{D:3}$	$-I^* \in M_{D:3}$	$-8 \in M_{D:3}$	$-S^* \in M_{D:3}$	$-14 \in M_{D:3}$
$\lambda = 1.00$	$\lambda = 0.98$	$\lambda = 0.58$	$\lambda = 0.98$	$\lambda = 0.56$	$\lambda = 0.99$

I^*: *INV* S^*: *SUM*

Fig. 10.15 Complete chain for all SMs in *Group 2* from Fig. 10.11

```
Input: T, optimalassignmentofCEsfromHungarianalgorithm
Output: K, ChainfornextiterationofN − Way
method generateChains()
    Add first CE ∈ T to the chain K
    while ThasunprocessedCEs do
        foreach CEinT do
            foreach CHEinK do
                if CEandCHEhaveblocksincommon then
                    if sim(CE) ≤ sim(CHE) ∧ sim(CHE ∪ CE) ≥ sim(CHE) then
                        MergeCHEandCE
                        RecalculateCHEssimilarity
                        BreakloopandproceedwithnextCE
                    end
                end
            end
            CreatenewCHEforCE
            AddCHEtoK, markCEasprocessed
        end
    end
return K;
```

Algorithm 10.1: Pseudo-code for chain generation in N-Way

we show the final chain produced by our adapted N-Way and detail all contained CHEs in Fig. 10.15. For instance, CHE_6 matches blocks from three models.

The CHEs comprise all functional blocks from *Group 2*, grouped together based on their similarity. CHE_6, for instance, contains all *Subsystem* blocks, labeled 14 in Fig. 10.3 and illustrates this block only to be present in some of the input variants. With a similarity value of $\lambda = 0.58$, CHE_2 also shows the *Integrator* blocks, labeled 6 in Fig. 10.3, as *alternative* to the *INV* block in M_D from the portfolio. We depict the chaining process in Algorithm 10.1.

10.5.3.3 Merging Individual Blocks

MATLAB/Simulink models comprise two basic entities as detailed in Sect. 10.2.1, *functional blocks* and *signals*. The merging process is twofold, processing functional blocks first and subsequently merging signals. The variability information of a signal is ultimately determined by its source block and its target block. Therefore, assigning a signal's variability requires knowledge about the variability of its corresponding blocks. Hence, blocks present within the final CHEs, having their associated signals as references, are merged first. Secondly, dangling signals are resolved and merged blocks are connected.

Omitting signals at first allows CHEs to be processed in an arbitrary order. Using the CHEs from Fig. 10.15, which contain functional blocks from SMs in Fig. 10.3, we illustrate the merged blocks in Fig. 10.16. Based on the similarity value of the CHEs and the respective thresholds stated in Sect. 10.5.2, variability categories are assigned to blocks. *Alternative* blocks are collapsed within a VP to illustrate them being *mutually exclusive*. Based on the CHEs from Fig. 10.15, four *mandatory* blocks exist as well as two *alternative groups*, here VPs. To show merged blocks holding signals as references to other blocks, Fig. 10.16 schematically depicts the signal targets for the *Input* block labeled 4 (cf. Fig. 10.3). Within the input model M_A, three outgoing signals exist with targets labeled 8, 6, and 7. For M_B, however, respective targets have changed, substituting target 7 with SUM.

After blocks have been merged, signals are added, exploiting the variability categories assigned to merged blocks. Figure 10.16 reveals the *mandatory* block labeled 4 to have partially different targets within M_A and M_B, here 7 and *SUM*, while also sharing common target blocks such as 8 and 6. Within the merged result shown in Fig. 10.16, the block labeled 8 is *mandatory*, while 6 is part of a VP and, thus, *alternative*. Consequently, the signal connecting 4 and 8 is also *mandatory*, while the signal targeting 6 is *alternative*. Specifically, it is alternative to the signal connecting 4 and *INV*, thereby forming the *alternative line group A1* in the merged result in Fig. 10.17.

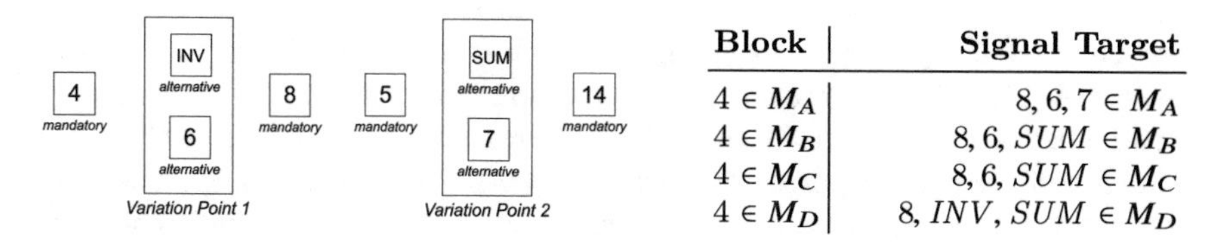

Block	Signal Target
$4 \in M_A$	$8, 6, 7 \in M_A$
$4 \in M_B$	$8, 6, SUM \in M_B$
$4 \in M_C$	$8, 6, SUM \in M_C$
$4 \in M_D$	$8, INV, SUM \in M_D$

Fig. 10.16 Merged blocks and their variability based on CHEs from Fig. 10.15

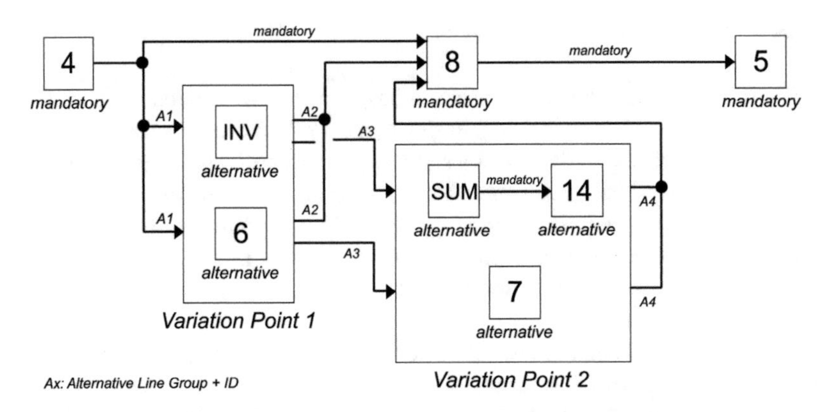

Fig. 10.17 Merged result with variability annotations for CHEs from Fig. 10.15

To this end, the internal *attributes* (cf. Sect. 10.2.1) of blocks have not been evaluated. Such attributes specify individual block behavior, and only small changes can render the systems overall behavior completely different during runtime. Consequently, varying attributes also constitute a VP, and their identification is crucial for a comprehensive family model. Moreover, the result in Fig. 10.17 does not contain details on other portfolio artifacts, that is, further groups (cf. Fig. 10.11), and, thus, it does not constitute a family model.

10.5.3.4 Post-processing Merged Results

With the procedure detailed in Sect. 10.5.3.3, functional blocks are merged solely on the basis of their containment within CHEs and their respective similarity values. Such a result may, however, be insufficient and may require post-processing. Comparing the *variation point 2* in Figs. 10.16 and 10.17 reveals a change with the latter comprising the block labeled 14. The CHEs shown in Fig. 10.15, specifically, CHE_5 and CHE_6, do not show any connection between the comprised blocks. However, evaluating the corresponding input models in Fig. 10.3 shows that the block labeled *SUM syntactically* requires the *Subsystem* block labeled 14. Consequently, VPs defined on the basis of CHEs may need to be reassessed and enriched with further data. Signals are the required information to do so, constructing any *syntactical* connection between functional blocks. During all phases of comparing, matching, and merging blocks, we preserve their referenced signals. Therefore, we post-process the result of every merging operation and may *break up* existing VPs if necessary. With the preliminary merging result given in Fig. 10.17, we add 14 to *variation point 2* as there is a direct syntactical connection between the blocks. Hence, because *SUM* always connects to 14, we categorize connection as *mandatory* but the block labeled 14 as alternative. The variability model we generate then shows the *Summation* block labeled *SUM* combined with the *Subsystem* block 14 to be alternative to the *XOR* block labeled 7 alone. The result as depicted in Fig. 10.17 is then *syntax-preserving* with respect to the evaluated input product portfolio as shown in Fig. 10.3.

After refining all VPs and finalizing variability categories (cf. Sect. 10.5.2), we can evaluate *mandatory* blocks to identify changes within their *internal behavior*. If such are not captured in the first place, they may account for inexplicable system behavior later on, hence requiring additional reengineering (Schlie et al. 2017a). Consequently, we compare all mandatory blocks regarding their internal properties. Such properties are specific to the *function* of a block. The *Switch* blocks, labeled 8 in Fig. 10.3, illustrate such change, specifically, in M_C the *Default Port* is different. Blocks exhibiting different functions, for instance, *Gain* labeled 12 in Fig. 10.3, have different internal properties. Here, we only compare mandatory blocks because they are of the same function, perforce given our metric and thresholds (cf. Table 10.3 and Sect. 10.5.2). For *alternative* and *optional* blocks, no information is collapsed, and, thus, we do not compare them further. For mandatory blocks in Fig. 10.17, labeled 4, 8, and 5, we identify the *Switch* block 8 (cf. Fig. 10.3) to exhibit varying internal properties. We enrich the merged results (cf. Fig. 10.17) with this information prior to reengineering the variability model, which represents the entire input portfolio.

10.5.4 Adaptations to N-Way for MATLAB/Simulink

N-Way describes the process of comparing and merging multiple UML class diagrams, consisting of three subsequential phases. Firstly, all UML classes of all input diagrams are retrieved and *compared* in a pairwise fashion, thereby calculating a similarity value. Secondly, with all UML classes compared, an *optimal matching* is found using the Hungarian algorithm (Kuhn 1955) and, thus, an assignment between compared UML classes with maximum weight regarding their similarity values. Precisely, the Hungarian algorithm operates on two data sets and matches an elements from one set with at most one element from the other set. However, with multiple input diagrams present, relations between more than two diagrams may be present. Thus, matched UML classes are subsequently subject to a *chaining* process, which further groups together similar elements from more than two input diagrams within *chain elements*. The resulting chain exhibits a *total weight*, that is, the sum of all similarity values of all chain elements, and, thus, grouped UML classes from multiple diagrams. N-Way then operates iteratively, using the chained elements as input, trying to increase the chains' total weight, thereby the quality of the result. N-Way comprises three sequential phases (cf. Fig. 10.12), *Comparison*, *Matching*, and *Merging*. Since UML class diagrams differ strongly from *MATLAB/Simulink*, we modified the comparison and merging phase accordingly. Rubin and Chechik 2013a calculate the similarity by evaluating attributes of UML classes. For t UML classes in n diagrams, they calculate it as follows.

$$similarity(t) = \frac{\sum_{2 \leq j \leq |t|} (j^2 \cdot n_j^p)}{n^2 \cdot |\pi(t)|}$$

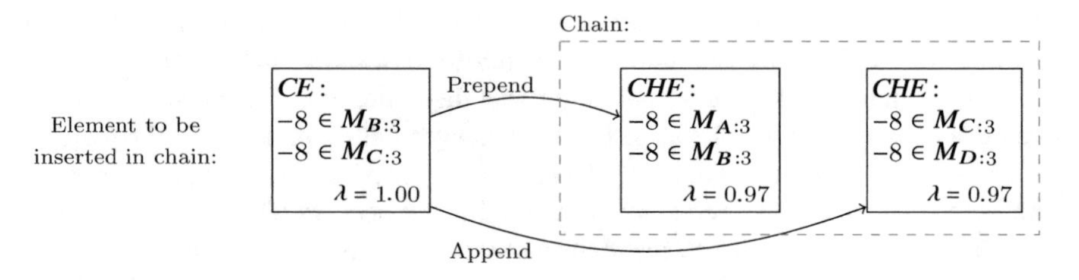

Fig. 10.18 Chain ambiguity with multiple possibilities to insert a CE

with $\pi(t)$ being the union of all distinct attributes from all classes in t and n_j^p being the distribution of attributes p appearing in j classes from t. In other words, the similarity value reflects the distribution of common attributes between multiple classes in relation to the total number of distinct attributes. Unfortunately, *MATLAB/Simulink* and function block diagrams in general exhibit a completely different syntax. First, a *functional block* is *not unique* within a model but can be present *many* times (cf. *Inports* in Fig. 10.3). Second, all blocks have the *same* attributes, that is, a *function* (cf. Table 10.1). Thus, attributes themselves do not differentiate blocks but their *specific value*. Therefore, we calculate the similarity for blocks as detailed in Sect. 10.5.3.2. Additionally, the chaining process itself had to be adapted. After processing all CEs in an arbitrary order, we found the final *chain* to be *invalid* as defined by Rubin and Chechik 2013a, as it contained duplicates of specific blocks. Scenarios exist, in which *multiple* possibilities exist to prepend or append a certain CE, rendering the chain *ambiguous* and ultimately invalid. Using the three assignments of the block labeled 8 from Fig. 10.14, each assignment being one CE, we illustrate the problem of chain ambiguity in Fig. 10.18.

The depicted CE, comprising the *Switch* blocks labeled 8 from $M_{B:3}$ and $M_{C:3}$, can *either* be prepended to the first CHE in the chain *or* can be appended to the second CHE. As a result, however, either $8 \in M_{C:3}$ or $8 \in M_{B:3}$ will be present *twice* in the chain, rendering it invalid. We addressed this by iterating over the chain and collapsing related CHEs further until no duplicates exist.

10.6 Reengineering Variability Models

In this section, we elaborate on how we combine results of our coarse-grained analysis procedure, SCMA, and our adapted N-Way procedure to create a *variability model*, containing merged artifacts and their variability information.

Every SM group generated by SCMA (cf. Sect. 10.4.2) is analyzed in greater detail using N-Way (cf. Sect. 10.5.3), each resulting in a merged model as depicted in Fig. 10.17. To this end, however, such individual results have not been aggregated to represent the entire product portfolio. With merged artifacts preserving references to their corresponding *parentsystem* or *childsystems*, respectively, detailed

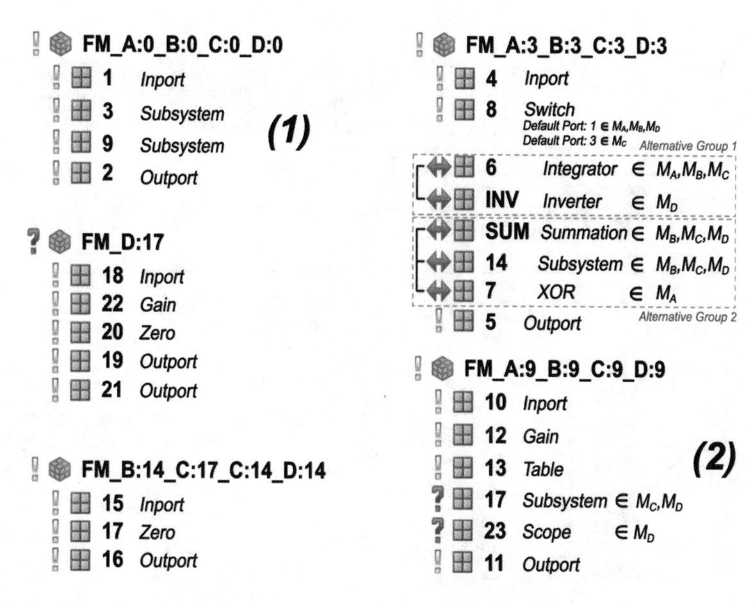

Fig. 10.19 Merged models created by N-Way for the portfolio from Fig. 10.3

results from N-Way can be combined to form a variability model, here a *family model*. For instance, within the *variation point 2* shown in Fig. 10.17, the block labeled 14 is a *Subsystem* block, comprising further functionality. Respective SMs (i.e., $M_{C:14}$) are in *Group 4* from Fig. 10.11, generated by SCMA, and CHE_6 from Fig. 10.15, produced by N-Way. Based on groups of similar SMs identified by SCMA (cf. Fig. 10.11) and their respective processing by N-Way, we provide all merged models in Fig. 10.19. For instance, the model shown at the top left contains all functional blocks from the SMs of *Group 1* from Fig. 10.11, thereby the top-level SMs from the portfolio (cf. Fig. 10.3). We state the blocks *variability*, *label*, and *function*. Our approach yields a *150% MATLAB/Simulink* model, thereby an SPL representation of the input portfolio. It allows for a better quality assurance and eased maintenance for the entire portfolio by collapsing reusable model parts and capturing variability-containing assets in a single platform. Such *150% solution* allows for the derivation of further products, the so-called *100% solutions* (Schulze et al. 2013).

150% models are an intuitive concept to model variability in flexible software-intensive systems, such as *MATLAB/Simulink* models (Kolassa et al. 2015). To create the *150% MALTAB/Simulink* model, in the following referred to as *150%ML*, we use the merged models from Fig. 10.19 and encapsulate every such model within a *Subsystem* block. We process models in a descending order based on their hierarchical depth. Consequently, we process the top-left model (cf. *(1)* in Fig. 10.19) first, as it contains all top-level elements (cf. Fig. 10.3). All blocks are mandatory and are migrated to the *Subsystem* we created in the *150%ML*. Subsequently, we process elements on the second hierarchical layer, here the model labeled *(2)* in Fig. 10.19.

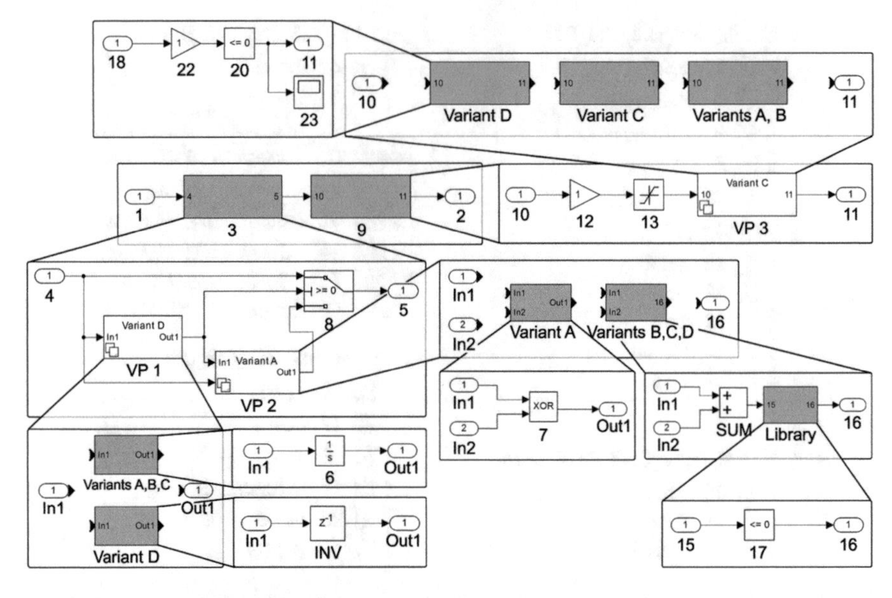

Fig. 10.20 *MATLAB/Simulink* SPL representation for the portfolio in Fig. 10.3

There exist a *parent-child* relation between the merged model *2* and the *Subsystem* block labeled 9 in the first merged model we processed. However, the merged model labeled (2) in Fig. 10.19 contains *optional* parts, which are only present in some variants of the input portfolio. First, we create a *Subsystem* block for the merged model, here *(2)* in the *150%ML*, and establish its parent-child relation to the existing *Subsystem* block labeled 9. Subsequently, variant-specific functionalities, here blocks labeled 17 and 23 in the merged model *(2)* from Fig. 10.19, are encapsulated within a *Variant Subsystem*, which explicitly supports the modeling of variability. We depict the final *150%ML* in Fig. 10.20, encapsulating common functionality in *Subsystems* and variant-specific functionality in *Variant Subsystems*. During simulation or code generation, all variant subsystems (i.e., the block labeled *VP 1*) are substituted by actual subsystems (highlighted gray), and, thus, they facilitate placeholders for concrete, variant-specific functionality.

10.7 Feasibility Study

In this section, we provide information about the product portfolio used to demonstrate the feasibility of our approach (cf. Sect. 10.7.1); elaborate on results produced by our approach, such as the *variability model* (cf. Sect. 10.7.2), and discuss the necessity for a two-step process (cf. Sect. 10.5.4).

10.7.1 Setup

The product portfolio used to demonstrate the feasibility of our holistic approach is depicted in Fig. 10.3. The portfolio comprises four model variants and a total of 17 individual SMs. We utilize the portfolio to specifically illustrate a *clone-and-own* scenario and the necessity for a holistic, multistep approach to reverseengineer variability information. However, we acknowledge that further such scenarios may exist, which we did not consider. Our coarse-grained analysis, SCMA, creates a total of 17 CMs for comparison. Each generated CM exhibits the same dimensions which correspond to the dictionary of all distinct block *functions* (cf. Sect. 10.2.1) present within the product portfolio. For the analyzed model variants, 12 distinct functions are used, and, therefore, every CM is of size 12×12. The feasibility study was evaluated on a Dual-Core i7 with 12 GB of RAM, running Windows on 64 bit.

10.7.2 Results

Comparing the product portfolio using our coarse-grained analysis procedure, SCMA yields 107 pairwise CM comparisons as depicted in Table 10.2. The identified hierarchical structures (cf. Fig. 10.9) allow for a precise grouping of similar CM, each representing an actual *MATLAB/Simulink* SM. Our SCMA generates four individual groups, each containing four CMs and, thus, referencing four SMs (cf. Fig. 10.11). Such groups are subsequently analyzed in greater detail to capture their implementation-specific variability. Our SCMA exhibits a computational complexity of $O(n^2)$ with n being the total number of CMs. Prior work has shown SCMA to scale for larger data sets, especially when using only those comparisons exceeding a certain similarity threshold (Schlie et al. 2018). Moreover, our SCMA shows that a detailed comparison can be omitted for the vast majority of model parts.

By that, SCMA specifically directs the subsequent fine-grained analysis procedures, here N-Way, to target relevant parts only. Hence, we illustrate the necessity of a coarse-grained analysis prior to any fine-grained comparisons. Directly applying the latter to an entire product portfolio is impractical by means of scalability but also by means of comprehensibility of produced results. If analyzing in detail all comparisons, which are identified by SCMA as omissible, the sheer amount of information would render their comprehensibility void. Furthermore, identification of reusable functionality would be more difficult as fine-grained analysis procedures would have to process more data. With groups of similar SMs identified by our coarse-grained analysis, thereby similar hierarchical structures between input models, we find SCMA to identify *all* such structures for our feasibility study. Consequently, we consider our coarse-grained analysis procedure applicable and SCMA feasible to identify and group together similar SMs structures between input variants. Although not detailed, we allow practitioners to filter results produced by SCMA and to focus on arbitrary variant subsets, for instance, when only a partial SPL migration is planned in the beginning. Moreover, scenarios may exist in which not all variants are relevant (Schulze et al. 2013) or when remaining variants are to be migrated incrementally. For the four groups generated by SCMA (cf. Fig. 10.11), we provide the corresponding merged models generated by our fine-grained analysis procedure, specifically our adapted merging algorithm. Specific CHEs generated by N-Way for the comparison of respective systems can be found online.[1]

With the generated merged models as illustrated in Fig. 10.19, we found all results to be comprehensible and, by that, all variability information to be correctly represented. *Mandatory* blocks are correctly collapsed, while *alternative* blocks are represented as mutually exclusive, and *optional* blocks are annotated correctly. We found no block missing in the family model and, thus, such model to comprise all implementation artifacts from all input variants. Therefore, we argue that the results produced by our approach, combining SCMA and our adapted N-Way, are complete by means of capturing all implementation artifacts for the considered feasibility study. To this end, the merged models in Fig. 10.19 each comprise only one hierarchical layer. We aggregate the merged models to derive one single family model, which represents all SMs and, by that, all VPs on all hierarchical layers (cf. Fig. 10.2).

For the feasibility study, our approach correctly finds all *variation point*. Combined with identical functionality between variants and reusable artifacts captured, we generate a new *MATLAB/Simulink* model, an SPL representation of all input variants, and we picture respective model in Fig. 10.20. Although not obvious, the generated *MATLAB/Simulink* SPL model achieves a drastic reduction in terms of complexity and size compared to the input portfolio from Fig. 10.3. Precisely, the illustrated SPL reduces the amount of individual block instances by half, lowering it from 75 in Fig. 10.3 to only 37 in Fig. 10.20. For instance, the *Subsystem* block labeled 3 on the top hierarchical layer, present four times in Fig. 10.3, has been collapsed to one single block within Fig. 10.20. Additionally, when generating the SPL, we utilize *VariantSubsystems*, a *MATLAB/Simulink*-specific block, which can reduce the number of blocks. For instance, the block labeled *In1* in the lower left corner of Fig. 10.20 is only *referenced* in the two *Subsystems* left of it. Here, the blocks labeled 6 and *INV* are the only actual block instances. With reusable artifacts identified as well as redundant artifacts collapsed, we reduce complexity and support overall software quality. By reverseengineering variability information, we allow for a precise assessment of individual artifacts, and, by that, we aim to mitigate security concerns for the entire portfolio.

10.7.3 Discussion

The need for a coarse-grained analysis prior to a fine-grained analysis is ultimately a question of scalability. The scalability of our approach is bound by N-Way's computational complexity being $O(n^4)$ with n being the number of blocks.

Although other work exists that incrementally compares variant pairs (Clements and Northrop 2001; Wille 2014), the effect of input order on the quality of the produced output, here variability models, remains unclear (Rubin and Chechik 2013a). N-Way and our adaptation for *MATLAB/Simulink* specifically addresses this and compares multiple models at once. Although assessing all four SMs in *Group 2* from Fig. 10.11 is feasible with 23 blocks contained in total and, thus, $O(23^4)$, it is evident that it quickly becomes impractical for a growing number of input elements. SCMA addresses this issue and abstracts *MATLAB/Simulink* models, reducing the overall size of the data set. Contrary to N-Way, the computational complexity of SCMA is quadratic, rendering it applicable to model portfolios of larger size. Furthermore, by showing the relations between all model variants with respect to their hierarchical structure, we allow practitioners to make strategic decisions and to tailor results prior to portfolio revamp (Schlie et al. 2018). In practice, however, practitioners may prefer an incremental transition, hence, focusing on certain variants first, rather than migrating the entire portfolio at once (Tang et al. 2010; Schulze et al. 2013). SCMA provides means to focus on any arbitrary variant subset, and, by that, it also supports an incremental transition of the portfolio.

Reengineering variability information is not limited to *MATLAB/Simulink* models but is of great importance for a variety of different domains. For instance, software-intensive systems in the automation domain, utilizing the IEC61131-3 language family, oftentimes lack a sophisticated variability management (Fischer et al. 2018). We argue that our approach is, in principle, applicable to other languages such as those of the IEC61131-3 language family or state charts (Wille et al. 2016). However, for our approach, such languages may exhibit peculiarities, which would need to be addressed accordingly.

10.8 Related Work

With our approach, we reverseengineer variability information from a product portfolio, thereby analyzing common and varying assets and relating them. In general, work related to our approach can be grouped in two main categories, which are *Clone Detection and Diffing* and *Variability Analysis*.

Clone Detection and Diffing

Kelter et al. 2005 propose an approach to compare pairs of UML class diagrams to derive their differences for semantic lifting, aiming to support comprehensibility. They extend their work in Kelter and Schmidt 2008, focusing on *State Charts*. With our work, we also aim to improve comprehensibility, but we target an entire model portfolio and, thus, multiple models rather than only two as by Kelter et al. 2005, Kelter and Schmidt 2008. A clone detection approach for *MATLAB/Simulink* models is proposed by Pham et al. 2009. We argue that our approach differs drastically from clone detection in general as our work is designed to cope with multiple models and to also capture implementation-specific variability information. Work proposed by Pham et al. 2009, Kelter et al. 2005, Kelter and Schmidt 2008 is limited to two models only. Liang et al. 2014 identify the longest *common-subsequence* within models. Contrary to our work, found results are not classified, and no variability information is determined. She et al. 2011 reverse engineer feature models from product maps. However, contrary to our work, which uses direct implementation artifacts, such maps are rather abstract and provide no direct insight into the implementation. Integrated within *ConQat* (CQSE – Continuous Quality in Software Engineering 2016; Deissenboeck et al. 2008) use a graph-based algorithm to find model clones. However, the specific variability of found elements is not determined, while our approach reengineers implementation-specific variability information. Multiple *MATLAB/Simulink* models are compared by Al-Batran et al. 2011 and evaluated for reusable, semantically equivalent, parts. Unlike our work, found parts remain isolated and are not aggregated. Extending on *NICAD* (Roy and Cordy 2008) and leveraged to cope with *MATLAB/Simulink* models, (Alalfi et al. 2012) introduce *SIMONE*. Unlike our approach, results are not aggregated while we combine results for further processing.

Variability Analysis

The majority of current work revolving around model comparison primarily focuses on two models (Martinez et al. 2014). Authors locate similarities between UML model variants in Martinez et al. 2015b. Extended by visualizing results in Martinez et al. 2015a, authors derive a feature model. However, domain knowledge is indispensable for the process, and with the variability models we create, such domain knowledge is not needed. Zhang et al. 2011 use *EMF Compare* to capture variability by means of the *Common Variability Language*. Sharing similarities with Martinez et al. 2014, explicit domain knowledge is required. Similar to Zhang et al. 2011, Font et al. 2015 use *CVL* to identify common and variable artifacts. Their work requires a *base model* to be defined, while our approach explicitly does not require a starting point to be defined. With the *ECCO tool* proposed by Fischer et al. 2014, Fischer et al. 2015, authors identify similar structures between system variants. Contrary to our work, which revolves around models, (Fischer et al. 2014; Fischer et al. 2015) target source code. *MATLAB/Simulink* models are *sliced* by Gold et al. 2017, and despite authors utilizing trees, they only focus on one model, whereas we target multiple models. Rumpe et al. 2015 transform *Stateflow State Machines* into normalized *State Machines* to assess their compatibility between versions, however, also limited to two models only. An extractive approach to create an SPL from a set of system variants is proposed by Alves et al. 2007. Unlike our work, Alves et al. (2007) focus on source code rather than models. Ryssel et al. 2010a also utilize matrices to identify variation points within *MATLAB/Simulink* models. However, their matrices are instance-specific and their approach is based on finding cliques of maximal size, an *NP-hard* problem. Our CMs are instance-agnostic, and our approach is specifically designed to cope with an entire portfolio. Proposing N-Way, (Rubin and Chechik 2013a) compare multiple UML model variants. We have shown their technique as proposed by Rubin and Chechik 2013a to be not applicable to *MATLAB/Simulink* and illustrated performed adaptations. Also, the approach in general is limited by means of scalability and the analysis being limited to one hierarchical layer. To allow for the merging of different products into one model, (Rubin and Chechik 2012) define a corresponding operator. Rubin and Chechik 2013b extend on their work and introduce measurable metrics. Although (Rubin and Chechik 2012; Rubin and Chechik 2013b) also aim to create an SPL, their proposed operator does not allow for the derivation of variability models.

10.9 Conclusion and Future Work

We propose a holistic approach to reengineer an entire *MATLAB/Simulink* model portfolio into one *variability model*, which collapsed redundant artifacts, captures reusable artifacts, and depicts the relation of model artifacts by means of their implementation-specific variability. Our approach includes an instance-agnostic, coarse-grained analysis procedure, denoted Static Connectivity Matrix Analysis. By that, we compare all input variants in their entirety *at once* and identify all similar structures between them. Such analysis allows for a preliminary assessment of relations between all model variants. Furthermore, SCMA allows for a precise identification of model parts, which need to be analyzed further in greater detail, while also indicating those for which such analysis can be omitted. With groups of similar model parts identified, we apply our adaptation of N-Way (Rubin and Chechik 2013a) to cope with multiple individual model parts simultaneously. Specifically, we compare respective systems at once and identify their implementation-specific variability. We demonstrate the feasibility of our approach using an industry-inspired running example, which reflects *clone-and-own* scenarios. We show our approach to be fast and precise and, by that, to capture all reusable parts and to produce understandable variability models. Thus, we conservatively consider our holistic approach applicable to reengineer *MATLAB/Simulink* product portfolios.

Assuring the quality and security of developed products in flexible and software-intensive systems is pivotal. In practice, however, practitioners face a product portfolio, which emerged adhoc using unstructured reuse. As a result, maintenance overhead is increased, and overall quality of the portfolio is adversely affected. SPL facilitate strategic reuse and promote better quality assurance for the entire portfolio. However, migrating an entire family of software-intensive systems, such as *MATLAB/Simulink* models, toward managed reuse, remains a challenging endeavor. Nevertheless, we argue that it is necessary to facilitate strategic reuse, ease maintenance, and, thereby, improve overall quality and mitigate security concerns for the entire portfolio.

In future work, we plan to compare our approach with other work on reverse-engineering *MATLAB/Simulink* models such as (Rubin and Chechik 2013b) and to identify drawbacks and benefits to improve our approach. We also plan to extend our feasibility study and apply our approach to model portfolios of industrial size. Although in principle capable of an incremental migration of individual variants, our approach at this point considers the entire portfolio at once. In practice, however, partial migration of certain variant subsets and subsequent inclusion of remaining variants may be desired. With our SCMA, we already provide means to focus on any arbitrary variant subset and to define groups for further analysis. To this end, we have not explored ways to incrementally include further variants without the need to run the entire analysis procedure again. We therefore plan to explore appropriate means to cope with subsequent inclusion of additional variants and to also ensure the consistency of previously created variability models.

Acknowledgements This work has been supported by the German Research Foundation (DFG) (SCHA 1635/12-1).

References

Al-Batran, B., Schätz, B., & Hummel, B. (2011). Semantic clone detection for model-based development of embedded systems. In *Proceedings of the International Conference on Model Driven Engineering Languages and Systems (MODELS)* (pp. 258–272). Berlin: Springer.

Alalfi, M., Cordy, J., Dean, T., Stephan, M., & Stevenson, A. (2012). Models are code too: Near-miss clone detection for Simulink models. In *Proceedings of the International Conference on Software Maintenance (ICSM)* (pp. 295–304). Piscataway: IEEE.

Alves, V., Matos, P., Cole, L., Vasconcelos, A., Borba, P., & Ramalho, G. (2007). *Extracting and evolving code in product lines with aspect-oriented programming* (pp. 117–142). Berlin: Springer.

Antoulas, A. (2005). *Approximation of large-scale dynamical systems. Advances in design and control*. Philadelphia: Society for Industrial and Applied Mathematics.

Benner, P., Mehrmann, V., & Sorensen, D. (2005). *Dimension reduction of large-scale systems. Lecture notes in computational science and engineering*. Berlin: Springer.

Clements, P., & Northrop, L. (2001). *Software product lines: Practices and patterns*. Boston: Addison-Wesley Longman Publishing Co., Inc.

Codabux, Z., & Williams, B. (2013). Managing technical debt: An industrial case study. In *Proceedings of the International Workshop on Managing Technical Debt (MTD)* (pp. 8–15). Piscataway: IEEE.

CQSE – Continuous Quality in Software Engineering (2016) Conqat. https://www.conqat.org

Cretu, L., & Dumitriu, F. (2014). *Model-driven engineering of information systems: Principles, techniques, and practice*. Palm Bay: Apple Academic Press.

Deissenboeck, F., Hummel, B., Juergens, E., Pfaehler, M., & Schaetz, B. (2010). Model clone detection in practice. In *Proceedings of the International Workshop on Software Clones (IWSC)* (pp. 57–64). New York: ACM.

Deissenboeck, F., Hummel, B., Jürgens, E., Schätz, B., Wagner, S., Girard, J. F., & Teuchert, S. (2008). Clone detection in automotive model-based development. In *Proceedings of the International Conference on Software Engineering (ICSE)* (pp. 603–612). New York: ACM.

Duan, G. (2010). Analysis and design of descriptor linear systems. In *Advances in mechanics and mathematics*. New York: Springer.

Dubinsky, Y., Rubin, J., Berger, T., Duszynski, S., Becker, M., & Czarnecki, K. (2013). An exploratory study of cloning in industrial software product lines. In *Proceedings of the European Conference on Software Maintenance and Reengineering (CSMR)* (pp. 25–34). Piscataway: IEEE.

Ernst, N., Bellomo, S., Ozkaya, I., Nord, R., & Gorton, I. (2015). Measure it? Manage it? Ignore it? Software practitioners and technical debt. In *Proceedings of the European Software Engineering Conference/Foundations of Software Engineering (ESEC/FSE)* (pp. 50–60). New York: ACM.

Fischer, J., Bougouffa, S., Schlie, A., Schaefer, I., & Vogel-Heuser, B. (2018). A qualitative study of variability management of control software for industrial automation systems. In *Proceedings of the International Conference on Software Maintenance and Evolution (ICSME)* (pp. 615–624)

Fischer, S., Linsbauer, L., Lopez-Herrejon, R. E., & Egyed, A. (2014). Enhancing clone-and-own with systematic reuse for developing software variants. In *Proceedings of the International Conference on Software Maintenance and Evolution (ICSME)* (pp. 391–400). Piscataway: IEEE.

Fischer, S., Linsbauer, L., Lopez-Herrejon, R. E., & Egyed, A. (2015). The ECCO tool: Extraction and composition for clone-and-own. In *Proceedings of the International Conference on Software Engineering (ICSE)* (pp. 665–668). Piscataway: IEEE.

Font, J., Ballarín, M., Haugen, Ø, & Cetina, C. (2015). Automating the variability formalization of a model family by means of common variability language. In *Proceedings of the International Software Product Line Conference (SPLC)* (pp. 411–418). New York: ACM.

Gold, N. E., Binkley, D., Harman, M., Islam, S., Krinke, J., & Yoo, S. (2017). Generalized observational slicing for tree-represented modelling languages. In *Proceedings of the European Software Engineering Conference/Foundations of Software Engineering (ESEC/FSE)* (pp. 547–558).

Grönniger, H., Krahn, H., Pinkernell, C., & Rumpe, B. (2014). Modeling variants of automotive systems using views. CoRR.

Haber, A., Kolassa, C., Manhart, P., Nazari, P. M. S., Rumpe, B., & Schaefer, I. (2013). First-class variability modeling in Matlab/Simulink. In *Proceedings of the International Workshop on Variability Modeling in Software-intensive Systems (VaMoS)* (pp. 4:1–4:8). New York: ACM.

Henley (ed.). (1973). *Graph theory in modern engineering: Computer aided design, control, optimization, reliability analysis. Mathematics in science and engineering*. Amsterdam: Elsevier.

Holthusen, S., Wille, D., Legat, C., Beddig, S., Schaefer, I., & Vogel-Heuser, B. (2014). Family model mining for function block diagrams in automation software. In *Proceedings of the International Workshop on Reverse Variability Engineering (REVE)* (pp. 36–43). New York: ACM.

Jeong, H., Obaidat, M., Yen, N., & Park, J. (2013). *Advances in computer science and its applications (CSA)*. Berlin: Springer.

Kastner, C., Dreiling, A., & Ostermann, K. (2014). Variability mining: Consistent semi-automatic detection of product-line features. *IEEE Transactions on Software Engineering, 40*, 67–82.

Kelter, U., & Schmidt, M. (2008). Comparing state machines. In *Proceedings of the International Workshop on Comparison and Versioning of Software Models (CVSM)* (pp. 1–6). New York: ACM.

Kelter, U., Wehren, J., & Niere, J. (2005). A generic difference algorithm for uml models. *Software Engineering, 64*(105–116), 4–9.

Kim, J. A. (2010). Case study of software product line engineering in insurance product. In *Proceedings of the International Software Product Line Conference (SPLC)* (pp. 495–495). Berlin: Springer.

Kolassa, C., Rendel, H., & Rumpe, B. (2015). Evaluation of variability concepts for simulink in the automotive domain. In *Hawaii International Conference on System Sciences (HICSS)* (pp. 5373–5382). Piscataway: IEEE.

Kuhn, H. (1955). The hungarian method for the assignment problem. *Naval Research Logistics Quarterly, 2*, 83–98.

Lapeña, R., Ballarin, M., & Cetina, C. (2016). Towards clone-and-own support: Locating relevant methods in legacy products. In *Proceedings of the International Software Product Line Conference (SPLC)* (pp. 194–203). New York: ACM.

Levenshtein, V. I. (1966). Binary codes capable of correcting deletions, insertions, and reversals. *Soviet Physics Doklady, 10*(8), 707–710.

Liang, Z., Cheng, Y., & Chen, J. (2014). A novel optimized path-based algorithm for model clone detection. *Journal of Software 9*(7), 1810–1817.

Martinez, J., Ziadi, T., Bissyandé, T. F., Klein, J., & Le Traon, Y. (2015a). Bottom-up adoption of software product lines: A generic and extensible approach. In *Proceedings of the International Software Product Line Conference (SPLC)* (pp. 101–110). New York: ACM.

Martinez, J., Ziadi, T., Bissyandé, T. F., Klein, J., & Traon, Yl. (2015b). Automating the extraction of model-based software product lines from model variants. In *Proceedings of the International Conference on Automated Software Engineering (ASE)* (pp. 396–406). Piscataway: IEEE.

Martinez, J., Ziadi, T., Klein, J., & le Traon, Y. (2014). Identifying and visualising commonality and variability in model variants. In *Proceedings of the European Conference on Modeling Foundations and Applications (ECMFA)* (pp. 117–131). Cham: Springer.

Merschen, D., Polzer, A., Botterweck, G., & Kowalewski, S. (2011). Experiences of applying model-based analysis to support the development of automotive software product lines. In *Proceedings of the International Workshop on Variability Modeling in Software-intensive Systems (VaMoS)* (pp. 141–150). New York: ACM.

Olshevsky, V. (2001). *Structured matrices in mathematics, computer science, and engineering I. Contemporary mathematics.* American Mathematical Society.

Pham, N. H., Nguyen, H. A., Nguyen, T. T., Al-Kofahi, J. M., & Nguyen, T. N. (2009). Complete and accurate clone detection in graph-based models. In *Proceedings of the International Conference on Software Engineering (ICSE)* (pp. 276–286). Piscataway: IEEE.

Pohl, K., Böckle, G., & van der Linden, F. J.. (2005). *Software product line engineering: Foundations, principles and techniques.* Berlin: Springer.

Pressman, R. (2005). *Software engineering: A practitioner's approach.* New York: McGraw-Hill Higher Education.

Ramasubbu, N., Kemerer, C. F., & Woodard, C. J. (2015). Managing technical debt: Insights from recent empirical evidence. *IEEE Software, 32*(2), 22–25.

Riva, C., & Rosso, C. D. (2003). Experiences with software product family evolution. In *Proceedings of the Joint Workshop on Software Evolution and International Workshop on Principles of Software Evolution (IWPSE-EVOL)* (pp. 161–169). Piscataway: IEEE.

Roy, C. K., & Cordy, J. R. (2008). NICAD: Accurate detection of near-miss intentional clones using flexible pretty-printing and code normalization. In *Proceedings of the International Conference on Program Comprehension (ICPC)* (pp. 172–181). Piscataway: IEEE.

Rubin, J., & Chechik, M. (2012). Combining related products into product lines. In *Proceedings of the International Conference on Fundamental Approaches to Software Engineering (FASE)* (pp. 285–300). Berlin: Springer.

Rubin, J., & Chechik, M. (2013a). N-way model merging. In *Proceedings of the European Software Engineering Conference/Foundations of Software Engineering (ESEC/FSE)* (pp. 301–311). New York: ACM.

Rubin, J., & Chechik, M. (2013b). Quality of merge-refactorings for product lines. In *Proceedings of the International Conference on Fundamental Approaches to Software Engineering (FASE)* (pp. 83–98). Basel: Springer.

Rumpe, B., Schulze, C., von Wenckstern, M., Ringert, J., & Manhart, P. (2015). Behavioral compatibility of simulink models for product line maintenance and evolution. In *Proceedings of the International Software Product Line Conference (SPLC)* (pp. 141–150). New York: ACM.

Ryssel, U., Ploennigs, J., & Kabitzsch, K. (2010a). Automatic variation-point identification in function-block-based models. In *Proceedings of the International Conference on Generative Programming and Component Engineering (GPCE)* (pp. 23–32). New York: ACM.

Ryssel, U., Ploennigs, J., & Kabitzsch, K. (2010b). Automatic variation-point identification in function-block-based models. In *Proceedings of the International Conference on Generative Programming and Component Engineering (GPCE)* (pp. 23–32). New York: ACM.

Schlie, A., Schulze, S., & Schaefer, I. (2018). Comparing multiple MATLAB/Simulink models using static connectivity matrix analysis. In *Proceeidngs of the International Conference on Software Maintenance and Evolution (ICSME)* (pp. 185–196). Piscataway: IEEE.

Schlie, A., Wille, D., Cleophas, L., & Schaefer, I. (2017a). Clustering variation points in matlab/simulink models using reverse signal propagation analysis. In *Proceedings of the International Conference on Software Reuse (ICSR)* (pp. 77–94). Cham: Springer.
Schlie, A., Wille, D., Schulze, S., Cleophas, L., & Schaefer, I. (2017b). Detecting variability in MATLAB/Simulink models: An industry-inspired technique and its evaluation. In *Proceedings of the International Software Product Line Conference (SPLC)* (pp. 215–224). New York: ACM.
Schulze, M., Mauersberger, J., & Beuche, D. (2013). Functional safety and variability: Can it be brought together? In *Proceedings of the International Software Product Line Conference (SPLC)*. New York: ACM.
She, S., Lotufo, R., Berger, T., Wasowski, A., & Czarnecki, K. (2011). Reverse engineering feature models. In *Proceedings of the International Conference on Software Engineering (ICSE)* (pp. 461–470). Piscataway: IEEE.
Sullivan, K. J., Griswold, W. G., Cai, Y., & Hallen, B. (2001). The structure and value of modularity in software design. In *Proceedings of the European Software Engineering Conference/Foundations of Software Engineering (ESEC/FSE)* (pp. 99–108). New York: ACM.
Tan, H. (2012). Knowledge discovery and data mining. In *Advances in intelligent and soft computing*. Berlin: Springer.
Tang, A., Couwenberg, W., Scheppink, E., Aan de Burgh, N., Deelstra, S., & Vliet, H. (2010). Spl migration tensions: An industry experience. In *Proceedings of the Workshop on Knowledge-Oriented Product Line Engineering (KOPLE)*
Tuytelaars, T., & Mikolajczyk, K. (2008). Local invariant feature detectors: A survey. *Foundations and Trends in Computer Graphics and Vision, 3*(3), 177–280.
van der Linden, F. J., Schmid, K., & Rommes, E. (2007). *Software product lines in action: The best industrial practice in product line engineering*. Berlin: Springer.
Wille, D. (2014). Managing lots of models: The famine approach. In *Proceedings of the International Symposium on the Foundations of Software Engineering (FSE)* (pp. 817–819). New York: ACM.
Wille, D., Holthusen, S., Schulze, S., & Schaefer, I. (2013). Interface variability in family model mining. In *Proceedings of the International Workshop on Model-Driven Approaches in Software Product Line Engineering (MAPLE)* (pp. 44–51). New York: ACM.
Wille, D., Schulze, S., Seidl, C., & Schaefer, I. (2016). Custom-tailored variability mining for block-based languages. In *Proceedings of the International Conference on Software Analysis, Evolution, and Reengineering (SANER)* (pp. 271–282). Piscataway: IEEE.
Zhang, D., & Lu, G. (2003). A comparative study of curvature scale space and fourier descriptors for shape-based image retrieval. *Journal of Visual Communication and Image Representation 14*(1), 39–57.
Zhang, X., Haugen, Ø, & Møller-Pedersen, B. (2011). Model comparison to synthesize a model-driven software product line. In *Proceedings of the International Software Product Line Conference (SPLC)* (pp. 90–99). Piscataway: IEEE.

Part III
Engineering Security Improvement

Chapter 11
Security Analysis and Improvement of Data Logistics in AutomationML-Based Engineering Networks

Bernhard Brenner and Edgar Weippl

Abstract The Automation Markup Language (AutomationML) is a concept developed in 2008 in order to provide a versatile data format for seamless exchangeability of engineering data, with the goal of simplifying the design and creation of cyber-physical production systems. Different software, such as CAD programs, shall be able to support this format. Especially in the case of collaborative work and data exchange, security can become an important issue as current approaches do not fulfill the essential security objectives necessary, meaning that authenticity, integrity, and confidentiality of the stored files are not ensured from the start of product design to the end product. This raises questions not only about the confidentiality of company information but also about the safety of production lines and end products. Leakage of confidential information (e.g., construction plans), leading to unintended spread of know-how, can be an expensive consequence. Unauthorized and undetected (malicious) modifications may even lead to faults in end products, availability issues, or serious accidents within the production line. This chapter focuses on the demonstration of open issues within AutomationML-based engineering project environments. We are going to demonstrate why some kind of security layer (i.e., layer ensuring access control and privileges, as well as ensuring data integrity) is crucial when using AutomationML. Therefore, we provide assumptions about potential attacks and their potential consequences. We introduce an approach to identify and analyze assets, potential threats and vulnerabilities, resulting risks, as well as countermeasures that are relevant for ensuring the

B. Brenner (✉)
SBA Research, Vienna, Austria
e-mail: bbrenner@sba-research.org

E. Weippl
Christian Doppler Laboratory for Security and Quality Improvement in the Production System Lifecycle (CDL-SQI), Institute of Information Systems Engineering, Technische Universität Wien, Vienna, Austria

SBA Research, Vienna, Austria
e-mail: edgar.weippl@tuwien.ac.at

S. Biffl et al. (eds.), *Security and Quality in Cyber-Physical Systems Engineering*,
https://doi.org/10.1007/978-3-030-25312-7_11

abovementioned properties: confidentiality of know-how, availability of the assets, and the integrity of relevant data.

Keywords AutomationML security · AutomationML-based data exchange · Access control for AutomationML

11.1 Introduction

The Automation Markup Language (AutomationML) is a state-of-the-art concept to provide a standardized scheme for exchanging engineering data and to provide a versatile data format to support engineers in the construction of cyber-physical production systems (Drath et al. 2008). How data shall be exchanged is still insufficiently defined, however. One reason is that there are usually no security policies implemented that cover data exchange. For example, many engineers produce their CAD drawings on their local computer, store the files locally, and use a USB drive or email program in order to transfer it to their colleagues, who then proceed to work on this version of their output until they get a new email with a new version (Biffl et al. 2014; Drath 2009).

There are practical problems with this kind of data exchange:

- Efficiency

  It can be difficult for an engineer to find out the newest version of her colleagues' work, leading her to repeated requests for newest versions. If there are many colleagues and many files, this leads to many unnecessary requests, which have a negative impact on the efficiency of their work.
- Unwanted information disclosure

  One is barely able to determine the set of people who will possess this file: emails are usually sent in plaintext, and a USB drive can get lost. In addition, employees might send them to third parties (e.g., TÜV or clients), making it a practical threat in terms of data (thus, know-how) leakage, which can have expensive consequences (Von Hippel 1989; Ahmad et al. 2014; Mohamed et al. 2006).
- Unauthorized modifications

  The authenticity and integrity of all construction files for production line and end product are crucial assets within cyber-physical production systems. Still, there is not yet an access control system that is fine-grained enough and easy to use in order to avoid unauthorized modifications of files on a per file or per component basis.

Using some kind of centralized versioning or data repository for project-related (AutomationML) data is a current approach to address the first of these problems. When including a centralized versioning system (e.g., SVN (Subversion) or git (Git-scm)), it is easier to collaborate on a set of files and to enable agile development within the creation of cyber-physical production systems. While git does already implement the functionality of signed commits (GitHub), it does not provide any AutomationML processing functionality, which is one motivation to develop repository software specifically tailored for AutomationML (AutomationMLe.V.). Furthermore, a concept for access control on a per component level that is built into an AML file has the big benefit of being independent (in terms of access protection) of any central AutomationML processing repository. Both systems could implement restricted views on a per component or per role class (AutomationMLe.V. c; Schyja et al. 2014) level instead of per file or per directory level. This functionality is important in the future because industrial espionage is still a practical threat that cannot be mitigated sufficiently by just implementing basic password authentication into version control (Thonnard et al. 2012; Tucker 1997; Tuptuk and Hailes 2018; Wangen 2015).

Considering authentication, there's another problem. Third parties, for example, to whom access to the repository is granted (e.g., TÜV or outsourcing partners), can become a cause of unintentional information disclosure as data leaves the boundaries of the company in an uncontrolled way. In other words: once a confidential file is stored on a device outside the company's premises, the company does not have full control of how this information is spread. Thus, there exist the same problems as with email and USB drives mentioned before.

There is furthermore not yet a way to identify unauthorized modifications on a per file basis. Adversaries who manage to break into a company network could therefore modify a file in an undetected way, leading to modified end products or modified hardware or software in production lines. These changes can remain undetected for a long period of time as cyber-physical production systems typically have very long life-spans and projects to create them can have very long durations. (A more detailed explanation on the impact of security vulnerabilities on the product, the company profit, and reputation as well as safety is given in Sect. 11.1.1).

In this chapter:

- We provide a systematic and general approach to introduce measures to improve the confidentiality and authenticity of central AutomationML data repositories
- We show how these measures can be introduced into productive environments with no or low negative impact on the usability of services or efficiency of the engineering teams
- We give advice based on common best practices and point out the most important mistakes to avoid

Furthermore, we strive to answer the following research questions:

1. Which use cases of AutomationML can lead to dangerous situations in terms of confidentiality of engineering data and integrity of product and production line design?
2. Which assumptions shall be made about potential attackers in a manufacturing company?
3. Which solutions for access protection can be derived?

11.1.1 The Relationship of Security to Profit, Reputation, and Safety

Let us define the terms "safety" and "security." We define safety according to the "Oxford Living Dictionary," as the condition of being protected from or unlikely to cause danger, risk, or injury (OxfordDictionaries) and furthermore as the measures to ensure physical integrity of people's health and security in this context as a state in which all defined rules are sufficiently enforced through technical means, such that the potential gains or benefits of breaking a rule are never worth the necessary effort (technical means) or the consequences (management and technical means).

Vulnerabilities related to security can have negative consequences for safety, namely, wherever there is a transition from concepts or software to hardware or physics. Safety measures shall ensure that even in case of unintentional wrongdoing, the impact on the physical integrity/health of employees and clients shall remain as small as possible. Security plays a role as soon as a system must be protected from malicious changes because these malicious changes may indeed have consequences in hardware components and the physical behavior of machines. One example is an end product that is entirely defined using AutomationML. Changes in parameters (thickness of a weld, lifetime, load stability, tensile stress, maximum torque, voltage/current/resistors, etc.) can have severe impact on product safety. This is why integrity and non-repudiation of data are requirements for engineering data. Furthermore, unauthorized and possibly malicious modifications can have severe consequences to the availability of the production line. Since the availability of the production line typically is of crucial importance for a company, negative impacts on it are a major threat. Due to the physical nature of these systems, the consequences of such malicious modifications could have severe consequences on the safety of employees working within the production line on one hand and have negative consequences even for the safety of clients dealing with the end product on the other hand. And, of course, as seen in the famous case of stuxnet,[1] production lines may also experience physical damage due to malware (Falliere et al. 2011; Langner 2011).

Just to give an example: imagine a manufacturer of production lines for car parts. In this case, the client would be a car manufacturer. Competitors could intentionally modify the construction plans of the production lines or the ones of the end product (i.e., modify the thickness or location of a weld, leading to end products that last only 3 years instead of the specified 10 years which have been negotiated with the client). About 3 years after the production line is already sold and operated by the client, a crushing majority of the manufacturer's clients (in this case, the car owners) complain that their cars broke or are at least less reliable than the ones of its competitors. The manufacturer then may complain about the quality of the

[1] Stuxnet is one of the first known malware that was designed to physically destroy a cyber-physical system by means of influencing the controller nodes' software parameters.

produced parts, and further contracts won't be negotiated with the manufacturer of the production line.

The attacker thus reached his/her goal to damage the company who built the production line (or the car manufacturer), *and* it can be extremely difficult in practice to trace the issues back to the attacker, who just changed a tiny parameter a few years ago. First, this could also have been a mistake by the responsible engineer, and second, a long timespan typically helps an attacker a lot; as in this case, it is likely that not even the responsible engineer can prove that this was not his/her mistake.

In order to mitigate these threats, our work targets the introduction of security-related measures into central AutomationML-based repositories and working environments.

11.2 Introduction to Access Control

The purpose of this section is to give a short introduction and background information about the topic. First, we present an introduction to access control, and then we provide a short introduction to authentication and authorization. The explanations are supplemented by examples.

The purpose of access control in general can be one or more of the following:

- Protect the confidentiality of information
- Protect the integrity of information
- Protect the use of resources (e.g., tools or machines)
- Protect the access to certain (virtual or physical) areas or services

Let us define an access control system to consist of an access control model and an access control policy. Access control models define *how* access *can be* distributed, while access control policies define to whom and *for what* access *is* distributed, meaning that it defines how an access control model is implemented to fit a certain purpose in a certain case.

An example: Imagine a family consisting of two parents (both older than 30) and two children in the age of 2 and 13 years.

Those four people are our subjects for now. Imagine that this family owns a chainsaw and a lawn mower. Parents are allowed to operate the lawn mower and the chainsaw, the 13-year-old is allowed to use the lawn mower but not the chainsaw, and the 2-year-old is not allowed to use either of them.

We have two sets then: a set of objects and a set of subjects.

Our simple access control model is defined as follows: For every object in the set of objects, it is explicitly defined which persons may operate them.

Our policy looks like this, then (see Table 11.1).

We thus have defined the creation of an access control list that, for our case, implements the policy we needed.

Table 11.1 Exemplary access policy for our exemplary family owning a lawn mower and a chainsaw

Object/subject	Parent 1	Parent 2	13-year-old	2-year-old
Lawn mower	Yes	Yes	Yes	No
Chainsaw	Yes	Yes	No	No

11.2.1 Discretionary and Mandatory Access Control

In the aforementioned example (Sect. 11.1), we have implemented an access control model that is referred to as discretionary access control (DAC) in the literature and has its theoretical roots for information technology before the 1960s. It defines the privileges on a per person (i.e., per identity) basis. Linux' and most other UNIX-based operating systems' file permission concepts are based on discretionary access control, with the additional feature of groups, and were designed in the late 1960s (TheLinuxFoundation). Also, Windows' "AccessCheck" works using discretionary access control and access control lists that are stored within the files themselves (Microsoft-Docs). However, Windows' Security Levels (Administrator > Local User > Guest) are, theoretically, an implementation of mandatory access control.

On the other hand, there is the paradigm of mandatory access control (MAC), which basically states that access restrictions are enforced through security levels as opposed to the discretionary access control model, where access is decided upon a user's identity. For the example of the family, the lawn mower, and the chainsaw, the rules could be the following: let there be three permission levels, 0, 1, and 2. The lawn mower is classified as level 1 device and the chainsaw as level 2 device. The 2-year-old has access level 0 and the 13-year-old access level 1, and the parents are assigned access level 2. The only rule necessary then is the following: For every object O and subjects S, access is granted if Level(S) >= Level (O).

Another important paradigm is the role-based access control model (Sandhu et al. 1996), in which policies are defined using the following sets:

- R. . .the set of all roles
- P. . .the set of all permissions a role may have
- S. . .the set of all subjects
- PA (Permission Assignment). . .the set of permissions a role may be assigned to
- SA (Subject Assignment). . .the set of subject assignments to roles
  RH. . .a partially ordered set defining the role hierarchy
- SE. . .the sessions, a mapping between S, R, and P—necessary if privileges are obtained session-wise

The analogy of role-based access control can be made quite well using a company as an example. Imagine a bank building having separate rooms that are dedicated to the specialization of their employees/clients. There is the entrance for the clients, leading to the client waiting area. Then, there is a vault, which only safekeepers are allowed to access. And there is the cashiers' area, in which only the client consultants

are allowed to be. The boss of the bank, of course, is allowed to access any area. Furthermore, all employees are allowed to access the client area.
That was our definition of the requirements. Let's start implementing it via a policy for a role-based access control model: In this case, S contains all subjects that are employed at this bank building. Imagine the following people:
$\{S = GenericPerson, Clara, Susan, Matthew, Anna, Alice\}$.
Then, there is our set of roles:
$R = \{client, safekeeper, clientconsultant, boss\}$.
There are furthermore the access permissions:
$P = \{access\ client\ area, access\ cashiers\ area, access\ vault\}$
These are our sets containing terminals. The next sets define mappings of these terminals to each other: our set SA (subject assignment), which maps subjects to roles, is defined as follows:

$$SA = \{Generic\ Person \rightarrow Client,\ Clara \rightarrow client\ consultant\} \cup \{Susan \rightarrow safe\ keeper,\ Matthew \rightarrow client\ consultant,\ Anna \rightarrow client\ consultant\} \cup \{Alice \rightarrow boss\}$$

Whereas "Generic Person" is the default identity for all individuals who cannot be identified otherwise.

And the set PA (Permission Assignment), which maps roles to permissions, looks like this:

$$PA = \{client \rightarrow client\ area, client\ consultant \rightarrow \{client\ area, cashiers\ area\}\} \cup \{safekeeper \rightarrow \{client\ area, vault\}, boss \rightarrow \{client\ area, cashiers\ area, vault\}\}$$

In addition, there are two optional sets. The first one is RH, which defines the role hierarchy—enabling roles to inherit permissions from other roles in order to make it easier to design and to understand the permission distribution.

And, there is the set SE, a set defining sessions, whereas a session can be defined as a temporal assignment of privileges for a resource to a role. As an example, we could assume that some special clients are, as soon as they are identified by the client consultants, assigned the role "VIP client" and get temporary access to the vault (e.g., as they may own a bank locker there).

Then, the set SE could look like this:
$SE = \{(Emil \rightarrow VIP\ client, EXPIRATIONTIMESTAMP)\}$,
and we extend our role set R by: $R = R \bigcup \{VIP\ client\}$
and finally, PA will be extended the following:
$PA = PA \bigcup \{VIP\ client \rightarrow access\ vault\}$

These were the basic about role-based access control (RBAC). To get a deeper insight into RBAC, take a look at the official NIST standard for role-based access control models (Sandhu et al. 2000).

11.2.2 Other/Special Forms of Access Control

In the case of *context-based access control*, permission is granted based not only on roles or identities but also on the specific context within an action that is to be performed. An example is a firewall blocking or passing through requests based on the layer 7 content of the packet. Another form is *relationship-based access control* (ReBAC), which is a relatively new access control model. It has, for example, been described by Aktoudianakis in his dissertation (Aktoudianakis 2016), in which he also describes many other forms (mainly mixed forms of the ones we have named before) of access control. The relationship-based access control model allows subjects to create binary relations to each other and to define relational security policies that regulate access to an owner's objects. The benefit of introducing relations into the access control model is that they can be combined to form more complex patterns allowing more abstract policy definitions that seamlessly adapt to "network" changes. The so generated policies may span any degree of separation between subjects and may reach an access control decision based on the relations involved as well as an optional variable, the "trust ratio" of a particular relation between two subjects. These optional trust ratios may be assigned individually by subjects for any relation they own and according to their individual decision (Aktoudianakis 2016).

11.2.3 Examples of Famous Access Control Models

Two well-known access control models have been developed in the 1970s, the Bell-LaPadula (Bell and LaPadula 1973) (or BLP) access control model and the BIBA (Biba 1977) access control model. The Bell-LaPadula model is a MAC-based access control model that has originally been invented to control the confidentiality of classified files within a military context, while the BIBA model has been published in order to be a modification of the LaPadula model applicable to control the integrity of information (in our case: files) within an organization.
There are plenty of literature describing these two models; however, to have a complete explanation in the book, we will explain these two in a short and concise way.

11.2.3.1 The LaPadula Model

It is a MAC (two rules) and DAC (one rule)-based model, meaning that there is a certain number of security levels (in this case: confidentiality levels) *and* an identity-based access matrix defined within the system. Let's make an example with three levels: 0 for public information, 1 for classified information, and 2 for top-secret information.

The LaPadula model describes three rules how information may flow between roles, whereas the word "up" means "to a higher level" and vice versa:

- Simple security Property ("no read up"): a subject of a certain confidentiality level may not have read access to information of a higher confidentiality level.
- The star (*) ("no write down") property: a subject of a certain confidentiality level may not have write access to information of a lower confidentiality level.
- The third property states that there is an access matrix defined (which makes it a discretionary access control property).

The star property may be replaced by the so-called strong star property, changing rule two to as follows: A subject of a certain confidentiality level may only have write access to objects at the same confidentiality level. Note that the word subject may refer to a person or a process.

11.2.3.2 The BIBA Model

This model is similar, but instead of focusing on the confidentiality of information, its focus lies on its integrity. Therefore, the rules look a bit different[2]:

- Simple security Property ("*read up*"): a subject of a certain integrity level may not have read access to information of a *lower* integrity level.
- The star (*) property ("*write down*"): a subject of a certain integrity level may not have write access to information of a higher integrity level.
- Invocation property: A subject of level S can only invoke other subjects, if they are of level S or lower.

11.2.4 Authentication and Authorization

According to current NIST definitions, digital authentication establishes that the subject with the intention to access additional service is in control of one or more valid proofs associated of that subject's digital identity. Such an identity is unique in the context of the digital service, but thus, however, not need to uniquely identify the subject in all contexts. That is, the real life identity does not have to be known in order to achieve authentication. Merely, that means that the subject must be able to prove that it takes in fact the role it claims to take. In this situation, the subject is the so-called prover and its counterpart is the verifier. In most cases, the prover is an actor, like a human being or a machine, and the verifier is a system, in particular a system's authentication routine.

[2]For a more detailed explanation of the Biba model, please refer to this presentation: nathanbalon.net/projects/cis576/Biba.ppt (last access: November 2018).

Current authentication techniques can depend on one or more factors, with the factors being one or more of "something you know" (e.g., a password, PIN code, OAuth token,. . . .), "something you have" (e.g., a token card), or "something you are" (e.g., a fingerprint or iris scan). In the abovementioned example (Sect. 11.1), it could be a fingerprint reader built into the entry of the room where the devices are stored.

Authorization is the process of granting access privileges to an entity while conveying an in this context official sanction to perform a security function or activity (Barker and Dang 2016). In the abovementioned example (Sect. 11.1), it could be part of the mechanism to open the door at the entry—e.g., a routine that checks if the according identity is allowed to enter the room by obtaining the privileges from a local storage or a database. Authorization is approved if the prover is in fact privileged to perform the desired action.

11.3 A Stepwise Approach

We will now introduce a systematic approach to harden a data exchange environment for further facilitating the development and construction of cyber-physical production systems. The purpose is to propose a systematic process one can follow in order to improve all necessary information security aspects of a productive engineering environment. We focus on AutomationML files and its exchange between peers and a central data repository.
The goals are to:

- Protect company secrets
- Keep the integrity of internal data
- Ensure clear records of responsibilities for files and their changes

. . . which are our most important (exemplary) assets. Reaching the following three goals of our chapter shall be the consequences of reaching the former three:

- ensure safety within the production line
- ensure safety for the use of the end product
- ensure the availability of important services and resources

11.3.1 Identifying Assets

In the first step, we will identify assets in order to get the basis for all further steps. We start with the definition of the word asset and then explain the aspects to consider. We propose a systematic approach to identify assets and provide examples.

Generally spoken, an asset is a subject/object or service that, to a certain extent, one's success builds upon (*Collins English Dictionary*). In the case of the production line, the prominent asset is the availability of its service and productivity (e.g., in units per hour).

Identifying assets is one of the most important, if not the most important, step to perform in order to set up a proper security policy for the company. It is the step which enables us to answer the question: *What really is important for our company to work trouble-free*? The answer not only may give a basis for the creation of security countermeasures but also may be the basis for future concepts of data safety and service availability (such as data backup and service redundancy concepts).
For now, we are only going to answer this one question. We are not answering any other questions, e.g., *how* are we going to protect them? It is too early to answer other questions yet, and we have to make sure that the answer to our "what?" is complete *before* we start answering the "how?" in the next steps.
Our main use case will be to set up a central data repository in order to store AutomationML files. We are now going to point out the most important data sets, services, and properties available within the repository's ecosystem and explain their importance.

We recommend to perform this step in a conscientious way, as it will form the basis for all further steps. How can you find your most important assets? The approach is relatively simple:

1. Create some kind of listing (list, table, etc.) of services or entities that your main company value proposition depends on
2. Prioritize (e.g., low-medium-high, 1–10, etc.)

Sometimes these entities/services are not obvious, and it is not always easy to find them. Still, the list should be as complete as possible. Examples for assets are secret recipes for durable metal alloys, availability of the telephone service to the client, availability of electricity supply, the availability of a certain server, the ability of a certain employee or role to work in an undisturbed way (e.g., imagine spamming), etc. Prioritization should be made on the negative impact the loss of an asset would have. From that point of view, e.g., the availability of electrical current may get a higher priority than the ability to work without email spam. We try to give a realistic example in Table 11.2. The company partners and experts we are collaborating with repeatedly claimed that their most important assets are mainly their know-how. The term know-how in this context describes company secrets, which can be methods, recipes or manufacturing best practices (e.g. for the manufacturing of certain alloys or surface treatment, certain ways to weld or glue, etc.). In fact, most of these secrets are probably stored in digital documents. Even if they are probably not stored in AutomationML format, the state can be saved at the central repository as one attachment file. Deducted from that, we can say that one of the most important assets is confidentiality of data; data does not only mean AutomationML files but also the so-called attachment files. Attachment files are all file types that are not AutomationML, meaning that they are treated by the repository as not transformable between AutomationML (one and two) formats and are therefore stored separately in a so-called container. The availability and confidentiality of both, the container and the AutomationML documents, are of crucial importance. Furthermore, the safety of

Table 11.2 Exemplary assets within an AutomationML data repository environment

Asset	Description	Priority
AutomationML documents (confidentiality)	It is important to prevent data leaks of any kind	Medium
Attachment documents/files (i.e. documents and files in other format than AutomationML) (confidentiality)	It is important to prevent data leaks of any kind	Medium
AutomationML documents (availability)	Data loss shall be prevented	Medium
Attachment documents/files (availability)	Data loss shall be prevented	Medium
The ability to store documents into the repository	This is part of the availability of the repository in general. One shall be able to store data into the repository if authorized	Low
The ability to read/retrieve documents from the repository	This is part of the availability of the repository in general. One shall be able to retrieve data from the repository if authorized	Medium
Engineer's and other user's credentials	It is crucial that all participants keep their credentials confidential For the case that they are lost or shared, they have to be renewed as soon as possible	Medium
Administrator's credentials	Even more important are the credentials of the administrator. There should always be more than one administrative account in order to be able to deactivate or renew an administrator's credentials for the case of loss, etc.	High
The safety of all employees	More important than any functionality or credentials is, of course, the safety of all participants	High
The safety of all clients	More important than any functionality or credentials is, of course, the safety of all participants	High

all participating people as an asset is of course of very high priority. Lastly, there are the credentials of normal users and the credentials of the administrator. If a normal user loses her credentials (smart card, password, etc.), they can be deactivated. However, in the meantime, someone gaining knowledge of these credentials could modify files. We assume that such unauthorized changes, if detected early enough, could be reverted using a backup of a previous state of the repository, thus assuming that the priority of these credentials is not as high as the priority of some other assets for now. Then, there are the administrator's credentials, which we consider of higher priority than those of normal users due to the higher privileges of administrators and the possibility for more severe consequences if unauthorized users gain knowledge of them.

All of the mentioned assets are listed in Table 11.2. Note: Such a table (or any other form of listing) has to be created individually per company as assets may differ.[3]

11.3.2 Analyzing Threats and Possible Attack(er)s

In this second step, we will again start with a definition of the word threat. In order to describe what a threat is, as well as to explain the relation to the terms vulnerability, and asset, we provide a comprehensive example. Analogous to the previous step, we will then propose a systematic approach to identify possible attackers and attacks.

A so-called threat, within the context of computer security, is anything with the potential to cause serious harm to your assets (Ross et al. 2016). A vulnerability is any type of flaw or weakness in an information system that may leave a system exposed to a threat (Ross et al. 2016). Last but not least, the so-called attack, within the context of computer security, is any kind of malicious activity with the motivation to obtain, alter, destroy, remove, insert, or reveal information without authorized access or permission (Ross et al. 2016). An attack may also have the motivation to disrupt the availability of the service.

Imagine an injection molding machine, for example (Osswald et al. 2008). Its main function is to produce plastic parts by molding raw plastic granulate and clamping the melted mass into a mold. It consists of a gas tank with the gas needed to heat the plastic, a heating unit, a worm shaft to guide the molded plastic through the injector, and a clamping unit in order to bring the product into the desired form. Considering this simple example, one can derive the following:

In this example, our asset is the functioning of the machine (i.e., the service availability). The vulnerabilities are the machine's dependencies on the availability of gas and plastic as well as regular inspection and maintenance. The threats are:

- A responsible person destroying the machine's availability by accident, e.g., by forgetting to ensure the availability of gas or plastic or neglecting to regularly inspect and maintain the machine.
- A person with the motivation to destroy the machine's availability intentionally (i.e., the asset), e.g., by causing an explosion of the tank or filling something other than the proper plastic into the plastic granulate tank that causes damage to the machine or other kinds of vandalism.

The risks are the facts that incidents negatively impacting the ability of the machine may occur. Examples are that the gas tank may explode or leak, the machine could come subject to vandalism, and the machine could run out of plastic or gas. The severity of the risk can be calculated by considering the two factors it consists of: the possibility for its occurrence and the damage caused. An exploding gas tank

[3]Source: Personal correspondence with manufacturing engineers in September 2018.

Table 11.3 Possible attacks in each layer of the OSI model, IT technology

Layer	Layer name	Protocols	Devices or technologies	Potential attacks
7	Application layer	DNS, HTTP, FTP, SMTP	OS, applications	Viruses, worms, Trojan horses, buffer overflow, OS exploits, web service vulnerabilities and SQL injections, Individual vulnerabilities of all applications, e.g., CAD, etc.
6	Presentation layer	MIME, ASN.1	–	–
5	Session layer	PPTP, RTP, SOCKS	–	Session hijacking (this is a layer 5–7 attack, however)
4	Transport layer	TCP, UDP	Ports and NAT/PAT (L3 and L4),	Port scans, Denial of Service by TCP syn flooding, UDP flooding, XMAS Attack, OS exploits
3	Network layer	IP, IPsec, AppleTalk	Routers and L3 switches	Man in the Middle (MITM) attack, DHCP attack, ICMP attack, OS exploits
2	Data link layer	ARP, MAC, PPP, IEEE 802.3	Switches, access points, mac addresses	ARP spoofing, sniffing, broadcast storms, misconfigured or malfunctioning NICs, spanning tree attack, MAC spoofing
1	Physical layer	IEEE 802.3, IEEE 802.11, DSL, RS232, Bluetooth	USB/Cables, devices, and drives	Passive sniffing (e.g., through tapping of copper (Blog) or fiber-optic cables (Thefoa)), vandalism—which is also a kind of denial of service attack, theft

would probably cause more damage than if the machine runs out of plastic; however, the latter is a lot more likely than the former.

It can furthermore be good practice to approach the search for types of attacks as systematically as possible. Therefore, we prepared a table that shows possible attacks on each layer of the OSI (Open Systems Interconnection) model.[4] Table 11.3 shows an example for IT systems (i.e., also IT systems within the production line infrastructure or the manufacturing company's offices and the data repository and its clients) (Cybersecuritynews; Cisconet; S21sec; Wikipedia).

[4]basics to the OSI model can be found in the according RFCs, 1122 and 1123, as well as, for a quick help in understanding, in the Wikipedia article (IETF a,b).

11.3.2.1 Possible Types of Attacks

Attackers that manage to break into a company's IT infrastructure could possibly do one or more of the following.
Attackers with read access to the repository's content may unveil and steal company secrets. Read access to metadata may reveal secrets to her as well. Attackers may furthermore eavesdrop communication channels between users and the repository, giving them also a kind of read access. Active attackers may eavesdrop the communication channel by sending messages to either user or repository (e.g., requests), while a passive attacker may just place himself as man in the middle somewhere into the communication channel.
Attackers with write access to the repository's files may insert or modify arbitrary files within the repository, for example, content that is stored on the repository or content that is stored on a user's computer. Modification may also include the deletion of arbitrary files and information within arbitrary files. Attackers may furthermore modify metadata on the repository, such as version control metadata or any other kind of usage, or change history protocols.
Attackers may furthermore influence the repository's service availability. Attackers may also modify metadata of arbitrary data within the repository. They may also hinder other users from using the service as intended or slow down the service (denial of service attack), by either attacking the service directly or by influencing the communication channel between user and repository.
Furthermore, an attacker may inject malware into the repository or the computer of one or more of its users. She may do this via the web interface but also may inject data into the communication channel between user and repository.
An attacker may impersonate an arbitrary user or change authenticity information of a certain file or metadata if she manages to alter authenticity information, resulting in non-repudiation attacks.

11.3.2.2 Identifying Possible Attacker Types

Let us start with modeling potential attackers, their motivations and situations, as well as possible consequences of successful attacks. There are quite some different possible attackers within the context of our central data repository. Probably the most obvious one is the competitor (refer to footnote 3), who may profit from gaining knowledge of company secrets or pricing, from disturbing the company's assets, or from damaging the company's reputation. A former employee may become an attacker (e.g., because of unresolved conflicts), as well as an employee that is still working at the company (Farahmand and Spafford 2013; Mohamed et al. 2006). Employees may improve their own reputation within the company or damage the reputation of others and, if there are unresolved conflicts, may want to damage the company in some way Within the set of possible attackers, there also is the potential criminal organization that can have quite diverse motivations. To damage the company in general (e.g., by mission of a competitor) or to decrease their revenues or its reputation may be one of them. They may also be interested in stealing secrets, e.g., in order to sell them to competitors. In addition, such an attacker may have a lot more resources than a former employee or a single activist—which can also be a potential attacker. Such activists or hackers may appear as single persons or in groups. Their motivation can be of political or idealistic nature, or just some form of (non-) directed aggression (e.g., imagine manufacturers of controversial products, like SUV with exceptionally high fuel consumption). In the case of many manufacturing companies, negative influences on the safety of their products may have serious injuries or deaths of operators, clients, and other involved people as consequence. Thus, an attacker's motivation could even be to kill.

At least since the Snowden revelations starting in 2012 (TheGuardian), we also have to take governmental agencies into the set of potential attackers. These are the strongest attackers in terms of resources and attack opportunities. In this attacker model, we assume them to have access to very high financial and computational resources and may also control Internet nodes, hardware devices that are used within the company, or, in the very worst case, even employees (e.g., through bribery or threatening). Their attacks may have political or economic motivations. Table 11.4 shows a summary of typical attackers.

Now that we have identified the attacks and attackers that can be dangerous for the repository, we want to identify the consequences of successful attacks. This is the last step of the process to identify threats. In order to prepare for the next step (in which we are going to rate risks), think about the consequences attacks may have and the potential cost of them.

Table 11.4 Attacker types, their typical motivations and typical resources

Attacker type	Internal/ext.	Motivations	Resources
Activist/hacker (single)	External	Political, idealistic, aggression, curiosity, ...	Low
Former employee	Both	E.g., unresolved conflicts	Low
Employee	Internal	Improve own reputation, damage or improve reputation of others	Low
Activist/hacker (group)	External	Same as single activist/hacker	Medium (e.g., through bundling of several nodes across the Internet)
Criminal organization	External	Damage in general, decrease revenues or reputation, steal information	Medium
Competitor	External	Steal information, damage reputation, sabotage	Medium
Governmental agency	Both	Steal information (industrial espionage), political	High

11.3.3 Identifying and Calculating Risks

We now found our valuable assets within the company and mentioned possible attacks, attackers, and their motivations. The next section will treat the derived risks. Again, we will start with a definition of the term and continue with a manual how to identify and especially rate risks. Based on the identification matrix, countermeasures can later on be defined.

11.3.3.1 What Is a Risk?

Generally spoken, a risk is the *possibility* of losing something of value. In the case of the injector molding machine, the prominent risk is the fact that a standstill is possible. A risk has two influence factors: the probability that this certain situation occurs and the damage that is caused if it happens. Based on the productivity loss, it is possible to estimate the loss of money that is caused by the downtime, as well as the maximum downtime that we can tolerate.

Often, one cannot decide upon the existence of certain risks. However, often, it is possible to mitigate the risk by lowering both factors. It follows that there are four possibilities to react to a risk: The first option is to deal with it, leaving both factors untouched ("I know it may occur, but what shall I do?"). The next two influence either of the factors, for example, by installing a second production line or hiring an emergency response team (thus reducing damage caused by the standstill of the production line) or by installing more reliable components (thus reducing the likelihood of such a standstill). The last option is to do both which is the most expensive but also the most effective one.

The goal of this section is to provide some concepts in order to help decide how much effort shall be put into mitigating a certain risk (of Queensland).

11.3.3.2 Rating and Calculating Risks

The VDI document (Ingenieure 2011) provides an approach to list and rate risks. In Table 4 of the document, a matrix has been drawn that shows how to rate a risk based on the value of both factors. Figure 11.1 shows an example of how such a matrix could look like.

This step is crucial because risks have been known in order to be countermeasured in later steps. At least risks that are rated as high or very high should be targeted and mitigated. If resources are available, we recommend to also mitigate risks rated as "medium." Ultimately, one can decide if the risk is worth being mitigated based upon the following calculation: What are the total practical costs (i.e., efforts, time, motivation, money) (e.g., per year) of mitigating the risk vs. what are the total practical costs if the according incident occurs (repeatedly)? The cheaper solution should always win. In many cases, countermeasuring threats is cheaper, as it often

Damage					
Client/employee injury	fatal	medium	very high	very high	very high
Confidentiality of repository content broken	serious	medium	high	high	very high
Theft of engineer's and other user's credentials	significant	low	medium	medium	high
Inability to store documents into the repository	neglectable	low	low	medium	medium
		low	medium	likely	very likely
		Probability of occurence			

Fig. 11.1 Exemplary risk matrix, following VDI/VDE 2812 guidelines (Ingenieure 2011)

requires only simple measures such as the introduction of a policy or a software update, while possible damages often require a lot of person-hours to be repaired. Another reason is that measures often do not have to be performed too frequently (e.g., the introduction of a policy), while the repair of the damage would have to be done every time the incident occurs. The risk analysis must be repeated in specified intervals in order to maximize its effectivity.

11.3.4 Countermeasuring Identified Risks

Based on the identified threats and their worst-case consequences, and based on the risk analysis and the costs of possible successful attacks or other damages that can arise due to the lack of security related measures, the next step is to decide upon countermeasures to apply. This section serves as explanation of this next step. We are going to mention and explain different technologies that can be applied and provide countermeasure examples in the form of a table.

It is recommended to apply countermeasures based on the risk analysis and for the risks that have been marked as "medium," "high," or "very high." Table 11.5 shows some example countermeasures. Note that also backups are countermeasures within the field of security. Imagine the famous example of WannaCry and other encrypting malware (Heise; Chen and Bridges 2017). Instead of trying to clean such malware from the hard disk(s), a system could just be wiped, and its states could be reverted, using a backup, to a previous state. A backup can also protect from data loss that has been caused, e.g., by theft of data carriers. Although this may not solve the problem of data being in the wrong hands, it still is a countermeasure against data loss—and can be a part of an effective mitigation plan against theft of data carriers, together with, e.g., disk encryption.

It is not possible to provide countermeasures for every possible vulnerability or attack, but for now, we want to provide some general rules in this chapter. There are some measures that normally mean low to no additional effort for the engineers, administrators, and managers while at the same time providing a dramatic increase

Table 11.5 Example: Potential attacks per layer and countermeasures

Category	Potential attacks	Possible countermeasures
OS and applications	Vulnerabilities of all applications, e.g., CAD software, etc.	Regular updates
Web attacks	Session hijacking (this is a layer 5–7 attack, however), web service vulnerabilities and SQL injections, Individual vulnerabilities of applications	The use of TLS (https), software updates for browsers and servers, security-aware web programming and regular updates of web software
Denial of service	Broadcast storms, misconfigured or malfunctioning NICs, spanning tree attack, MAC spoofing, vandalism, ICMP attack, Rogue DHCP, DHCP Starvation Attack, port scans, TCP syn flooding, UDP flooding, viruses, worms, Trojan horses	Smart multilayer switches, firewall, IDS/IPS
Unauthorized write	Viruses, worms, Trojan horses, buffer overflow, OS exploits	Antivirus solution, IDS/IPS, update policy and regular OS updates
Unauthorized read or reconnaissance	Passive sniffing, wire cutting and slicing of copper or fiber-optic cables, theft (of data carriers), Man in the Middle (MITM) attack, XMAS packet attack (TCP stack fingerprinting), port scans, viruses, worms, Trojan horses, social engineering, Covert Channels	Use of Bend-Insensitive (BI) Fiber cables, OTDR and similar techniques (Execsecurity), Firewall, IDS/IPS, policies, awareness and trainings, CCTV, mechanical measures (locks, security doors, etc.)

in practical security.[5] One of these measures is the supply with recent updates. Probably one of the most effective and at the same time one of the easiest measures an individual can do in order to harden a software application is simply update supply. Of course, it depends on the practical reaction time of software suppliers to known malware and known vulnerabilities. And although this is a measure that is not as easy on operation technology, it will probably become a lot easier in the future—especially in connected ("always online") IoT (Internet of Things)-based operation technology. Antivirus software can be used to protect against known malware—of course only to a certain extent and most of the available antivirus software concepts also rely upon recent updates of their virus definition database. IDS (intrusion detection systems) can support the detection of anomalies in a network and even detect malware trying to spread itself, while IPS (intrusion prevention systems) may even actively react to those kinds of threats without making additional interactions necessary. Policies can be a very powerful tool in order to harden data safety (through regular backups),

[5]Refer to the famous MELTDOWN and SPECTRE attacks, against which Microsoft already provided a patch before they started becoming so popular (Meltdownattack; Support).

confidentiality (rules how to identify and react to phishing or social engineering attacks), and authenticity (e.g., by the rule that users may not share their smart card and may immediately tell it to an administrator if it is lost). Cryptographic means can help protect confidentiality (e.g., through drive encryption, secure channels between nodes and VPN endpoints). Cryptographic means may also help in enforcing access control mechanisms and to protect the authenticity and even integrity of data (e.g. through cryptographic hash functions, MACs, and signatures).
Furthermore, we want to clearly state that we recommend the implementation of a smart card-based public key infrastructure into a company. Modern EC card-based systems are often secure enough in terms of the inability to derive credentials (i.e., private keys) or abuse the card without knowledge of the correct pin code (Alliance 2015) and can be the basis for providing sophisticated (maybe even cryptography-based) access control and the ability to create signatures and encrypt drives and data with relatively low effort.
Cryptography-based access control may be a powerful approach to ensure integrity, confidentiality, and authenticity of files even when these files leave the borders of the company. Approaches may be developed in order to implement fine-grained access control enabling also some kind of "security restricted view" in order to limit the view on certain files for external institutions, targeting a famous problem many production companies have.[6]

11.3.5 Policies

The next step will handle the topic of policies. We will, again, start with a definition and then explain the term as well as good and bad choices for policies with the help of examples.

11.3.5.1 Definition

A policy can be described as a statement of intent, implemented into a set of rules. Its purpose is to assist in particular decisions (user, manager) and technical implementations (engineer). Within the context of computer security, a policy's purpose is furthermore to define the "border" to decide if a system's state is "secure" or "insecure." Within this context, "secure" means that the system is working as intended—and "insecure" means that something is not working as intended.

[6]personal correspondence with engineers from such companies revealed that it is a common problem to restrict views on, e.g., construction plans in order for external institutions such as TÜV or through outsourcing to get insights into the construction, without confidential information leaving the company. As of now, since most of this information is stored within MS Excel files and CAD documents, it is not possible to restrict the view, leaving them access to a lot more information than is needed and endangering the confidentiality of company secrets.

A policy can be enforced either by management directions or by technical implementations. A lot of time/thoughts should be invested into the policy (and how it shall be enforced) as it is (similar to the asset list) also a helpful definition of a company's goals and intentions and the basis of any set of security-related measures. Note: The following guidelines are important to define a policy that is useful:

1. A policy should always be defined in a positive way (i.e., state what you want instead of what you don't want).
2. Furthermore, a policy should always be as complete as possible (i.e., we want the management to be satisfied if the policy is kept, i.e., no "unofficial" rules should exist)
3. The rules of a policy shall be described as concise as possible.
   A good example: backups shall be made every week, and the backup medium has to be stored in a safe outside the server room
   A bad example: backups have to be made often, and we want our data to be safe
4. All rules shall be realistic. Keeping them should be within reach of all people involved.
   A bad example: it is forbidden to make mistakes when writing a text
   A good example: every text has to be proofread twice by independent people. The mistakes found during these two proofreads have to be corrected prior to hand-in.

The first guideline is important because it is a lot easier to know what to do if it is written in a positive way (e.g., similar to the ideal formulation of goals). The second guideline is important in order to avoid additional rules to be kept for the employees or other users. It would take a lot more effort if there were additional rules, and it would question the meaning of the policy if the state cannot be defined as correct or secure even if all rules are kept. The fourth guideline is crucial as a policy simply does not make any sense if it demands unrealistic efforts from involved people.

A backup policy for an employee's home office could look like this:

- The purpose of the backup concept is to minimize the probability for data loss in cases of unintended deletions, unintended modifications that have to be reverted, viruses, fire, and burglaries.
- Daily backup: every day, an incremental (Searchdatabackup) backup is made to an external hard disk.
- Weekly backup: every Friday, this external hard drive is mirrored on to another external hard drive
- This hard disk is then stored in another room in which the computer is not located.
- Every first Monday of the month, this external hard drive is mirrored on to an encrypted external hard drive which is stored at another location at least 500 m away.

Let's define an exemplary policy for our repository. This policy defines the handling of smart cards within the company. Note that the use of a public key infrastructure utilizing smart cards in order to store the public-private key pairs is strongly recommended for security and convenience (i.e., usability) reasons.

- Every employee gets a smart card, which is the only way to authenticate within the (logical) borders of the company.
- Every user authenticates with her smart card.
- Smart cards (including public/private key pairs) are renewed every 36 months.
- Every user is fully responsible for her smart card.
- In order to access files on the repository, authentication using a smart card is needed. That means that also partner institutions get a card in order to authenticate if they need any access to data on the repository.
- Access to rooms may only be assigned on the basis of a white list and with a valid smart card.
- The card owner is fully responsible for her card. Loss must be reported immediately. Furthermore, sharing or lending the card in any way is forbidden.

Furthermore, in order to ensure reliable operation of the production line and the repository, continuous audits are crucial.

11.3.6 Reaction Plan

The last step is about the creation of reaction plans, a way to handle risks whose probability of occurrence should not be neglected. We explain the need for such plans, provide guidelines how to create them, and give an example.

Once assets, risks (including potential damage and their costs), and countermeasures have been specified, a reaction plan should be created. The reason why a reaction plan can be very helpful in practice is that the countermeasures that may be accomplished in advance are not always enough. Imagine possible incidents that occur in spite of countermeasures such as IDS/IPS, firewalls, access control (physical and logical) measures, antivirus applications, and operating system updates. A practical example is malware. No matter how often backups and updates are prepared, there is still the possibility for recent or specifically tailored malware that finds its way through all barriers (Bilge and Dumitraş 2012). A reaction plan is a guideline (in particular, a structured set of guidelines) with the aim of telling *what to do if a certain incident occurs*. It is important to have one in order to be prepared for these incidents, even in worst-case scenarios. Such scenarios can be overcome a lot better if having thought of them in advance.
We now want to give an example reaction plan in the form of a table: Let's stick to one of the worst-case scenarios: malware trying to steal company secrets.

Table 11.6 shows a response plan for the scenario of malware successfully stealing information from the central data repository. Two cases have been covered in this table. The first being the less catastrophic one: An IDS detects unusual information flow (e.g., if someone uploads hundreds of gigabytes from the repository to an online storage). The second case is, so to say, the worst case: the company finds out that

Table 11.6 Example for a response plan for the scenario: Malware steals secret information from repository

Type	Way of detection	Guide
Malware to steal information	IDS alarm	1. Cut dataflow 2. Find origin of unintended spread 3. Clean (i.e., remove malware) origin If that is not possible: Shut down and replace (or reset) origin
Malware to steal information	Competitor is aware of secret	1. Cut data flow 2. Analyze data flow 3. Find origin 4. Update all systems 5. Search for: – Suspicious third party hardware or software (begin with recently installed ones) – If found: clean device from malware – If not found/not possible: 7. Isolate important data sets 8. Clean/clear whole system step by step 9. Return back to normal operation 10. If possible: define characteristics for this certain incident, and install intrusion detection measures in order for this case to not occur ever again

competitors may know information that they should not know and develops the suspicion of having malware installed within the control systems or the repository and its users. Note that all guidelines must be written in a way that is easy to understand and to follow. It is not sufficient to write down "find a solution" as a step, for example. Instead, it must give clear instructions how the state can be returned back to normal. It is important that the guidelines are complete, meaning that following the instructions should be sufficient in order to return to the normal state. If wished, another column could be added stating the approximate cost (in terms of time and of money) of the measures. Furthermore, incidents may be classified and categorized. They may also be stored in a database instead of a table, for example.

11.4 Mistakes to Avoid

The goal of our approach is to provide a systematic way of hardening the data exchange environment in AutomationML-based engineering networks. Based on experiences and references from the fields of project management and network security, we describe mistakes that may occur in practice and explain how they can be avoided.

One common mistake is the subtle creeping of implicit assumptions, which often has a fostering effect on misunderstandings and, in this case, also on security-related vulnerabilities. In fact, it is in many cases almost unavoidable to make assumptions when building a (security) concept. Typical assumptions are the unbroken-ness of used primitives (i.e., cryptographic algorithms). Further assumptions are, for example, the awareness to and correct execution of policies by involved employees, the probability of untrustworthy employees being low, etc. Making these assumptions explicit is important because one can develop countermeasures if (and only if) the assumptions are known. Imagine an administrator who is silently assuming that people use passwords longer than four characters, for example, but in reality some of them don't. The whole password security is not at the required state then, and no one is aware of this fact.

Another common mistake in implementing security measures is the order in which important questions are being answered. Try to make sure to complete the answer of all "what?" questions (assets, threats, attacks, attackers) before answering any of the "how?" questions (countermeasures, policy). It is crucial to keep this order as any confusion leads to a dramatic increase of required effort in order to obtain feasible results.

One should be aware that some security measures may on one hand solve a problem (e.g., authentication of the user in a certain scenario) but also have great drawbacks (e.g., multifactor authentication with too many factors or too complex password restrictions may be cumbersome to use in practice). Always consider whether the benefits of a measure outweigh its costs (in terms of time, energy/concentration, and money). One should keep in mind that the productivity of employees is a very valuable asset and should therefore be treated with care. Therefore, security measures should ideally work in background without its users being able to notice any slowdown or additional effort. If that is not possible (e.g., because of the use of smart card with a pin code), the extra effort should be kept as low as possible (e.g., through a four-digit pin code and wireless capabilities of the card and the cardreader).

Continuous audits have to be performed in order for the system to stay in a secure state. It is not a "fire and forget"-like goal that one can accomplish once. Instead, the state of the environment has to be verified within continuous security audits.

11.5 Related Work

In this section, we briefly highlight existing work regarding access control, access control in outsourced information repositories, XML encryption, and related fields. We will furthermore elaborate on previous work about security and access control and explain publications about AutomationML itself, AutomationML processing, and AutomationML repositories. Lastly, we will highlight selected related publications for data management solutions.

11.5.1 AutomationML and Related Schemes

The Automation Markup Language (AutomationML) has been developed in 2008 by Drath et al. as a modeling language to function as "glue for seamless automation engineering" (Drath et al. 2008). It uses the CAEX (Computer Aided Engineering Exchange) standard's syntax for system-independent exchange of engineering data (Schleipen et al. 2008; Drath et al. 2008; AutomationMLe.V. a). The goal of AutomationML is to fulfill the growing demand for efficient exchange of engineering data within the whole lifecycle of production system engineering. Its focus lies on concepts for system modeling, following object-oriented paradigms in order to enable versatile, efficient exchange of engineering data and centralized management of engineering artifacts (AutomationMLe.V. d). COLLADA is a data format developed in a cooperative effort between companies of the fields of game-, platform and application development in order to provide a format for efficient exchange of 3D models (Arnaud and Barnes 2006). Since both languages are aiming and efficient exchange of artifacts, AutomationML follows a similar goal as the one of COLLADA. The difference between AutomationML and CAEX can be explained best by examining the definition of both: CAEX has been invented in order to provide modeling functionality for engineering applications, while AutomationML integrates CAEX (object and relation syntax and topology), COLLADA (geometry and kinematics), and PLCOpen XML (Plcopen) for modelling of component logic (Drath 2009).

11.5.2 Current Version Systems

The most famous examples of version control software are git (Blischak et al. 2016; Loeliger and McCullough 2012; Git-scm) and svn (Pilato et al. 2008). Such version control systems are widely deployed in software development, for example. However, they are the current best practice tools and are widely in use in other fields, such as research and engineering, and they can be very useful as soon as teamwork on the same directory or data set is required. The software can also be used in order to exchange AutomationML files. However, the creation of a specific AutomationML data exchange system can have benefits, such as AutomationML processing and server-side graphical web-based views including advanced access control on a per component (or her role class) level (see Chapter 8, "Engineering Data Logistics for Agile Automation Systems Engineering"). That way, it can also be possible to implement desirable properties such as full traceability of changes, integrity control measures, and non-repudiation on an AutomationML content-based granularity.

11.5.3 Access Control

As mentioned in the Introduction (Sect. 11.2.3), the first theoretical principles of access control have their origin in the 1960s (TheLinuxFoundation) and 1970s (Bell and LaPadula 1976; Biba 1977). Access control on the level of operating systems and file system security therefore can be considered a very old and thoroughly researched topic. However, at least since the trend of outsourcing and cloud storage, demands for access control have changed (Takabi et al. 2010). Concepts have been introduced (Miklau and Suciu 2003; Smart 2003; Wang et al. 2006; Zhou et al. 2013), even on XML level (Zhou et al. 2011). Furthermore, some service providers already offer some kind of access protected cloud data storage in order to ensure access control files without being dependent on any file or operating system security measures. The reason is that these measures are, in the case of cloud storage, often not in the hand of the data owner.

11.6 Conclusion and Future Work

This section provides a short discussion of achieved results, answers to the aforementioned research questions, and an outlook for future projects and demands.

In this chapter, we provided a high level guide of how to harden a central AutomationML data repository environment for engineering of cyber-physical systems. We provided an introduction to access control and historic approaches and models. We furthermore provided a guide divided into six steps on how to harden the network around and the central data repository itself and included practical examples. The most important steps being first finding and identifying the assets in order to create an asset list and prioritize them and second the creation of a policy as one of the main countermeasure. This chapter ends with some tips and a list of common mistakes that should be avoided when following this approach.

In the following, we want to answer the research questions that we have listed in the Introduction:

RQ1:Which use cases of AutomationML can lead to dangerous situations in terms of confidentiality of engineering data and integrity of product and production line design? The most important situations in which data integrity and confidentiality are in danger are two situations: the first situation is whenever files leave the boundaries of the company and thus also the boundaries of control. Section 11.1 describes two possible situations; however, there is an infinite number of possibilities for such scenarios in practice. The second situation is when intruders actively or passively steal such information from a company. The second situation has been described in Sects. 11.3.2 and 11.3.2.2, in which attackers were identified.

RQ2: Which assumptions shall be made about potential attackers in a manufacturing company? We also made assumptions about potential attackers, attacks, their consequences, and attacker motivations. It is good to be careful with such assumptions; careful in this context means that one has to assume the strongest possible attacker attacking at the weakest available point. In order to describe what these weak points are, the context and the termini, we provided an example (Sect. 11.3.2) by taking a gas injection molding machine that produces plastic parts as example. We have described potential attackers, their motivations, and arising risks in Sects. 11.3.2.2 and 11.3.3.

RQ3: Which solutions for access protection can be derived? The countermeasures that can be applied in order to prevent unintended information leakage on one hand and to ensure data integrity on the other hand are twofold. The first category are the more general IT security measures (Sect. 11.3.4) like recent updates, a public key infrastructure utilizing smart cards (Sect. 11.3.5.1), and a reaction plan (Sect. 11.3.6). Another category however of access protection is providing cryptographically enforced access control that is independent of OS measures. The latter, however, is a matter of future work. However, this chapter does not cover the whole topic, and a lot of future work is being left.

It would be useful to have a realistic use case of a real company. A use case including a policy, a list of assets, a complete risk analysis, etc. However, this may be hard to get as it could reveal secret information of this company.

Apart from that, we think it would be useful to spend more thoughts on some concepts. Examples are the file-based access control that is independent from the data storage system. We would like to implement and test this concept as part of our future work. We will therefore work on such a concept and practical test implementation. Future work can also elaborate more on potential OT (operation technology)-specific attacks and their consequences. Furthermore, guides could be created that are specifically tailored and offer concrete solutions for more specific areas, a guide that is specifically tailored for the car supply industry, for example.

Acknowledgements This material is based on the work partially supported by (1) the Christian-Doppler-Laboratory for Security and Quality Improvement in the Production System Lifecycle; the financial support by the Austrian Federal Ministry for Digital and Economic Affairs and the Nation Foundation for Research, Technology and Development is gratefully acknowledged; (2) SBA Research; the competence center SBA Research (SBA-K1) is funded within the framework of COMET—Competence Centers for Excellent Technologies by BMVIT, BMDW, and the federal state of Vienna, managed by the FFG.

References

Ahmad, A., Bosua, R., & Scheepers, R. (2014). Protecting organizational competitive advantage: A knowledge leakage perspective. *Computers & Security, 42*, 27–39.

Aktoudianakis, E. (2016). Relationship based access control. Ph.D. Thesis, University of Surrey.

Alliance, S. C. (2015). Smart card alliance. In *INSIDE Contactless Offers Free, Downloadable, Open NFC API and Source Code on SourceForge*.

Arnaud, R., & Barnes, M. C. (2006). *COLLADA: sailing the gulf of 3D digital content creation*. Natick, Massachusetts: AK Peters/CRC Press.

AutomationMLe.V. Automationml FAQ. https://www.automationml.org/o.red.c/faq.html, a. Non peer-reviewed reference. Accessed Jan 2019.

AutomationMLe.V. AutomationML – first steps. https://www.automationml.org/o.red.c/erste-schritte.html, b. Non peer-reviewed reference. Accessed 24 Jan 2019.

AutomationMLe.V. Whitepaper automationmlpart 1 – architecture and general requirements. https://www.automationml.org/o.red/uploads/dateien/1542621846-Whitepaper, c. Non peer-reviewed reference. Accessed 22 Feb 2019.

AutomationMLe.V. AutomationML in a Nutshell. http://www.unserebroschuere.de/automationml/WebView/, d. Non peer-reviewed reference. Accessed 13 Feb 2019.

Barker, E., & Dang, Q. (2016). Nist special publication 800-57 part 1, revision 4. *NIST, Tech. Rep*.

Bell, D. E., & LaPadula, L. J. (1973). *Secure computer systems: Mathematical foundations*. Technical report. Bedford, MA: MITRE Corp.

Bell, D. E., & LaPadula, L. J. (1976). *Secure computer system: Unified exposition and multics interpretation*. Technical report. Bedford, MA: MITRE Corp.

Biba, K. J. (1977). *Integrity considerations for secure computer systems*. Technical report. Bedford, MA: MITRE Corp.

Biffl, S., Winkler, D., Mordinyi, R., Scheiber, S., & Holl, G. (2014). Efficient monitoring of multi-disciplinary engineering constraints with semantic data integration in the multi-model dashboard process. In *2014 IEEE, emerging technology and factory automation (ETFA)* (pp. 1–10). Piscataway: IEEE.

Bilge, L., & Dumitraş, T. (2012). Before we knew it: an empirical study of zero-day attacks in the real world. In *Proceedings of the 2012 ACM Conference on Computer and Communications Security* (pp. 833–844). New York: ACM.

Blischak, J. D., Davenport, E. R., & Wilson, G. (2016). A quick introduction to version control with git and github. *PLoS Computational Biology, 12*(1), e1004668.

Blog, O. Open security research: Sniffing traffic on the wire with a hardware tap. http://blog.opensecurityresearch.com/2013/03/sniffing-traffic-on-wire-with-hardware.html. Non peer-reviewed reference. Accessed Dec 2018.

Chen, Q., & Bridges, R. A. (2017). Automated behavioral analysis of malware a case study of wannacry ransomware. arXiv preprint arXiv:1709.08753.

Cisconet. Preventing security attacks from all osi 7 layer. http://cisconet.com/security/security-general/140-preventing-security-attacks-from-all-osi-7-layer.html. Non peer-reviewed reference. Accessed Dec 2018.

Collinsdictionary. Asset definition und bedeutung | collins wörterbuch. https://www.collinsdictionary.com/de/worterbuch/englisch/asset. Non peer-reviewed reference. Accessed Nov 2018.

Cybersecuritynews. Network vulnerabilities and the osi model; cyber security news. https://cybersecuritynews.co.uk/network-vulnerabilities-and-the-osi-model/. Non peer-reviewed reference. Accessed Dec 2018.

Drath, R. (2009). *Datenaustausch in der Anlagenplanung mit AutomationML: Integration von CAEX, PLCopen XML und COLLADA*. Berlin: Springer.

Drath, R., Lüder, A., Peschke, J., & Hundt, L. (2008). Automationml-the glue for seamless automation engineering. In *IEEE International Conference on Emerging Technologies and Factory Automation, 2008. ETFA 2008* (pp. 616–623). Piscataway: IEEE.

Execsecurity. Wiretap detection and telecom threats to businesses. https://execsecurity.com/wiretap-detection/. Non peer-reviewed reference. Accessed Dec 2018.

Falliere, N., Murchu, L. O., & Chien, E. (2011). W32. Stuxnet dossier. *White Paper, Symantec Corporation, Security Response, 5*(6), 29.

Farahmand, F., & Spafford, E. H. (2013). Understanding insiders: An analysis of risk-taking behavior. *Information Systems Frontiers, 15*(1), 5–15.

GitHub. About commit signature verification – github help. https://help.github.com/articles/about-commit-signature-verification/. Non peer-reviewed reference. Accessed Jan 2019.

Git-scm. About – git. https://git-scm.com/about. Non peer-reviewed reference. Accessed Nov 2018.

Heise. Cyberangriff: Kraussmaffei von hackern erpresst | heise online. https://www.heise.de/newsticker/meldung/Cyberangriff-KraussMaffei-von-Hackern-erpresst-4244880.html. Non peer-reviewed reference. Accessed Dec 2018.

IETF. RFC 1122. https://www.ietf.org/rfc/rfc1122.txt, a. Accessed Nov 2018.

IETF. Rfc 1123 – requirements for internet hosts – application and support. https://tools.ietf.org/html/rfc1123, b. Accessed Nov 2018.

Ingenieure, V. D. (2011). Vdi/vde 2182 vdi/vde 2182 blatt 1:2011-01 informationssicherheit in der industriellen automatisierung – allgemeines vorgehensmodell. www.beuth.de.

Khronos. Main page – collada public wiki. https://www.khronos.org/collada/wiki/Main_page, a. Non peer-reviewed reference. Accessed 24 Jan 2019.

Khronos. COLLADA schema version 1.5.0. http://www.khronos.org/files/collada_schema_1_5, b. Non peer-reviewed reference. Accessed 24 Jan 2019.

Langner, R. (2011). Stuxnet: Dissecting a cyberwarfare weapon. *IEEE Security & Privacy, 9*(3), 49–51

Loeliger, J., & McCullough, M. (2012). *Version Control with Git: Powerful tools and techniques for collaborative software development*. Sebastopol: O'Reilly Media.

Meltdownattack. Meltdown and spectre. https://meltdownattack.com/. Non peer-reviewed reference. Accessed Dec 2018.

Microsoft-Docs. How accesscheck works – windows applications | microsoft docs. https://docs.microsoft.com/en-us/windows/desktop/secauthz/how-dacls-control-access-to-an-object. Non peer-reviewed reference. Accessed Nov 2018.

Miklau, G., & Suciu, D. (2003). Controlling access to published data using cryptography. In *Proceedings of the 29th International Conference on Very Large Data Bases* (Vol. 29, pp. 898–909). Los Angeles, CA: VLDB Endowment.

Mohamed, S., Mynors, D., Grantham, A., Walsh, K., & Chan, P. (2006). Understanding one aspect of the knowledge leakage concept: people. In *Proceedings of the European and Mediterranean Conference on Information Systems (EMCIS)* (pp. 6–7). Alicante: EMCIS.

Osswald, T. A., Turng, L.-S., & Gramann, P. J. (2008). *Injection molding handbook*. Munich: Hanser Verlag.

OxfordDictionaries. Safety | definition of safety in english by oxford dictionaries. https://en.oxforddictionaries.com/definition/safety. Non peer-reviewed reference. Accessed Feb 2019.

Pilato, C. M., Collins-Sussman, B., & Fitzpatrick, B. W. (2008). *Version control with subversion: next generation open source version control*. Sebastopol: O'Reilly Media.

Plcopen. PLCopen xml. http://www.plcopen.org/pages/tc6_xml/. Non peer-reviewed reference. Accessed 13 Feb 2019.

Ross, R., McEvilley, M., & Oren, J. (2016). Nist sp 800-160 systems security engineering: Considerations for a multidisciplinary approach in the engineering of trustworthy secure systems. In *National Institute of Standards Technology, US Department of Commerce, Gaithersburg, MD, USA, Tech Report NIST SP* (pp. 800–160).

S21sec. Attacks on layer two of the osi model (i) – s21sec. https://www.s21sec.com/en/attacks-on-layer-two-of-the-osi-model-i/. Non peer-reviewed reference. Accessed Dec 2018.

Sandhu, R., Ferraiolo, D., Kuhn, R. (2000). The NIST model for role-based access control: towards a unified standard. In *ACM workshop on role-based access control* (Vol. 2000, pp. 1–11).

Sandhu, R. S., Coyne, E. J., Feinstein, H. L., & Youman, C. E. (1996). Role-based access control models. *Computer, 29*(2), 38–47.

Schleipen, M., Drath, R., Sauer, O. (2008). The system-independent data exchange format caex for supporting an automatic configuration of a production monitoring and control system. In *IEEE International Symposium on Industrial Electronics, 2008. ISIE 2008* (pp. 1786–1791). Piscataway: IEEE.

Schyja, A., Bartelt, M., & Kuhlenkötter, B. (2014). From conception phase up to virtual verification using automationml. *Procedia CIRP, 23*, 171–177.

Searchdatabackup. Full, incremental or differential: How to choose the correct backup type. https://searchdatabackup.techtarget.com/feature/Full-incremental-or-differential-How-to-choose-the-correct-backup-type. Non peer-reviewed reference. Accessed Dec 2018.

Smart, N. P. (2003). Access control using pairing based cryptography. In *Cryptographers Track at the RSA Conference* (pp. 111–121). Berlin: Springer.
Subversion. Apache subversion. https://subversion.apache.org/. Non peer-reviewed reference. Accessed Nov 2018.
Support, M. Protect windows against spectre and meltdown. https://support.microsoft.com/en-us/help/4073757/protect-your-windows-devices-against-spectre-meltdown. Non peer-reviewed reference. Accessed Dec 2018.
Takabi, H., Joshi, J. B., & Ahn, G.-J. (2010). Security and privacy challenges in cloud computing environments. *IEEE Security & Privacy* (6), 24–31.
Thefoa. The foa reference for fiber optics – how to tap fiber optic cables-. http://www.thefoa.org/tech/ref/appln/tap-fiber.html. Non peer-reviewed reference. Accessed Dec 2018.
TheGuardian. Nsa files decoded: Edward snowden's surveillance revelations explained | us news | theguardian.com. https://www.theguardian.com/world/interactive/2013/nov/01/snowden-nsa-files-surveillance-revelations-decoded. Non peer-reviewed reference. Accessed Aug 2018.
TheLinuxFoundation. Overview of linux kernel security features | linux.com | the source for linux information. https://www.linux.com/learn/overview-linux-kernel-security-features. Non peer-reviewed reference. Accessed Nov 2018.
Thonnard, O., Bilge, L., O'Gorman, G., Kiernan, S., & Lee, M. (2012). Industrial espionage and targeted attacks: Understanding the characteristics of an escalating threat. In *International workshop on recent advances in intrusion detection* (pp. 64–85). Berlin: Springer.
T. S. of Queensland. Managing information technology risks | business queensland. https://www.business.qld.gov.au/running-business/protecting-business/risk-management/it-risk-management/managing. Non peer-reviewed reference. Accessed Dec 2018.
Tucker, R. L. (1997). Industrial espionage as unfair competition. *University of Toledo Law Review. University of Toledo. College of Law, 29*, 245.
Tuptuk, N., & Hailes, S. (2018). Security of smart manufacturing systems. *Journal of Manufacturing Systems, 47*, 93–106.
Von Hippel, E. (1989). Cooperation between rivals: Informal know-how trading. In *Industrial Dynamics* (pp. 157–175). Berlin: Springer.
Wang, H., Sheng, B., & Li, Q. (2006). Elliptic curve cryptography-based access control in sensor networks. *International Journal of Security and Networks, 1*(3–4), 127–137.
Wangen, G. (2015). The role of malware in reported cyber espionage: a review of the impact and mechanism. *Information, 6*(2), 183–211.
Wikipedia. Spritzgießmaschine – wikipedia. https://de.wikipedia.org/wiki/Spritzgie%C3%9Fmaschine. Non peer-reviewed reference. Accessed 22 Feb 2019.
Zhou, L., Varadharajan, V., & Hitchens, M. (2011). Enforcing role-based access control for secure data storage in the cloud. *The Computer Journal, 54*(10), 1675–1687.
Zhou, L., Varadharajan, V., & Hitchens, M. (2013). Achieving secure role-based access control on encrypted data in cloud storage. *IEEE Transactions on Information Forensics and Security, 8*(12), 1947–1960.

Chapter 12
Securing Information Against Manipulation in the Production Systems Engineering Process

Peter Kieseberg and Edgar Weippl

Abstract Modern engineering projects often include extensive cooperation with partners as well as external experts, either due to specific knowledge required that cannot be acquired otherwise or even due to rules and regulations that have to be obeyed to enter a specific market. Still, Production Systems Engineering (PSE) processes contain significant intrinsic and explicit knowledge that is a key resource of a partner. Therefore, the partners in such a collaborative process need to protect their vital knowledge assets while still being forced to share much of the information, thus rendering proactive solutions for information protection infeasible. Information fingerprinting has been used as a reactive measure in many data-based information processes. While fingerprinting does not hinder unsolicited information exchange, fingerprinting techniques can be used to prove ownership of information and to determine the leaking partner. In addition, expert information is integrated into the overall process, requiring means to hold single participants responsible for errors and/or other issues. Still, in current environments, manipulation of information is largely possible. This becomes especially problematic in cases where the expert information is used as input in intelligent algorithms, thus rendering any chance of simple detection impossible, even for the expert originally entering the information. In this chapter, we adopt an approach for providing information integrity in the so-called *doctor in the loop* Holzinger (Brain Inform 3(2):119–131, 2016) systems in order to fit the PSE process and its special requirements and combine it with fingerprinting methods for protecting the ownership of vital information assets. Furthermore, we extend this approach to not only control data manipulation but also access to sensitive

P. Kieseberg (✉)
St. Pölten University of Applied Sciences, St. Pölten, Austria
e-mail: peter.kieseberg@fhstp.ac.at

E. Weippl
Christian Doppler Laboratory for Security and Quality Improvement in the Production System Lifecycle (CDL-SQI), Institute of Information Systems Engineering, Technische Universität Wien, Vienna, Austria

SBA Research, Vienna, Austria
e-mail: edgar.weippl@tuwien.ac.at

S. Biffl et al. (eds.), *Security and Quality in Cyber-Physical Systems Engineering*,
https://doi.org/10.1007/978-3-030-25312-7_12

information. In order to further mitigate attacks targeting data exfiltration, we provide two new approaches for logging SELECT-queries in a way that cannot be manipulated even by attacks in the possession of administrator privileges.

Keywords Data protection · Audit and control · Exfiltration detection · PSE databases

12.1 Introduction and Motivation

During the last years, data science has become ubiquitous in many parts of the economy and scientific workforce, not only spawning novel user-centered applications but also new approaches in, for example, security research or production systems. Data has become the new oil[1] and as such possesses a lot of value. This has been known in "classical" industry even before, with know-how being a key factor for being able to compete on the worldwide market. Still, while this is certainly true with respect to pure data protection as is common in the producing industry, the integration of additional features like machine learning approaches requires the in-depth cooperation with other companies that could easily result in lost or stolen intellectual property. On the other hand, in the new world of Industry 4.0 and data-driven approaches, machines and factories are required to work together very closely; thus, shunning the cooperation with other partners is not an option. Furthermore, many emerging markets and especially China often enforce cooperation with local companies from a political side, that is, in order to be able to compete on these markets, the incorporation of local experts is required, thus further complicating the protection of critical know-how.This problem of data protection becomes especially apparent in cases where experts from various partners work closely together and exchange information, as well as make changes to assets originally introduced by other partners. In PSE systems, the types of information that require protection is manifold, ranging from construction information over control information to other (semi-)structured data. Moreover, with the advent of data science in most industries, it will only be a matter of time for these techniques to also appear in PSE processes, that is, the integration of machine learning results in this process, with its problem of verifiability regarding subsequent manipulation. In this chapter, we will discuss how centralized information exchange platforms that allow for a cooperative PSE process can be secured using techniques like data fingerprinting, as well as taking care of manipulation detection. Furthermore, we provide a novel approach for detecting unsolicited data extraction through a novel logging mechanism withstanding even attackers with administrator privileges. Furthermore, we provide a new fingerprinting algorithm that is capable of detecting and proving data leakage based on information steganographically hidden in the primary indices of the database.

[1]The Economist, 6.5.2017: "The worlds most valuable resource is no longer oil, but data", available through https://www.economist.com/leaders/2017/05/06/the-worlds-most-valuable-resource-is-no-longer-oil-but-data.

12.2 Background and Related Work

In this section, we will discuss some prerequisites for our approach, namely, an approach for fingerprinting information, as well as an approach for securing know-how in expert in the loop (Holzinger et al. 2016) systems.

12.2.1 Chained Witnesses for Manipulation Detection

While digital forensics has seen a surge in scientific work in recent years, the topic of forensics on databases has not been in the focus, which is a huge problem when considering that most larger applications are built upon a database management system (DBMS). Typically, database forensics is often reduced to the topic of effectively managing and searching the database log files, which is of course very important but does leave out an important attacker, the database administrator himself/herself, who is typically perfectly capable of manipulating said log files or at least turning off the logging before carrying out malicious behavior. In their work (Frühwirt et al. 2012), the authors provided an approach for using the database transaction log for forensic purposes. While the name suggests otherwise, the transaction log is not a log file meant for human control but an internal mechanism used for carrying out rollbacks and crash recovery; thus, human readability is low on the one side, but any kind of manipulation of this mechanism might result in unrecoverable crashes of the database. While this should be motivation enough for not fiddling with this mechanism, an administrator could still try to delete entries after he/she forced a commit statement, thus reducing the danger of corrupting the whole database. In order to thwart this, Frühwirt et al. (2014) introduced the concept of a chained witness into the transaction log that allows for the detection of deleted and/or manipulated log entries. In the basic approach, the log files are used solely for incident response, that is, in the aftermath of an incident, the information is filtered for its informative value. It must be pointed out that both approaches were only tested for the InnoDB storage engine (Zaitsev 2009) of MySQL, still most of the techniques can be adapted for closed-source DBMSs when switching to the data replication mechanism instead of the transaction mechanisms, so we do not see a major drawback in using this approach for securing the PSE process. The major advantage of this primitive approach is that is does not rely on any changes made to the DBMS, which would be impossible in case of closed-source products (except when using the database replication log, which was discussed in (Frühwirt et al. 2014)). With the extended approach, the author changed the method for writing data to the database; more precisely, they changed the procedure for writing the changes to the transaction log files (there are two redundant files in InnoDB, sporting a well-devised mechanism for ensuring log correctness). Instead of simply writing the log block information as in the original approach, the new logging mechanism adds a

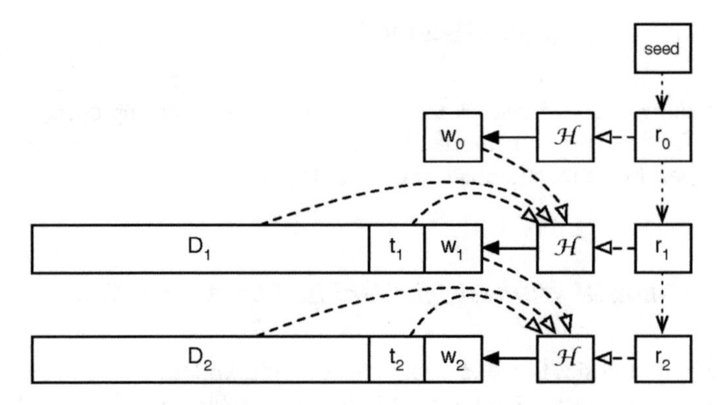

Fig. 12.1 Construction of the chained witness Frühwirt et al. (2014)

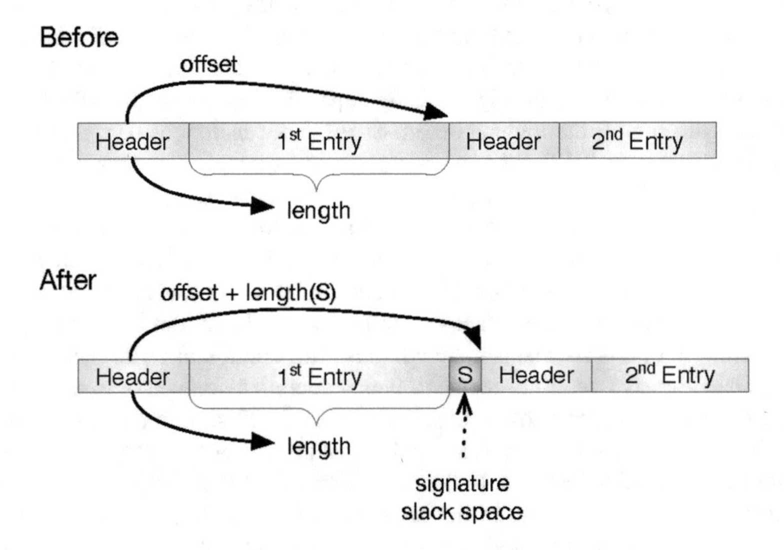

Fig. 12.2 Integration of the witness into the log entries (Frühwirt et al. 2014)

so-called chained witness to each entry. Figure 12.1 shows a graphical representation of the approach, while Fig. 12.2 shows how it is included into the transaction log.

The main principle is similar to many other hash chains, be it blockchain mechanisms or simple log chaining as already introduced by Schneier and Kelsey (1999). The so-called witness is constructed by hashing[2] the current log data block xored with the current timestamp, a pseudorandom number, and the witness of the last block. The witness is then added to the log data together with the timestamp. This addition, as also shown in Fig. 12.2, is made possible by the structure of the

[2]Of course using a cryptographic hash function (Rogaway and Shrimpton 2004).

transaction log in InnoDB. Instead of a fixed format, InnoDB does not always assume fixed formatted fields but allows for a flexible number of descriptors. This is done in order to be able to log INSERT-statements with their various numbers of attributes efficiently. Thus, the log works based on offsets, which can be manipulated in order to accommodate for the previously constructed witness without interfering with any restore or rollback mechanisms. In fact, the only part requiring any changes is the routine that handles writing to the log in order to calculate and add the witness, as well as the introduction of the pseudorandom number generator that is used for generating the numbers r_i, which are required in order to add a (pseudo-)random element into the chain. Also, compared to the size of the rest of the blocks, the newly generated signature slack space is rather small, thus not blowing up the log file too much.

In order to be able to verify the correctness of the witnesses, simple calculations are sufficient for low-level attacks (i.e., changes without meddling in the transaction log). For adversaries that consider removal of large parts of the log and recalculation after initial manipulation, for provable detection, an old, trusted state is required, against which, all transactions in the log are executed and compared with respect to witness calculation and data inside the database.[3] Since pure read access as facilitated through SELECT-statements are not sent to the transaction log, read access cannot be controlled using this techniques, an issue that we will take care of later in this work.

12.2.2 Data Fingerprinting and Watermarking

Fingerprinting and watermarking are two closely related techniques for reactively protecting information. While they slightly differ in definition, the names for these techniques are often used interchangeably; still we will differentiate between them: Watermarking methods provide techniques for identifying ownership or authenticity, that is, in our context, a watermark can be used to prove that information originally belonged to a specific person. Fingerprinting extends this a step further and not only proves ownership of the data but also indicates who the original recipient, and therefore the data leak was in a provable manner. The latter is especially important in case information is leaked and has become publicly available in order to identify the data leak and take (e.g., legal) measures. Most approaches rely on adding visible or invisible marks to the data. This was most prominently explored for multimedia data, where early on concepts for fingerprinting and watermarking have been devised (Willenborg and de Waal 1996) in order to enable copyright infringement lawsuits. When relating to traditional microdata as it is stored in databases, many approaches from the multimedia world are not applicable anymore.

[3]In addition, other approaches for storing the tree structure based on the so-called signature have been devised in (Kieseberg et al. 2013) and can be used for relaxing this prerequisite.

Thus, approaches,for example, (Willenborg 1999), construct the fingerprints from combinations of selected records. Other approaches like (Bertino et al. 2005) that target medical fingerprinting utilize data binning for construction. Regarding watermarking and (in further extensions) fingerprinting information in general relational databases, a technique was introduced in (Agrawal and Kiernan 2002) that uses a private key only known to the data owner for constructing the watermark. Still, these typical approaches have one severe drawback when related to the PSE process scenario: They all target watermarking/fingerprinting sets of information, that is, not single data records but a relatively large amount of records. They typically utilize either the addition (or removal) of selected "marker" records for the identification of the culprit or the introduction of statistical distortion in an attribute, thus making detection highly improbable in case only single data records are leaked. While this is typically not a problem in case of medical databases, where large amounts of patient data are hosted, extracted and required for meaningful illegal use, in case of a database holding small but extremely valuable information particles, this does not hold up anymore. For illustration, let us assume that in our case, the database contains valuable steering data or simply construction details for machinery parts. Here, the theft of a single asset can pose significant damage to the original owner. Furthermore, most watermarking and fingerprinting approaches rely on changing the underlying information in an unforeseeable manner, thus introduce distortion that might interfere with the PSE process. Last, but not least, a major issue lies in the wide variety of different data that needs to be stored in such a central repository. While structured data can be fingerprinted using typical approaches, other data, like construction information, requires different approaches. Most modern fingerprinting approaches require the leakage of at least a significant amount of data records in order to allow for detection, not working on the level of single record leakage detection. In order to mitigate this shortcoming, an approach based on data anonymization was developed for structured table data in (Schrittwieser et al. 2011). The fundamental idea of this approach lies in anonymizing the data differently for each data recipient, thus making the resulting data look slightly different for each recipient. Since the approach is based on k-anonymity (Sweeney 2002), an anonymization algorithm that works on record level, each anonymized record possesses the intrinsic trace of the so-called anonymization strategy (Emam et al. 2009) explicitly used for this single user, thus allowing for single record detection. It must be noted though that this approach has two major drawbacks, one of them critical for the application in PSE processes: (1) The number of possible recipients is rather low when compared to other fingerprinting mechanisms, and (2) the approach solely works for structured table data. The first limitation is not problematic in PSE processes, as the number of recipients is low compared to medical environments, where each doctor and lab worker is seen as an individual recipient. Still, the second drawback needs further investigation; thus we will outline another idea for fingerprinting construction data in PSE processes.

12.2.3 Shared Analysis Environments for Experts in the Loop

Experts in the loop systems try to integrate the power and speed of modern computing environments with the versatile thinking of humans and their expertise in certain fields (Holzinger et al. 2019). A good example is medical diagnosis, where combining efforts of human expertise with machine learning categorizers has led to unprecedented detection rates (Girardi et al. 2016). Still, a major issue of these systems lies in the area of manipulation detection. This becomes especially apparent when considering the possible effects manipulation attacks can have on the stability of a system. One major example for such an attack would be the STUXNET malware (Langner 2011), where sensor information was manipulated in order to let the enrichment centrifuges run outside their specifications. Providing a centralized platform for managing assets in the PSE process would offer a valuable target for any attacker, ranging from pure data exfiltration to intelligent targeted manipulation attacks. In their work (Kieseberg et al. 2015), the authors provided a solution for a relevant similar problem: Their work focused on doctor in the loop systems (Holzinger 2016), where doctors interact with intelligent analysis algorithms in ways that makes fast and simple control of the results difficult if not impossible. Especially when considering modern machine learning environments, while we do understand the fundamental math behind these mechanisms (as we are the ones who program them), explaining why,for example, a classifier took a certain decision is often impossible (Gunning 2017). Thus, in order for doctors to participate in such a platform, it is of the utmost importance to assure that their input is not manipulated, for example, in order to cover up for mis-implemented algorithms or downright failure. With the machine learning part not trusted by the doctor in the loop, this also includes dealing with a potentially malicious database administrator. The solution of this problem was based on the chained witnesses approach already outlined and provides a flexible and secure solution to this problem; still, it does not consider information theft as an issue, as this did not make any sense in the context of a doctor in the loop.

12.2.4 Database Steganography Through Slack Space

The term "slack space" typically defines the problem that most file systems need to reserve whole clusters when writing files to the disc. In case the data stored inside the file does not fill up the whole allocated space, the remaining space is lost. Still, direct addressing can be used either for retrieving non-overwritten old data from this slack space within a forensic investigation (Kent et al. 2006) or even for using it to store data (Garfinkel 2009). In their work (Frühwirt et al. 2015), the authors have extended the notion of slack space to MySQL more precisely using the InnoDB storage engine in order to hide data. While other authors (Pieterse and Olivier 2012) have already provided ideas for hiding data inside databases, as they offer multiple advantages like high volatility and typically large size, the authors in (Frühwirt et al. 2015) used the index tree in order to generate this free space instead of hiding data inside other records. The basic idea of the concept is that all data inside InnoDB is structured along the primary key of a table which is therefore always present and constitutes the primary index of the table. On deletion, the database like many file systems does not overwrite the data but simply removes the key and the address from the primary index, as well as updates all secondary indices accordingly. The free address is then added to the garbage collection for future reuse. Thus, the only modification to the database source required to generate large amounts of slack space hindering the garbage collection from accessing the space we want to utilize for data storage. This can be done by simple modification of the deletion function, which will result in no space being added to the garbage collection with the drawback of potentially opening up a problematic data leak. The major advantage of slack space generated through this method over other methods of database steganography is the complete loss of any control of the database and especially the database administrator over this space. Neither can the database access this space nor can it be overwritten by database reorganization methods; no board tools allow access short of using file carving (Pal and Memon 2009) to retrieve and change the information in the underlying database storage files, which is a tedious, dangerous, and noticeable task. Thus, in order to actually use this space, either the user himself/herself needs to use file carving, or specific additional read functions need to be implemented into the database.

12.3 Protecting Information Inside a Centralized PSE Platform

As already stated in the introduction, one major issue when protecting the information flow in many current PSE processes is the reliance of exchanging information. This creates multiple issues when done in a more traditional waterfall style of environment:

- No control over the actual accessed (and possibly reused) content
- No methods for proving data leaks
- No defenses against manipulation attacks

Thus, centralized information platforms, as outlined in the working proposal of the CD-lab SQI, seem to be a good solution to these problems. Still, establishing such a system can result in serious issues, as a lot of know-how and data is put into a central place that makes this platform become an interesting target for attackers. Thus, in order to detail our approach, we will define a basic platform architecture based on the doctor in the loop concept. While we certainly believe this approach has its merits, we leave it to other researchers to come up with platforms more likely to be actually used in the industry. Still, the basic features that we require to

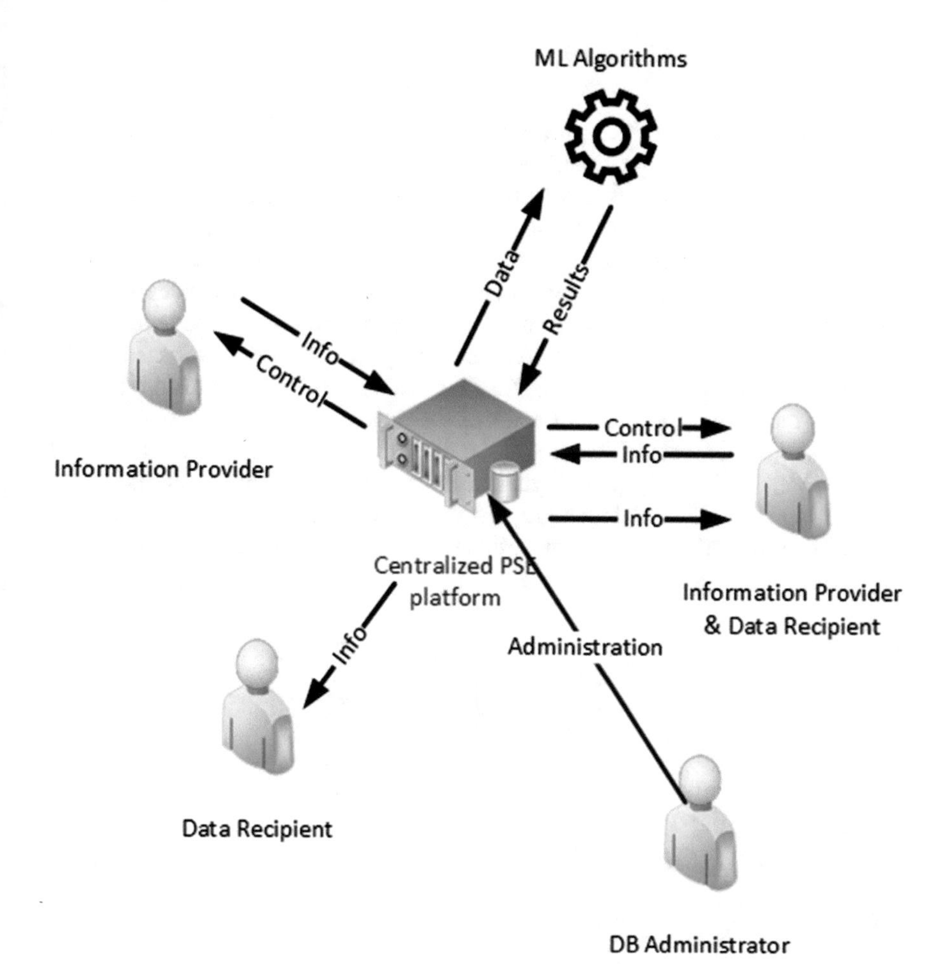

Fig. 12.3 The structure of the PSE platform

demonstrate our approach for information protection will still be in place in these platforms (Fig. 12.3):

- An ACID-compliant database at its core (Haerder and Reuter 1983), thus including mechanisms for transaction safety, rollbacks, and crash recovery
- Interfaces to various (human) participants with different interests
- A platform provider that typically plays fair, that is, the administrator of the underlying database does not plan on manipulation at the time the system is set up (else, the chained witnesses approach will not work in a provable manner)
- Mechanisms that work on the data in a manner that is not easily conceivable, for example, advanced machine learning algorithms

12.3.1 Basic Architecture

In our approach, we use the approach for providing an audit and control system for the doctor in the loop as outlined in (Kieseberg et al. 2016) and extend it to suit the purpose of a distributed PSE environment. One major difference to the original approach is that we do not focus that heavily on the pure integration of human expertise into machine learning environments but solely look at different partners, called "workers" from here on, that work on data residing inside the shared resource database. Figure 12.4 gives an overview on the principle architecture of the system, where each transaction on the underlying database is protected by the chained witnesses approach. In addition, each worker is assigned his/her/its own signature that is used in order to mark any changes made to the data with their originator. Another major difference to the doctor in the loop approach as outlined in (Kieseberg et al. 2016) are the types of information stored in the database. The original approach only cared for highly structured information like patient data, with a lot of records holding simply structured information like timestamps, numbers, or

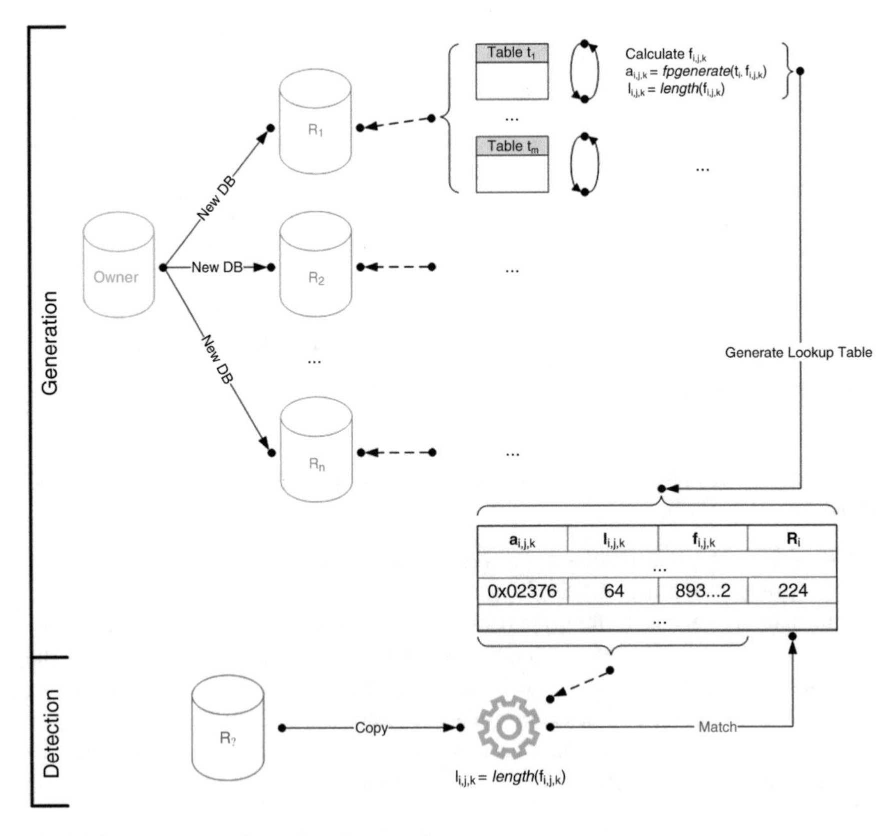

Fig. 12.4 Overview on the proposed approach

other. This might not be the case for a database required for exchanging information required in a PSE process: Especially when considering planning the facility with respect to placing, machineries, and electrical/hydraulic assets, data will be highly complex in nature; thus the data model of the underlying database needs to be reasonably flexible to cater for such complex structures. This is also another major benefit of using the chained witnesses approach, as it does not concern itself with the exact way data and information is modelled in the data model, but lets this be done by the DBMS and the underlying models. Since it only cares for the actual transactions carried out on the database, it does not need to decipher the actual meaning behind them in order to be able to prove manipulations.

Our system model consists of the following workers, where some roles can be combined, especially considering the role of the information provider and the one of the data recipient:

Information Provider: These are the owners of information entered into the centralized PSE platform and the primary proliferators of data, aside from the results of the ML algorithms. They are also the ones who have most to lose with respect to know-how and industrial espionage; still they could pose as potential attackers with respect to trying to provide faulty information and later covering up or by unrightfully accusing other workers.

Data Recipient: These are the typical data and information consumers, for example, experts who rely on the data that other partners enter into the system. Especially with experts, and even more with experts in the loop, these workers will often take over a combination of the data recipient and the information provider roles.

ML Algorithms: Since machine learning algorithms have become ubiquitous in many research and development heavy industries, it is, in our opinion, just a matter of (short) time until they will be facilitated inside PSE processes. Thus, our approach already includes the integration of intelligent workers that take input data at a certain point in time and return results that are not easily controlled and verifiable by the other workers.

DB Administrator: The database administrator is seen as an own worker, as he/she is in the possession of high privileges and thus exceptionally interesting from an attacker perspective. Many of the techniques outlined in this chapter allow for protection against attacks from this side, which is especially important in order to motivate information providers to participate in the centralized PSE platform at all.

Centralized PSE platform: This participant symbolizes the centralized platform with all its technicalities and especially personal involved. Of course, the centralized PSE platform can belong to one of the participating workers, most likely an information provider. Accordingly, the platform cannot be trusted; still we assume that it is trustworthy on setup, that is, the platform provider did not plan to cheat the partners when setting up the platform but might start playing foul later on.

12.3.2 Thwarting Information Leakage

The original approach of using witnesses for securing database content as outlined in this section has one severe limitation regarding data leak detection and the detection of unsolicited information extraction, as it was solely concerned with thwarting manipulation of data, especially by high-profile attackers possessing administrator privileges. This is certainly of the utmost importance for any data-driven process using sensitive (or at least valuable) information, as it guarantees the integrity of the information stored in the database to some extent. There are many foreseeable scenarios where attacks on the integrity of data assets are an interesting factor, also including methods like specialized malware that do not need to infect the machinery on the premises but are already implemented at the system level, thus available at each update and running with administrator privileges (see, e.g., Thonnard et al. 2012 for APTs in industrial environments). Still, the original approach does not thwart any information theft, as read access to the database is not integrated into the transaction mechanisms (Frühwirt et al. 2010). It is rather simple to deal with this issue in case of normal database users (as it was also the case with respect to data manipulation), as there are many logging and tracing mechanisms in place; still, these are under the direct control of the database administrator. Challenging the trustworthiness of the database administrator is fundamental for providing lasting security in such an environment: Many attacks launched against companies come from the inside (Huth et al. 2013), and a disgruntled database administrator is surely a valid concern (Claycomb and Nicoll 2012). Still, often attacks from the inside are the result of a malware infection of the user in question, ranging from simple mass spam infection to targeted attacks explicitly designed against this very system (Sood and Enbody 2013). Thus, we need an equivalent technique regarding information retrieval comparable to the witness for read access. We will thus look into two different approaches for storing this read information, namely, by

1. Making read access to the database show up in the transaction mechanisms as well
2. Using database steganographic methods for storing read and write information into the database

For the first approach, that is, letting SELECT-statements issued toward the database show up in the transaction log, we present two approaches: While the naive approach would follow the general approach on changing the logging strategy by modifying the code of the SELECT-statement in the underlying DBMS in a way it writes its content to the log, this poses a major problem. Currently in InnoDB, there exists no log block structure defined for SELECT-statements, and other blocks are not suitable to hold such a transaction, since this would confuse the rollback and crash recovery mechanisms. The actual definition of the log is far more complex than just writing a single data block per transaction to the files, and the definition of the transaction in question is spread over several different supporting blocks (Frühwirt et al. 2010). While it is of course possible to change this and add a new block type with supporting block structure into the log, these changes would need to be implemented consistently throughout all internal database mechanisms and would require updates every time the basic InnoDB code is changed. Thus, we do not consider this strategy to be helpful, especially as it does not work automatically for closed-source DBMSs and as this would mean changing the replication log functions as well. But, a simpler way of introducing the content of SELECT-statements into the transaction and replication logs can be defined straightforward: We could change the function that handles the SELECT-statement just a bit in order for it to carry out an additional INSERT into a predefined table, thus writing the content of the SELECT-statement to the database. This would result in automatic writing of the statement into the transaction log. Of course it is possible for the database administrator to change the content of this table, as it offers no additional protection over any other table in the database; thus the table itself does not have much value in manipulation detection, but since an INSERT-statement was triggered, the relevant part of the data is also inserted into the chaining of the witnesses. Since the data in this table is of no relevance, it can be deleted right after the INSERT is finished. The major issue with this technique lies in the vast amount of additional information inserted into the log files, thus exceeding the typically reserved space quite soon. Since the files holding the log information are structured like ring storage (Frühwirt et al. 2010), this would result in the first parts to be overwritten. While this does not pose a problem per se for the witness algorithm, it vastly increases the efforts required for manipulation detection. The structure of the additional table that is required to hold the SELECT information is extremely simple: It just requires a field of type varchar that holds the statement.

The other approach for detecting data extraction/theft that we want to present in this section relies on the concept of database steganography as described in Sect. 12.2.4. The basic idea is to generate a suitable amount of database slack space right at the beginning at the setup of the database by providing an additional DELETE-routine that, in addition to deleting information like the original DELETE-function, also generates slack space by saving the deleted space from the garbage collection. In order to thwart file carving, we will spread the slack space over a larger region of the database files by generating a large amount of data throughout the setup phase of the database and delete it, most of it by using the standard deletion route and some with the help of the slack space generating deletion command. The deletion will not be done right after the insertion but at random time in the INSERT/DELETE cycle; thus, the database will be filled up at first, and then random space will be allocated to the slack space, thus spreading it over large parts of the underlying database files.

The structure of the slack space looks as follows: The space starts at a predefined address that acts like the garbage collector by pointing to the next free address. In addition to the pointer, it contains the following data: (1) size of the slack space, (2) remaining free space in the slack, (3) a pointer to the latest entry in the slack, and (4) the last address of the slack. This information is required in order to extend the slack space when needed. Another modification to the database environment that needs to be done at compile time (like the implementation of the new DELETE methods) is a way of writing information to and retrieving it from the slack space. Writing should take place at every SELECT-statement; thus we define a log block as the concatenation of a timestamp, an offset, user credentials, and the statement itself:

$$block := timestamp||userID||offset||statement||witness$$

The timestamp contains the exact time of the issuing of the SELECT-statement, and the userID is required in order to store the user issuing the statement. It is retrieved from the internal access control system that is invoked by each SELECT-statement in order to check whether a user possesses the credentials to actually access the information declared by the SELECT-statement. This information is retrieved inside the search routines from database internals and cannot be changed by the administrator for obvious reasons. The offset simply contains the length of the statement that needs to be stored. In order to additionally secure the data inside the database slack, we will apply the chained witnesses approach as we have done for the transaction mechanism. Before adding the statement to the database slack, the routine inside the SELECT-method calculates the required space and checks whether the slack space is still big enough. If not, the extension routine is called, which basically creates some arbitrary random data and deletes it using the slack-generating DELETE-routine and adds the required pointers from the previously last address of the slack. In any case, the routine jumps to the latest entry in the slack and adds the information in the subsequent slack space. Afterward, it changes the last entry address and the remaining space counter in the slack space header. In order to mitigate the issue of having to split statements over several addresses, we assume that at generation time each data particle generated for slack generation was larger than the longest SELECT-statement. In case this does not hold true, the data needs to be split over several addresses, which does not pose any problem as the data length can be calculated using the offset, thus making perfect use of all slack space available. For reading the log inside the slack space, another routine needs to be provided. The definition is quite simple, and the routine takes the address of the slack space start as an input parameter and generates the list of all SELECT-statements executed against the database. Since we know the address of the latest entry based on the respective pointer in the slack space, the enumeration can be done either forward or backward in time. For the latter, it must be noted though that the witnesses cannot be verified backward.

In comparison, the approach writing the information into the slack space has two vital benefits over using the transaction mechanism: First, since the transaction log is designed like a ring storage, that is, when it is full, it starts writing from the start again and typically has a size of just several megabytes; the second approach can store far more data as its size is basically limited by the amount of disc space allocated to the total database. Furthermore, data will not be overwritten, in case the mechanism runs out of space. On the other hand, it needs special consideration in case the log cannot be extended any more, which is no problem for the mechanism based on the transaction log. Also the mechanism based on the transaction log is especially well suited for small installations, as the hidden data in the steganographic approach is not meant to be deleted, thus growing with the age of the database.

12.3.3 ExPost Protection Through Fingerprinting

With the protection measures in place through manipulation detection and controlling read access for being able to detect mass information exfiltration of unsolicited access, one major problem with respect to data leaks is still present: The one of a legitimate data recipient taking the data it legitimately received and further using it in other projects or even publishing it. This is a serious problem, as within a more complex project, it is not unreasonable for several partners receiving the same information, thus making data leaks hard to detect by themselves. Still, the data stored inside the system will often constitute a non-negligible value to its owners, thus protection of data theft with reactive (legal) measures needs to be in place for companies to be willing to accept such a shared environment in the first place. Thus, we will add fingerprinting to our approach in order to thwart these kinds of leakage attacks. More precisely, contrary to many other systems that purely deal with high volumes of rather simply structured information, our system needs to be able to take care of a multitude of different types of information. A lot of different approaches toward fingerprinting have been devised in the past, mostly concerning the record level, that is, the data is extracted through the SQLinterface or shared via tables and is changed with respect to some secret markers in order to generate a fingerprinted copy ready for detection. While this is perfect for many data driven use cases and some of these approaches can surely be added to the database interfaces of our approach, we want to take care of another attacker model: copying the whole PSE database, either through a malicious administrator or simply a copy actually planned for backup purposes. Thus, in this section, we will provide a novel database fingerprinting algorithm that can be used for arbitrary database structures holding arbitrary forms of data, thus being perfect for shared PSE environments. Furthermore, this novel fingerprint does not change any information in the data parts, thus making it even more an interesting choice for PSE data.

In Sect. 12.2.4 we already referenced some of our earlier work regarding information hiding in index trees, especially considering manipulating the primary index in order to generate highly stable slack space that cannot be accessed by the database anymore. The approach required the deletion function of the database management system to be changed in order to not add deleted records to the garbage collection for further reuse. In case the data was to be read again, special "hidden" functions were required to be implemented into the DBMS as well. While surely feasible, this opens the problem that the system needs to run a special version of the DBMS in question, which might make users (or at least administrators) suspicious, and introduces additional overhead and programming effort. In our fingerprinting approach, the scenario is far more simple: While the original owner of the DBMS will need to have his/her own version of the DBMS running, the data recipient is not required to be in the possession of anything unorthodox, that is, he/she can run the "normal" version of the DBMS.

DBMS Modifications The data owner is required to run a DBMS that he/she has full control of at the source code level in order to introduce some changes, as two new functions *fpgenerate (table, fingerprint)* and *fpextract (address, length)* (see below) need to be introduced.

Function fpgenerate is a duplication of the normal deletion routine with the exception that it inserts the parameter *fingerprint* into the table provided in parameter *table* using the normal insertion routine provided by the database and then executes an incomplete deletion that does not invoke the addition of the record to the garbage collection, thus effectively removing the information from the reach of normal database functions. Since the transaction log is stored in external files in MySQL, we can simply use the original INSERT routines, as we won't copy these files to the recipient, thus not leaking information through these files and still maintaining transaction safety even in the fingerprinting process. Of course, in order to guarantee this (and maybe in case of other DBMSs), also the INSERT routine might be changed in order to not leave any traces.

1. Insert a new record into the table that contains the fingerprint specifically constructed for the data recipient.
2. Unlink the record from the neighboring records in the $\mathcal{B}^+$-Tree leaf node, but do not add it to the garbage collection pointer.
3. Check secondary indices for the record—in case of positive retrieval, unlink it from the respective secondary indices.
4. Check the dictionary of the table—in case the record can be retrieved from it, unlink it.
5. Depending on the actual DBMS in use, return the absolute address or the offset in the DBMS file of the fingerprint location

Function fpextract takes an absolute address or an offset in the DBMS file (depending on the actual implementation and the DBMS in question), as well as the length of the information requiring extraction, and returns the data at the given address/offset.

Fingerprint Generation and Verification One major benefit of this technique is the possibility to introduce an arbitrary amount of fingerprints to the database extract, that is, the data owner does not need to resort to a single fingerprint but can spread multiple instances across the database files.

For the construction of a fingerprint, any reasonable technique might be used. In our basic approach, we assume that the fingerprint is a salted hash of data identifying the respective data recipient together with the table name in order to thwart detection through duplicate entries. This fingerprint will be stored in the primary key of the table, if suitable, and the other columns will be filled with default values, depending on the actual structure of the table. Thus, in real-life implementations, it might be practical to provide standardized functions for fingerprint generation for various types of data. We further assume to use a different salt for each fingerprint in the same table:

$$f_{i,j,k}(u_i, t_j, s_k) := \mathfrak{H}(u_i \oplus t_j \oplus salt_k),$$

with u_i denoting the i-th data recipient, t_j the j-th table in the database, and s_k the k-th entry of the set of salt values.

For adding the fingerprints, the data owner simply selects the table in the database that should be used as carrier for the fingerprint and generates the fingerprinting information according to the table structure as outlined above. Then, the fingerprint is added by invoking the *fpgenerate* function. The resulting address needs to be stored in a table outside the database, together with the recipient ID and the fingerprint.

Detection is straightforward: In case an illegal copy of the database is found, the data owner needs to obtain a full version of it and check all addresses or offsets (again depending on the DBMS) using the function fpextract. The results of fpextract are then checked against the respective fingerprinting values in the table holding all fingerprints for all copies of the database in question. Figure 12.4 gives an overview on the proposed approach.

Evaluation and Limitations In this section, we will discuss the benefits and limitations of the outlined approach.

Performance Generation of the fingerprints is rather simple and fast as we require the calculation of one hash and two xors, as well as one database insert and one deletion per fingerprint, which is negligible compared to the computational effort introduced by requiring to copy all database files. For verification, the worst case scenario can be constructed as follows: Let n be the number of data recipients and $m_i, i = 1, \ldots, n$ the number of fingerprints for the i-th recipient. Then, in the worst case scenario, $\sum_{i=1}^{n} m_i$ fingerprints need to be checked. Given that this only constitutes the extraction of information at a single address/offset and comparison with pre-calculated hashes, this is feasible even for higher numbers of n and $m_i, i = 1, \ldots, n$, respectively.

Variability The construction of the fingerprint allows for quasi-endless variations, depending on the actual size of the carrier database.

Stability Within the defined side parameters, that is, the copying of whole databases on a file basis, the fingerprint is extraordinarily stable: As it is not encoded within the actual data in the database but in meta-structure, it is possible to change every single data record completely and still being able to tell who was the original data recipient of the database instance in question. This notion is so strong that it must be discussed though whether this still constitutes a desirable amount of stability, or is simply too strong, as the fingerprint will also "detect" databases that, due to not possessing any original information any more, most surely do not constitute a case of data leakage.

Detectability The detectability is difficult to answer: While it is stronger than, for example, the approach outlined in Sect. 12.2.2 in the sense that it does not require any leaked record to be present for detecting a copied database, it does require the presence of the original files with their meta-information, most precisely the database slack space. Furthermore, it is discussible whether the files without any of the original data left still constitutes a data leak, still, it might be interesting to know that a copy of the database files was reused somewhere else.

Limitations While this technique is practical in order to detect copies of whole databases, even in case all the data in the tables have changed, it does have drawbacks: First, it can easily be circumvented by not copying the database files but by inserting full table scans into another freshly setup database or by extracting (part of) the data to files. This major drawback is due to the fact that the fingerprint is not provided within the information that we want to protect itself but within the meta-structure of the database. Furthermore, attackers could still try to remove the fingerprint by (1) finding all addresses in the database files that are not referenced by any primary index (and are not used by other database assets like procedures, views, secondary indices, etc.) and (2) overwriting the information at the said addresses by using file carving techniques. While this is feasible form a theoretical point of view, depending on the size of the database, it does involve a lot of effort.

Susceptibility to Collusion Attacks In case two data recipients work together for breaking the fingerprint in two fresh copies of the shared database, they could calculate diffs for the received files and thus extract the data holding the fingerprints. This can be circumvented by selecting the fingerprints in a way that each subset of the m recipients R_i has at least one fingerprint that they all share, but none of the other recipients possesses. This leads to the introduction of at least $2^m - 1$ fingerprints (the empty subset does, of course, not require a fingerprint, thus the "-1"). While this seems to be infeasible for large numbers of data recipients, again, for attackers working at this level, regarding removal of the fingerprint, it would far more easy to simply extract all the information from the tables to a file and load these files into a fresh database instance, thus removing the fingerprint entirely at the cost of not being in the possession of a forensic copy anymore.

12.4 Limitations

Our approach possesses several limitations that need to be addressed in any real-life implementation:

1. The fundamental basic principle of the chained witness approach only works in case of a database administrator that did not plan on going rogue when already setting up the environment, that is, the very initialization of the witnesses approach requires a trusted third party that does not play any further part in the system, especially not cooperating with the attacker.
2. While the original chained witnesses approach did work with closed-source DBMSs to some extent, this cannot be stated for the adaption to logging read access. Both methods outlined in the approach require the modification of parts of the database, as neither the database transaction nor the replication mechanisms catch read access in any database management system analyzed by us.
3. The system still requires a neutral player that is open for proofs regarding data manipulation, that is, either a central authority within the system rules on the correctness of the individual behavior of the partner, or courts must be made fit in order to be able to understand and validate the information presented to them by the system and its methods.
4. Another issue that requires attention with this approach is the issue of using too much of the storage capacity for the slack space. In the current approach, an attacker could issue large amounts of fast SELECT-statements, thus filling the slack space up with information that requires a long time for verification. Still, this would be obvious for any analyst, thus making this not a big limitation.
5. In order to facilitate the slack space construction, the source code of the underlying DBMS in question needs to be available; thus, this approach cannot be used for closed-source DBMSs. This also holds true for logging read access to the transaction log through writing the statements into a table, as this requires changes in the routine parsing and executing the SELECT-statements.
6. In case for highly structured data, the fingerprinting approach based on k-anonymity that we propose to use in the approach does not work for a large number of data recipients. More precisely, in case the number of usable attributes exceeds the number of participants, collaborating attackers cannot be identified in a provable manner.
7. As we already outlined in the relevant section, new fingerprinting methods need to be discovered that cater for the very specific requirements put forth by the special nature of data required within the PSE process: no distortion of the important information while still being hard to detect and remove without altering the information and rendering it useless at best. We hinted at some initial ideas for providing such fingerprint, still currently the best approach lies in providing hidden markers inside this assets, as well as using standard fingerprinting approaches aiming at the file formats these assets are shared in.

Still, with all these limitations in place, we do consider this platform approach to be reasonable for implementation inside a working environment.

12.5 Conclusion

In this chapter, we proposed methods for securing shared asset databases for PSE processes against data exfiltration, manipulation, as well as data leakage. In order to provide context for these techniques to be applied, we proposed a basic shared environment sporting different workers. From an attacker model point of view, we consider the partner providing the platform to be honest at the start but allow him/her to join the attacker later on while still providing the security features outlined above. Our main contribution in this chapter lies in the manipulation secure collection of information on SELECT-statement executed against the database. We proposed two different methods for reaching this target: One is based on a simple change in the routine implementing the SELECT-statement in order to write the statement into a (arbitrarily defined) table, thus sending the data to the transaction log that is protected by the chained witnesses approach. The other works by generating database slack space in order to store the SELECT-statements in a more structured manner but requires far more changes to the underlying DBMS. Finally, our approach includes fingerprinting of all sensitive information in order to protect it against data leakage through legit data recipients and other project partners participating in a joint development project. Here, we provided a novel fingerprinting solution based on database steganography that can deal with the issue of very diverse data types and structures.

Acknowledgements The financial support by the Christian Doppler Research Association, the Austrian Federal Ministry for Digital and Economic Affairs, and the National Foundation for Research, Technology, and Development is gratefully acknowledged.

References

Agrawal, R., & Kiernan, J. (2002). Watermarking relational databases. In *Proceedings of the 28th International Conference on Very Large Databases* (pp. 155–166).

Bertino, E., Ooi, B. C., Yang, Y., & Deng, R. H. (2005). Privacy and ownership preserving of outsourced medical data. In: *21st International Conference on Data Engineering (ICDE'05)* (pp. 521–532).

Claycomb, W. R., & Nicoll, A. (2012). Insider threats to cloud computing: Directions for new research challenges. In: *2012 IEEE 36th Annual Computer Software and Applications Conference* (pp. 387–394). Piscataway, NJ: IEEE

Emam, K. E., Dankar, F. K., Issa, R., Jonker, E., Amyot, D., Cogo, E., et al. (2009). A globally optimal k-anonymity method for the de-identification of health data. *Journal of the American Medical Informatics Association, 16*(5), 670–682.

Frühwirt, P., Huber, M., Mulazzani, M., & Weippl, E. (2010). InnoDB database forensics. In: *2010 24th IEEE International Conference on Advanced Information Networking and Applications (AINA)* (pp. 1028–1036). Piscataway, NJ: IEEE.

Frühwirt, P., Kieseberg, P., Krombholz, K., & Weippl, E. R. (2014). Towards a forensic-aware database solution. *Digital Investigation, 11*(4), 336–348.

Frühwirt, P., Kieseberg, P., Schrittwieser, S., Huber, M., & Weippl, E. R. (2012), Innodb database forensics: Reconstructing data manipulation queries from redo logs. In *2012 Seventh International Conference on Availability, Reliability and Security* (pp. 625–633).

Frühwirt, P., Kieseberg, P., & Weippl, E. (2015). Using internal mysql/innodb b-tree index navigation for data hiding. In *IFIP International Conference on Digital Forensics* (pp. 179–194).

Garfinkel, S. L. (2009). Automating disk forensic processing with sleuthkit, xml and python. In. *2009 Fourth International IEEE Workshop on Systematic Approaches to Digital Forensic Engineering* (pp. 73–84).

Girardi, D., Küng, J., Kleiser, R., Sonnberger, M., Csillag, D., Trenkler, J., et al. (2016). Interactive knowledge discovery with the doctor-in-the-loop: A practical example of cerebral aneurysms research. *Brain Informatics, 3*(3), 133–143.

Gunning, D. (2017). Explainable artificial intelligence (XAI). In *Defense Advanced Research Projects Agency (DARPA), nd Web* .

Haerder, T., & Reuter, A. (1983). Principles of transaction-oriented database recovery. *ACM Computing Surveys, 15*(4), 287–317.

Holzinger, A. (2016). Interactive machine learning for health informatics: when do we need the human-in-the-loop? *Brain Informatics, 3*(2), 119–131.

Holzinger, A., Plass, M., Holzinger, K., Crişan, G. C., Pintea, C.-M., & Palade, V. (2016), Towards interactive machine learning (IML): applying ant colony algorithms to solve the traveling salesman problem with the human-in-the-loop approach. In *International Conference on Availability, Reliability, and Security* (pp. 81–95). Berlin: Springer.

Holzinger, A., Plass, M., Kickmeier-Rust, M., Holzinger, K., Crisan, G. C., Pintea, C. M. et al. (2019). Interactive machine learning: Experimental evidence for the human in the algorithmic loop: A case study on ant colony optimization. *Applied Intelligence, 49*(7), 2401–2414.

Huth, C. L., Chadwick, D. W., Claycomb, W. R., & You, I. (2013). Guest editorial: A brief overview of data leakage and insider threats. *Information Systems Frontiers, 15*(1), 1–4.

Kent, K., Chevalier, S., Grance, T., & Dang, H. (2006). *Guide to integrating forensic techniques into incident response*. (No. Special Publication (NIST SP)-800-86).

Kieseberg, P., Malle, B., Frühwirt, P., Weippl, E. R., & Holzinger, A. (2016). A tamper-proof audit and control system for the doctor in the loop. *Brain Informatics, 3*(4), 269–279.

Kieseberg, P., Schantl, J., Frühwirt, P., Weippl, E. R., & Holzinger, A. (2015). Witnesses for the doctor in the loop. In *International Conference on Brain Informatics and Health* (pp. 369–378). Berlin: Springer.

Kieseberg, P., Schrittwieser, S., Morgan, L., Mulazzani, M., Huber, M., & Weippl, E. (2013). Using the structure of B+-trees for enhancing logging mechanisms of databases. *International Journal of Web Information Systems, 9*(1), 53–68.

Langner, R. (2011). Stuxnet: Dissecting a cyberwarfare weapon. *IEEE Symposium on Security and Privacy, 9*(3), 49–51.

Pal, A., & Memon, N. (2009). The evolution of file carving. *IEEE Signal Processing Magazine, 26*(2), 59–71.

Pieterse, H., & Olivier, M. S. (2012). Data hiding techniques for database environments. In *8th International Conference on Digital Forensics (DF)* (pp. 289–301).

Rogaway, P., & Shrimpton, T. (2004). Cryptographic hash-function basics: Definitions, implications, and separations for preimage resistance, second-preimage resistance, and collision resistance. In *International Workshop on Fast Software Encryption* (pp. 371–388). Berlin: Springer.

Schneier, B., & Kelsey, J. (1999). Secure audit logs to support computer forensics. *ACM Transactions on Information and System Security, 2*(2), 159–176.

Schrittwieser, S., Kieseberg, P., Echizen, I., Wohlgemuth, S., & Sonehara, N. (2011), Using generalization patterns for fingerprinting sets of partially anonymized microdata in the course of disasters. In *2011 Sixth International Conference on Availability, Reliability and Security* (pp. 645–649).

Sood, A. K., & Enbody, R. J. (2013). Targeted cyberattacks: A superset of advanced persistent threats. *IEEE Security & Privacy, 11*(1), 54–61.

Sweeney, L. (2002). k -anonymity: a model for protecting privacy. *International Journal of Uncertainty, Fuzziness and Knowledge-Based Systems, 10*(5), 557–570.

Thonnard, O., Bilge, L., O'Gorman, G., Kiernan, S., & Lee, M. (2012), Industrial espionage and targeted attacks: Understanding the characteristics of an escalating threat. In *International Workshop on Recent Advances in Intrusion Detection* (pp. 64–85). Berlin: Springer.

Willenborg, L. (1999). Fingerprints in microdata sets. In *Joint ECE-Eurostat Work Session on Statistical Data Confidentiality, Thessaloniki*.

Willenborg, L. C. R. J., & de Waal, T. (1996). *Statistical disclosure control in practice*. Berlin: Springer.

Zaitsev, P. (2009). Innodb architecture and performance optimization. In *O'Reilly MySQLConference and Expo*.

Chapter 13
Design and Run-Time Aspects of Secure Cyber-Physical Systems

Apostolos P. Fournaris, Andreas Komninos, Aris S. Lalos, Athanasios P. Kalogeras, Christos Koulamas, and Dimitrios Serpanos

Abstract Cyber-Physical Systems (CPSs) combine computational and physical components enabling real-world interaction. Digitization, decentralization, and high connectivity, as well as incorporation of various enabling technologies, raise various security issues. These security concerns may affect safety, endangering assets and even human lives. This is especially true for CPS utilization in different sectors of great significance, including manufacturing or critical infrastructures, creating a need for efficiently handling relevant security issues. Including security as part of a software-intensive technical system (i.e., the CPS) that can be distributed and highly resilient highlights the need for appropriate security methodologies to be applied on the CPS from the engineering stage during CPS design. The efficient security-related processes that are implemented at design time have an impact on security monitoring during the CPS operational phase (at run-time). Efficient and accurate security monitoring that follows security-by-design principles can be a potent tool in the hands of the CPS manager for detecting and mitigating cyber threats. Monitoring traffic and activity at the system boundaries, detecting changes to device status and configuration, detecting suspicious activity indicating attacks, detecting unauthorized activity that is suspicious or violates security policies, and timely responding to security incidents and recovering from them are issues that need to be efficiently tackled with by security monitoring. In this chapter, we explore the various CPS cybersecurity threats and discuss how adding security as a parameter at the CPS design phase can provide a well-structured and efficient approach on providing strong security CPS foundations. New technologies on CPS security design are presented and emerging security directions are discussed. Furthermore, in the chapter, the different aspects of security monitoring are presented with a special emphasis on CPSs, discussing the various existing monitoring approaches that are followed in order to detect security issues at run-time. Specific use cases of CPSs in the manufacturing domain and with reference to critical infrastructures are also

A. P. Fournaris (✉) · A. Komninos · A. S. Lalos · A. P. Kalogeras · C. Koulamas · D. Serpanos
Industrial Systems Institute, ATHENA Research Center, Patras, Greece
e-mail: fournaris@isi.gr

S. Biffl et al. (eds.), *Security and Quality in Cyber-Physical Systems Engineering*,
https://doi.org/10.1007/978-3-030-25312-7_13

detailed and security requirements like confidentiality, integrity, and availability are discussed.

Keywords Security by design · Security run-time monitoring · Cyber-physical systems security · Cybersecurity · Digital Twins

13.1 Introduction

Cyber-physical systems (CPS) are characterized by the tight integration of computing, communication, and control technologies (Rajkumar et al. 2010). They transverse a number of application areas ranging from manufacturing to transportation, to energy and healthcare, to name just a few. They are associated with several research domains including real-time networking, real-time computing, hybrid systems, wireless sensor networks, model-driven development, and security (Kim and Kumar 2012).

Cyber-Physical Systems (CPSs) constitute a disruptive technology applicable in many industrial domains and present strong impact on economies and social processes, on many fronts, including robotics, security, safety, and military, and across industries and applications (Serpanos 2018), as they aim to bridge the cyber world of computing and communications with the physical world (Rajkumar et al. 2010). CPSs represent complex engineered physical systems, centered on Information and Communication Technologies (ICT) to integrate, control, monitor, and coordinate their operations. The advances of ICT toward interacting with the physical world actually make CPS rather ICT systems integrated into the physical world processes and applications (Gollmann 2012; Humayed et al. 2017).

The application of CPSs is wide, covering different domains. Such domains include critical infrastructure monitoring, control and protection (energy, water resources, communication systems, and transportation infrastructures), manufacturing, factory automation and control, building management and control, environmental monitoring, automotive systems, healthcare, and defense and military systems. Identification of end user needs, challenges, and opportunities in the different application domains can advance research in CPS. Multidisciplinary collaborative research can lead toward high confidence systems characterized by compatibility, synergistic behavior, and integration at all scales between cyber and physical designs (Baheti and Gill 2011). This confidence cannot be attained unless it stems from the careful integration of security and, by extension, reliability considerations in the design of such systems.

The inherent characteristics of CPSs raise significant security challenges. CPSs are characterized by wide geographical distribution comprising different types of sensing, actuating, computing, and control devices, usually without physical security. In several cases, those devices are left unattended and unsupervised in "hostile" environments where they can be easily attacked. They have requirements with a need to react in a real-time manner. They have different communication channels that may be exploited by adversaries due to the CPS feedback from the physical environment. They are characterized by distribution of management and control usually involving multiple parties. They present *System of Systems* control characteristics (Neuman 2009).

The peculiarities of CPSs with reference to their characteristics and the high economic and societal impact stemming from potential security attacks make reliability and security integral to their operation. In order to guarantee such properties in complex systems comprising heterogeneous devices interconnected with different communication technologies, it is imperative to have a deep understanding of related threats and vulnerabilities at the outset of the design phase. This approach leads to formal specifications describing the CPS implementation. However, this task is quite challenging as it needs to address both discrete and continuous CPS behavior. There is a strong need for security operations integration to CPS hardware structure as well as CPS software-intensive operations beginning from the CPS engineering (design) phase and progressing to CPS operation phase in order to be able to achieve a high level of protection against cybersecurity attacks and to support a broad range of security activities. Applying security-by-design principles is the best approach in order to structure CPSs that have the capability to support intrinsically strong security at operational time with minimal number vulnerabilities. This also reduces the fault tolerance of the CPS system, thus making it more resilient to support safety critical applications.

This security-by-design approach that leads to designs with several supported security features, however, has no impact on security if such features are not put in good use throughout the full lifecycle of the CPS. More specifically, at the CPS regular operation, security protocols must be efficiently established at the communication level and specialized security mechanisms must be deployed and executed in order to monitor the system for cyberthreats.

Run-time security monitoring utilizes program monitors in order to examine deviations between the expected and real behavior of a system. The formal specification describing the CPS behavioral model as well as machine learning algorithms can be used for the determination of the expected "good behavior" of the system. Building such program monitors is quite complex and challenging when it comes to large-scale CPSs of high complexity involving physical processes. Enabling technologies in the context of Industry 4.0 and the Industrial Internet of Things can be used to enhance such monitoring mechanisms. Digital Twin (Tao et al. 2018) by representing a digital model that accurately mimics its physical counterpart and evolves, is continuously updated to reflect changes in the physical world (Maurer 2017) becomes fertile ground for detecting deviations between the physical twin and its digital counterpart.

In this chapter, we discuss emergent security and reliability challenges pertinent to CPSs focusing on the mechanisms, approaches, and solutions on achieving security-by-design goals in CPSs in order to support security-based software-intensive functions at the CPS's operational level. We discuss the design principles that need to be applied in order to protect communications and to achieve resilience, robustness in CPSs. We also present and analyze existing and emerging approaches on how to constantly monitor the CPS's security level through appropriate cyber-threat and anomaly detection schemes considering also novel concepts like the Digital Twins technology. Finally, we also present a use-case scenario that targets security in industrial/critical infrastructure CPSs and give some future research directions on the chapter's concepts.

The rest of the chapter is organized as follows: Sect. 13.2 presents CPS security challenges and deals with reliability and security by design, Sect. 13.3 addresses Run-Time Security Monitoring in CPSs. Section 13.4 deals with the use case of Industrial Control Systems (ICS), a subclass of CPS, their threats and vulnerabilities. Finally, Sect. 13.4 presents relevant research challenges and concludes the chapter.

13.2 Reliability and Security by Design

Reliability and Security are two equivalent aspects of a system's operation. They are related with two different domains of thinking. While security ensures that a system is doing the *right thing*, system reliability safeguards that the system is doing the thing *in the right way*. To be characterized as secure, a system must exhibit the following three properties: confidentiality, integrity, and availability. Confidentiality represents a set of rules limiting access to information. It thus ascertains that sensitive information does not reach inappropriate parties, while guaranteeing that it is accessible by the right parties. Integrity is the assurance of information consistency, trustworthiness, and accuracy throughout its entire lifecycle. It thus implies that information is only modified by authorized parties. Availability is a guarantee of authorized parties' reliable access to information.

13.2.1 Threats to Cyber-Physical Systems

In the past, CPS' safety and reliability have been the foremost concern, even though their security had always been recognized as important. Despite this acknowledgment, CPS security has traditionally been treated as an "afterthought" in the engineering phases of systems (Mouratidis et al. 2003), owing mostly to the fact that in traditional CPSs security dangers were limited, since these systems were operated in isolation from the rest of the world. With the use of open networks, wireless technologies, the Internet and the IoT, and the cloud, the isolation of CPSs has ended, resulting in Internet-based attacks being the majority of attacks after 2001 (Byres and Lowe 2004). Unfortunately, despite the increasing connectivity of systems and the realization that the connectedness requirements of modern systems also imply an increased need for system security, the application of security-conscious design practices in industrial systems remains fragmented. Modern systems are still designed without a consistent integration of security practices or requirements into the core functional requirements identified through design (Ruiz et al. 2015). In a sense, this is not overly surprising—after all, with the ever-changing nature of cyber threats, it would be impossible to design a completely secure system that could be robust against any future type of attack. However, to ensure CPS reliability and security, it is imperative that these systems are designed from the outset with a thorough understanding and consideration of the threats and challenges that at least a *current* adversary may pose.

Although the precise form of a CPS security threat may differ, and indeed, even though it is impossible to predict the novel and ingenious types of attack that may emerge in the future, the fundamental nature of a threat remains unchanged. A security threat represents a set of circumstances with the potential to cause loss or harm (Pfleeger and Pfleeger 2006). The US National Institute of Standards and Technology defines threats more analytically as "Any circumstance or event with the potential to adversely impact organizational operations (including mission, functions, image, or reputation), organizational assets, or individuals through an information system via unauthorized access, destruction, disclosure, modification of information, and/or denial of service" (Ross et al. 2006). In a CPS, system vulnerabilities that can cause such circumstances or events might be distinguished into cyber, physical, or cyber-physical. Physical vulnerabilities include physical sabotage of equipment or jamming but also fault injection and side channel attacks (Fournaris et al. 2017a). Cyber vulnerability types include communications and communication protocols, software, and web-based attacks. Cyber-physical vulnerabilities include interconnected devices, insecure protocols, insecure Operating Systems, Software, Replay, and Injection attacks.

Summarizing, a threat is a situation where a given system is vulnerable to one or multiple *attackers*, attempting to gain access to one or more components of the *system*, in order to carry out an unauthorized *objective*. Through a review of existing literature, Lei et al. (2018) proposed a concise taxonomy of threats to the security (and by extension, reliability) of CPS, in which a specific attack can be categorized from three major perspectives, namely, its origination, purpose, and target. Furthermore, in Humayed et al. (2017) we also find a consideration about the *consequences* of a successful attack.

These definitions may seem somewhat abstract, but understanding the fundamental components of a threat allows a system engineer to begin to integrate security aspects as an intrinsic part of the requirements capture process and subsequent design choices. As a result, while designing around the requirements of a system, an engineer might begin to consider security by asking *Who* (i.e., which entities), *Why* (i.e., what objective these persons might have) and *How* (i.e., what tools or exploits will they attempt to use). The estimated consequences of an attack (we could call this the *What* aspect), allows an engineer to identify the individual components that merit security considerations, classify the severity of each threat, and therefore prioritize and inform the design work required to counter these threats (Fig.13.1). Such consequences may affect the operation of a system asset (technical consequence), business aspects (e.g., loss of confidential information, inability to respond to customer demands) and also psychological consequences both within the organization operating the CPS (in-house morale) and outside it (e.g., investor relationships, public sentiment).

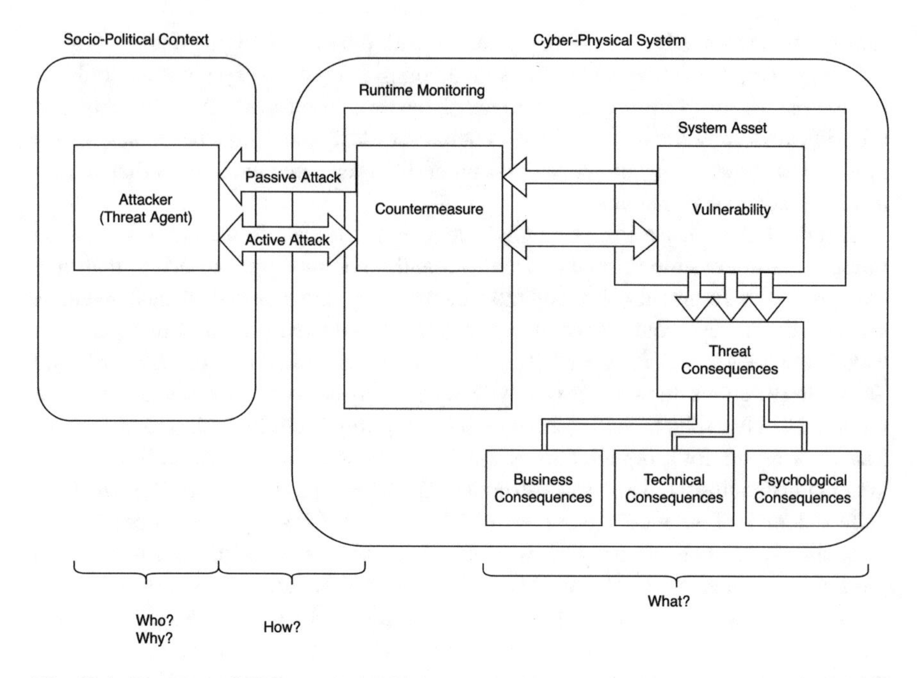

Fig. 13.1 Overview of CPS security and run-time monitoring environment. At the design phase, engineers should factor the *Who*, *Why*, *How*, and *What* aspects in the specification of system requirements. Answers to these questions feed into the design and implementation of run-time countermeasures

Concerning the "*Who*" aspect, a CPS engineer should be aware that attacks can be adversarial, initiated by malicious insiders and outsiders, that is, entities (persons or organizations) with authorized or unauthorized (illegitimate) access to the CPS. We often think of other humans as "attackers" of a system, and indeed this is a significant source of anxiety since we cannot always predict who might be interested in initiating an attack, but humans are not the only source of threats. Accidental damage caused by legitimate components (e.g., a buggy software update that may inadvertently wipe, or corrupt data under very specific and rare contexts), environmental factors including natural or man-made disasters ("act-of-God" phenomena), or simple failures, can also be the source of a threat that can affect security and reliability. Great heterogeneity of CPS components is also a source of vulnerability as they generally come from different vendors, following different paths of specification, design, implementation, and integration, involving different entities. This greater level of integration of components that CPSs are composed of brings forth the inherent vulnerabilities of all constituent components (Ericsson 2010). Thankfully, these nonhuman sources are easier to predict (and thus guard against).

In respect to the "*Why*" aspect, the motivation of attackers can be classified into cybercrime, cyber espionage, cyber terrorism, or cyber war (Setola 2011). Further threats move beyond the realm of the technical and into the sociopolitical domain (e.g., societal and legal reactions toward increased automation, surveillance, data gathering). As such, motives might include personal factors (e.g., disgruntled employees), political factors (e.g., protest against government actions, retaliation against physical attacks), and sociocultural factors (e.g., anniversaries of important or historic events). Answers to the "Why" question help a system engineer understand the context of attacks, and perhaps even anticipate their timings (Gandhi et al. 2011).

The "*How*" aspect defines the type of attack, through the medium, tools, and techniques that are used to carry out the attack. This is more formally defined as the *attack vector*. Attacks can be categorized as passive or active, in the sense that passive attacks attempt to obtain access to sensitive information without affecting the operation of the CPS, while active attacks explicitly aim to affect the CPS resources and modify the operation of the system (e.g. bringing nodes offline, or maliciously modifying their behavior). Alternatively, an attack can be classified according to the system asset that it targets. Interception attacks aim to obtain unauthorized access to information (e.g., via keylogging, packet sniffing, side channel leakage exploitation) and are, by their very nature, passive. Modification attacks target the system's integrity, by modifying control signals or sensor values for example. Interruption attacks target the system resource availability (e.g., Denial of Service Attacks), by disrupting communication, modifying software or deleting data. Finally, fabrication attacks target the system's operation by inserting non-authenticated operations, for example, fake control signals and transactions. These three latter types of attack are all considered as active attacks. These active attacks can further be viewed under two typical scenarios, random attacks (in which, e.g., an intruder blindly injects or modifies information into the CPS) and targeted attacks, where the attacker attempts to affect specific components or states of the CPS (Ding et al. 2017).

13.2.2 Approaches to Preventing and Detecting CPS Attacks

As discussed previously, compared to pure software systems, the design of a CPS needs to be concerned with a mixture of physical, cyber, and cyber-physical threats. To mitigate the effects of such attacks, a CPS may be designed so as to implement preventative measures at both the physical level (e.g., restricting physical access to certain assets, monitoring employee, contractor or asset presence, maintaining backup copies in physically remote locations) and also the cyber levels (e.g., enforcing strong cryptography, software certification and user/service authentication measures). Some measures involve a hybrid cyber-physical approach, for example, securing physical access to assets can be implemented using both physical equipment (e.g., gates, door locks) and cyber equipment (e.g., RFID employee badges, smart locks, sensor equipment). Other examples might include the physical location of networking equipment, which in turn, dictates the type of networking equipment that can be used, and thus the type of cybersecurity mechanisms that need to be implemented on top of the network layer. For example, at the design phase, the engineer may have to choose between wired and wireless communication technologies for certain assets, depending not just on the physical properties of the environment where these assets might be located, but also on the likelihood of potential types of attack that exploit weaknesses of the network types. In the next sections, we discuss a few core considerations during the design of secure CPSs and present common approaches that the engineer might consider during the design phase.

13.2.2.1 Prevention Strategies

Communication networks are perhaps the most obvious attack target for connected CPSs. The obvious approach is to prevent such attacks by strictly securing all communications, data, and access to various subcomponents, but this approach introduces problems at scale, given the required complexity and generated overheads. This is because many of the participating IoT devices (such as sensors) do not have the computing capability required to implement such measures, or at the very least, not with the required lack of delay (Zhang et al. 2016). Moreover, given the heterogeneity of large-scale CPSs, the different operating characteristics and configurations (often left at "defaults") of commercial off-the-shelf components that make up parts of a CPS can be easily exploited for attacks, or equally may make it difficult to integrate them into a comprehensive security framework (Humayed et al. 2017). To circumvent the problem, it has been demonstrated that securing intelligently selected subsets of data, or strategically introducing secure infrastructure at key areas of a CPS, like hardware security tokens (Fournaris et al. 2017b), may act as a strong deterrent, especially in large-scale systems such as power grids, since such approaches make it extremely hard and impractical for an attacker to effect more than small and inconsequential compromises to the system's integrity (Kim and Poor 2011).

13.2.2.2 Detection Strategies

The design and integration of preventive strategies can add a strong deterrent to the security of a networked CPS, but it is not enough to guarantee reliable operation of a system. For example, an insider attack from a corrupted employee can completely override any network security measures and is not preventable with such means. Organized, motivated, and well-funded attackers may be able to find ways to penetrate the security of a CPS network, but there is a range of attacks that do not require penetration of a network's security (e.g., DoS through flooding), and which are able to wreak havoc on even the most heavily secured system. To secure a CPS and ensure its reliability, engineers must integrate detection mechanisms at the design phase, to continuously monitor the operation of the system and identify potential breaches of security, or threats to the reliable operation of the system, ideally in real time.

In the design of CPS, mathematical modeling of the effects of various attack types can be useful in order to derive monitoring schemes that can detect these attacks in real time. For example, network performance degradation, such as observed in denial of service attacks, is well studied in terms of its modeling and is therefore easily detectable (Pang et al. 2011; Amin et al. 2013; Befekadu et al. 2015). Other types of network attack are harder to detect and defend against, for example, in Lee et al. (2014) a framework is presented for detecting wormhole attacks. These attacks target the control signals relayed between distant sites over wireless networks, by establishing links that appear to be authentic, but in reality, work by delaying signals, or replaying previous signals. This type of attack is not preventable by cryptography, since the delayed or replayed messages are all valid.

When attackers are able to bypass network security and gain access to a networked CPS, there are still measures that can be implemented from the design phase, which can help ensure reliable operation of the system. A frequent objective of attack after penetration is to enact interruption or fabrication types of attack, threatening the integrity of CPS data. It is possible to implement simple yet effective detectors of bad data, either using simple thresholds (which the attacker cannot know in advance) or by detecting significant deviations from the expected reported states (Mo et al. 2010). Even with such measures, small changes effected by attackers may incrementally mount to large consequences in the operation of CPSs, and still, an attacker might adopt conservative strategies to minimize the chances of being detected. As such, it is apparent that real-time detection should ideally be paired with longitudinal monitoring of system behavior, in order to detect such cumulative effects on the system.

13.2.2.3 Implementing CPS Security Strategies

More recently, the rise of popularity (and accessibility) of machine learning tools has led to the recommendation for applying such techniques (especially deep learning neural networks) to detect reliability or security issues (Kriebel et al. 2018). One drawback of these approaches is that although a trained classifier can work in real time to detect threats, on the cloud, fog, or even edge level (Mamdouh et al. 2018), the training process has to be performed typically offline, and particularly so when the training data consists a large volume. Hence, such classifiers cannot be re-trained online and require multitier architectures (Khorshed et al. 2015) for their implementation (e.g., online for detection, near-line for model tuning, offline for training).

Other approaches, such as Singh et al. (2016) are more concerned not with detecting attacks, but mitigating the results of attacks by implementing a better control mechanism that is robust to a variety of network effects, whether these effects occur naturally (e.g., dynamic inadvertent changes in network operating conditions) or occur maliciously (e.g., certain types of attack).

The introduction of trusted computing as part of the security-by-design approach can also provide a proactive countermeasure against possible attacks on CPS devices. Latest processor technologies provide trusted execution environment (TEE) generation that can be used for security sensitive software execution. Such execution environments cannot be accessed by attackers to install malicious code or alter existing software code since all activities are monitored. For example, ARM offers the ARMTrustzone TEE for all its cortex A family and in some of its cortex M processor family. Dedicated hardware tokens can also be placed in non-embedded system CPS devices like Trusted Platform Modules (TPMs) in order to instill security and trust on control management subsystems of a CPS (Fournaris and Sklavos 2014).

13.2.3 Challenges to Securing Cyber-Physical Systems

There are several open problems and research challenges associated with security of CPS. First of all, there is a need to consider CPS security aspects throughout their lifecycle. This means that security should be an intrinsic property of CPS right from its design phase, rather than an issue to be dealt with at a later stage, for example, for mitigating a potential attack. Secondly, CPS security design has to consider both cyber and physical system aspects. The combination of cybersecurity and physical system theoretic security, results in better guaranteeing CPS security as both approaches are incomplete and present drawbacks (Mo et al. 2012). Yet, this approach requires a change in design principles so as to address both the cyber and physical worlds. Thirdly, the real-time nature of CPS security needs to be addressed. Real-timeliness of decision making is critical in ascertaining survivability of CPS in the case of an attack. Ensuring attack resilience of CPS mandates taking into account the physical and cyber world interactions, during the design phase (Cárdenas et al. 2011). A fourth research challenge is associated with CPS change management. CPSs represent complex systems that comprise a big number of components and systems. Changes in CPS, for instance changes in hardware, device mobility, software updating and patching or addition of capabilities, often result in different overall systems. It is a challenge to ascertain that no new vulnerabilities are introduced and that the CPS security assumptions remain valid as well as ascertain that updates are not tampered and are transmitted to CPS devices in a secure way.

Despite technical progress such as reported above, and as can generally be found in the domain of cryptography, system security and network security, a CPS introduces fundamental challenges to the holistic aspect of security and reliability, given its very nature. Starting off with the fact that we should not forget the *physical* aspect of CPSs, it is imperative that physical security measures (e.g., gates, barriers, locks, security personnel, access protocols, and policies) are designed, tested, and implemented. These physical security measures are by no means perfect and undefeatable, and their compromise can quickly lead to mounting threats in the *cyber* aspect of CPSs (e.g., manual installation of rootkits that compromise cybersecurity). Still, one challenge to be faced in the transition from classical industrial control systems to dynamic CPSs is the fact that older CPSs assumed isolation from the external world (hence burden was more heavily focused on physical, rather than cybersecurity). These legacy systems will inadvertently form part of extended, interconnected CPSs (until such time at least as they are upgraded or replaced) and therefore present an inherent security risk, since they rely on unsecure software and connection protocols (since the principle of isolation did not mandate more sophisticated approaches) (Humayed et al. 2017).

In this regard, it should be considered that a CPS may be subject to not just single types of attack, but needs to be secured against multiple types of simultaneous attacks. Given its scale, it is plausible that attackers might choose to adopt a low-detectability, low-consequence form of attack, whose effects might however amount to large cumulative problems in the operation of CPSs. Scale also aggravates the problem of attack detection, since network effects and the nonlinear dynamics of networked, adaptive, and self-configurable operations might signal unexpected deviations from normal operating parameters, which might be confused for attacks. The latter aspect of networked, adaptive, and self-configurable operation of CPSs (especially in the context of Industry 4.0 and distributed manufacturing) demonstrates the difficulty in developing robust, accurate, and integral formal models for the simulation of such systems. Quite simply, the dynamic configuration of available components, operating modes, and context-sensitive goals of CPSs make the modeling of any nontrivial such system a daunting prospect.

On the positive side, this dynamic complexity of adaptive and self-configurable CPSs makes it hard for attackers to carry out targeted attacks, since this typically requires an intimate knowledge of the system's configuration and operations (Cárdenas et al. 2011). This aspect can work well to the advantage of defending such systems, since it has been demonstrated that prior knowledge about a system's configuration (physical model) can significantly help in identifying the most critical sensors and attacks (Cárdenas et al. 2011), and of course, a priori knowledge of this configuration can help to dynamically identify and control security aspects during operation.

13.2.4 Designing Security and Reliability into CPSs Using Digital Twins

At this time, given the relative infancy of large-scale networked CPSs, it is not surprising that attempts at defining engineering and design processes for incorporating security into the design of such systems are very limited. In Neuman (2009) it is argued that for CPSs, all communication channels within an application are enumerated and analyzed for security constraints, under a domain-specific understanding that includes both physical (sensor) and external (human operator or third-party control) process channels. This analysis feeds into the relatively modern concept of a "Digital Twin": an online, cyber model of the physical processes taking place in a CPS, which can be used to simulate outcomes at decision points, or monitor the operation of a CPS in real time. In this sense, a Digital Twin can incorporate many of the online threat detection and prevention measures discussed above. To become a successful tool toward this end, a Digital Twin cannot be statically defined; instead, it needs to be dynamically generated and adapted using environment specification.

Multiple such twins can be produced for various purposes: control, monitoring, and testing, the latter particularly important for security engineers, who can exploit these models to test CPS resilience against a variety of attack vectors, without fear of risking the real CPS in operation (Eckhart and Ekelhart 2018a). Even better, the Digital Twin is something that can evolve along with the design of a CPS. As various options are explored in the design phases, they can be modeled and trialed in a partial Digital Twin model, thereby informing the design process and help engineers make better decisions in the context of security and reliability, as they work toward designing the full system.

Although the concept of a Digital Twin is promising, considering the assistance it may offer to engineers throughout the whole lifecycle of a CPS (design, operation, maintenance), there are some issues that need to be considered. The main challenge in producing a Digital Twin for a CPS is that engineers need to fully document the subsystems that comprise the CPS using a parseable formal specification language (e.g., AutomationML). Changes to existing components must also be reflected in their specification, to ensure model validity. This challenge is significant, as it requires considerable additional effort at the design stage of a CPS, and also there is no guarantee that components (e.g., third-party services, off-the-shelf assets) that are later added to the CPS will come with such documentation; therefore, it might be challenging to integrate these new components into an existing Digital Twin model. Even if a Digital Twin model was able to fully capture the operation of a CPS, the model cannot display complete fidelity to the real world, since observed effects in real-world operation are often nonlinear and chaotic, hence may not be so accurately modeled (e.g., signal noise, network latency, power grid fluctuations).

Further from preliminary steps toward the specific use of Digital Twins as tools for security and reliability analysis, little other progress has been made in generalizing the use of Digital Twins as engineering design and lifecycle monitoring tools, as acknowledged in Bécue et al. (2018). However, it is envisioned that digital twins that act as a testbed for real large-scale CPSs can enhance the potential of identifying and correctly applying strategic security approaches, possibly through carrying out adversarial exercises (red team vs. blue team, attackers vs. defenders) across the security engineers and personnel involved in the operation of a CPS. In that sense, the CPS users can be trained on the security aspects of their CPS using trial and error without affecting the reliability of the actual system. Also, CPS Digital Twins can be used in order to test new security policy effectiveness before such policies are applied to the real system, thus considerably helping the construction of a tailored-made, concrete security policy for each CPS at hand.

The Digital Twin concept will undoubtedly feedback results of such exercises into the design of CPSs, helping to improve their configuration. At a time where it is uniformly accepted that our knowledge about how to best configure and secure large-scale adaptive CPSs is rather limited, the digital twin can be seen not just as a tool for posthoc evaluations of system security and reliability, but potentially as an integral part in the incremental design process of robust CPSs, from their very core conceptual phases.

13.3 Run-Time Security Monitoring

Considering the security threats and challenges that CPS have, as those described in the previous section, it becomes obvious that there is a considerable need to continuously monitor a CPS during its regular operation for security anomalies that can result to some security attack. Typical ICT systems have a series of well-developed tools that by combining a wide range of technologies and methods can detect, respond, and mitigate security attacks. The generic category of run-time monitoring systems may comprise various components like intrusion detection systems (IDS), zero-vulnerability malware detectors, and anomaly detectors that are all interconnected under a security information and event management (SIEM) system. The SIEM is usually responsible for the correlation between various events and logs to extract security alerts and make attack mitigation suggestions. However, a CPS run-time security monitoring system must consider the CPS specificities that, in several cases, are distinctly different than those of a typical ICT system.

According to Mitchell and Chen (2014), there are four basic characteristics that distinguish CPSs from typical ICT systems in terms of run-time security intrusion detection: physical process monitoring, Machine-to-Machine communications, heterogeneity, and legacy system interactions. Due to their connection between the cyber and the physical world, the CPS devices measure physical phenomena and perform physical processes that are governed by the laws of physics. Thus, a CPS security monitoring system must perform physical process monitoring, using physical laws as a control mechanism to model and predict valid instructions and outcomes. Furthermore, many CPSs application scenarios are highly focused on automation and time-driven processes that realize closed control loops, which do not require human intervention (and its associated unpredictability). This kind of behavior focused on Machine-to-Machine communications, increasing the regularity and predictability of the CPSs' activities. The CPS security monitoring system should be able to monitor regularly closed control loops. Thirdly, the attack surface of a CPS is considerably broader than that of an ICT system. CPSs consist of many heterogeneous subsystems and devices while they follow a broad range of different, not ICT related, control protocols like ISA 100, Modbus, CAN, etc. Some of these devices and protocols have proprietary software or standards that may constitute ICT attacks unfitting. This characteristic, along with the fact that a successful CPS attack has high impact and thus high payoff, attracts very skilled attackers that can mount very sophisticated attacks on CPSs (Fournaris et al. 2017a). Such attacks are usually very hard to discover and document since typical ICT intrusion detection software cannot identify them (e.g., the attacks may not be IT related but rather OT related). Attackers exploit CPS zero-day vulnerabilities, which would render many ICT security monitoring toolsets useless (e.g., knowledge-based ones (Mitchell and Chen 2014)).

Lastly, many CPSs include legacy hardware that is difficult to modify or physically access. Such components may be partially analog, have very limited installed software resources and be dictated by physical processes. The challenge here is how to install security monitoring sensors on such devices and how to predict/model their behavior correctly in order to detect possible anomalies. It needs also to be considered that legacy devices do not have many computational resources and it becomes hard for the monitoring system to retain its real-time responsiveness when collecting security metrics from them.

Run-time security monitoring in the CPS domain, considering the above specificities can take various forms. However, they all rely on two core functions, the collection of data from various CPS sources and the analysis of data in a dedicated run-time security monitoring subsystem. To achieve appropriate data collection, the security monitoring system must deploy security agent software/hardware (Fournaris et al. 2018) on the monitored CPS devices, or introduce virtual entities (Virtual Machines) for data collection (Eckhart and Ekelhart 2018a) within the CPS infrastructure. Examples of collected data can be Syslog log events, system call logs, traffic recordings from network interfaces, reputation scores, processing loads, connection/communication failures, etc. All collected data are analyzed in the CPS run-time security monitoring system that uses data mining, machine learning, pattern recognition or statistical data analysis to extract metrics on security issues that may take place inside the CPS at run-time. Such issues may be possible incidents detected via data that can be binarily characterized as bad/good, or continuously characterized by a specific significance grade. The performance of the security monitoring system is measured by the False Positive Rate (FPR), the False Negative Rate (FNR), and the True Positive Rate (TPR). The system is also measured in terms of incident detection latency and consumed resources number, computational overhead, excessive network traffic, and power consumption (Mitchell and Chen 2014).

To better understand the monitoring/detection approach that run-time security monitoring systems follow, we can broadly identify two approach categories: knowledge-based detection and behavioral-based approaches. In a knowledge-based security monitoring system run-time, features that are extracted from collected data are matched with a specific profile pattern or model. Alarms are raised when there is a behavior mismatch with the existing profiles or models. This approach may lead to low FPR, but needs a very well-described profile or model to be effective (e.g., an attack dictionary, a CPS device functionality pattern) since it relies on identifying a specific pattern/model.

On the other hand, behavior-based security monitoring systems do not rely on a specific prescribed knowledge but rather look for run-time features that seem out of the ordinary and act as outlier values to the expected behavior of a CPS. Supervised, semi-supervised, or unsupervised machine learning algorithms can be employed on this approach. As expected, in supervised and semi-supervised algorithms a predefined training set must be constructed in such a way that it reflects accurately the expected CPS behavior. Given the CPS specificities, this is a nontrivial task. It takes a lot of time and effort to structure such a dataset (e.g., using state-of-the-art feature analysis, discovery, and engineering techniques) and still the behavior-based monitoring may result in high FPR. Unsupervised behavior-based monitoring does not need a prestructured training set and creates the dataset using CPS live data (Mitchell and Chen 2014). The behavior pattern that the above approaches evaluate can be a deviation from good behavior or a match to bad behavior (Khan et al. 2016). Bad behavior matching monitors detect attacks by building profile(s) of known bad system behavior, such as statistical profiles of attacks (Hodge and Austin 2004; Paxson 1998). Such monitors are robust since machine learning techniques tend to generalize from the presented data (Khan et al. 2016). On the other hand, good behavior deviation monitors build a statistical profile of normal (good) behavior and detect deviations from this profile (Lakhina et al. 2005; Watterson and Heffernan 2007). Their robustness is better than that of bad behavior monitors since their employed machine learning techniques do not rely on historical knowledge of possible attacks (Khan et al. 2016).

There are several CPS security monitoring systems that consider some of the distinguishing CPS characteristics in their design, like the works in Kane (2015) and Koopman and Wagner (2016), which are focused on closed control loop monitoring in autonomous computing systems and on traditional network traffic monitoring. Specifically, for industrial network run-time security monitors, there are solutions that take advantage of the physical process measuring taking place in an industrial site as well as the closed control loop processes (Qin 2012), but they still use techniques based on traditional network traffic monitoring. For example, the ARMET (Khan et al. 2018) system can identify good behavior deviations in a reliable way that has very low FPR and FNR. ARMET can observe at run-time an application's execution, compare it against the predicted execution behavior, and identify deviations.

When it comes to security run-time monitoring based on knowledge-based approaches using models, there is a need for some model description language that can take into account CPS characteristics like real-time responsiveness (Blum and Wasserman 1994). Barnett et al. (2003) proposed the use of AsmL as an executable specification language for run-time monitoring. AsmL, an extension of Abstract State Machines (ASM), is based on the formalism of a transition system whose states are first-order algebras (Börger and Stärk 2012). In Chupilko and Kamkin (2013), a full framework for executing specifications of real-time systems is proposed. This proposal can be used for security run-time monitoring in CPS timed systems.

There are very few CPS security run-time monitoring systems that provide efficiency metrics as are specified at the beginning of this section (Kane 2015). There exist works where such results are provided but only for CPS monitoring subsystems like IDSs (Khan et al. 2016).

What needs also to be mentioned is the fact that existing solutions on CPS security monitoring are primarily focused on detecting computational and network security incidents happening in a CPS. However, since a CPS implements closed control loops that rely on collected data for autonomic decision-making, malicious attacks on the collected data can also constitute a very serious threat. Recently, effort has been invested in detecting false data injection (FDI) attacks that aim to maliciously alter the CPS control loops. Research works aiming to provide protection against FDI are focused on making efficient vulnerability analysis like the work in Khan et al. (2016) where vulnerability to FDI is expressed as a satisfiability problem and solved using a solver that supports functions over real numbers (Gao et al. 2013) or focused on utilizing appropriately FDI fault diagnosis techniques (Rigatos 2015, 2016).

It is important here to note that the full potentials of a run-time security monitoring system are not unrelated with the security-by-design principles described in the previous section. A very important aspect of any monitoring tool is the mechanism that provides input to such a tool. As mentioned, in security monitoring tools inputs are provided by event data collection points (security agents or sensors) that are installed in various parts of a CPS. It is of prime importance that these security sensors are designed and realized in the CPS architecture during the engineering phase (design time) and that they are fully integrated with the CPS architecture. Only then can such sensors maximize their efficiency (in terms of speed but also in terms of impact) on collecting all security-related information that may trigger run-time security anomalies.

As will be explored in detail in the following chapters of this book, Digital Twins can enhance CPS security monitoring mechanisms. It has already been proposed as a tool to provide additional security in a CPS system by testing security components in complex CPSs (Eckhart and Ekelhart 2018b; Tauber and Schmittner 2018; Damjanovic-Behrendt 2018). A framework providing a security-aware environment for Digital Twins is described in Eckhart and Ekelhart (2018a, b), demonstrating, among others, how security and safety rules can be monitored in security-relevant use cases. The framework is extended in Eckhart and Ekelhart (2018c) by a specification-based, physical device state replication approach, by passively monitoring their inputs and outputs, showing successful detections of attacks against a CPS testbed. However, there is still no concrete proposal on how to use Digital Twins of a CPS as part of a run-time security monitoring system since Digital Twins is a relatively new modeling approach and it has considerable complexity making it hard to be integrated in a monitoring tool. However, conceptually, Digital Twins can be an integral part of security monitoring since it provides a trusted environment for testing real inputs without the possible presence of malicious entities. In that sense, a Digital Twin of a CPS can act as a trusted replica of the actual system where good behavior can be modeled and evaluated as well as device patterns can be described and trusted. Following the knowledge-based or behavior-based CPS run-time monitoring approaches, we can use the Digital Twin virtual environment to match collected data and its behavioral patterns with the known good behavior of the Digital Twin. Alternatively, Digital Twins can be used in order to construct in a safe, virtual environment, training datasets for machine learning algorithms that are used during CPS run-time security monitoring.

The industrial sector is beginning to understand the security benefits and potentials of Digital Twins in the industrial CPSs. As such, large companies have announced their plans to launch relevant products based on the Digital Twins concept. The announcement of General Electric's (GE) Digital Ghost, which is a combination of GE's Digital Twin efforts and industrial control technologies, as a means to thwart cyber-attacks (Dignan 2017) constitutes an indication of the above interest.

13.4 CPS Security Use Cases

In this section, we focus more specifically on a CPS use case, targeting Industrial Control Systems (ICS) and describe state-of-the-art approaches in the application of CPS security and reliability design for such systems, in order to concretely demonstrate the integration of the aspects discussed in the preceding chapters. An ICS is a subclass of CPS that is associated primarily with the manufacturing sector, yet is increasingly utilized for control and management of critical infrastructures. It thus covers a wide area of application use cases such as energy smart grids, transport systems, water management systems apart from the pure industrial manufacturing domain as its name implies. ICS constitute the infrastructure of the so-called Operational Technology (OT), comprising control equipment (Programmable Logic Controllers (PLCs), Network Controllers (NCs), and robot controllers), supervisory control and data acquisition systems (SCADA), and their industrial networking infrastructure. OT has a different path of evolution with reference to Information Technology (IT) systems, as it addresses quite different end user needs. The interoperable convergence of OT and IT is an emerging challenge, as ICSs are viewed in the context of the emergence of the Industrial Internet of Things (IIoT) both in the pure industrial manufacturing domain, and other application domains taking advantage of OT like critical infrastructures or healthcare (Serpanos and Wolf 2017).

Industry 4.0 (Schweichhart n.d.) is a high-tech strategy of the German government that promotes computerization in manufacturing. It tries to bridge the two worlds of OT and IT to enable higher, more flexible and efficient productivity in the manufacturing sector and more services. Its reference architecture RAMI 4.0 comprises three axes related to Hierarchy, Architecture, and Product Lifecycle (Fig. 13.2). The hierarchical axis actually dissolves the traditional multilevel hierarchy in the manufacturing domain to a flat and flexible hierarchy that distributes functionalities to devices and equipment in a Smart Factory producing smart products and being connected to the world.

The Industrial Internet Consortium (Lin et al. 2017a) has also developed a reference architecture, the Industrial Internet Reference Architecture (IIRA) (Fig. 13.3). IIRA is applicable to ICS that are related to different domains ranging from manufacturing to transportation to energy and healthcare. IIRA Functional Viewpoint focuses on functional components, structure and interrelation, interfaces and interactions. There are efforts for mapping between IIRA and RAMI 4.0 recognizing the commonalities between the two reference architectures (Lin et al. 2017a, b) (Fig. 13.4).

Different types of threats are relevant for ICSs: criminal, financial, political, and physical. For each threat five factors may be identified, namely, source, target, motive, attack vector, and consequences (Humayed et al. 2017). A criminal threat has as a potential consequence ICS application disruption of operation through its remote control utilizing wireless connectivity (vector) by an attacker (source). A financial threat leads to financial losses (consequence) of a utility (target) through false data

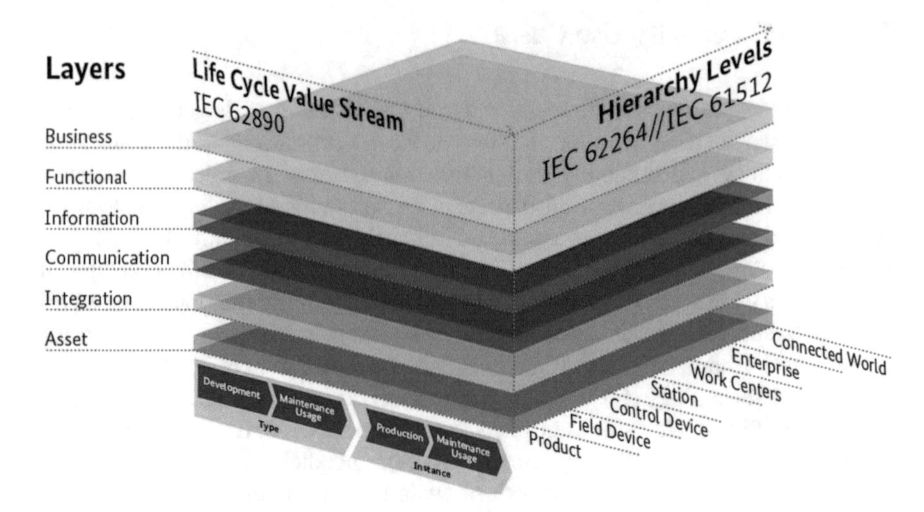

Fig. 13.2 Reference Architectural Model Industry 4.0 (RAMI 4.0) (Schweichhart n.d.)

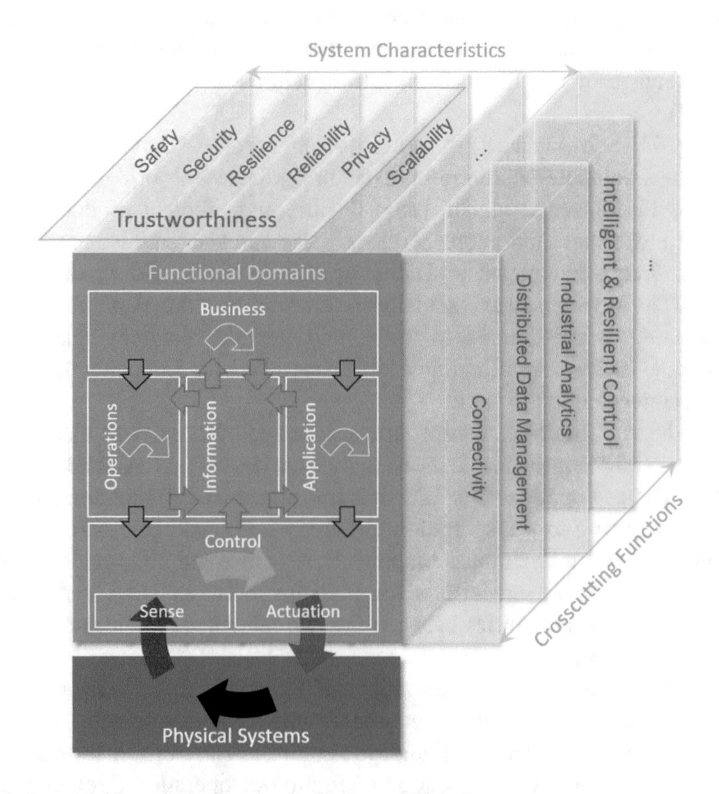

Fig. 13.3 Industrial Internet Reference Architecture (IIRA) (Lin et al. 2017a)

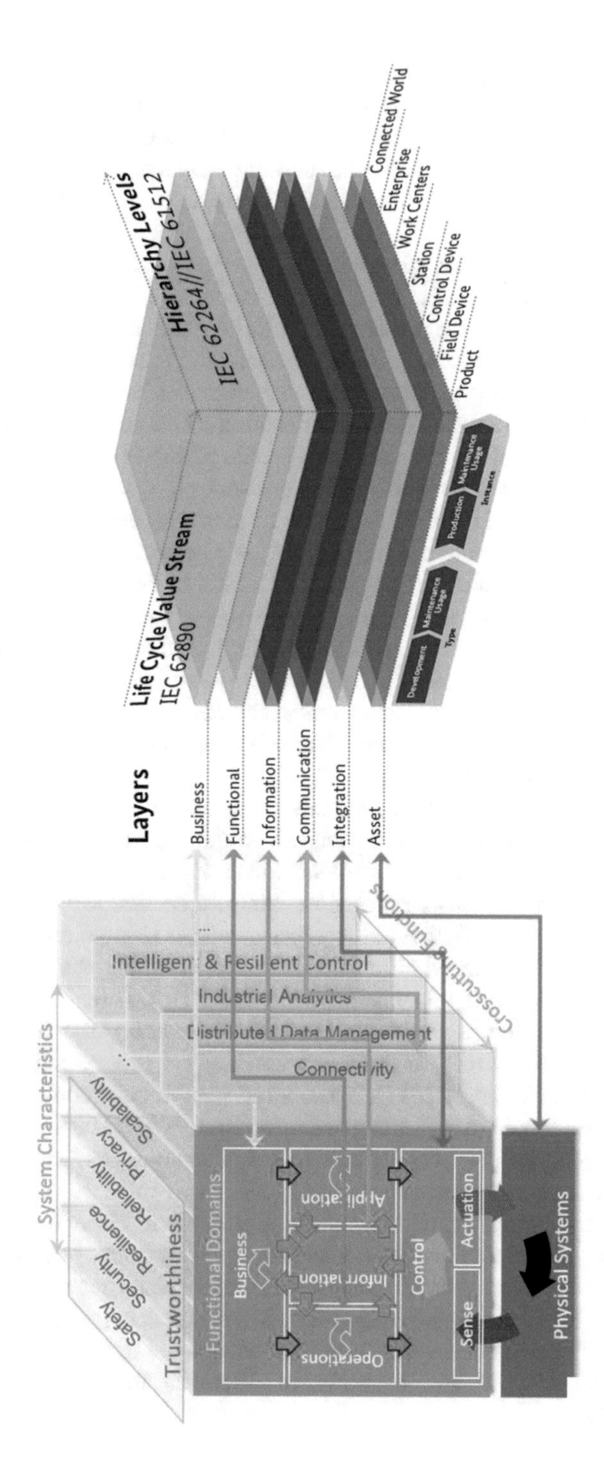

Fig. 13.4 Architecture Alignment between RAMI 4.0 and IIRA (Lin et al. 2017a, b)

injection or tampering (vector) by a customer (source) (Turk et al. 2005). Critical infrastructures of a targeted nation are usually attacked by a hostile nation (source) in political threats utilizing to this end access to ICS devices or malware and leading to sabotage actions or environmental destruction (Ryu et al. 2009). A physical threat is of the type of sensor input spoofing like the spoofing of optical flow cameras (target) in UAVs possible even for not sophisticated adversaries (source) (Davidson et al. 2016).

ICSs present different *cyber*, *physical*, and *cyber-physical* vulnerabilities. The geographically distributed nature of ICS especially with reference to Critical Infrastructure applications, for example, road infrastructure, and physical exposure of its different components creates a physical vulnerability, as inability to physically secure each and every component makes them vulnerable to tampering or sabotage.

ICS cyber vulnerabilities include communication and software vulnerabilities. Communication vulnerabilities are related to specific protocols utilized in ICSs and their inherent vulnerabilities, as well as the communication physical medium. Use of open protocols, like TCP/IP, not intended to be secure by design, raises security issues (Bellovin 1989). Remote procedure call (RPC) protocol vulnerabilities have contributed to the Stuxnet attack (Langner 2011). Man in the Middle attacks are a vulnerability of both wired and wireless systems (Hwang et al. 2008; Francia III et al. 2012). False data injection, capturing traffic, attacking employee probably unsafe personal devices connected to ICS network are just a few of the vulnerabilities. A taxonomy for wireless communication vulnerabilities is presented in Welch and Lathrop (2003). Software vulnerabilities include SQL injection (Halfond et al. 2006) and email-based attacks (Fovino et al. 2009).

ICS cyber-physical vulnerabilities include communication, OS, and software vulnerabilities. Communication vulnerabilities comprise vulnerabilities of the protocols used for ICS component communications. For instance, widely used in ICSs, the Modbus protocol lacks encryption, integrity checks, and authentication measures, making it vulnerable to different types of attacks (Byres and Lowe 2004). MTF-Storm fuzzer has evaluated different Modbus implementations and identified issues with all of them (Katsigiannis and Serpanos 2018). Direct access to ICS devices, left with default passwords (Mo et al. 2012), or being directly connected to the Internet (Leverett 2011), or through secondary emergency communication channels (Alcaraz and Zeadally 2013), is also an ICS communication vulnerability. OS vulnerabilities comprise real-time operating system (RTOS), used by ICS devices, and general-purpose OS, where ICS applications run, vulnerabilities. Absence of access control mechanisms in RTOS makes them vulnerable (Igure et al. 2006) to authentication and confidentiality failures. General-purpose OS vulnerabilities are exploitable for attacks as proven in the case of Stuxnet attack, exploiting Windows Print Spooler Service and Server Service vulnerabilities (Chen and Abu-Nimeh 2011). Software vulnerabilities are associated with ICS devices reduced computational resources that limit their capacity to enforce cryptographic measures (Langner 2011). Existing backdoors in ICS devices further facilitate attacks (Santamarta 2012).

A solution toward ascertaining that ICS applications are free from vulnerabilities while meeting their requirements is developing reliable and secure applications by design. The challenge is to derive in a formal way the ICS application implementation starting from a declarative specification. This challenge is quite ambitious as it has to take into account both continuous and discrete system behaviors. Different approaches in literature address security by design (Yang et al. 2012; Zhang et al. 2015; Martinelli and Matteucci 2007; Matteucci 2007) or reliability by design (Soulat 2014). An approach for reliable and secure by design ICS applications (Khan et al. 2018) based on deductive synthesis (Delaware et al. 2015) utilizes stepwise refinements of declarative specifications. In this context, an initial nondeterministic specification is refined into a fully deterministic efficient, correct, and secure implementation. As an example, the Coq proof assistant (Barras et al. 1997) is used to encode ICS behavior declarative specification based on an abstraction relation specification (Hoare 1978).

The reliable and secure by design ICS applications is coupled by run-time security monitoring utilizing to this end both the specification and implementation of the ICS application (Khan et al. 2016). The specification models normal behavior of ICS resources, data, and control flow between their submodules as well as misbehavior (bad behavior). The divergence between anticipated normal behavior and misbehavior of submodules of an ICS is an indication of potential attacks. Hypothetical attacks can be used for diagnostic reasoning, increasing robustness of the approach.

13.5 Research Directions

Part of the current challenges in certain dimensions have been initially discussed above, referring mainly to the evolution path of core CPS engineering disciplines that should widely adopt and adapt modern model-based and formal methods that should incorporate security aspects from the system specification and design phases. Looking into the core aspects of existing approaches, there are two main research directions identified as of paramount importance.

The first direction involves the formal description methods for the principal, as well as for the emerging composable system properties and the exploitation of automatic executable model transformation and component code generation methods and tools in order to create security monitor(s) from a distributed system specification. While there is a significant progress on the available knowledge and tools, these are typically applicable to provable correct autonomous computing nodes for the detection of specific attack types (Khan et al. 2018). Their applicability though to complex distributed systems is not straightforward, considering the introduction of additional system properties that are contributing to the system-wide correctness condition, as well as the introduction of additional attack surfaces that need to be handled altogether, and not on an isolated basis, both contributing to an exponential complexity increase for the detection of multiple, concurrent attacks. Furthermore, the existence of other nonfunctional system properties like the timeliness constraints and critical dependencies in a distributed CPS, pose substantial challenges to the engineering of robust program synthesis and code generation techniques and modules. Needless to say, the unpredictability of human-in-the-loop scenarios (Folds 2015) adds to the complexity of formally modeling CPSs, and while CPS literature has, so far, largely avoided the subject, it is hard to see how it can be completely disregarded in many CPS use cases, where human operators play a vital role to critical and safety-sensitive systems' behavior (e.g., intelligent transportation and aviation).

Then, there is the quest for the consolidation, or even more, the standardization of basic run-time frameworks, component libraries, and subsystem interfaces that will ease the deployment of interoperable customized components into generic, domain-specific solutions and architectural frameworks (Koulamas and Kalogeras 2018). These include the prevalent reference architectures of Industry 4.0 and IIC, and the challenge is enlarged after considering the widening of the Digital Twins deployment alternatives, stemming from the expected evolution of AI capabilities in embedded devices at the edge (Koulamas and Lazarescu 2018), in contrast to the typical cloud-based paradigm of today. That is, the security designing and run-time monitoring becoming a distributed system procedure on its own terms, opening then other research challenges on the necessary computing hardware and networking hardware and software support for the isolation of the two concurrently executed distributed systems, such as the exploitation and integration of results from trusted execution environments and trusted platform modules (TPM) or from time-sensitive networking (TSN) research.

Acknowledgments This work is supported by the project "I3T—Innovative Application of Industrial Internet of Things (IIoT) in Smart Environments" (MIS 5002434) implemented under the "Action for the Strategic Development on the Research and Technological Sector," funded by the Operational Programme "Competitiveness, Entrepreneurship and Innovation" (NSRF 2014–2020) and co-financed by Greece and the European Union (European Regional Development Fund).

References

Alcaraz, C., & Zeadally, S. (2013). Critical control system protection in the 21st century. *Computer, 46*, 74–83.

Amin, S., Schwartz, G. A., & Shankar Sastry, S. (2013). Security of interdependent and identical networked control systems. *Automatica, 49*, 186–192. https://doi.org/10.1016/j.automatica.2012.09.007.

Baheti, R., & Gill, H. (2011). Cyber-physical systems. *The Impact of Control Technology, 12*, 161–166.

Barnett, M., & Schulte, W. (2003). Runtime verification of net contracts. *Journal of Systems and Software, 65*, 199–208.

Barras, B., Boutin, S., Cornes, C., Courant, J., Filliatre, J.-C., Gimenez, E., et al. (1997). *The Coq proof assistant reference manual: Version 6.1* (PhD Thesis). Inria.

Bécue, A., Fourastier, Y., Praça, I., Savarit, A., Baron, C., Gradussofs, B., et al. (2018). CyberFactory#1—Securing the industry 4.0 with cyber-ranges and digital twins. In *2018*

14th IEEE International Workshop on Factory Communication Systems (WFCS) (pp. 1–4). https://doi.org/10.1109/WFCS.2018.8402377.

Befekadu, G. K., Gupta, V., & Antsaklis, P. J. (2015). Risk-sensitive control under Markov modulated denial-of-service (DoS) attack strategies. *IEEE Transactions on Automatic Control, 60*, 3299–3304. https://doi.org/10.1109/TAC.2015.2416926.

Bellovin, S. M. (1989). Security problems in the TCP/IP protocol suite. *ACM SIGCOMM Computer Communication Review, 19*, 32–48.

Blum, M., & Wasserman, H. (1994). Software reliability via run-time result-checking. *Journal of the ACM*. Citeseer.

Börger, E., & Stärk, R. (2012). *Abstract state machines: A method for high-level system design and analysis*. Cham: Springer.

Byres, E., & Lowe, J. (2004). The myths and facts behind cyber security risks for industrial control systems. In *Proceedings of the VDE Kongress* (pp. 213–218). Citeseer.

Cárdenas, A. A., Amin, S., Lin, Z.-S., Huang, Y.-L., Huang, C.-Y., & Sastry, S. (2011). Attacks against process control systems: Risk assessment, detection, and response. In *Proceedings of the 6th ACM Symposium on Information, Computer and Communications Security, ASIACCS'11* (pp. 355–366). New York: ACM. https://doi.org/10.1145/1966913.1966959.

Chen, T., & Abu-Nimeh, S. (2011). Lessons from stuxnet. *Computer, 44*, 91–93.

Chupilko, M., & Kamkin, A. (2013). Runtime verification based on executable models: On-the-fly matching of timed traces. *ArXivPrepr*. ArXiv13031010.

Damjanovic-Behrendt, V. (2018). A digital twin architecture for security, privacy and safety. *ERCIM NEWS, 115*, 25–26.

Davidson, D., Wu, H., Jellinek, R., Singh, V., & Ristenpart, T. (2016). Controlling UAVs with sensor input spoofing attacks. In *10th USENIX Workshop on Offensive Technologies (WOOT 16)*.

Delaware, B., Pit-Claudel, C., Gross, J., & Chlipala, A. (2015). Fiat: Deductive synthesis of abstract data types in a proof assistant. In *ACM SIGPLAN notices* (pp. 689–700). New York: ACM.

Dignan, L. (2017). *GE aims to replicate digital twin success with security-focused digital ghost*. ZDNet.

Ding, D., Wei, G., Zhang, S., Liu, Y., & Alsaadi, F. E. (2017). On scheduling of deception attacks for discrete-time networked systems equipped with attack detectors. *Neurocomputing, 219*, 99–106. https://doi.org/10.1016/j.neucom.2016.09.009.

Eckhart, M., & Ekelhart, A. (2018a). Towards security-aware virtual environments for digital twins. In *Proceedings of the 4th ACM Workshop on Cyber-Physical System Security, CPSS'18* (pp. 61–72). New York: ACM. https://doi.org/10.1145/3198458.3198464.

Eckhart, M., & Ekelhart, A. (2018b). Securing cyber-physical systems through digital twins. *ERCIM NEWS, 115*, 22–23.

Eckhart, M., & Ekelhart, A. (2018c). Aspecification-based state replication approach for digital twins. In *Proceedings of the 2018 Workshop on Cyber-Physical Systems Security and Privacy* (pp. 36–47). New York: ACM.

Ericsson, G. N. (2010). Cyber security and power system communication—Essential parts of a smart grid infrastructure. *IEEE Transactions on Power Delivery, 25*, 1501–1507.

Folds, D. J. (2015). Human in the loop simulation. In *Modeling and simulation in the systems engineering lifecycle* (pp. 175–183). London: Springer.

Fournaris, A. P., & Sklavos, N. (2014). Secure embedded system hardware design–a flexible security and trust enhanced approach. *Computers and Electrical Engineering, 40*, 121–133.

Fournaris, A. P., Pocero Fraile, L., & Koufopavlou, O. (2017a). Exploiting hardware vulnerabilities to attack embedded system devices: A survey of potent microarchitectural attacks. *Electronics, 6*, 52.

Fournaris, A. P., Lampropoulos, K., & Koufopavlou, O. (2017b). Hardware security for critical infrastructures-the CIPSEC project approach. In *2017 IEEE Computer Society Annual Symposium on VLSI (ISVLSI)* (pp. 356–361). IEEE.

Fournaris, A. P., Lampropoulos, K., & Koufopavlou, O. (2018). Trusted hardware sensors for anomaly detection in critical infrastructure systems. In *Modern Circuits and Systems Technologies (MOCAST), 2018 7th International Conference* (pp. 1–4). IEEE.
Fovino, I. N., Carcano, A., Masera, M., & Trombetta, A. (2009). An experimental investigation of malware attacks on SCADA systems. *International Journal of Critical Infrastructure Protection, 2*, 139–145.
Francia, G., III, Thornton, D., & Brookshire, T. (2012). Cyberattacks on SCADA systems. In *Proceeding of the 16th colloquium for Information Systems Education* (pp. 9–14).
Gandhi, R., Sharma, A., Mahoney, W., Sousan, W., Zhu, Q., & Laplante, P. (2011). Dimensions of cyber-attacks: Cultural, social, economic, and political. *IEEE Technology and Society Magazine, 30*, 28–38. https://doi.org/10.1109/MTS.2011.940293.
Gao, S., Kong, S., & Clarke, E. M. (2013). dReal: An SMT solver for nonlinear theories over the reals. In *International Conference on Automated Deduction* (pp. 208–214). New York: Springer.
Gollmann, D. (2012). Security for cyber-physical systems. In *International doctoral workshop on Mathematical and Engineering Methods in Computer Science* (pp. 12–14). New York: Springer.
Halfond, W. G., Viegas, J., Orso, A., et al. (2006). A classification of SQL-injection attacks and countermeasures. In *Proceedings of the IEEE International Symposium on Secure Software Engineering* (pp. 13–15). IEEE.
Hoare, C. A. R. (1978). Proof of correctness of data representations. In *Programming methodology* (pp. 269–281). Springer.
Hodge, V., & Austin, J. (2004). A survey of outlier detection methodologies. *Artificial Intelligence Review, 22*, 85–126.
Humayed, A., Lin, J., Li, F., & Luo, B. (2017). Cyber-physical systems security—A survey. *IEEE Internet of Things Journal, 4*, 1802–1831. https://doi.org/10.1109/JIOT.2017.2703172.
Hwang, H., Jung, G., Sohn, K., & Park, S. (2008). A study on MITM (man in the middle) vulnerability in wireless network using 802.1 X and EAP. In *Information Science and Security, 2008. ICISS. International Conference* (pp. 164–170). IEEE.
Igure, V. M., Laughter, S. A., & Williams, R. D. (2006). Security issues in SCADA networks. *Computers & Security, 25*, 498–506.
Kane, A. (2015). *Runtime monitoring for safety-critical embedded systems.*
Katsigiannis, K., & Serpanos, D. (2018). MTF-storm: A high performance fuzzer for Modbus/TCP. In *2018 IEEE 23rd International Conference on Emerging Technologies and Factory Automation (ETFA)* (pp. 926–931). IEEE.
Khan, M. T., Serpanos, D., & Shrobe, H. (2016). A rigorous and efficient run-time security monitor for real-time critical embedded system applications. In *Internet of Things (WF-IoT), 2016 IEEE 3rd World Forum* (pp. 100–105). IEEE.
Khan, M. T., Serpanos, D., & Shrobe, H. (2018). ARMET: Behavior-based secure and resilient industrial control systems. *Proceedings of the IEEE, 106*, 129–143.
Khorshed, M. T., Sharma, N. A., Kumar, K., Prasad, M., Ali, A. B. M. S., & Xiang, Y. (2015). Integrating internet-of-things with the power of cloud computing and the intelligence of big data analytics—A three layered approach. In *2015 2nd Asia-Pacific World Congress on Computer Science and Engineering (APWC on CSE)* (pp. 1–8). https://doi.org/10.1109/APWCCSE.2015.7476124.
Kim, K.-D., & Kumar, P. R. (2012). Cyber-physical systems: A perspective at the centennial. *Proceedings of the IEEE, 100*, 1287–1308.
Kim, T. T., & Poor, H. V. (2011). Strategic protection againstdata injection attacks on power grids. *IEEE Transactions on Smart Grid, 2*, 326–333. https://doi.org/10.1109/TSG.2011.2119336.
Koopman, P., & Wagner, M. (2016). Challenges in autonomous vehicle testing and validation. *SAE International Journal of Transportation Safety, 4*, 15–24.
Koulamas, C., & Kalogeras, A. (2018). Cyber-physical systems and digital twins in the industrial IoT. *Computer, 51*(11), 95–98.
Koulamas, C., & Lazarescu, M. T. (2018). Real-time embedded systems: Present and future. *MDPI Electronics, 7*.

Kriebel, F., Rehman, S., Hanif, M. A., Khalid, F., & Shafique, M. (2018). Robustness for smart cyber physical systems and internet-of-things: From adaptive robustness methods to reliability and security for machine learning. In *2018 IEEE Computer Society Annual Symposium on VLSI (ISVLSI)* (pp. 581–586). https://doi.org/10.1109/ISVLSI.2018.00111.

Lakhina, A., Crovella, M., & Diot, C. (2005). Mining anomalies using traffic feature distributions. In *ACM SIGCOMM Computer Communication Review* (pp. 217–228). New York: ACM.

Langner, R. (2011). Stuxnet: Dissecting a cyberwarfare weapon. *IEEE Security and Privacy, 9*, 49–51.

Lee, P., Clark, A., Bushnell, L., & Poovendran, R. (2014). A passivity framework for modeling and mitigating wormhole attacks on networked control systems. *IEEE Transactions on Automatic Control, 59*, 3224–3237. https://doi.org/10.1109/TAC.2014.2351871.

Lei, H., Chen, B., Butler-Purry, K. L., & Singh, C. (2018). Security and reliability perspectives in cyber-physical smart grids. In *2018 IEEE Innovative Smart Grid Technologies - Asia (ISGT Asia)* (pp. 42–47). https://doi.org/10.1109/ISGT-Asia.2018.8467794.

Leverett, E. P. (2011). Quantitatively assessing and visualising industrial system attack surfaces. *University of Cambridge, Darwin College, 7*.

Lin, S.-W., Crawford, M., & Mellor, S. (2017a). *The industrial internet of things, volume G1: Reference architecture*. Industrial Internet Consortium.

Lin, S.-W., Murphy, B., Clauer, E., Loewen, U., Neubert, R., Bachmann, G., et al. (2017b). Architecture alignment and interoperability - An industrial internet consortium and platform industrie 4.0 joint whitepaper (No. IIC:WHT: IN3: V1.0:PB: 2017120 5).

Mamdouh, M., Elrukhsi, M. A. I., & Khattab, A. (2018). Securing the internet of things and wireless sensornetworks via machine learning: A survey. In *2018 International Conference on Computer and Applications (ICCA)* (pp. 215–218). https://doi.org/10.1109/COMAPP.2018.8460440.

Martinelli, F., & Matteucci, I. (2007). An approach for the specification, verification and synthesis of secure systems. *Electronic Notes in Theoretical Computer Science, 168*, 29–43.

Matteucci, I. (2007). Automated synthesis of enforcing mechanisms for security properties in a timed setting. *Electronic Notes in Theoretical Computer Science, 186*, 101–120.

Maurer, T. (2017). What is a digital twin? Siemens. https://community.plm.automation.siemens.com/t5/Digital-Twin-Knowledge-Base/What-is-a-digital-twin/ta-p/432960.

Mitchell, R., & Chen, I.-R. (2014). A survey of intrusion detection techniques for cyber-physical systems. *ACM Computing Surveys(CSUR), 46*, 55.

Mo, Y., Garone, E., Casavola, A., & Sinopoli, B. (2010). False data injection attacks against state estimation in wireless sensor networks. In *49th IEEE Conference on Decision and Control (CDC)* (pp. 5967–5972). https://doi.org/10.1109/CDC.2010.5718158.

Mo, Y., Kim, T. H.-J., Brancik, K., Dickinson, D., Lee, H., Perrig, A., & Sinopoli, B. (2012). Cyber-physical security of a smart grid infrastructure. *Proceedings of the IEEE, 100*, 195–209.

Mouratidis, H., Giorgini, P., & Manson, G. (2003). Integrating security and systems engineering: Towards the modelling of secure information systems. In J. Eder & M. Missikoff (Eds.), *Advanced information systems engineering* (pp. 63–78). Berlin: Springer.

Neuman, D. C. (2009). Challenges in security for cyber-physical systems. In *DHS workshop on future directions in cyber-physical systems security*.

Pang, Z. H., Liu, G. P., & Dong, Z. (2011). Secure networked control systems under denial of service attacks. In *IFAC proceedings volumes, 18th IFAC World Congress 44*, 8908–8913. https://doi.org/10.3182/20110828-6-IT-1002.02862.

Paxson, V. (1998). Bro. A system for detecting network intruders in real-time. In *Proceedings of the 7th USENIX security symposium*.

Pfleeger, C. P., & Pfleeger, S. L. (2006). *Security in computing* (4th ed.). Upper Saddle River, NJ: Prentice Hall.

Qin, S. J. (2012). Survey on data-driven industrial process monitoring and diagnosis. *Annual Reviews in Control, 36*, 220–234.

Rajkumar, R., Lee, I., Sha, L., & Stankovic, J. (2010). Cyber-physical systems: The next computing revolution. In *Design Automation Conference (DAC), 2010 47th ACM/IEEE* (pp. 731–736). IEEE.

Rigatos, G. (2015). *Differential flatness approaches to nonlinear filtering and control: Applications to electromechanical systems*. New York: Springer.

Rigatos, G. (2016). *Intelligent renewable energy systems: Modelling and control*. Cham: Springer.

Ross, R. S., Katzke, S. W., & Johnson, L. A. (2006). *Minimum security requirements for federal information and information systems*.

Ruiz, J. F., Maña, A., & Rudolph, C. (2015). An integrated security and systems engineering process and modelling framework. *The Computer Journal, 58*, 2328–2350.

Ryu, D. H., Kim, H., & Um, K. (2009). Reducing security vulnerabilities for critical infrastructure. *Journal of Loss Prevention in the Process Industries, 22*, 1020–1024.

Santamarta, R. (2012). *Here be backdoors: A journey into the secrets of industrial firmware*. Black Hat USA.

Schweichhart, K. (n.d.). *Reference architectural model industrie 4.0 (RAMI 4.0) - An introduction*.

Serpanos, D. (2018). The cyber-physical systems revolution. *Computer, 51*, 70–73.

Serpanos, D., & Wolf, M. (2017). *Internet-of-things (IoT) systems: Architectures, algorithms, methodologies*. Cham: Springer.

Setola, R. (2011). *Cyber threats to SCADA systems*.

Singh, V. P., Kishor, N., & Samuel, P. (2016). Load frequency control with communication topology changes in smart grid. *IEEE Transactions on Industrial Informatics, 12*, 1943–1952. https://doi.org/10.1109/TII.2016.2574242.

Soulat, R. (2014). *Synthesis of correct-by-design schedulers for hybrid systems* (PhD Thesis). École normale supérieure de Cachan-ENS Cachan.

Tao, F., Zhang, H., Liu, A., & Nee, A. (2018). Digital twin in industry: State-of-the-art. *IEEE Transactions on Industrial Informatics, 15*(4), 2405–2415.

Tauber, M., & Schmittner, C. (2018). Enabling security and safety evaluation in industry 4.0 use cases with digital twins. *ERCIM News*.

Turk, R. J., et al. (2005). *Cyber incidents involving control systems*. New York: CiteSeer.

Watterson, C., & Heffernan, D. (2007). Runtime verification and monitoring of embedded systems. *IET Software, 1*, 172–179.

Welch, D., & Lathrop, S. (2003). Wireless security threat taxonomy. In *Information assurance workshop, 2003. IEEE systems, man and cybernetics society* (pp. 76–83). IEEE.

Yang, J., Yessenov, K., & Solar-Lezama, A. (2012). A language for automatically enforcing privacy policies. In *ACM SIGPLAN notices* (pp. 85–96). New York: ACM.

Zhang, M., Duan, Y., Feng, Q., & Yin, H. (2015). Towards automatic generation of security-centric descriptions for android apps. In *Proceedings of the 22nd ACM SIGSAC Conference on Computer and Communications Security* (pp. 518–529). New York: ACM.

Zhang, H., Shu, Y., Cheng, P., & Chen, J. (2016). Privacy and performance trade-off in cyber-physical systems. *IEEE Network, 30*, 62–66. https://doi.org/10.1109/MNET.2016.7437026.

Chapter 14
Digital Twins for Cyber-Physical Systems Security: State of the Art and Outlook

Matthias Eckhart and Andreas Ekelhart

Abstract Digital twins refer to virtual replicas of physical objects that, inter alia, enable to monitor, visualize, and predict states of cyber-physical systems (CPSs). These capabilities yield efficiency gains and quality improvements in manufacturing processes. In addition, the concept of digital twins can also be leveraged to advance the security of the smart factory. More precisely, this concept can be applied as early as in the design phase by providing engineers the means to spot security flaws in the specification of the CPS. Security testing or intrusion detection are other security-enhancing technical use cases of digital twins that can be realized in systems engineering or during plant operation. In this chapter, we will discuss how digital twins can accompany their physical counterparts throughout the entire lifecycle and thereby strengthen the security of CPSs. The findings of this chapter indicate that the concept of digital twins will open up new paths to secure CPSs. However, efficiently creating, maintaining, and running digital twins still represents a major research challenge, as the overhead costs hinder the adoption of this concept. We believe that these insights are valuable to shape future research in this emerging research area at the intersection of digital twins and information security.

Keywords Digital twin · Information security · Cyber-physical systems · Industrial control systems · Digital thread

14.1 Introduction

Cyber-physical systems (CPSs) are essential for the realization of the Industry 4.0 vision (Kagermann et al. 2013), owing to their capabilities that blend physical and virtual components in order to interface both worlds (Baheti and Gill 2011). While

M. Eckhart (✉) · A. Ekelhart
Christian Doppler Laboratory for Security and Quality Improvement in the Production System Lifecycle (CDL-SQI), Institute of Information Systems Engineering, Technische Universität Wien, Vienna, Austria

SBA Research, Vienna, Austria
e-mail: matthias.eckhart@tuwien.ac.at; andreas.ekelhart@sba-research.org

S. Biffl et al. (eds.), *Security and Quality in Cyber-Physical Systems Engineering*,
https://doi.org/10.1007/978-3-030-25312-7_14

these systems interact through sensors and actuators with the physical (real) world, the computational and networking elements allow them to function in the digital (cyber) space (Baheti and Gill 2011). In this way, physical processes in a variety of sectors (e.g., health care, energy, transportation (Shi et al. 2011)) can be fully automated but also operated in an intelligent fashion, leading to the emergence of multiple *smart* applications, e.g., *smart grid* and *smart factory*. In fact, CPSs are even considered the next computing revolution (Rajkumar et al. 2010).

Given that CPSs can impact the physical as well as the digital world, ensuring that these systems operate in a *secure* and *safe* manner is paramount. Multiple prominent cyber attacks against industrial control systems (ICSs), which we consider a subset of CPSs, have demonstrated how severe the consequences of these incidents can be. To give an example, an attack launched against the Ukrainian power grid in 2015 disconnected several substations, causing a power outage that affected approx. 225,000 households (Lee et al. 2016). As a result, successfully attacking ICSs, due to a lack of adequate security measures, can even represent a threat to public safety.

As the interconnectivity of ICSs increases in light of Industry 4.0 (Kagermann et al. 2013) and Information Technology (IT) and Operational Technology (OT) gradually converge (Hahn 2016), the attack surface expands substantially. This is also reflected in the past annual reports published by ICS-CERT (2017, 2015, 2013), as the reported incidents increased significantly over the past years.[1] The main reason for the increased susceptibility to security issues of ICSs is the fact that IT and OT are driven by different challenges and, in further consequence, pursue different objectives. While the typical (business) IT systems tend to place more weight on the confidentiality and integrity of data, OT systems (i.e., ICSs) primarily focus on the availability of industrial operations (Knowles et al. 2015). For instance, most industrial network protocols have not been designed with security in mind but rather focus on reliability and meeting real-time requirements (Knapp and Langill 2014). Thus, ICSs often rely on the *security through obscurity* principle (McLaughlin et al. 2016). A recent study conducted by Dragos, Inc. (2018) supports this claim, as they have found that 64% of the patches for vulnerabilities discovered in ICSs, which have been released in 2017, cannot completely remedy the found weaknesses, due to an insecure design of these systems. Consequently, security aspects must be taken into account when engineering CPSs but then also be considered in subsequent phases of the systems' lifecycle.

Recently, researchers started to explore the concept of digital twins in order to implement security-enhancing technical use cases for CPSs (Bécue et al. 2018; Bitton et al. 2018; Eckhart and Ekelhart 2018c,b,a; Tauber and Schmittner 2018; Damjanovic-Behrendt 2018a; Damjanovic-Behrendt 2018b), suggesting that it may even qualify for the realization of a holistic approach to CPS security. Most of these works present a specific technical use case, such as privacy

[1]More specifically, the following numbers of ICS incidents were recorded by fiscal year, starting from 2010 to 2016: 39, 140, 197, 257, 245, 295, 290 (ICS-CERT 2017, 2015, 2013).

enhancement (Damjanovic-Behrendt 2018b), even though Eckhart and Ekelhart (2018c) give a general, brief overview of the applicability of the digital-twin concept in the CPS security context. However, little is known about the concept's full potential relating to information security as well as the research challenges that need to be addressed in order to overcome barriers to adoption. This chapter aims to fill this gap.

The contribution of this chapter is twofold and can be summarized as follows:

- We introduce the concept of digital twins for the purpose of enhancing the security of CPSs. First, we describe the origins of the digital-twin concept, discuss its use cases in the manufacturing domain, explain the term *digital thread* (Lubell et al. 2013; Singh and Willcox 2018), and clarify how it connects to digital twins. Second, we attempt to establish a coherent definition of the term *digital twin* in the context of information security and map the traditional use cases of the concept to security-related applications.
- We provide a comprehensive outlook on possible research directions worth pursuing. More precisely, we study existing work in the field and explore how current security challenges related to CPSs may be overcome by adopting the digital-twin concept.

The remainder of this chapter is structured as follows. First, in Sect. 14.2, we provide background information on the concept of digital twins and the digital thread. In Sect. 14.3, we propose a definition of the term *digital twin* in the context of information security and present technical use cases of this concept that aim to strengthen the security of CPSs. Section 14.4 suggests future research directions based on existing works in the literature. Finally, Sect. 14.5 concludes the chapter by summarizing the main findings of this work.

14.2 Background

This section introduces the concept of digital twins by first describing its origins and then explaining the concept's manifestations. Furthermore, traditional use cases of digital twins in the manufacturing domain are presented. A brief discussion on digital threads and how they relate to digital twins completes this section.

14.2.1 The Digital Twin

The concept of digital twins has attracted significant attention from both academia and industry in the past few years. In fact, Gartner has even recognized digital twins as a top strategic technology trend for 2019, ranking on place four (Panetta 2018). Upon first glance, it may seem that this term has been introduced merely for marketing purposes in order to revamp a long-established concept, namely, the use of virtual models of systems during various phases of their lifecycle (e.g., engineering). The following subsections attempt to demystify this technology buzzword.

14.2.1.1 Origins of the Concept of Digital Twins

According to Rosen et al. (2015), the concept of digital twins has its origins in NASA's Apollo program, as a twin of a spacecraft was built for two purposes, viz., (i) training before the mission and (ii) supporting the mission by mirroring flight conditions based on data coming from the spacecraft in operation. However, owing to the technological progress concerning simulations and connectivity that has been achieved in the past decades, creating twins has evolved from building physical copies to virtual models of systems (Schleich et al. 2017). As stated in (Rosen et al. 2015; Schleich et al. 2017; Negri et al. 2017), the term *digital twin* was coined by Shafto et al. (2010), who published a report that includes the following definition of the term: "A digital twin is an integrated multiphysics, multiscale simulation of a vehicle or system that uses the best available physical models, sensor updates, fleet history, etc., to mirror the life of its corresponding flying twin." While in this seminal work and a few subsequent papers (e.g., Glaessgen and Stargel 2012; Tuegel et al. 2011; Gockel et al. 2012; Reifsnider and Majumdar 2013) the focus of the digital-twin concept was on mirroring the life of air vehicles, Lee et al. (2013) introduced it to the manufacturing sector in 2013 (Negri et al. 2017). Motivated by the need to utilize machine or process data for the purpose of prognostics, Lee et al. (2013) propose to run digital twins of production systems in the cloud that simulate the conditions of their physical counterparts based on physical models. With the advent of digital twins in the manufacturing domain, the concept expanded to health monitoring, systems engineering (e.g., optimizing the development of control algorithms (Grinshpun et al. 2016)), and managing other phases of the systems' lifecycle (e.g., virtual commissioning (Schluse and Rossmann 2016)) (Negri et al. 2017). Furthermore, Ríos et al. (2015) also investigate the role of digital twins in the Product Lifecycle Management (PLM) and how the digital-twin concept relates to *product avatars* (Hribernik et al. 2006, 2013), i.e., virtual counterparts of products (Negri et al. 2017).

Given the variety of applications for digital twins, multiple interpretations of the concept exist, which is also clearly reflected by the plethora of definitions that can be found in the literature. To clear up the confusion, Negri et al. (2017) provide a comprehensive overview of definitions of the term *digital twin* that appeared in existing works. Interestingly, the authors of (Negri et al. 2017) found that papers related to the digital-twin concept, which do not touch on the simulation aspects, exist, even though it originally emerged from research in this area. Moreover, although several digital-twin-related proofs of concept have been developed (e.g., Haag and Anderl 2018; Alam and Saddik 2017; Schroeder et al. 2016a; Uhlemann et al. 2017; Vachálek et al. 2017) and some solutions are already available on the market, there seems to be still a lack of clarity about what constitutes a digital twin. Durão et al. (2018) attempt to address this issue in their recent paper by gathering requirements for the development of digital twins based on a literature review and interviews with professionals from the industry. Their findings indicate that the requirements (i) real-time data, (ii) integration, and (iii) fidelity have been addressed by most of the reviewed works, while at the same time these are the ones that are the most desired properties of industry solutions according to the interviewees. The reason for this is that real-time data that is fed into a digital twin would reflect the actual state of its physical counterpart; thus, making a seamless data integration also a crucial component of digital twins (Durão et al. 2018). Furthermore, the fidelity of digital twins indicates how precisely they mirror their physical counterparts (Durão et al. 2018). However, simulations without real-time data flows still seem to be the state of practice concerning digital twins, even though the adoption of high-fidelity simulations that are able to integrate data in real-time is envisioned for the future (Durão et al. 2018).

14.2.1.2 Types of Digital Twins

Due to the fact that the interpretation of the digital-twin concept varies among scholars as well as industry professionals and considering that the concept can be applied to solve different problems, several types of digital twins have been proposed so far. As pointed out by Kritzinger et al. (2018), a digital-twin solution is characterized by (i) its intended areas of application, (ii) the used technologies, and (iii) the data integration level. In the following, we focus on the technological characteristics and levels of data integration, as the next subsection, Sect. 14.2.1.3, is devoted to the use cases of the digital-twin concept in the manufacturing domain.

As already discussed in Sect. 14.2.1.1, the digital-twin concept emerged from advances in the field of modeling and simulation. Boschert and Rosen (2016) even declare digital twins as "the next wave in simulation technology". Over the past 50 years, the number of papers published related to simulation has steadily increased, reaching its peak between 2010 and 2014 with 5,677 published works (Mourtzis et al. 2014). Interestingly, literature analyses of simulation technology in the manufacturing domain (Negahban and Smith 2014; Polenghi et al. 2018) indicate that simulation applications for operational aspects (i.e., middle-of-life phase) attracted increasing research interest from 2002 to 2013, while interest in its applications in the beginning-of-life phase appeared to decline over the same period of time. To give a few examples of simulation applications, *system design*, *facility design/layout*, and *material handling system design* appear to be among the most used in the beginning-of-life phase of manufacturing systems (Polenghi et al. 2018). On the other hand, *operations planning*, *scheduling*, and *real-time control* are among the most used simulation applications in the middle-of-life phase (Polenghi et al. 2018). It is also worth noting that the use of simulation technology plays a little role in the end-of-life phase, even though specialized simulation applications may be vital when decommissioning entails high risks, e.g., as is the case with nuclear power plants (Polenghi et al. 2018). Considering that a plethora of simulation applications have been studied for both the design and operation phase, adopting a holistic view on how digital twins (i.e., simulation applications) can be leveraged along the systems' lifecycle represents a reasonable next step to take in the light of Industry 4.0. In fact, several works (e.g., Boschert and Rosen 2016; Schluse and Rossmann 2016; Grieves and Vickers 2017) suggest that the digital twin of a system evolves

Table 14.1 Classification based on the level of data integration according to Kritzinger et al. (2018)

Level of integration	Dataflow	
	Physical → Digital	Digital → Physical
Digital model	Manual	Manual
Digital shadow	Automatic	Manual
Digital twin	Automatic	Automatic

with its physical counterpart, meaning that fidelity tends to increase as the lifecycle progresses and, by implication, complexity too. An example of a digital twin's lifecycle is given by Schluse and Rossmann (2016), where animations of the system are created in the design phase and a discrete event simulation is then developed for examining the system's performance, followed by a rigid body simulation and a finite element method (FEM) simulation, which are used for further analysis. The authors of (Schluse and Rossmann 2016) expand their idea by proposing *experimentable digital twins*, i.e., interactive virtual replicas of systems that function in a *virtual testbed*, enabling engineers to interactively analyze the system in the environment in which it operates. In this context, 3D simulations play an important role, as accurate visual representations may facilitate certain engineering tasks. Besides adopting simulation technology for realizing the concept of digital twins, there are also a few works that do not associate it with simulation applications (Negri et al. 2017), even though the digital-twin concept evidently has its roots in this field. For example, mere visualizations (e.g., realized by utilizing augmented reality (Schroeder et al. 2016b)) or data-driven models based on machine learning methods (e.g., Jaensch et al. (2018)) are also regarded as implementations of the digital-twin concept.

Integrating data, acquired either from past lifecycles or in real-time from live systems, into virtual replicas is a cornerstone of the concept of digital twins. However, in the literature, there appears to be no consensus concerning the minimum level of data integration required for qualifying as an actual implementation of the concept. Consequently, Kritzinger et al. (2018) proposed a classification of the digital-twin concept based on how the data exchange between the virtual replica and its physical counterpart is realized. As shown in Table 14.1, the authors introduced the terms *digital model* and *digital shadow*, in addition to *digital twin*, which are defined on the basis of the data flows to and from the virtual replica. For instance, according to the definitions proposed by Kritzinger et al. (2018), a digital twin is characterized by an automated, bidirectional data exchange between the real system and its digital representation.[2]

Now that an overview of types of digital twins, which have been covered in the literature, has been provided, views from industry professionals on this topic remain

[2]In this work, we do not adopt the classification proposed in (Kritzinger et al. 2018) for the sake of simplicity, as the level of data integration plays only a secondary role for the security-related use cases.

to be discussed. As indicated in Sect. 14.2.1.1, Durão et al. (2018) conducted, inter alia, interviews with six companies to gather requirements related to the digital-twin concept. Their study reveals that, from the point of view of industry professionals, the digital-twin concept appears to be regarded as a simulation model of a physical object that does not receive data instantly or continuously (Durão et al. 2018).

14.2.1.3 Use Cases of Digital Twins in the Manufacturing Domain

Based on the literature reviews conducted by Negri et al. (2017) and Kritzinger et al. (2018), as well as the works published by Rosen et al. (2015) and Grieves and Vickers (2017), we determined the areas of application in the manufacturing domain of the digital-twin concept. In particular, Grieves and Vickers (2017) describe in detail how the concept of digital twins can be utilized in a variety of ways throughout the systems' lifecycle. Furthermore, Negri et al. (2017) identify the following three categories for use cases: (i) *monitoring* (e.g., health assessment), (ii) *mirroring the systems' life* (e.g., lifecycle management), and (iii) *decision support* (e.g., modeling, visualization, simulation, optimization). The works by Kritzinger et al. (2018) and Rosen et al. (2015) provide further details regarding the use cases of the digital-twin concept for cyber-physical production systems (CPPSs) and were therefore used supplementary to gather the areas of application in the manufacturing domain. Figure 14.1 depicts a CPPS-centric view on the areas of application without considering the product lifecycle (e.g., Ríos et al. 2015). In the following, we briefly review the role of the digital-twin concept within the three high-level phases of the CPPSs lifecycle, viz., (i) engineering, (ii) operation, and (iii) end-of-life.

In (Grieves and Vickers 2017), the authors explain how the system evolves virtually during engineering until the fabrication of its physical twin. Owing to the

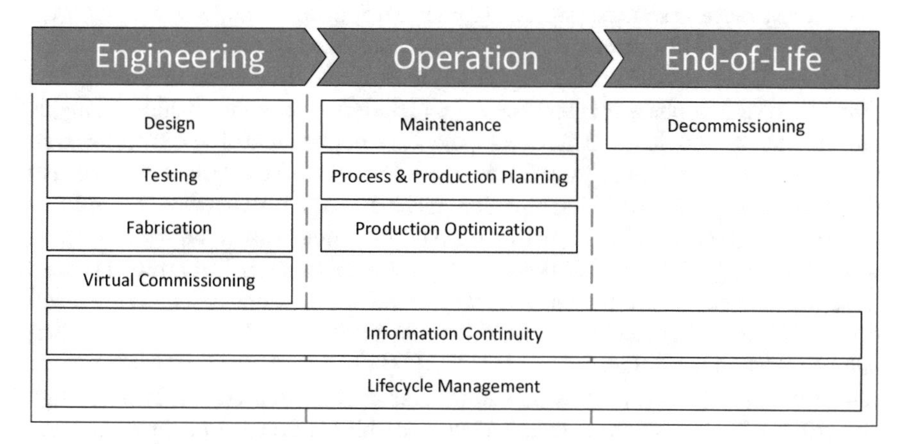

Fig. 14.1 Areas of application of the digital-twin concept based on (Negri et al. 2017; Rosen et al. 2015; Kritzinger et al. 2018; Grieves and Vickers 2017) within the lifecycle of CPPSs (inspired by Lüder et al. (2017))

use of 2D/3D models as well as physical models to simulate the behavior of systems, the efficiency of the engineering process can be drastically increased (Grieves and Vickers 2017). As already indicated in the previous section, this practice itself is not new per se, but the technological progress made in the past decades opened up new methods to develop realistic, high-fidelity models that facilitate the design, testing, fabrication, and commissioning of systems. On top of that, these models lay the foundation for supporting subsequent activities in the lifecycle (Rosen et al. 2015), making the data model an integral component of digital twins (Negri et al. 2017). Thus, efforts have been made by Schroeder et al. (2016a) to improve the modeling and exchange of digital-twin-related data by utilizing AutomationML (AML) (Drath et al. 2008), i.e., an engineering data exchange format.

Use cases of digital twins that belong to the operation phase typically rely on data coming from real systems. For instance, the health of the system can be continuously assessed by analyzing data collected during operation on the basis of physical models (e.g., as discussed by Glaessgen and Stargel (2012) in the context of air vehicles) in order to prevent failures, reduce downtime, and optimize maintenance. Besides monitoring the health of CPPSs, digital twins have also been adopted for the purpose of optimizing production processes (Uhlemann et al. 2017; Rosen et al. 2015).

Finally, when the end-of-life is reached and the CPPS is decommissioned, the respective digital twin can be of use in two different ways, viz., to retain knowledge about the system's life for reuse and to properly dispose of its materials (Grieves and Vickers 2017).

14.2.2 The Digital Thread

According to several sources (West and Pyster 2015; Boschert and Rosen 2016; West and Blackburn 2017), the term *digital thread* has been introduced by the United States Air Force (USAF) (Maybury 2013) to describe the notion of linking data throughout various phases of the lifecycle (e.g., design, processing, manufacturing) in order to increase efficiency in the development and deployment of systems. However, as indicated in (West and Pyster 2015), there appears to be a lack of a consistent definition of this term in the literature. Some scholars (e.g., Boschert and Rosen 2016) only see a negligible difference between the digital-thread and digital-twin concept, while others (e.g., Singh and Willcox 2018; West and Blackburn 2017) prefer to keep these two concepts apart. In this work, we adopt the definitions proposed by Lubell et al. (2013) and Singh and Willcox (2018) who describe the digital thread as "[the] unbroken data link through the lifecycle [. . .]" (Lubell et al. 2013) of a system that can be utilized "[. . .] [to] generate and provide updates to a Digital Twin" (Singh and Willcox 2018). In this context, the interoperability of tools used throughout the lifecycle represents a prerequisite for the implementation of the digital thread. As a result, technologies that foster semantic interoperability (e.g., OPC UA, AML) may become even more important with wider adoption of this concept. Although the works (Eckhart and Ekelhart 2018c,b; Schroeder et al. 2016a) do not explicitly mention the digital-thread concept, they provide valuable insights into how AML supports the exchange of data for realizing digital twins.

Although the digital thread can be considered as an enabler for digital twins, which in turn may be leveraged to improve the security of CPSs, the digital thread represents an attractive target for attacks, as it links various assets that are high in value (e.g., design artifacts) (Glavach et al. 2007). Due to the fact that a compromised digital thread may lead to severe consequences (e.g., manipulated updates to put the digital twin into a malicious state), adequate security measures to protect each link within the digital thread are paramount.

14.3 Digital Twins in the Information Security Domain

In this section, we review the definitions given in earlier works that deal with security aspects of CPSs in conjunction with the concept of digital twins and attempt to make one step toward a coherent definition of the term *digital twin* in the context of information security. Furthermore, we extend the use cases presented in (Eckhart and Ekelhart 2018c) (viz., (i) intrusion detection, (ii) system testing and simulation, (iii) detecting misconfigurations, and (iv) penetration testing) in order to provide a more comprehensive view of the significance of the digital-twin concept for the information security community. Besides extending the research conducted by Eckhart and Ekelhart (2018c), we expand on the use cases that have been proposed in other previous works, viz., (Bécue et al. 2018; Bitton et al. 2018; Tauber and Schmittner 2018; Damjanovic-Behrendt 2018a; Damjanovic-Behrendt 2018b). Thus, this section shows the state of the art in using the concept of digital twins to increase the security of CPSs.

14.3.1 Definitions

In the recent past, a few works have appeared that explore how the concept of digital twins can be applied to secure CPSs. Table 14.2 provides an overview of the *digital twin* definitions given in these works and thereby extends the view of definitions presented in (Negri et al. 2017).

As can be seen in Table 14.2, the definitions overlap to some extent yet include aspects that are relevant to the respective use cases presented in these papers. For instance, Bécue et al. (2018); Damjanovic-Behrendt (2018b); Tauber and Schmittner (2018) express that a digital twin is not only composed of a system's virtual model but also includes historical information thereof. Furthermore, the definition given by Bécue et al. (2018) explicitly includes physical processes, which may be useful for implementing process-aware intrusion detection systems (IDSs) (e.g., Nivethan and Papa 2016; Chromik et al. 2016).

Table 14.2 Definitions of the term *digital twin* in papers published on information security

Reference	Definition of the term *Digital Twin*
Bécue et al. (2018)	"[...] [An] evolving digital profile of the historical and current behavior of a physical object or process."
Bitton et al. (2018)	"[...] [A] replica of a specific ICS; i.e., a model that consists of all of the components from the original industrial environment."
Damjanovic-Behrendt (2018b)	"[...] [A] virtual counterpart to actual physical devices (entities) that combines many Artificial Intelligence (AI)-based technologies and methods, real-time predictive analyses, and forecasting algorithms performing on top of Big Data derived from the Internet of Things (IoT) sensors and acquired historical data."
Eckhart and Ekelhart (2018c)	Refers to the definition proposed by Shafto et al. (2010), namely "[...] the use of holistic simulations to virtually mirror a physical system."
Eckhart and Ekelhart (2018a)	"[...] virtual replicas of the network and the logic layer of physical devices, closely matching the physical devices' behavior on these layers."
Eckhart and Ekelhart (2018b)	Semantically equivalent to the definition given in (Eckhart and Ekelhart 2018a).
Tauber and Schmittner (2018)	"[...] [A] digital representation of a real system, with the history of all changes and developments."

To foster a common understanding of the term *digital twin* in the context of information security, we propose a definition that reflects the recent research progress made in this field. In particular, in the following, we introduce a uniform definition based on the synthesized interpretations from works cited in Table 14.2: A digital twin, which is used for the purpose of enhancing the security of a cyber-physical system, is a *virtual replica of a system that accompanies its physical counterpart during phases of its lifecycle, consumes real-time and historical data if required, and has sufficient fidelity to allow the implementation of the desired security measure*. It is worth noting that we assume that the knowledge about the process can be contained in a digital twin, depending on the implemented use case. For instance, the digital twins in (Eckhart and Ekelhart 2018c,b,a) represent simulated or emulated devices that can accurately mirror the physical counterparts on the logic and network layer, meaning that process knowledge is readily accessible through them. On the other hand, in (Damjanovic-Behrendt 2018b), the digital twins are composed of machine learning methods that learn security- and privacy-relevant aspects based on sensor data. Thus, process knowledge can merely be learned but not obtained directly through digital twins, as they are not aware of any control logic per se.

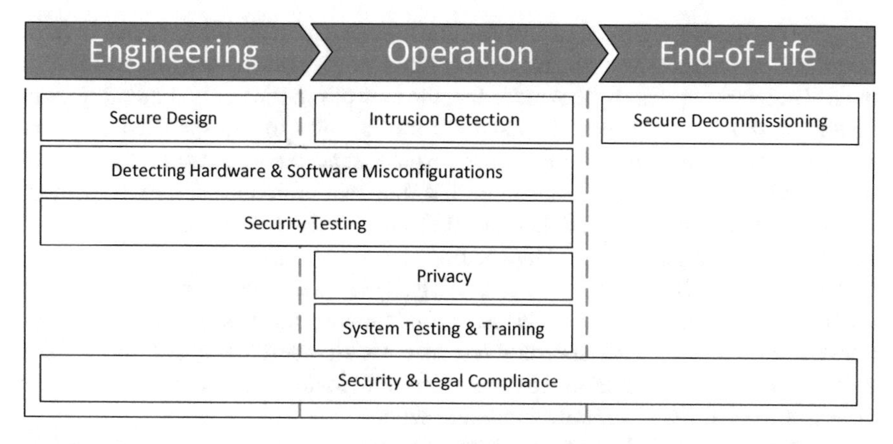

Fig. 14.2 Security-relevant use cases of the digital-twin concept based on (Bécue et al. 2018; Bitton et al. 2018; Eckhart and Ekelhart 2018c,b,a; Tauber and Schmittner 2018; Damjanovic-Behrendt 2018a; Damjanovic-Behrendt 2018b) within the lifecycle of CPSs (inspired by Lüder et al. (2017))

14.3.2 Security Use Cases of Digital Twins

Similarly to Sect. 14.2.1.3, we assigned the security-relevant use cases to the phases of the CPS lifecycle (cf. Fig. 14.2). The following subsections discuss each of these uses cases in detail.

14.3.2.1 Secure Design of Cyber-Physical Systems

Digital twins that gradually evolve over the course of the engineering may support engineers in designing more secure CPSs.

For instance, Bécue et al. (2018) suggest to use digital twins in combination with a cyber range[3] to analyze how the system to be engineered behaves under attack. The authors state that this method would allow engineers to estimate potential damages, which may facilitate designing the security and safety mechanisms of CPSs. As a result, this security activity may yield more robust and fault-tolerant designs of CPSs.

Besides simulating attacks to evaluate whether the system fails securely and safely, a virtual representation of the CPSs may also support reducing the attack surface. In particular, security analyses conducted on the basis of digital twins can reveal weak spots in the architecture, unnecessary functionality of devices or even unprotected services that would allow an adversary to gain a foothold in the system.

[3]Bécue et al. (2018) do not provide a definition of the term *cyber range*, but they indicate that it represents a virtual environment that provides the means to interact with the digital twins, e.g., to execute attacks against them.

To give an example, digital twins that have been automatically generated from specification (Eckhart and Ekelhart 2018c) may allow security analysts to identify unused network services by first recording the network traffic while simulating plant operation and then mapping the captured traffic flows to the specified services. As a consequence, this activity would expose superfluous network services in the specification of the CPS, meaning that they can be removed entirely without restraining plant operation and thereby minimize the attack surface.

Additionally, digital twins that are equipped with logic and network features (Eckhart and Ekelhart 2018c) may aid in realizing a defense in depth strategy, as network security controls can be thoroughly tested by simulating attack scenarios layer-wise. For example, security analysts could test whether an attacker is able to pivot from a compromised data historian to a programmable logic controller (PLC) with the objective to steer the plant into an insecure state.

Another viable technical use case in this context is the evaluation of how damages can be limited in the event of a compromise. In particular, simulating attack scenarios may help in preparing a containment strategy for compromised devices and thereby facilitate incident handling in the operation phase.

14.3.2.2 Intrusion Detection

In 2017, Rubio et al. (2017) published a survey paper on IDSs for ICSs. In this paper, the authors discuss, inter alia, the role of IDSs in the context of Industry 4.0 and suggest that the concept of digital twins provides promising opportunities in this area.

To the best of our knowledge, only two papers, namely, (Eckhart and Ekelhart 2018c,b), have been published thus far that demonstrate how the concept of digital twins can be leveraged to implement IDSs.[4]

In the first work by Eckhart and Ekelhart (2018c), the authors show how a knowledge-based intrusion detection system can be implemented. This particular intrusion detection technique relies on certain misuse patterns that the system would exhibit upon a compromise (Mitchell and Chen 2014). In (Eckhart and Ekelhart 2018c), these patterns have been specified with AML and are part of the specification of the CPS. More specifically, the authors defined two rules, namely, safety and security rules, that specific digital twins must adhere to. The safety rule specifies a threshold for a tag of a PLC (maximum velocity of a motor that the PLC controls), whereas the security rule defines a consistency check between a tag of a PLC and a tag of an human-machine interface (HMI) (the velocity of a motor can be set by using the HMI, as it can send a request to the PLC that controls the motor; thus, it can be assumed that the respective tags on these two devices should match). During the operation of the CPS, the digital twins are checked continuously for any rule

[4]In this work, we adopt the classification of intrusion detection techniques proposed by Mitchell and Chen (2014).

violations. However, the authors of Eckhart and Ekelhart (2018c) do not touch on the aspects concerning the replication of states to digital twins and merely evaluate the implemented IDS in simulation mode, i.e., without incorporating real-time data from live systems into digital twins so that they do not mirror the behavior of their physical counterparts during plant operation. Furthermore, although this intrusion detection technique generally yields a low false-positive rate, it is limited to detecting known misbehavior (Mitchell and Chen 2014).

Their second work (Eckhart and Ekelhart 2018b) builds on (Eckhart and Ekelhart 2018c), as the authors introduce a passive state replication approach that aims to replicate the program states from physical devices to the corresponding digital twins. Based on this state replication approach, the digital twins follow the states of their physical counterparts and thereby allow to virtually mirror the behavior of the real CPS during operation. It is self-evident that the implementation of such a state replication approach represents a fundamental requirement for realizing the intrusion detection use case, as the digital twins are utilized for detecting abnormal behavior that the real CPS may exhibit. To demonstrate the viability of the proposed state replication approach, Eckhart and Ekelhart (2018b) implemented a behavior-specification-based IDS and evaluated the effectiveness thereof by launching a man-in-the-middle (MITM) and an insider attack against a real CPS. This intrusion detection technique requires that the correct, benign behavior of the system is defined, as this specification is utilized to determine whether the system's behavior during runtime diverges from it due to an intrusion (Mitchell and Chen 2014). The beauty of this intrusion detection technique is that it generally yields a low false-negative rate while also being capable of detecting attacks that were unknown at the time of defining the legitimate behavior (Mitchell and Chen 2014). On the contrary, creating the specification of the system's correct behavior typically requires effort (Mitchell and Chen 2014). In (Eckhart and Ekelhart 2018b), the authors evade this issue intentionally by making the assumption that the specification of the CPS is readily available, as it has been developed in the course of the engineering phase. The specification of the CPS can then be used to automatically generate the digital twins, which model the correct behavior of their physical counterparts. During operation, the states of the physical devices are passively observed in the real environment and then replicated virtually in order to ensure that the digital twins receive the same inputs (e.g., network packet, simulated digital input, user input) as their physical counterparts. If, for example, a programmer performs an insider attack by manipulating the source code of a PLC, its behavior will deviate from that of the corresponding digital twin, provided that the adversary was not able to tamper with the specification. As a result, an intrusion can simply be detected by comparing the inputs and outputs of physical devices and those of the digital twins.

14.3.2.3 Detecting Hardware and Software Misconfigurations

Assuming that the hardware and software of devices are simulated or emulated to form digital twins, these virtual representations should mimic the functionality of corresponding devices to a certain level of detail. For example, the digital twin of a PLC may have a similar (virtualized) communications interface and I/O modules for the hardware layer, while the software layer may be replicated by executing the control logic. Thus, it can be expected to observe the common features of a digital twin and its physical counterpart. If hardware and software configurations of real devices have been manipulated, the digital twin should exhibit noticeable differences in terms of its characteristics, which would be indicative of malicious actions. As a matter of fact, this technical use case is similar to implementing a behavior-specification-based IDS based on digital twins in the sense that any deviation between the virtual replicas and their physical counterparts may indicate an attack.

Moreover, detecting manipulated software configurations can also be achieved by comparing configuration data (e.g., parameterization) of physical devices to their corresponding digital twins (Eckhart and Ekelhart 2018c). Yet, instead of checking whether the behavior of the physical devices deviates from that of their corresponding digital twins, only the software configuration settings are checked.

It is also worth highlighting that for realizing this technical use case, we have to assume that the digital twins run in an isolated environment protected against malicious acts. Otherwise, an adversary could tamper with the digital twins' configurations to ensure that any manipulations of the physical devices' configurations go unnoticed.

As can be seen in Fig. 14.2, this use case can be applied in two different phases of the CPSs' lifecycle. First, in the course of the commissioning of CPSs, the digital twins can be used to test if the devices have been set up according to their virtual replicas. Since security controls may be completely or partially deactivated during commissioning in order to ease the start-up phase, external or internal (i.e., commissioning staff) threat actors may be able to launch attacks even before the actual operation of the CPS. Thus, running final security checks to test the systems' configurations on the basis of their virtual replicas prior to the final acceptance may be worthwhile. Second, running these checks can be continued after commissioning in order to ensure that the integrity of configuration data is maintained throughout the operation phase. Evidently, if any legitimate changes to the physical devices' configurations are made during the operation phase, the configurations of the respective digital twins have to be adjusted.

14.3.2.4 Security Testing

Conducting security tests in OT environments represents a critical activity, especially when these tests are carried out during the operation of the CPS. In the past, multiple incidents occurred due to penetration tests that were carried out on live systems, causing severe physical damages and business interruption (Duggan et al. 2005). Thus, a testbed may be used in order to avoid any interference with live systems. However, building and maintaining a testbed can be time- and cost-intensive, in particular, when it should accurately reflect the actual CPS in operation (Eckhart and Ekelhart 2018c; Bitton et al. 2018). The adoption of digital twins has been proposed to address this issue (Eckhart and Ekelhart 2018c; Bitton et al. 2018; Bécue et al. 2018). In essence, digital twins enable penetration testers to perform security tests virtually, i.e., on the digital twins instead of on real systems. In this way, it can be ensured that the execution of these tests does not negatively affect the operation of live systems while also sparing operators from having to deal with the costs associated with testbeds. However, in this context, the challenge is to balance the fidelity of digital twins and the costs involved in creating them, so that the conducted security tests still yield useful results while keeping expenses low. In the work published by Bitton et al. (2018), the authors attempt to solve this problem by proposing a method for developing a cost-effective specification of a digital twin that would support the execution of specific security tests under a certain budget.

Besides performing security assessments in the operation phase, this approach can be likewise applied during engineering in order to fix vulnerabilities early on in the lifecycle of the CPS.

14.3.2.5 Privacy

Damjanovic-Behrendt (2018b) studied how the concept of digital twins can be applied to protect the privacy of smart car drivers. In particular, this work explores how automated privacy assessments can be carried out based on a virtual replica of a smart car that continuously receives data (e.g., from on-board sensors) in real time. The author provides an exemplary use case in which an insurer offers a usage-based insurance product based on the data obtained from the digital twins of smart cars. Since the digital twins integrate machine learning methods to classify personal data that can then be anonymized prior to the data transfer to the insurer, the customers' privacy rights are preserved. In this way, the concept of digital twins assists *controllers* or *processors* in meeting General Data Protection Regulation (GDPR) requirements.

Although the work published by Damjanovic-Behrendt (2018b) focuses on smart cars, the presented approach appears to be also applicable to other types of CPSs. Nevertheless, privacy-enhancing techniques based on the digital-twin concept for smart grids, transportation systems, and, in particular, medical CPSs may be worth exploring in greater detail.

14.3.2.6 System Testing and Training

Due to the fact that digital twins only exist virtually and are typically running in an environment that is isolated from live systems, they may also qualify to be used as a testing and training platform. Similarly to a cyber range, users could test new defenses before putting them into production or train how to respond to cyber incidents.

In (Bécue et al. 2018), the authors propose to adopt digital twins in combination with a cyber range to realize this use case. In particular, their work suggests launching attacks against digital twins from the cyber range for training and testing purposes.

Eckhart and Ekelhart (2018c) describe system testing as a use case for their proposed digital-twin framework. For example, similar to hardware-in-the-loop (HIL) simulation, real devices may be interfaced with the digital-twin framework for the purpose of testing. The authors of (Eckhart and Ekelhart 2018c) also present a proof of concept, named *CPS Twinning*,[5] which may provide rudimentary support for testing the network and logic layer of the CPS. This reason behind this claim is that the framework provides a virtual environment based on *Mininet* (Lantz et al. 2010) to emulate the network layer of the CPS but also supports a variety of device types (e.g., PLC, HMI, motor) whose logic can be virtually replicated to some extent. Based on this, it seems to be that their proposed digital-twin framework can also serve as a training tool, even though the authors do not explicitly mention this use case. Taking this idea one step further, *CPS Twinning* may also be suitable for carrying out red vs. blue team exercises that involve the network and logic layer of the CPS. Besides uncovering weaknesses resulting from the attacks launched by the red team, these exercises can also be used for training information security personnel (i.e., blue team) to implement adequate defenses in response to attacks. Collecting data over the course of such events, which may be helpful for risk assessments, can be a side benefit of these exercises (Sommestad and Hallberg 2012; Cook et al. 2016).

14.3.2.7 Secure Decommissioning

CPSs and, in particular, ICSs tend to have a long lifecycle, which can be up to 30 years (Macaulay and Singer 2016) or even longer. Yet, when the end-of-life phase is eventually reached, it must be ensured that components are disposed of in a secure manner.

In addition to supporting the proper disposal of materials (Grieves and Vickers 2017) (cf. Sect. 14.2.1.3), digital twins may also help to answer questions related to media sanitization. For instance, the NIST SP 800-88 (Kissel et al. 2014) guideline suggests considering, inter alia, confidentiality requirements of data as well as the costs associated with the sanitization process.

While the digital twins and the digital thread may facilitate secure disposal of physical devices, they can be equally affected by unauthorized access, if data security requirements are not met when disposing of them. Thus, it must be ensured that the digital thread is not only cut off but also properly archived and that the digital twins are finally laid to rest securely.

14.3.2.8 Security and Legal Compliance

Recently, Tauber and Schmittner (2018) published an article that highlights the importance of monitoring the CPS's security and safety posture during operation.

[5]https://github.com/sbaresearch/cps-twinning.

The authors emphasize that this activity could provide evidence of meeting security standards (e.g., IEC 62443 (IEC 2009)), which would, in turn, assist organizations in complying with legal requirements. In particular, Tauber and Schmittner (2018) suggest that the digital twins may provide an accurate reflection of CPSs throughout their entire lifecycle and thereby allow continuous monitoring and documentation of security and safety aspects. Considering that regulatory requirements for operators of CPSs appears to be increasing (e.g., the NIS Directive (European Parliament and the Council of the European Union 2016) for critical infrastructure providers), integrating security and legal compliance support into digital twins seems worthwhile.

14.4 Future Research Directions

This chapter has discussed several security-enhancing use cases for digital twins that may be worth researching in depth. Besides these use cases, we identified a variety of interesting questions in need of further investigation. In particular, we derived research directions as well as gaps from these questions and determined relevant work that may serve as a starting point for future studies. Furthermore, we classified the research directions according to their applicability in the three high-level lifecycle phases, viz., engineering, operation, and end-of-life. Table 14.3 summarizes the results of this assessment. In the following, we briefly discuss the identified research directions.

Practical Aspects

Examining the practicality of applying the digital-twin concept for securing CPSs focuses on answering fundamental research questions related to efficiently creating, maintaining, and running digital twins. These research topics are motivated by cost-benefit considerations, as implementing a digital-twin framework that supports the use cases presented in Sect. 14.3.2 seems to require substantial effort. Although such a digital-twin framework could leverage existing open-source tools (cf. Eckhart and Ekelhart 2018c), there is still significant work required to achieve an implementation of digital twins that provides an adequate level of detail for the desired use cases. In fact, this issue appears to be a major barrier to adopting the digital-twin concept, as other non-digital-twin approaches to implementing these security-enhancing use cases (e.g., intrusion detection) may incur less overhead in terms of effort required for implementation and maintenance (in the CPS's operation phase). Thus, a necessary first step would be to determine the required fidelity of digital twins for realizing the use cases discussed in Sect. 14.3.2. Note that creating identical digital representations that replicate the CPS in its entirety would defeat the concept's purpose, as the digital twin should merely provide support instead of a redundancy gain for protecting against failures of the real system. The work by Bitton et al. (2018) represents a valuable contribution toward the cost-efficient development of digital twins. However, it is still unknown how accurately the digital twins are required to

Table 14.3 Overview of research directions related to digital twins and information security

Research direction	Research gaps	Phase		
		E	O	D
Practical aspects	• Limited understanding of the required fidelity for use cases • Accuracy and performance requirements for use cases are unknown • Evaluation in a real-world setting is required	●	●	●
Legacy systems	• Automated generation of digital twins despite non-existent specification • Dealing with proprietary hardware and software of CPSs		●	
Risk assessment	• Unknown how cyber risks can be (automatically) identified, quantified, and (re-)evaluated based on digital twins	●	●	●
Resilience improvements	• Little is known how resilience can be measured based on digital twins • Unknown how to simulate attacks against digital twins	●	●	
Automated security testing	• Little is known how security tests for CPSs can be generated and executed in the digital-twin environment	●	●	
Intrusion detection	• Monitoring the physics of CPSs based on digital twins to detect intrusions is unexplored thus far		●	
Intrusion prevention	• Feasibility is unknown • Introduced latency is an obstacle, especially when real-time requirements must be met		●	
Honeypots	• Questionable how the behavior of digital twins can be altered to avoid disclosing valuable information while ensuring that the honeypot is still realistic		●	
Incident response training	• Attack simulation is an obstacle • Unknown whether digital twins can be exploited as a cost-effective training environment		●	

(continued)

Table 14.3 (continued)

Research direction	Research gaps	Phase		
		E	O	D
Attacks based on digital twins	• Unknown how digital twins or the digital thread can be exploited for launching advanced, covert attacks	●	●	
Attacks against digital twins	• Consequences of attacks against digital twins are unknown	●	●	

E = Engineering, O = Operation, D = Decommissioning (End-of-Life)

follow the states of their physical counterparts. In this context, achieving sufficient performance of the digital twins represents an obstacle (Eckhart and Ekelhart 2018b). As a result, identifying the optimal balance between budget and the required fidelity as well as state replication accuracy is still a research direction worth pursuing.

Legacy Systems

Considering the typical long lifecycle of CPSs, implementing the digital-twin concept for brownfield sites will become increasingly important. These legacy systems tend to be insufficiently documented, and detailed knowledge of their inner workings is rare. This, however, affects the accuracy of the virtual models to be developed, as a lack of understanding of the legacy system may lead to a flawed digital representation thereof. In (Eckhart and Ekelhart 2018c), the authors present a rudimentary prototype that allows to automatically generate digital twins based on the specification of the CPS. In their paper, the authors make the strong assumption that the specification is complete to the extent that the presented use case (i.e., intrusion detection) can be realized and that it is available in the engineering data exchange format AML. However, in a real-world setting, the specification of the CPS may be nonexistent or incomplete, at least for realizing the security-enhancing use cases discussed in Sect. 14.3.2. Nevertheless, this challenge may be overcome by first determining the information required to realize a specific use case (i.e., abstraction level of the digital twin) and then mining the specification from existing resources (e.g., monitoring systems, extracting data from other related artifacts). For example, Caselli et al. (2016) propose a specification mining approach for the implementation of an intrusion detection system used in building automation systems. Their work may be a starting point for researching mining methods capable of yielding a specification that can then be used to generate digital twins for the purpose of intrusion detection. On the other hand, if legacy virtual models are indeed available, research is required on how they can be retrofitted for digital-twin applications.

Risk Assessment

Cook et al. (2016) indicate the need for a CPS simulation environment, allowing the execution of attack scenarios that could then be factored into the risk assessment. The authors propose to adapt simulations of physical processes in a way that would allow consideration of boundary conditions caused by attacks, provided that these simulations already exist. In this way, the severity of potential cyber incidents would become apparent. Cook et al. (2016) also suggest that this could be realized by blending virtualized and physical devices, taking into account that such an environment must also support the representation of threat scenarios and potential consequences (e.g., financial loss) thereof. Thus, in the context of the digital-twin concept, this would mean that digital twins must be equipped with (i) accurate knowledge about the process under control (i.e., simulating the physical process) and not just replicating the control systems' logic (i.e., executing the programs that are running on their physical counterparts) and (ii) features to describe and simulate cyber risks. Both topics have been covered already in the literature, albeit not associated with digital twins. For instance, Krotofil et al. (2015) present a framework named *Damn Vulnerable Chemical Process (DVCP)* that leverages the Tennessee Eastman (Downs and Vogel 1993) and Vinyl Acetate (Chen et al. 2003) process models, enabling users to simulate attacks on the physical layer. Moreover, a considerable amount of literature has been published on simulating network attacks (e.g., Chabukswar et al. 2010) and assessing the impact of (simulated) threats to CPSs (e.g., Bracho et al. 2018).

It is also worth mentioning that a digital-twin approach to risk assessment may be suitable to deal with the dynamic nature of cyber risks. As a side note, both the probability of an attack and its impact can vary throughout the operation phase of the CPSs, meaning that risk mitigation strategies must be adapted accordingly. If digital twins run in parallel to their physical counterparts (i.e., they continuously mirror the behavior of real devices), this may be a viable approach to dynamic security risk assessment.

Resilience Improvements

In the context of ICSs, and presumably, also CPSs, (cyber) resilience refers to the systems' ability to maintain an adequate level of control of the physical process despite facing undesirable incidents (e.g., being under attack) (Wei and Ji 2010). As proposed by Wei and Ji (2010), improving the resilience of ICSs may be achieved by following a four-step process, which consists of (i) risk assessment, (ii) resilience engineering, (iii) resilience operation, and (iv) resilience enhancement. In essence, these four steps aim to minimize the probability of incidents occurring, their impacts, and the time required to recover from them, albeit at different phases of the ICSs's lifecycle. The concept of digital twins may support activities of this four-step process, as it may enable users to systematically introduce chaos (e.g., by simulating cyber attacks) into virtualized environments reflecting the real systems used for process control. In this way, users can determine the potential loss incurred (e.g., in terms of service degradation) and, in further consequence, mitigate these incidents.

A few works have been published on improving the (cyber) resilience of CPSs, which also give pointers for this future research direction of the digital-twin concept. For example, the work by Krotofil and Cárdenas (2013) investigates how the resilience of physical processes against manipulations of sensor readings can be increased. Their work shows that a well-versed adversary could maximize the economic and safety impact of malicious acts by strategically targeting specific sensors and manipulating readings at different points in time, depending on the process dynamics. This, by implication, means that the control system may be designed in a way that could make the physical process more resilient to certain kinds of attacks (Krotofil and Cárdenas 2013). Krotofil and Cárdenas (2013) leverage a simulation of the Tennessee Eastman process (Downs and Vogel 1993) for conducting their experiments to analyze process resilience. If such process simulations provide an interface for digital twins, a more comprehensive analysis of plant resilience may be feasible, also allowing to examine resilience at the system level.

Automated Security Testing

Automating security analyses of CPSs is an emerging research area. Several works, such as (Lemaire et al. 2017) and (Depamelaere et al. 2018), propose methodologies that aim to automate the identification of vulnerabilities of CPSs based on system models, which, for example, have been created in SysML during engineering. Extending this idea to the concept of digital twins, security tests may be automatically executed against virtual models reflecting either early versions of the systems to be engineered or the actual system during operation. Put differently, instead of automatically analyzing the systems' specifications to spot weaknesses, automated security tests are run continuously aiming to discover newly introduced flaws in digital twins. The beauty of this approach is that a replica of the actual system's implementation (i.e., the digital twin) can be tested, rather than, or in addition to, verifying that its specification does not have security weaknesses. Furthermore, depending on the fidelity of digital twins, certain types of security tests may be feasible. To give an example, digital twins that mirror the network and logic layer of devices may allow performing automated vulnerability scanning of the CPS's infrastructure.

In general, automated security testing based on the concept of digital twins may be beneficial for both the engineering and operation phase of CPSs. In the engineering phase, this use case may be applied on low- to medium-fidelity digital twins to check for potential attack vectors after certain engineering activities have been performed. On the other hand, in the operation phase of CPS, automated security tests may be executed against high-fidelity digital twins when adaptations to the CPS are made.

Although this security-enhancing use case may appear far-fetched at the present state of digital-twin research, in particular, the work by Eckhart and Ekelhart (2018c) already provides initial insights into how a digital-twin framework may be realized, which seems to be also extensible to support automated security testing.

Intrusion Detection and Intrusion Prevention

As indicated in Sect. 14.3.2.2, the first steps in this research direction have already been taken, as a knowledge- and behavior-specification-based IDS, which both build on the digital-twin concept, is presented in (Eckhart and Ekelhart 2018c,b). These works primarily focus on mirroring the logic and network layer of real devices, leaving the CPS's physical properties out. However, due to the fact that CPSs interact with the real world (e.g., for the purpose of controlling a physical process), it is possible to take advantage of the physical properties of these systems and use them as another dimension for detecting intrusions. In recent years, researchers have shown an increased interest in physics-based intrusion detection techniques (Giraldo et al. 2018). According to Giraldo et al. (2018), these techniques are characterized by the use of models of the physical system (e.g., autoregressive or linear dynamical state-space models) in order to predict system behavior. The predictions are then used to determine whether the sensor readings deviate from what is expected and whether the system reaches an unsafe state (Giraldo et al. 2018). Although digital twins may already include models that represent the physical properties of the system, researchers have not yet demonstrated how they can be utilized to detect intrusions. To date, only one paper (Eckhart and Ekelhart 2018c) mentions physics-based IDSs in the context of digital twins, albeit the work lacks further explanation on how this approach can be implemented. Thus, future research needs to be conducted in order to examine how digital twins that are composed of physical models can be leveraged for physics-based intrusion detection.

Besides investigating how physics-based IDSs can be implemented based on the digital-twin concept, realizing behavior-based IDSs by using data-driven digital twins may be another possible area of future research. Although no work has been published on this subject matter to date, we believe that the research conducted by Damjanovic-Behrendt (2018b) could represent the first step toward this direction, as this work covers digital twins that integrate machine learning methods to detect privacy-related anomalies.

Investigating new approaches to detect intrusions accurately is a major area of interest within the field of CPSs security. Yet, the mere detection of intrusions is of limited use if countermeasures cannot be taken in a timely manner, since the launched attacks may have already caused damages to equipment, environment, or human health. Therefore, intrusion prevention systems (IPSs) may be required, as they provide the means to take active security measures (e.g., by blocking malicious control commands) before incidents occur. However, as, for example, indicated in (Cárdenas et al. 2011), developing IPSs for CPSs represents a challenging task, due to the fact that false alarms (e.g., dropped packets of benign control commands) may raise safety concerns. Overcoming this challenge seems to be also relevant for digital-twin research in general, since data flows from a digital twin (back) to its physical counterpart can serve as a response mechanism (Kritzinger et al. 2018). Thus, further research regarding the role of digital twins for realizing IPSs for CPSs would be worthwhile.

Honeypots

Honeypots are systems that are installed for the sole purpose of being attacked (Spitzner 2002). These systems have several advantages, for example, (i) detecting intrusions, (ii) deterring attackers, or (iii) capturing malicious actions (e.g., attack patterns) for subsequent analysis (Spitzner 2002). Thus, honeypots can be used by defenders as a security measure and by security researchers as a means to develop novel countermeasures.

If honeypots are deployed with the objective to lure adversaries who launch targeted attacks, they should be as realistic (in terms of mimicking the real systems) and attractive (i.e., worthwhile to attack) as possible. Physical honeypots are composed of real devices, therefore, representing the most realistic form of honeypots (Antonioli et al. 2016). In recent years, a few works have been published that demonstrate how these honeypots can be used for CPSs (e.g., *HoneyTrain* (Fichtner and Krammel 2015), *SIPHON* (Guarnizo et al. 2017)). Although physical honeypots may allow defenders to gain a deep understanding of attacks, the development and maintenance costs associated with them may be too high, especially when used for CPSs (Antonioli et al. 2016). To alleviate this problem, the systems designated to lure attackers can also be virtualized. Depending on the achieved fidelity or realism of virtual honeypots, they can be categorized into low- and high-interaction honeypots (Fan et al. 2015). Past research has explored low-interaction (e.g., Vasilomanolakis et al. 2016; Rist et al. 2019) as well as high-interaction (e.g., Antonioli et al. 2016; Zhao and Qin 2017) virtual honeypots for CPSs, meaning that future work can build on a considerable body of research that deals with both types of honeypots.

Since digital twins can be considered as virtual replicas of physical devices, it appears that digital twins and virtual honeypots can also share commonalities in terms of their implementation. Thus, digital twins may also be exploited as a honeypot or, more precisely, honeynet (i.e., a network of honeypots (Fan et al. 2015)) solution. Implementation-wise, this similarity can already be observed between the works (Antonioli et al. 2016) and (Eckhart and Ekelhart 2018c), as both of the therein presented prototypes are based on *Mininet* (Lantz et al. 2010) to emulate the network layer, albeit they are unrelated to each other. If digital twins accurately reflect physical devices, except for the vulnerabilities that have been introduced deliberately, and follow the states of their physical counterparts, a significant increase of the honeynet's level of realism may be achieved. As a result, the simulated plant behavior may spark the adversary's interest in attacking the honeynet.

The primary issue of exploiting digital twins as honeypots is that defenders would give adversaries a detailed picture of the real plant upfront, making attacks against the real systems significantly easier, provided that adversaries are able to detect the trap. Based on this, we can derive the following research question: How can existing digital twins be modified in a cost-effective manner so that they can still mimic plausible plant behavior while ensuring that attackers do not gain valuable information about the real systems when they fall for these honeypots? Answering this research question will provide insights into the feasibility and applicability of realizing honeypots based on the concept of digital twins.

Incident Response Training

Section 14.3.2.6 discusses the idea of utilizing digital twins as a testing and incident response training platform, which resembles the notion of cyber ranges. Similar to traditional training environments for CPSs and, in particular, ICSs (Plumley et al. 2017), the supported training scenarios vary depending on digital twins' fidelity. In this context, the cost-effectiveness of digital twins seems to be a major research challenge. Although Bitton et al. (2018) already made the first steps toward a cost-effective digital twin for the purpose of conducting security analyses, it is unknown whether exploiting the digital-twin concept for certain training purposes, which would require an advanced fidelity, is financially worthwhile. For instance, the cost associated with achieving the fidelity required to support forensic investigation training scenarios may potentially exceed the cost of the real device. Thus, further research regarding the cost-effectiveness of digital twins for incident response training would be interesting. The work by Plumley et al. (2017) may be used as a starting point, as they provide a categorization of ICSs training environments that aids in determining the required level of realism based on training needs and budget constraints.

Covert Attacks Based on Digital Twins and Attacks Against Digital Twins
Stuxnet, one of the most prominent examples of ICS-tailored malware, aimed to cause significant equipment damage at the nuclear facility at Natanz by covertly manipulating the speed of centrifuge rotors (Langner 2013). According to Langner (2013), the attackers behind Stuxnet had a deep understanding of the plant design, which enabled them to tailor the malware to the target plant. The discovery of the Stuxnet malware led to an increased interest in such covert attacks against CPSs, i.e., attacks that are executed based on in-depth knowledge about the physical process and corresponding control devices in order to manipulate plant behavior in a covert manner. Due to the fact that digital twins may constitute accurate virtual replicas of physical devices, they represent valuable knowledge that might be misused for launching covert attacks if they were to fall into unfriendly hands. Building upon existing research in the area of covert attacks (e.g., Smith 2015; de Sá et al. 2017), it would be interesting to analyze the level of covertness that can be achieved based on digital twins, which have been obtained by attackers beforehand.

Another possible abuse case of digital twins is to launch targeted attacks against them in order to sabotage (security-enhancing) use cases and potentially also the behavior of their physical counterparts, provided that backflows to physical devices exist. Taking the example of intrusion detection (cf. Sect. 14.3.2.2), if attackers are able to manipulate the behavior of digital twins, they can ensure that the digital twins do not exhibit the defined pattern of misbehavior (to delude knowledge-based IDSs) nor deviate from their physical counterparts (to delude behavior-specification-based IDSs), hence allowing them to remain undetected when attacking the real systems. Furthermore, if digital twins directly affect plant operation (e.g., via an automatic data flow to field devices for optimizing manufacturing processes), attacks launched against them may have similar consequences as direct attacks against real devices.

To sum up, more research is definitely needed to better understand the threats posed by unsecured digital twins and to investigate how to mitigate them.

14.5 Conclusion

This chapter set out to provide a comprehensive overview of how the concept of digital twins can be applied to strengthen the security of CPSs. In particular, we have (i) provided relevant background information about the digital-twin concept, (ii) proposed a definition of the term *digital twin* in the context of information security, (iii) described security-enhancing use cases of the concept, and (iv) suggested future research directions.

The concept of digital twins appears to be an emergent stream of research in the information security field. Thus far, only a few papers have been published that merely scratch the surface of what seems to be possible with this concept. While some of the reviewed work only describe use cases and give general recommendations on how to realize them, there are also a few papers that discuss details regarding the implementation or even provide a proof of concept (Eckhart and Ekelhart 2018c,b; Bitton et al. 2018; Damjanovic-Behrendt 2018b).

Despite the fact that the chapter at hand reveals the state of the art of present approaches related to digital twins and CPS security, our work is limited in the following ways: First, we analyzed only papers that discuss the digital-twin concept in the context of information security. There may also be other existing works, which do not explicitly mention digital twins per se but still propose to use virtual models or simulations in a way that would have a positive effect on the security of CPSs. Second, our analysis lacks consideration of what the commercial market currently has to offer. Companies may already provide digital-twin solutions adaptable or extensible for realizing some of the use cases discussed in Sect. 14.3.2.

Nevertheless, we believe our work could be the basis for ongoing research, as the presented findings enhance our understanding of the term *digital twin* and envision what role the concept can take on when securing CPSs. In the future, more research is definitely required to investigate the practicality of the concept for security-enhancing use cases.

Acknowledgements The financial support by the Christian Doppler Research Association; the Austrian Federal Ministry for Digital and Economic Affairs; and the National Foundation for Research, Technology, and Development and COMET K1, FFG—Austrian Research Promotion Agency is gratefully acknowledged. Furthermore, this work was supported by the Austrian Science Fund (FWF) and netidee SCIENCE under grant P30437-N31.

References

Alam, K. M., & Saddik, A. E. (2017). C2PS: A digital twin architecture reference model for the cloud-based cyber-physical systems. *IEEE Access, 5*, 2050–2062.

Antonioli, D., Agrawal, A., & Tippenhauer, N. O. (2016). Towards high-interaction virtual ICS honeypots-in-a-box. In *Proceedings of the 2Nd ACM Workshop on Cyber-Physical Systems Security and Privacy', CPS-SPC '16* (pp. 13–22). New York, NY: ACM.

Baheti, R., & Gill, H. (2011). Cyber-physical systems. *The Impact of Control Technology, 12*, 161–166.

Bécue, A., Fourastier, Y., Praça, I., Savarit, A., Baron, C., Gradussofs, B., et al., (2018). Cyberfactory#1 — securing the industry 4.0 with cyber-ranges and digital twins. In *2018 14th IEEE International Workshop on Factory Communication Systems (WFCS)* (pp. 1–4)

Bitton, R., Gluck, T., Stan, O., Inokuchi, M., Ohta, Y., Yamada, Y., et al., (2018). Deriving a cost-effective digital twin of an ICS to facilitate security evaluation. In J. Lopez, J. Zhou & M. Soriano (Eds.), *Computer Security* (pp. 533–554). Cham: Springer.

Boschert, S., & Rosen, R. (2016), *Digital twin—the simulation aspect* (pp. 59–74). Cham: Springer.

Bracho, A., Saygin, C., Wan, H., Lee, Y., & Zarreh, A. (2018). A simulation-based platform for assessing the impact of cyber-threats on smart manufacturing systems. *Procedia Manufacturing, 26*, 1116–1127. 46th SME North American Manufacturing Research Conference, NAMRC 46, Texas, USA.

Cárdenas, A. A., Amin, S., Lin, Z.-S., Huang, Y.-L., Huang, C.-Y., & Sastry, S. (2011). Attacks against process control systems: Risk assessment, detection, and response. In *Proceedings of the 6th ACM Symposium on Information, Computer and Communications Security, ASIACCS '11* (pp. 355–366). New York, NY: ACM.

Caselli, M., Zambon, E., Amann, J., Sommer, R., & Kargl, F. (2016). *Specification mining for intrusion detection in networked control systems* (pp. 791–806), Berkeley: USENIX Association.

Chabukswar, R., Sinopoli, B., Karsai, G., Giani, A., Neema, H., & Davis, A. (2010). Simulation of network attacks on SCADA systems. In *First workshop on secure control systems, cyber physical systems week 2010.*

Chen, R., Dave, K., McAvoy, T. J., & Luyben, M. (2003). A nonlinear dynamic model of a vinyl acetate process. *Industrial & Engineering Chemistry Research, 42*(20), 4478–4487.

Chromik, J., Remke, A., & Haverkort, B. (2016). *What's under the hood? Improving SCADA security with process awareness*. Piscataway: IEEE.

Cook, A., Smith, R., Maglaras, L., & Janicke, H. (2016). Measuring the risk of cyber attack in industrial control systems. In *Proceedings of the 4th International Symposium for ICS & SCADA Cyber Security Research 2016', ICS-CSR '16* (pp. 1–11). Swindon, UK: BCS Learning & Development.

Damjanovic-Behrendt, V. (2018a). A digital twin architecture for security, privacy and safety. *ERCIM News, 2018*(115).

Damjanovic-Behrendt, V. (2018b). A digital twin-based privacy enhancement mechanism for the automotive industry. In *Proceedings of the 9th International Conference on Intelligent Systems: Theory, Research and Innovation in Applications.*

Depamelaere, W., Lemaire, L., Vossaert, J., & Naessens, V. (2018). CPS security assessment using automatically generated attack trees. In *Proceedings of the 5th International Symposium for ICS & SCADA Cyber Security Research 2018*. London: British Computer Society (BCS).

de Sá, A. O., d. C. Carmo, L. F. R., & Machado, R. C. S. (2017). Covert attacks in cyber-physical control systems. *IEEE Transactions on Industrial Informatics, 13*(4), 1641–1651.

Downs, J., & Vogel, E. (1993). A plant-wide industrial process control problem. *Computers & Chemical Engineering, 17*(3), 245–255. Industrial challenge problems in process control.

Dragos, Inc. (2018). *Industrial Control Vulnerabilities: 2017 in Review*, Tech report. Hanover: Dragos, Inc.

Drath, R., Luder, A., Peschke, J., & Hundt, L. (2008). AutomationML – the glue for seamless automation engineering. In *2008 IEEE International Conference on Emerging Technologies and Factory Automation* (pp. 616–623).

Duggan, D., Berg, M., Dillinger, J., & Stamp, J. (2005). *Penetration testing of industrial control systems*. Albuquerque: Sandia National Laboratories.

Durão, L. F. C. S., Haag, S., Anderl, R., Schützer, K., & Zancul, E. (2018). Digital twin requirements in the context of industry 4.0. In P. Chiabert, A. Bouras, F. Noël & J. Ríos, (Eds.), *Product Lifecycle Management to Support Industry 4.0* (pp. 204–214). Cham: Springer.

Eckhart, M., & Ekelhart, A. (2018a). Securing cyber-physical systems through digital twins. *ERCIM News, 2018*(115).

Eckhart, M., & Ekelhart, A. (2018b). A specification-based state replication approach for digital twins. In *Proceedings of the 2018 Workshop on Cyber-Physical Systems Security and PrivaCy, CPS-SPC '18* (pp. 36–47). New York, NY: ACM.

Eckhart, M., & Ekelhart, A. (2018c). Towards security-aware virtual environments for digital twins. In *Proceedings of the 4th ACM Workshop on Cyber-Physical System Security, CPSS '18* (pp. 61–72). New York, NY: ACM.

European Parliament and the Council of the European Union (2016), Directive (EU) 2016/1148 of the European Parliament and of the Council of 6 July 2016 concerning measures for a high common level of security of network and information systems across the Union, https://eur-lex.europa.eu/legal-content/EN/TXT/?uri=uriserv:OJ.L_.2016.194.01.0001.01.ENG. Accessed 11 Feb 2019.

Fan, W., Du, Z., & Fernández, D. (2015). Taxonomy of honeynet solutions. In *2015 SAI Intelligent Systems Conference (IntelliSys)* (pp. 1002–1009).

Fichtner, H.-P., & Krammel, M. (2015). *Project HoneyTrain*, Techreport. Saarbrücken: Koramis GmbH.

Giraldo, J., Urbina, D., Cardenas, A., Valente, J., Faisal, M., Ruths, J., et al. (2018). A survey of physics-based attack detection in cyber-physical systems. *ACM Computing Surveys, 51*(4), 76:1–76:36.

Glaessgen, E. H., & Stargel, D. (2012). The digital twin paradigm for future NASA and U.S. air force vehicles. In *53rd AIAA/ASME/ASCE/AHS/ASC Structures, Structural Dynamics and Materials Conference* (pp. 1–14).

Glavach, D., LaSalle-DeSantis, J., & Zimmerman, S. (2017). *Applying and assessing cybersecurity controls for direct digital manufacturing (DDM) systems* (pp. 173–194). Cham: Springer.

Gockel, B., Tudor, A., Brandyberry, M., Penmetsa, R., & Tuegel, E. (2012). Challenges with structural life forecasting using realistic mission profiles. In *53rd AIAA/ASME/ASCE/AHS/ASC Structures, Structural Dynamics and Materials Conference*. Reston: American Institute of Aeronautics and Astronautics.

Grieves, M., & Vickers, J. (2017). *Digital twin: mitigating unpredictable, undesirable emergent behavior in complex systems* (pp. 85–113). Cham: Springer.

Grinshpun, G., Cichon, T., Dipika, D., & Rossmann, J. (2016). From virtual testbeds to real lightweight robots: Development and deployment of control algorithms for soft robots, with particular reference to industrial peg-in-hole insertion tasks. In *Proceedings of ISR 2016: 47st International Symposium on Robotics* (pp. 1–7).

Guarnizo, J. D., Tambe, A., Bhunia, S. S., Ochoa, M., Tippenhauer, N. O., Shabtai, A., et al. (2017). Siphon: Towards scalable high-interaction physical honeypots. In *Proceedings of the 3rd ACM Workshop on Cyber-Physical System Security, CPSS '17* (pp. 57–68). New York, NY: ACM.

Haag, S., & Anderl, R. (2018). Digital twin – proof of concept. *Manufacturing Letters, 15*, 64–66. Industry 4.0 and Smart Manufacturing.

Hahn, A. (2016). *Operational Technology and Information Technology in Industrial Control Systems* (pp. 51–68). Cham: Springer.

Hribernik, K. A., Rabe, L., Thoben, K., & Schumacher, J. (2006). The product avatar as a product-instance-centric information management concept. *International Journal of Product Lifecycle Management, 1*(4), 367–379.

Hribernik, K., Wuest, T., & Thoben, K.-D. (2013). Towards product avatars representing middle-of-life information for improving design, development and manufacturing processes. In G. L. Kovács & D. Kochan (Eds.), *6th Programming Languages for Manufacturing (PROLAMAT), Digital Product and Process Development Systems* (Vol. AICT-411, pp. 85–96). Dresden, Germany: Springer. Part 2: Digital Product- and Process- Development.

ICS-CERT (2013), Year in review 2012, Technical report, Department of Homeland Security.

ICS-CERT (2015), Year in review 2014, Technical report, Department of Homeland Security.

ICS-CERT (2017), Year in review 2016, Technical report, Department of Homeland Security.

IEC (2009). 62443: Industrial communication networks – network and system security. *International Standard, First Edition, International Electrotechnical Commission, Geneva, 1*, 170.

Jaensch, F., Csiszar, A., Scheifele, C., & Verl, A. (2018), Digital twins of manufacturing systems as a base for machine learning. In *2018 25th International Conference on Mechatronics and Machine Vision in Practice (M2VIP)* (pp. 1–6).

Kagermann, H., Helbig, J., Hellinger, A., & Wahlster, W. (2013). *Recommendations for implementing the strategic initiative industrie 4.0 – securing the future of german manufacturing industry, Final report of the industrie 4.0 working group, acatech*. München: National Academy of Science and Engineering.

Kissel, R. L., Regenscheid, A. R., Scholl, M. A., & Stine, K. M. (2014). Guidelines for media sanitization. *NIST Special Publication, 800*(88r1).

Knapp, E. D., & Langill, J. T. (2014). *Industrial Network Security: Securing critical infrastructure networks for smart grid, SCADA, and other Industrial Control Systems*. Rockland: Syngress.
Knowles, W., Prince, D., Hutchison, D., Disso, J. F. P., & Jones, K. (2015). A survey of cyber security management in industrial control systems. *International Journal of Critical Infrastructure Protection, 9*, 52–80.
Kritzinger, W., Karner, M., Traar, G., Henjes, J., & Sihn, W. (2018). Digital twin in manufacturing: A categorical literature review and classification. *IFAC-PapersOnLine, 51*(11), 1016–1022. 16th IFAC Symposium on Information Control Problems in Manufacturing INCOM 2018.
Krotofil, M., & Cárdenas, A. A. (2013). Resilience of process control systems to cyber-physical attacks. In H. Riis Nielson & D. Gollmann (Eds.), *Secure IT Systems* (pp. 166–182). Berlin: Springer.
Krotofil, M., Isakov, A., Winnicki, A., Gollmann, D., Larsen, J., & Gurikov, P. (2015). Rocking the pocket book: Hacking chemical plants for competition and extortion, resreport, Black Hat.
Langner, R. (2013). *To kill a centrifuge: A technical analysis of what stuxnet's creators tried to achieve*. Arlington: The Langner Group.
Lantz, B., Heller, B., & McKeown, N. (2010). A network in a laptop: Rapid prototyping for software-defined networks. In *Proceedings of the 9th ACM SIGCOMM Workshop on Hot Topics in Networks, Hotnets-IX* (pp. 19:1–19:6). New York, NY: ACM.
Lee, J., Lapira, E., Bagheri, B., & an Kao, H. (2013). Recent advances and trends in predictive manufacturing systems in big data environment. *Manufacturing Letters, 1*(1), 38–41.
Lee, R. M., Assante, M. J., & Conway, T. (2016). Analysis of the cyber attack on the ukrainian power grid, techreport, SANS Institute.
Lemaire, L., Vossaert, J., Jansen, J., & Naessens, V. (2017). A logic-based framework for the security analysis of industrial control systems. *Automatic Control and Computer Sciences, 51*(2), 114–123.
Lubell, J., Frechette, S. P., Lipman, R. R., Proctor, F. M., Horst, J. A., Carlisle, M., et al. (2013). Model based enterprise summit report, Technical Report 1820, National Institute of Standards and Technology.
Lüder, A., Schmidt, N., Hell, K., Röpke, H., & Zawisza, J. (2017). *Fundamentals of artifact reuse in CPPS* (pp. 113–138). Cham: Springer.
Macaulay, T., & Singer, B. (2016). *Cybersecurity for industrial control systems: SCADA, DCS, PLC, HMI, and SIS*. Boca Raton: CRC Press.
Maybury, M. T. (2013). Global horizons: Final report, resreport AF/ST TR 13-01; Air Force/Small Business Technology Transer 13-01, United States Air Force.
McLaughlin, S., Konstantinou, C., Wang, X., Davi, L., Sadeghi, A. R., Maniatakos, M., et al. (2016). The cybersecurity landscape in industrial control systems. *Proceedings of the IEEE, 104*(5), 1039–1057.
Mitchell, R., & Chen, I.-R. (2014). A survey of intrusion detection techniques for cyber-physical systems. *ACM Computing Surveys, 46*(4), 55:1–55:29.
Mourtzis, D., Doukas, M., & Bernidaki, D. (2014). Simulation in manufacturing: Review and challenges. *Procedia CIRP, 25*, 213–229. 8th International Conference on Digital Enterprise Technology – DET 2014 Disruptive Innovation in Manufacturing Engineering towards the 4th Industrial Revolution.
Negahban, A., & Smith, J. S. (2014). Simulation for manufacturing system design and operation: Literature review and analysis. *Journal of Manufacturing Systems, 33*(2), 241–261.
Negri, E., Fumagalli, L., & Macchi, M. (2017). A review of the roles of digital twin in CPS-based production systems. *Procedia Manufacturing, 11*, 939–948. 27th International Conference on Flexible Automation and Intelligent Manufacturing, FAIM2017, 27–30 June 2017, Modena, Italy.
Nivethan, J., & Papa, M. (2016). A SCADA intrusion detection framework that incorporates process semantics. In *Proceedings of the 11th Annual Cyber and Information Security Research Conference, CISRC '16* (pp. 6:1–6:5). New York, NY: ACM.

Panetta, K. (2018). Gartner top 10 strategic technology trends for 2019, https://www.gartner.com/smarterwithgartner/gartner-top-10-strategic-technology-trends-for-2019/. Accessed 12 Dec 2018.

Plumley, E., Rice, M., Dunlap, S., & Pecarina, J. (2017). Categorization of cyber training environments for industrial control systems. In M. Rice & S. Shenoi (Eds.), *Critical Infrastructure Protection XI* (pp. 243–271). Cham: Springer.

Polenghi, A., Fumagalli, L., & Roda, I. (2018). Role of simulation in industrial engineering: Focus on manufacturing systems. *IFAC-PapersOnLine, 51*(11), 496–501. 16th IFAC Symposium on Information Control Problems in Manufacturing INCOM 2018.

Rajkumar, R., Lee, I., Sha, L., & Stankovic, J. (2010). Cyber-physical systems: The next computing revolution. In *Design Automation Conference* (pp. 731–736).

Reifsnider, K., & Majumdar, P. (2013). Multiphysics stimulated simulation digital twin methods for fleet management. In *54th AIAA/ASME/ASCE/AHS/ASC Structures, Structural Dynamics, and Materials Conference*. Reston: American Institute of Aeronautics and Astronautics.

Ríos, J., Hernández, J. C., Oliva, M., & Mas, F. (2015). Product avatar as digital counterpart of a physical individual product: Literature review and implications in an aircraft. In *ISPE CE* (pp. 657–666).

Rist, L., Vestergaard, J., Haslinger, D., Pasquale, A., & Smith, J. (2019). Conpot ICS/SCADA Honeypot. http://conpot.org/. Accessed 11 Feb 2019.

Rosen, R., von Wichert, G., Lo, G., & Bettenhausen, K. D. (2015). About the importance of autonomy and digital twins for the future of manufacturing. *IFAC-PapersOnLine, 48*(3), 567–572. 15th IFAC Symposium on Information Control Problems in Manufacturing INCOM 2015.

Rubio, J. E., Alcaraz, C., Roman, R., & Lopez, J. (2017). Analysis of intrusion detection systems in industrial ecosystems. In *14th International Conference on Security and Cryptography (SECRYPT 2017)*.

Schleich, B., Anwer, N., Mathieu, L., & Wartzack, S. (2017). Shaping the digital twin for design and production engineering. CIRP Annals, 66(1), 141–144.

Schluse, M., & Rossmann, J. (2016). From simulation to experimentable digital twins: Simulation-based development and operation of complex technical systems. In *2016 IEEE International Symposium on Systems Engineering (ISSE)* (pp. 1–6).

Schroeder, G., Steinmetz, C., Pereira, C. E., Muller, I., Garcia, N., Espindola, D., & Rodrigues, R. (2016). Visualising the digital twin using web services and augmented reality. In *2016 IEEE 14th International Conference on Industrial Informatics (INDIN)* (pp. 522–527).

Schroeder, G. N., Steinmetz, C., Pereira, C. E., & Espindola, D. B. (2016). Digital twin data modeling with AutomationML and a communication methodology for data exchange. *IFAC-PapersOnLine, 49*(30), 12–17. 4th IFAC Symposium on Telematics Applications TA 2016.

Shafto, M., Conroy, M., Doyle, R., Glaessgen, E., Kemp, C., LeMoigne, J., et al. (2010). Draft modeling, simulation, information technology & processing roadmap. *Technology Area, 11*. NASA

Shi, J., Wan, J., Yan, H., & Suo, H. (2011). A survey of cyber-physical systems. In *2011 International Conference on Wireless Communications and Signal Processing (WCSP)* (pp. 1–6).

Singh, V., & Willcox, K. E. (2018). Engineering design with digital thread. *AIAA Journal, 56*(11), 4515–4528.

Smith, R. S. (2015). Covert misappropriation of networked control systems: Presenting a feedback structure. *IEEE Control Systems Magazine, 35*(1), 82–92.

Sommestad, T., & Hallberg, J. (2012). Cyber security exercises and competitions as a platform for cyber security experiments. In A. Jøsang & B. Carlsson (Eds.), *Secure IT Systems* (pp. 47–60). Berlin: Springer.

Spitzner, L. (2002). *Honeypots: tracking hackers*. Boston, MA: Addison-Wesley Longman Publishing.

Tauber, M., & Schmittner, C. (2018). Enabling security and safety evaluation in industry 4.0 use cases with digital twins. *ERCIM News, 2018*(115).

Tuegel, E. J., Ingraffea, A. R., Eason, T. G., & Spottswood, S. M. (2011). Reengineering aircraft structural life prediction using a digital twin. *International Journal of Aerospace Engineering, 2011*, 14. Article ID 154798.

Uhlemann, T. H.-J., Lehmann, C., & Steinhilper, R. (2017). The digital twin: Realizing the cyber-physical production system for industry 4.0. *Procedia CIRP, 61*(Supplement C), 335–340. The 24th CIRP Conference on Life Cycle Engineering.

Vachálek, J., Bartalský, L., Rovný, O., Šišmišová, D., Morháč, M., & Lokšík, M. (2017). The digital twin of an industrial production line within the industry 4.0 concept. In *2017 21st International Conference on Process Control (PC)* (pp. 258–262).

Vasilomanolakis, E., Srinivasa, S., Cordero, C. G., & Mühlhäuser, M. (2016). Multi-stage attack detection and signature generation with ICS honeypots. In *NOMS 2016 – 2016 IEEE/IFIP Network Operations and Management Symposium* (pp. 1227–1232).

Wei, D., & Ji, K. (2010). Resilient industrial control system (RICS): Concepts, formulation, metrics, and insights. In *2010 3rd International Symposium on Resilient Control Systems* (pp. 15–22).

West, T. D., & Blackburn, M. (2017). Is digital thread/digital twin affordable? A systemic assessment of the cost of dod's latest manhattan project. *Procedia Computer Science, 114*, 47–56. Complex Adaptive Systems Conference with Theme: Engineering Cyber Physical Systems, CAS October 30 – November 1, 2017, Chicago, Illinois, USA.

West, T. D., & Pyster, A. (2015). Untangling the digital thread: The challenge and promise of model-based engineering in defense acquisition. *INSIGHT, 18*(2), 45–55.

Zhao, C., & Qin, S. (2017). A research for high interactive honepot based on industrial service. In *2017 3rd IEEE International Conference on Computer and Communications (ICCC)* (pp. 2935–2939).

Chapter 15
Radio Frequency (RF) Security in Industrial Engineering Processes

Martin Fruhmann and Klaus Gebeshuber

Abstract Interconnection and information transparency are major players when it comes to the 4th Industrial Revolution, also known as the Industry 4.0 (I4.0). Hence, wireless transmission systems have a growing potential in the engineering of new industrial machines. In fact, Radio Frequency (RF) technologies have already found their ways into the engineering process. This trend, however, goes hand in hand with a rising awareness for IT security. Since industrial machines are known to have a great lifetime, it is inevitable to not consider security from the very first development phase onward. To improve the security of any industrial system findings from industrial penetration tests as well as possible mitigations should be already considered at an early stage of the design and development process. This chapter therefore discusses use cases and security measures of wireless systems in industrial facilities. Based on an overview of RF technologies in the industrial field, several devices and software products (software-defined radios) for the analysis of such systems are introduced. Furthermore, the feasibility for Penetration Testing of these devices is addressed to strengthen the security aspect when it comes to the I4.0.

Keywords RF · Software-defined radio · Information security

15.1 Introduction

Many components in industrial settings can benefit from wireless technologies. This may be due to cost saving, performance, or monitoring reasons (Caro et al. 2014). But new technologies implicate new challenges in terms of security. Industrial machines are known to have a long lifetime which emphasizes an early identification of possible security flaws. This is especially true considering that industrial components controlled via wireless modules may be responsible for safety tasks. If flaws are

M. Fruhmann (✉) · K. Gebeshuber
FH JOANNEUM GmbH, Institute of Internet Technologies & Applications, Kapfenberg, Austria
e-mail: martin.fruhmann2@fh-joanneum.at; klaus.gebeshuber@fh-joanneum.at

S. Biffl et al. (eds.), *Security and Quality in Cyber-Physical Systems Engineering*,
https://doi.org/10.1007/978-3-030-25312-7_15

introduced in the engineering process, a worst case result can be injuries, or even human death. However, also financial losses have to be considered in a security incident due to wireless protocol flaws. Furthermore, security tests on machines in operation are very costly, because of possible downtimes or damage caused during the test. Therefore, this chapter highlights the importance of security considerations, especially early in an engineering process.

In the first part of the chapter, different fields of application of wireless technologies are presented. Additionally, common wireless protocols are briefly described. The next part shows basic attack techniques like replay, jamming, or reverse engineering and possible mitigation strategies. The chapter continues with strategies of integrating security measures into an early process of the engineering process. Therefore, Threat Modeling and Penetration Testing are discussed. Regarding analysis of wireless technologies, software-defined radios (SDR) are proposed as tools for security tests during and after the engineering process. Lastly an example for reverse engineering a simple wireless signal is shown in form of a remote garage door opener.

15.2 Industrial RF Applications

Compared to traditional industrial automation processes, wireless technologies in this sector are still very young. Therefore only a few different standards are in use. This section first provides an overview about differences and challenges related to classical wired data transmission. Next, some fields of application are mentioned, followed by state-of-the-art standards in industrial wireless transmission. Lastly different known vulnerabilities of the presented technologies are briefly described.

15.2.1 Wired vs. Wireless Data Transmission

Sensor and actuator technologies in industrial machines need any kind of data transmission to provide status information to their base stations. Before wireless technologies have existed, wired connections from every component were in use. This resulted in high costs for cables, low expandability, and sometimes in major maintenance costs, especially on moving parts. Wireless technologies had the potential to mitigate or reduce these problems.

With the absence of cables, the installation and maintenance costs can be reduced. Furthermore wireless systems are more likely to be easily expandable, and there is no problem with moving parts without cables. This might sound like a perfect replacement for wired data transmission. However, also wireless transmission brings its own challenges and problems.

When it comes to cables, most wireless systems are not truly wireless. Every component at least needs some type of power supply. Even when the power supply is managed via battery, the wireless network often needs to be wired to transfer collected data to a different network or computer. Therefore the costs for cables may be reduced compared to wired systems, but in terms of maintenance, the costs may be even higher due to degrading batteries.

Also new challenges are introduced by wireless systems due to the different transportation medium. Industrial wireless applications are free to operate in the industrial, scientific, and medical (ISM) frequency band. Since this band is restricted to a certain frequency (e.g., 2.4 GHz–2.5 GHz) and no access control mechanism is available across the different standards, one thing to consider is interference (Frotzscher et al. 2014). Not only different productive wireless systems may be a problem, also malicious actors have to be considered. The act of maliciously interfering with wireless signals is known as jamming and is covered in Sect. 15.4.

Another challenge in wireless transmission is the signal behavior itself. Dead spots is a term for areas where no wireless reception is possible. This might be due to interference as mentioned before, multipath behavior where the signal is reflected and interferes with the original signal, or due to attenuating building materials. More unlikely but still to mention are sunspots, which are electromagnetic waves emitted by the sun which can interfere with radio signals on earth.

Lastly, as the transmission medium can be accessed by nearly anyone with the proper equipment, privacy and data integrity concerns arise. Compared to wired communications, there is no need to physically intercept the transmission medium with, for example, a wiretap. Therefore wireless transmission systems are more prone to security incidents (Caro et al. 2014).

15.2.2 Fields of Application of Wireless Technologies

Wireless communication in industrial applications is mainly used at the field and sensor level, shown in Fig. 15.1. As mentioned before, one operation area is on movable parts. Traditionally components like sensors on those parts where connected via trailing cable systems or sliding contacts. These are expensive to maintain due to their high abrasion.

On the field level, different technologies from different manufacturers are available. Possible operation areas are connecting human-machine interfaces (HMI) with their programmable logic controller (PLC) counterparts. Another field may be connecting computers on hardly reachable areas with an adaption of wireless LAN (WLAN).

Furthermore Radio Frequency Identification (RFID) or Near Field Communication (NFC) is often used in transport, logistics, or material management (Frotzscher et al. 2014).

On the sensor level, the term wireless sensor and actuator network (WSAN) is often used. Due to the introduction of wireless sensors and actuators, they can now be placed on various parts of industrial machines. This may be used for condition monitoring. With this technique it is possible to use the data to plan maintenance

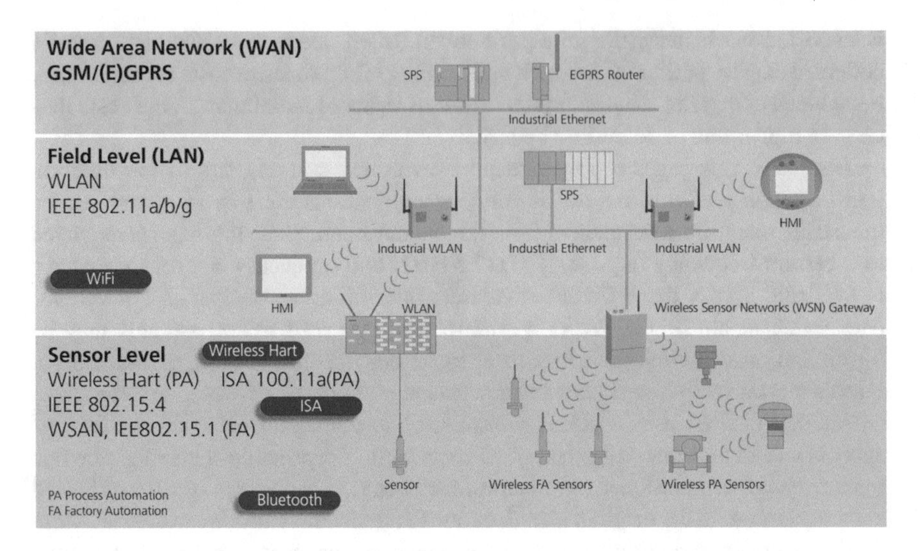

Fig. 15.1 Industrial wireless communication systems overview (Frotzscher et al. 2014)

windows and therefore improve the availability of the machine itself (Akerberg et al. 2011).

15.2.3 Industrial Wireless Standards

Starting with the mostly well-known technology Wi-Fi, which is standardized in IEEE 802.11. It is very popular because of its wide range of off-the-shelf products on the market. The main usage of Wi-Fi in industrial settings is connecting computers used machines or cameras for operation controlling.

Traditional Wi-Fi operates at 2.4 GHz but has been extended to another frequency band at 5 GHz with the introduction of IEEE 802.11n. Another improvement is a technique called multiple input-multiple output (MIMO), which provides a stronger and more reliable signal. This is managed with an array of antennas to collectively send and receive Wi-Fi signals. Single-channel Wi-Fi has a theoretical data rate up to 54 MBit/s, whereas with multiple channel usage and 802.11ac compatible devices, up to 1 GBit/s is possible (Caro et al. 2014).

The next well-known standard is Bluetooth, specified in IEEE 802.15.1. Similar to Wi-Fi, Bluetooth has a lot of different products already on the market and can be adopted for different usages fairly easy. Bluetooth nodes can be interconnected and therefore form a piconet with a maximum of eight nodes. One node can be connected to two networks, which allows routing information between different piconets. Bluetooth operated at the same frequency band as traditional Wi-Fi, which opens the possibility of interference between these technologies.

Table 15.1 Wireless system usage in industrial control systems (Reaves and Morris 2012)

Wireless system	ICS prevalence (%)
IEEE 802.11	51.5
Proprietary systems	34.0
WirelessHART	23.0
Bluetooth	18.2
ISA 100-11a	4.4
ZigBee	1.5

An expansion to common Bluetooth, Bluetooth Low Energy (BLE) was introduced to provide a low-end variant of the technology for the use in battery-powered devices. It operates at the same frequency band but with significant lower performance which also impacts the power usage (Caro et al. 2014).

Coming to the next technology, ZigBee is defined in the IEEE 802.15.4 standard. ZigBee was developed as a personal area network (PAN) and is mainly used in home automation. There are various other protocols which extend the IEEE 802.15.4 standards like WirelessHART, ISA100, or WIA-PA, which are more specialized for the use in industrial application. One thing they have in common is the goal of using as little power as possible, to be used with battery-powered devices. This is accomplished by a sleep mode, which allows the device to only be powered when data has to be sent.

To provide an overview of the use of these technologies in real-world scenarios, Reaves and Morris (2012) presents in Table 15.1 their prevalence respectively.

15.2.4 Known Vulnerabilities

Wi-Fi has a long history of technology changes and therefore also different vulnerabilities. One rather recently discovered vulnerability is the Key Reinstallation Attack (Vanhoef 2017; Vanhoef and Piessens 2017). It targets the four-way handshake of the Wi-Fi Protected Access 2 (WPA2) Authentication in modern Wi-Fi networks. The author discovered that it is possible to manipulate this handshake to trick the devices into reusing an already used encryption key.

Wi-Fi standards have, similar to industrial machines, a long lifetime. WPA2 was introduced in 2004 and was mandatory for new devices in 2006 (Alliance 2006), whereas WPA3, the successor technology, was only introduced in 2018 (Alliance 2018). This also emphasizes the importance of penetration tests during the lifespan of industrial machines, because new attack vectors are published even for apparently secure technologies.

Regarding Bluetooth, different attack techniques are still present today. Qu and Chan (2016) describe the different categories of attacks against Bluetooth and BLE devices. One of the techniques is Bluesniffing, where an attacker can extract unauthorized data from a target device. Bluejacking allows to send unsolicited messages to a target, whereas Bluebugging enables the attacker to fully control the victim device. Lastly Bluesmacking is the technique to DoS a device via jamming on all frequency bands of Bluetooth.

Another very prominent example of Bluetooth attacks is described in Chap. 15.3, where a man-in-the-middle attack is explained.

ZigBee and the variations of IEEE 802.15.4 for the use in industrial settings have a lot of security features in place (Sastry and Wagner 2004).

Bowers (2012) describes different ways these protocols may be exploited. One major attack vector are physical attacks where the encryption key for the secure communication is extracted to manipulate the traffic of the device. One example is the attack by Chapman (2014), attacking wireless light bulbs. Other vulnerabilities are arise due to implementation bugs in specific devices, like Ronen et al. (2017), who again attacks wireless light bulbs via malicious over-the-air (OTA) updates.

15.3 Threats in RF Systems

To get a feeling for possible attack vectors, this section provides different ideas on how wireless systems may be penetrated. Therefore the first part describes basic attack types, followed by specialized attacks against more modern approaches of radios.

15.3.1 Replay Attacks

The most basic attack type is the replay attack. This attack consists of four simple steps. First, the victim signal has to be detected. For that purpose, the documentation of the target device can be used. If none of this information is available, spectrum analysis of common frequency bands is the way to go. When the signal is identified, the next step is recording the signal. This record is then used to send the signal to the victim. Lastly the attack has to be verified, wherefore the response of the target has to be captured or any desired consequences have to be identified.

The major advantage of this technique is that the attacker does not need deep knowledge of the target system. The prerequisite to conduct a successful attack is that the target system does not implement security measures like rolling codes (Chernyshev 2013; Ossmann 2016). Rolling codes ensure that every signal is only used once, for example, by introducing a cryptographic hash into every signal.

15.3.2 Jamming Attacks

As mentioned before, jamming is the process of stressing a target to make any data transmission for the device impossible. This behavior classifies the attack as a denial of service (DoS). There are different types of jamming attacks, namely, constant, random, reactive, and pilot jamming. Constant jammers transmit radio signals continuously which make them very effective. The downside of constant jamming is the high energy consumption and therefore high energy cost. Next, random jammers transmit signals on a random pattern. These types of jammers are more energy-efficient, but not as powerful as constant jammers. Reactive jamming is a more creative way of the attack. Radio signals are only transmitted when a certain power level on a specific channel is detected. This makes it the most efficient method, compared to the previous mentioned ones. Another positive side effect is that the attack is harder to detect, because the jammer signal interferes with the signal of the target. The last, pilot jammers transmit signal in synchronization with the target system, to have the best interference (Punal et al. 2012).

Jamming is often used in combination with other techniques. Some of these combinations are described below.

Jamming and Replay
For example, vehicle key fobs often use rolling codes as a security measure against replay attacks. Therefore every lock and unlock operation uses a different code which is synchronized with the vehicle itself. With a combination of jamming and replaying recorded signals, this security method can be overcome.

The vehicle receives signals on a specific frequency range. The range of the transmitted signal is often much more narrow. This behavior is due to a better performance of the whole system. When an attacker now jams the receiver on a different frequency than the transmitter is operating, she can record the transmitted signal with an additional device. Now the attacker holds signal with a valid rolling code (Kamkar 2018).

Jamming and Man-in-the-Middle
Another combination of two attack methods is using jamming and a man-in-the-middle attack. An example for this approach is an attack on Bluetooth, shown in Fig. 15.2. The attacker jams the target to force the user to reconnect his devices. Now the man-in-the-middle device imitates the communication partner of the user, which is in this case a printer. It provides malicious information about the two devices to enter a lower security mode, which can be easily intercepted (Haataja and Hypponen 2008).

This attack is possible due to the "Just Works" association mode in the Secure Simple Pairing (SSP) mechanism of Bluetooth. This mode is used, for example, for printers, where it is often not possible to interact as a user with the printer during a connection attempt.

15.3.3 Reverse Engineering

If none of the above mentioned attack techniques are successful, reverse engineering is a way to get deeper understanding of the system. This allows specialized attacks

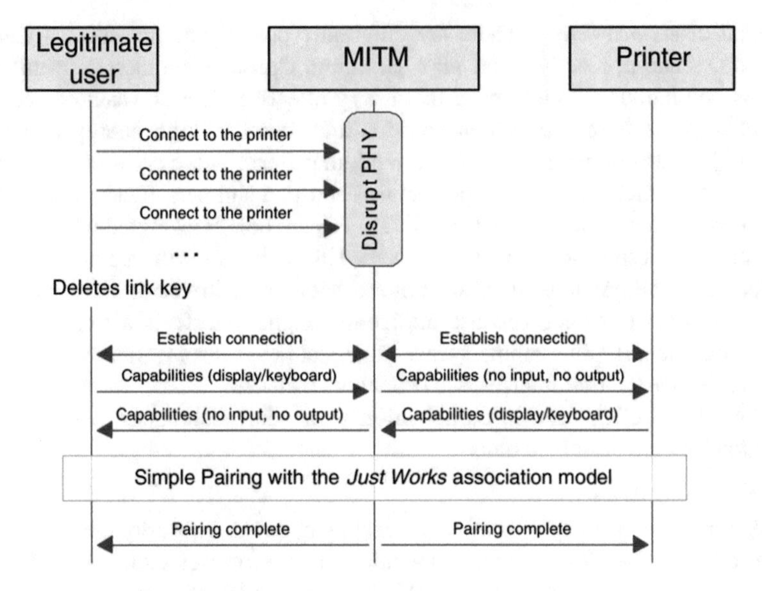

Fig. 15.2 Main idea of the man-in-the-middle attack against Bluetooth (Haataja and Hypponen 2008)

against the target. In general there are three types of reverse engineering in the RF domain. The first is hardware-based analysis, where the attacker has to have access to the target device. Next, also requiring access to the device, is firmware analysis. The last one is radio-based analysis which can be performed remotely (Klostermeier and Deeg 2016).

Hardware-Based Analysis

In hardware-based analysis, the components of the target device itself are investigated. This process is structured in the following tasks:

- Open device to get access to the internal printed circuit boards (PCB)
- Analysis of PCBs, especially identifying chips
- Reading documentation about discovered components
- Analysis of memory via Serial Peripheral Interface (SPI)

Firmware Analysis

Firmware analysis targets the software of the device. In most cases the software of proprietary radio systems is not open source and therefore has to be dumped from the device. For disassembling, applications like IDA Pro, Hopper, or Radare2 can be used (Hex-Rays 2018; Cryptic Apps 2018; pancake 2018).

Radio-Based Analysis

The last type is radio-based analysis, which is the method used in the practical example later in the chapter. Basically the process is similar to the steps a receiver has to perform to work with the transmitted signal.

- *Signal Acquisition:* Depending on the target device the signal can be detected in two different ways. The first is to use publicly available information like the Federal Communications Commission (FCC) identifier[1] of the device. The FCC stores data like frequency range, signal tests, and much more. The second way is to analyze common spectrum bands for the target signal, which is much more time-consuming.
- *Signal Analysis:* After the signal has been found, it can be recorded and further analyzed. This step includes detection of the modulation technique, as well as the investigation of the data encoding method.
  - *Demodulation:* Demodulation is the process of isolating the data signal from the carrier signal. Sometimes it is really easy to identify the used modulation technique. For example, Binary Frequency Shift Keying (BFSK) with its two distinct frequencies can be easily distinguished from a simple on-off keying (OOK) modulation.
  - *Decoding:* This step requires the identification of the used encoding method to extract the raw bits from the data signal. It is useful to refer to common encoding types specifically for the used modulation technique.
- *Data Analysis:* After the raw bits are at your fingertips, the data itself can be analyzed. Therefore a lot of data has to be collected to compare different signals. A big part of the data analysis is also to identify a preamble or synchronization sequence possibly present in the data. This is then further used to perform direct attacks against the system (Ossmann 2016).

15.3.4 Mitigation of Basic Attack Vectors

Considering the mentioned attacks, the first protection which comes to mind is against replay attacks. Different security improvements can be deployed to eliminate this type of attacks. First, the most obvious measure is to introduce rolling codes in the communication. Li et al. (2011) shows a cryptographic approach to include rolling codes into implantable medical devices (IMD). The main idea is to embed a sequence counter in the transmitter and in the receiver which is used during encryption and decryption of the data to be sent. In the decryption process, this counter value is checked. If it is in a certain range, the data is accepted.

A similar process is described by Alrabady and Mahmud (2005), as an industry standard for keyless entry systems. As rolling codes are also prone to be vulnerable, a solution to this problem is introducing authentication to RF systems. Again Alrabady and Mahmud (2005) mentions a challenge-response mechanism where

[1]The FCC is a US regulatory commission for radio, television, wire, satellite, and cable. Every device which may be used in the USA is therefore to be registered at this commission to be legally utilized (FCC 2018).

the communication partners share a secret encryption key. This key is used during initialization to verify the authenticity of the device. In an example of a car using a keyless entry system, the vehicle sends a random challenge to the identification device. This challenge is then encrypted using the secret key and sent back to the car. This encrypted message is checked by the car with its own secret key.

Regarding jamming attacks, mitigation is a difficult topic due to the destructive nature of these attacks. With enough transmission power, it is almost impossible to perform normal communication. One way to handle jamming attacks is proposed by Kar et al. (2014). They describe a detection system for vehicle GPS information. A more effective solution is called frequency hopping. An example is shown by Liechti et al. (2015). Frequency hopping is a technique where the transmission and receiving frequency of a certain RF system is permanently changed. The communication partners know the defined hopping sequence to perform their operations. Key factor is the randomization of the sequence. It has to be unpredictable to a possible attacker. Another example of frequency hopping is implemented in the well-known Bluetooth protocol. They use at least 20 to a maximum of 79 channels (Bluetooth 2016).

Lastly, the introduction of cryptography is a powerful security measure against radio-based reverse engineering. Similar to the authentication model by Shafagh and Hithnawi (2014); Li et al. (2006) propose a way to generate private keys using the special parameters of multi-channel behavior of RF communication. A more high level approach is presented by Owor et al. (2007). They are using elliptic curve cryptography (ECC), which provides high confidentiality together with strong performance compared to traditional asymmetric cryptography systems. This matches the limited performance of most RF systems perfectly.

15.3.5 Cognitive Radio Threats and Mitigation

Cognitive radio (CR) is an extension to the functionality SDR. A definition describes it as: "A cognitive radio is a radio that can change its transmitter parameters based on interaction with the environment it operates" (Fragkiadakis et al. 2013). This means the radio is capable of changing its transmission, as well as receiving configuration to increase the systems performance. This is accomplished by the use of artificial intelligence (AI) and SDR.

This technology brings its own security considerations and vulnerabilities. Clancy and Goergen (2008) describe three different attack scenarios on CR, namely, sensory manipulation attacks, belief manipulation attacks, and cognitive radio viruses. These vulnerabilities cannot be mitigated as easily as those which already exist for SDR. The first vulnerability focuses on the fact that policy radios use sensors to change their RF configuration in a positive manner. Therefore, manipulated sensor data may result in faulty behavior. Learning radios are affected by the same problem. Due to its learning characteristics, such an attack is much more powerful and can result in link degradation and lower data rates. This technique is therefore called belief manipulation attack. The final detected weakness, cognitive radio viruses, is also based on the changing behavior. In this case one infected radio may infect others in the same area.

Clancy and Goergen (2008) also present different mitigation to these threats. Improving sensor input, better sensing algorithms, constant reevaluation of learned behavior, or more complex techniques like particle swarm optimization are only a few mentioned upgrades.

Fragkiadakis et al. (2013) also address the topic of CR threats and detection mechanisms. They conducted a survey containing literature to different attack and detection scenarios. The first category of attacks described are primary user emulation attacks, which focus on the change of the transmission frequency of CRs. Next, spectrum sensing data falsification attacks use interference with other networks to cause DoS. MAC layer threats basically consist of three attack types, spoofing, flooding, and jamming attacks. Combined attacks are described in the cross-layer attacks chapter.

A different protection approach is presented by Fadlullah et al. (2013), which introduces an intrusion detection system (IDS) into a CR networks (CRN). IDSs are usually divided into two categories. Misuse detection is based on attack signatures which are already known. The other category is an anomaly-based attack detection, where the IDS is capable of detecting abnormal behavior. The latter has the benefit of identifying unknown attacks. The proposed IDS uses anomaly-based attack detection. It gathers the required information in a learning phase, where a profile during normal conditions is created. This profile is then used to discover different attacks in the detection phase. Fadlullah et al. (2013) introduce an example in which a jamming attack on a CRN is performed and, as a result of the attack detection, an alert is being generated.

15.4 Integration of RF Security in Engineering Life Cycle

This section proposes two ways to integrate security measure against mentioned attacks into the engineering process of industrial machines. The first is in the form of Threat Modeling, and as a second measure Penetration Testing may be used.

15.4.1 Threat Modeling

Threat Modeling is an approach to identify design flaws in any engineering process. This process can be basically integrated in two phases: first at the design phase of a project, where the main components and technologies have been chosen, and second as part of a penetration test at the end of the engineering process. A general best practice is to use both possibilities to get the most out of the threat model.

The output of a Threat Model should be a document containing a high level model, for example, realized with a data flow diagram (DFD), a list with assets

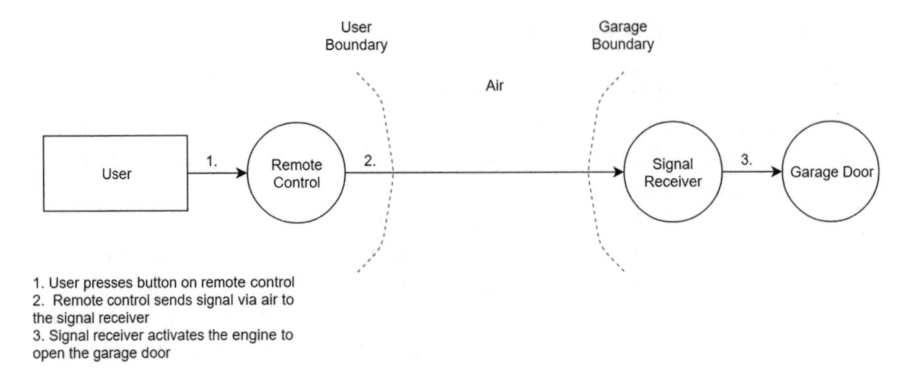

Fig. 15.3 DFD of a scenario from a wireless remote control

which have to be protected, threats ranked by their impact, and possible mitigation for those threats.

For the creation of a Threat Model, different techniques are available. One very popular is STRIDE (Spoofing, Tampering, Repudiation, Information Disclosure, DoS, Elevation of Privileges) in combination with DREAD (Damage, Reproducibility, Exploitability, Affected Users, Discoverability). STRIDE is a technique to identify threats in a model, whereas DREAD is a method for ranking the impact of flaws (Howard and Lipner 2006).

The basic approach in Threat Modeling is structured in the following steps (Howard and Lipner 2006):

- Define use scenarios (including external dependencies and security assumptions)
- Create DFDs for the defined scenarios
- Detect threats (e.g., via STRIDE)
- Determine risk (e.g., via DREAD)
- Plan mitigations

To provide an example of a Threat Model with regard to wireless transmission mechanisms, Fig. 15.3 shows a scenario from the garage door opener later shown in the case study.

On the basis of this scenario, the STRIDE model is used to find possible threats. The individual threats are described in the following list.

- **S**poofing: Given that the system is not capable of diversifying user identities, spoofing is not possible.
- **T**ampering: In this scenario, tampering is the most probable threat, which is also shown in the case study later in this chapter.
- **R**epudiation: The system does not record any signal history; therefore repudiation is not a possible threat.
- **I**nformation Disclosure: Similar to tampering, information disclosure is very likely due to the open transport medium.
- **D**oS: Another probable threat is DoS, again due to the transportation medium.

Table 15.2 DREAD risk rating for scenario defined in Fig. 15.3

Threat	Damage	Reproducibility	Exploitability	Affected users	Discoverability	Score
Tampering	10	10	5	10	10	9
Information disclosure	5	10	10	0	10	7
DoS	10	10	10	10	10	10
Overall score	8,7					

- Elevation of Privileges: Similar to spoofing, elevation of privileges is not possible due to missing authorization.

With the threats identified, it is now possible to rate the particular threats with the DREAD model. One thing to mention is that the risk rating is a very subjective task. Therefore, it is best to perform a risk rating in a group to discuss different opinions.

Table 15.2 shows how a risk rating with DREAD can be performed. Each factor is rated from 0 to 10, which then leads to a score for each threat. These scores may also be categorized into different categories like critical, important, moderate, or low. The overall score shows the impact of all threats combined for a specific scenario.

Threat Modeling encourages the identification of security threats in an early development stage and should be part of every engineering process. System integrator and vendors of wireless solutions can benefit by integrating this step by exposing vulnerabilities before they are present in their products and have to be patched in a comparable costly way.

15.4.2 Penetration Testing

Penetration Testing is nowadays a widely known term and has many different definitions. Basically a penetration test, or short pentest, is an authorized attack on a computer system to find possible vulnerabilities (Heinaearo 2015). Due to the enormous scope of this definition, three main testing methods have evolved. A vulnerability assessment is a way to find vulnerabilities in a given environment. Therefore a pentester scan targets to find well-known vulnerabilities by checking the operating system (OS), or version numbers. However, Penetration Testing goes a step further, so that found vulnerabilities are being actively exploited to provide a proof of concept (PoC). The last method is red teaming. It simulates a real attack where targeted systems already have countermeasures in place and eventually a defensive "blue team" is present. In contrast to the aforementioned methods, the goal of red teaming is not to find as many vulnerabilities as possible but to test the defensive measures instead. The only difference to a real attack is the permission of the customer (Hayes 2016).

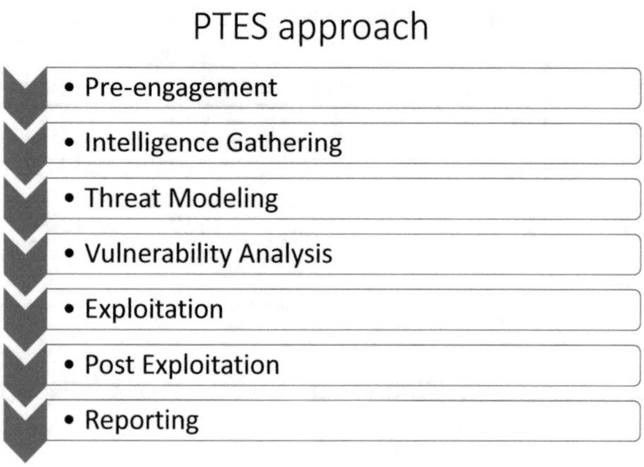

Fig. 15.4 Workflow of Penetration Testing phases (The Penetration Testing Standard 2014)

Additionally to this permission, a scope for the test has to be defined. This scope covers all details regarding what to be tested. The main parts are the time period in which the test takes place and the targets. The target can be anything from a whole company network (different IP ranges), including social engineering, wireless, and physical pentesting, to a simple web application test. A part of the target scope also includes the type of the attack scenario. Black box testing refers to a test method where nothing about the internal logic of a target is present to the attacker. In contrast, a white box test includes documentation about a given target. There are also hybrid approaches which are called gray box tests. Further the location of the testing team has to be defined. This depends on the customer's requests. In a typical scenario, a test is performed from outside of the company network, from inside, or sometimes both (The Penetration Testing Standard 2014).

The preparation of this scope is performed in the pre-engagement phase of a penetration test, according to the Penetration Testing Standard (PTES).[2] Figure 15.4 shows the seven phases during a penetration test defined by the PTES.

The second section covers intelligence gathering (IG), where information about the target is collected. PTES defines three levels of intensity during this step.

- **Level One:** automated IG
- **Level Two:** Level One and manual analysis
- **Level Three:** Level One and Two and heavy manual analysis

An example for level three IG is when automated collected information is further investigated to find relationships between companies or even connecting to employees

[2]The Penetration Testing Standard is a collaborative Standard for customers and service providers regarding penetration tests, to define a baseline of knowledge for both parties. It is developed by a group of security specialists, which is open for contribution The Penetration Testing Standard (2014).

on social media. Levels two and three basically only differ in the time invested into IG.

Additionally to these levels, three types of IG are presented.

- **Passive Information Gathering:** No direct interaction with the target itself is performed. Only stored information can be used, e.g., Shodan[3] and Google.
- **Semi-passive Information Gathering:** Normal interaction with the target, like basic web requests, query published nameservers, etc.
- **Active Information Gathering:** Suspicious or malicious interaction, including full port scans, vulnerability scanning, searching for unpublished files and servers, etc.

Next, Threat Modeling defines the approach to analyze assets provided by the customer to identify and categorize known threats. It should be differentiated between business assets (what can be targeted) and business processes (how to attack them).

The third step is vulnerability analysis. As the name indicates, this method is used to find flaws in the target system. They can vary from an insecure configuration to a known exploit vulnerability. The result of this phase is highly dependent on the infrastructure under test. Again, vulnerability analysis is split into active and passive techniques. Active methods include network port scanning and web application scanning, whereas with a passive process, only traffic monitoring and metadata analysis is possible.

With the information from step three, it is now possible to continue with exploitation, the next phase according to PTES. There are different ways to bypass security restrictions in a given environment. Often vulnerabilities are already well known and public exploits can be used. If this is not the case, it is common to use a zero-day approach. This means using different techniques like fuzzing or source code analysis to find unknown exploits in a given environment.

After exploitation, post-exploitation is performed. In general this represents the action of injecting different ways to access a compromised machine for further investigation. This step has to be specified by a set of rules, so that the target is not in risk by consequences of this phase. An example of such a risk is when a system in a production line from a company is being compromised and further investigation via network attacks shuts down the machine. This would cause, on the one hand, a financial damage for the company and, on the other hand, possible legal consequences for the tester.

If post-exploitation was used in a test, a cleanup has to be performed afterward. Therefore all executables and scripts have to be removed, and the original system settings have to be restored.

The last but most important step is the reporting of found vulnerabilities and threats. The report should be broken into two main parts: the executive summary, which represents a high level review, and the technical report, which should contain all technical details.

---

[3] Shodan is a search engine developed specifically for Internet-connected devices (Shodan 2018).

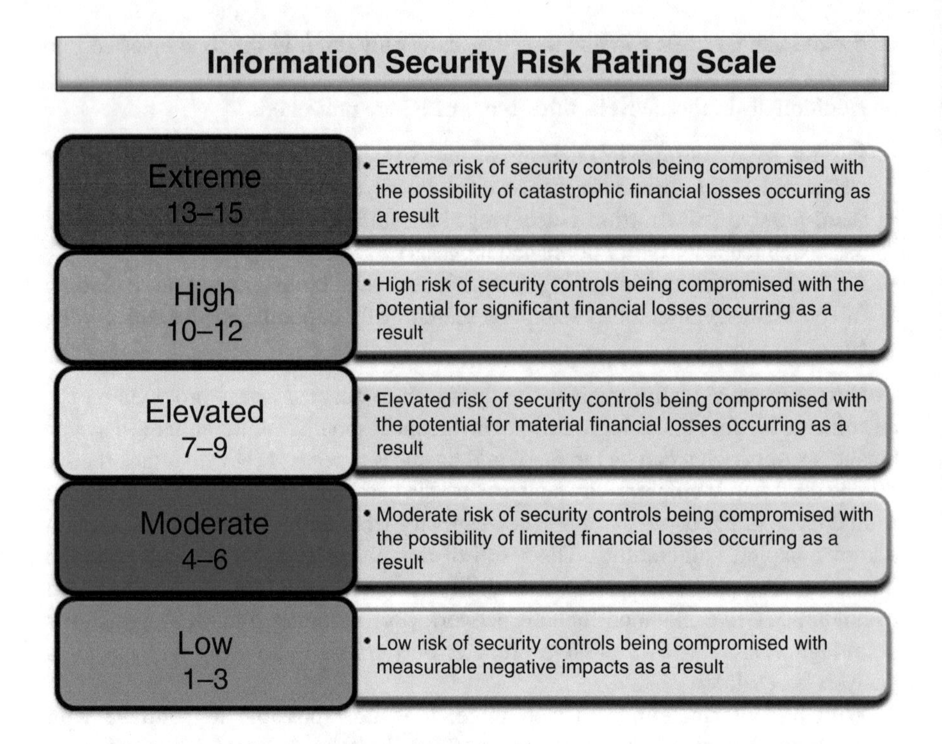

Fig. 15.5 Risk rating scale suggested by The Penetration Testing Standard (2014)

The executive summary should contain a background section, where basic information about the test itself is summarized. It should contain the reason for the test, goals, objectives, and relative results. For the covered vulnerabilities, a risk rating should be introduced. PTES presents an example, shown in Fig. 15.5.

Additionally, an overall risk score should be integrated to present the results with one rating and to further compare the rating to future test results. For every investigated security flaw, a recommendation to resolve the vulnerability should be given. The technical report should include all details of the test, including the raw data like screenshots of compromised web pages.

In industrial Penetration Testing setting, this approach is not the same as in standard IT environment. This is due to the fact that systems may fail in response to some testing techniques. This may result in downtimes, damage, or in the worst case human injuries due to safety-related failures. Therefore it is recommended to be more careful, especially with operational components like PLCs or in the wireless spectrum with connected actuators and sensors.

Also, to minimize the probability of interference between devices during a test, it is recommended to restrict transmit power and bandwidth of the attacking device. For critical devices the interference can be reduced to a minimum with the use of a faraday cage.

Regarding the pre-engagement phase, wireless testing settings stand a benefit from white box tests. This is due to the time-consuming nature of the information gathering phase of RF devices. Valuable information is the used technology, frequency spectrum, data rate, and if it is known, possible security measures in place. This information kick-starts the testing procedure.

In black box scenarios, a lot of resource have to be assigned to the identification of the data transmission via spectrum analysis. A helpful technique to accelerate this process is to isolate the signal as much as possible, to distinguish from other wireless traffic in place.

15.4.3 Software-Defined Radio (SDR) as Penetration Testing Tool

This section provides the fundamental concepts of SDR, as well as different hardware and software tools available at the moment. Traditional radio communication was realized with specialized hardware, which was only capable of receiving signals on a specific frequency band, with a fixed modulation and fixed filter mechanism. With the development of digital technology, especially analog to digital (AD) and digital to analog (DA) converters, it was possible to perform these steps in the digital domain. Furthermore the use of digital signal processing (DSP) and the invention of field-programmable gate arrays (FPGA) enabled the development of the first SDRs. In contrast to the traditional approach, it is now possible to create very flexible and therefore cheap hardware to discover RF communication (Machado and Wyglinski 2015).

15.4.3.1 Architecture of SDR

As shown in Fig. 15.6, SDR has two major domains. The analog domain consists of an RF front-end, usually an antenna with an RF amplifier, and the channel, which is air. The conversion of the analog signal is accomplished by the AD/DA converters. The steps of the digital domain are now controlled by software components (Machado and Wyglinski 2015).

The AD conversion is also known as sampling. Sampling is the process of using the analog waveform to produce a stream of digital data. Therefore periodical measurements are taken from the received signal every T_s seconds, which is also referred to as sampling period. The sampling frequency is calculated as $f_s = 1/T_s$. To successfully reconstruct the original signal from the sampled signal, the Nyquist theorem implies that the sampling frequency has to be more than twice the frequency of the original signal (Machado and Wyglinski 2015).

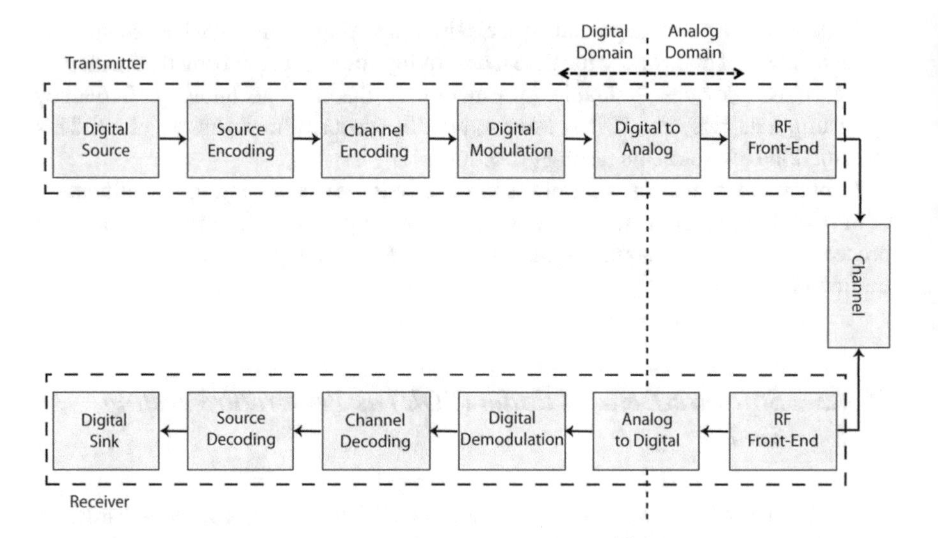

Fig. 15.6 Block diagram showing SDR architecture(Machado and Wyglinski 2015)

Table 15.3 SDR hardware with specifications (Wright and Cache 2015)

Device	Max. sample rate	Tuning range	ADC resolution	Transmit capability	Cost
RTL-SDR	2.5 MS/s	50 MHz–1.7 GHz	8 bits	None	20$
HackRF	20 MS/s	10 MHz–6 GHz	8 bits	Half duplex	330$
BladeRF	40 MS/s	300 MHz–3.8 GHz	12 bits	Full duplex	420$
Ettus USRP B200	61.44 MS/s	70 MHz–6 GHz	12 bits	Full duplex	675$

15.4.3.2 SDR—Hardware

Nowadays there are a lot of different SDR platforms available. They mainly differ in their transmission capability, sample rate, and tuning range. Table 15.3 shows four popular devices with their hardware specification.

RTL-SDR

RTL-SDR originally was not intended to be a SDR device. In fact, it is a conventional DVB-T TV tuner dongle. A group of specialists found that it is possible to access the digital signal of the RTL2832U chipset directly via special crafted software. This makes it a really cheap SDR for professionals and hobbyists. The only downside is the missing transmission capability (RTL-SDR 2018). It costs about 20€ and works with Linux, as well as with Windows (NooElec 2018).

HackRF

The HackRF One is a more advanced SDR platform, created by Great Scott Gadgets. It is very versatile with a sample rate up to 20 MSPS and a tuning range of 10 MHz to 6 GHz. The hardware and software components of the HackRF are open source and available on GitHub (Great Scott Gadgets 2016).

As antennas for radio communication are traditionally built only for one frequency, the HackRF team developed an add-on board, called Opera Cake, to switch between different antennas during operation (Ossmann and Spill 2017).

BladeRF

The first board, in this list, which can handle full-duplex communication is the BladeRF. This functionality enables the board to perform attacks on cellular mobile stations. Further the USB 3.0 support enables a sample rate up to 40 MSPS. Due to its high performance, it is more expensive than the previously mentioned SDR platforms (Nuand 2018).

Universal Software Radio Peripheral (USRP)

Another SDR capable of full-duplex communication is the USRP. It is the most expensive platform compared to the others mentioned before. With its very broad tuning range and very high sample rate, it is, for example, also able to receive and transmit 5G Wi-Fi signals (Ettus 2018).

Adalm-Pluto SDR

A quite new platform was developed by Analog Devices in 2017. It is called Adalm-Pluto and is marketed as an active learning module for students. It is not meant to be a replacement for professional grade SDR, due to some limitations like missing support for frequencies below 325 MHz, missing RF shielding, and a limited bandwidth due to the USB2.0 connection to name a few (Analog Devices Inc. 2018c).

In regard to hardware specification, the Adalm-Pluto has a tuning range from 325 MHz to 3.8 GHz, and it can communicate half- and full-duplex and has a 12-bit ADC resolution like the HackRF One and the BladeRF. The price of the device is about 99$ (Analog Devices Inc. 2018a).

The major limitation is due to its short time on the market and that a lot of the available software does not support the Adalm-Pluto platform; however GNU Radio support is present (Analog Devices Inc. 2018b).

YARD Stick One

The YARD Stick One is another product developed by Great Scott Gadgets. It features a transceiver IC and a hardware-based modem, which makes it not totally software controllable and therefore technically not a SDR. The modem only has a few modulation techniques built in (Ossmann 2016).

In combination with a RTL-SDR, the YARD Stick is a great tool to attack low-cost devices, like key fobs, wireless power adapters, and so on. The YARD Stick compensates the unsupported transmit functionality of the RTL-SDR. This combination is cheaper than a HackRF One, hence a great toolset for beginners (Ossmann 2016).

15.4.3.3 SDR—Software

SDR has a great open-source community, and therefore a variety of software tools with diverse functionality are available. This chapter provides an overview of common SDR software and their advantages and disadvantages. The presented tools are mainly available for Linux with one exception which is SDR#.

GNU Radio

GNU Radio is an open-source toolkit to implement SDRs. It provides basic blocks to perform different steps of signal processing, for example, filters, decoders, demodulators, and many more. It works with all of the mentioned SDR hardware platforms. The major benefit is the huge extensibility of the framework. It is possible to write blocks in C++, or Python (The GNU Radio Foundation 2018).

Osmocom-FFT

Osmocom-FFT is a spectrum analyzer included in the Osmocom GNU Radio blocks, which is an expansion of GNU Radio. Osmocom-FFT is part of the gr-osmosdr package, a GNU Radio library with various standalone tools. The package is available via GitHub,[4] or common Linux repositories.

Osmocom-FFT is also capable of recording raw signals and saving them to a file. This file can then be further investigated by a tool called inspectrum, which will be described next (Stolnikov 2018).

Inspectrum

As an offline radio signal analyzer, inspectrum can visualize recorded signals for analysis. It was also developed by an open-source community via GitHub.

A very useful feature is time selection, where it is possible to show a cursor for time measurement (miek 2018).

GQRX

GQRX is another spectrum analyzer including common demodulators like AM or FM. In the background the application also works with GNU Radio. Due to the demodulation functionality, it is possible to record demodulated signal streams which can be further analyzed by Audacity, for example (GQRX 2018).

Audacity

The main functionality of Audacity is multichannel audio editing, but in combination with GQRX, it can be used as a radio signal analyzer. Audacity is being developed open source and is available for all common OS (Audacity 2018).

Similar to inspectrum, Audacity accepts recorded signals; however the signal has to be demodulated, like GQRX files are.

Universal Radio Hacker (URH)

As the name indicates, URH is a multifunctional radio investigation tool. The project is, as the previously mentioned tools, open source and was developed using GNU

[4]GitHub is a collaborative platform for open-source software development (GitHub 2018).

```
from rflib import *

d.setFreq(434044000)
d.setMdmModulation(MOD_ASK_OOK)
d.setMdmDRate(int(1.0/0.000400))
d.RFxmit("\x8E\x8E\x8E\x88\x8E\x8E\x8E\x80\x00\x00\x00\x00"*30)
```

Fig. 15.7 Example RFCat script

Radio as a foundation. It is designed to automate all steps in the radio analysis workflow, from spectrum analysis to sending manipulated signals. URH is capable of recognizing modulation types and provides automatic decoding. For manual inspection, a differential view of received bit streams is available. On the sending side, the tool can send previously received signals and implements a fuzzing mode (Pohl and Noack 2018).

Scapy-Radio
Scapy-radio is an extension to Scapy, an open-source network packet manipulation tool, written in Python. This extension uses Scapy as a back end for radio packet manipulation. As the gateway from Scapy to the SDR device, GNU Radio is used (Picod et al. 2014).

RFCat
RFCat is a tool for interaction with compatible radio transceivers like the YARD Stick One. To tune a transceiver, an interactive Python shell is used (atlas0fd00m 2018).

The following Fig. 15.7 shows an example program for sending a data stream modulated with OOK on 434 MHz.

SDR#
SDR# offers the same functionality than GQRX but is also supported on Windows PCs. It is developed by Airspy, which also produces different SDR hardware for aerospace monitoring (Airspy 2018).

15.4.4 Wireless Transmission Attack—Case Study

To show an example of a wireless reverse engineering process, a garage door remote was used. Remotely controllable garage doors are very common in industrial facilities, as well as in home automation scenarios. The used garage door is about 15 years old, which is in a typical life cycle of such devices.

The following approach is commonly used for proprietary systems and black box tests where the used technology is not known. As shown in Table 15.1, proprietary protocols are very common in industrial settings. Figure 15.8 shows the remote control. The remote itself has two buttons, where only one of them is used to open

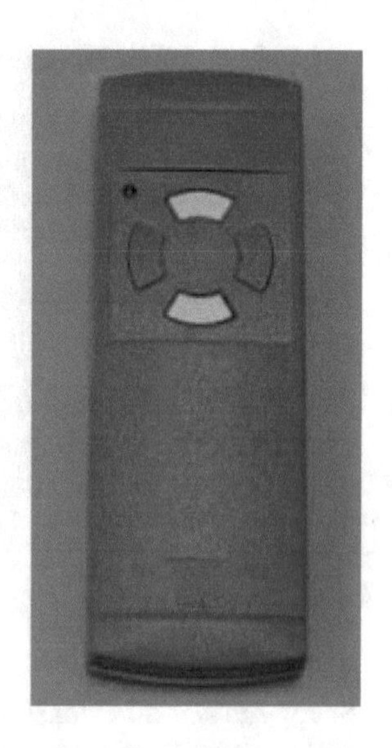

Fig. 15.8 Garage door opener

and close the garage door. The other button can be used for a different door. This makes the attack even easier because only one signal is used for both actions. The online research showed that the transmitter operates at 27 MHz. Interestingly the signal which was received at this frequency was quite weak.

The first tool that was used is GQRX. As mentioned before, it is a spectrum analyzer with a lot of functionality. It is able to record demodulated signals from the SDR, which then can be viewed as an audio file in Audacity.

Figure 15.9 shows GQRX tuned to 40.699 MHz. Due to the multipath behavior of RF systems, the signal at 27 MHz was apparently only a reflection.

To verify the assumption, osmocom_fft was used. Instead of recording demodulated signals like GQRX, osmocom_fft is able to record the raw signal data with the help of an SDR. With this raw signal, replay attacks are possible. Figure 15.10 shows the signal detection with osmocom_fft.

Another way of detecting the signal is directly with GNU Radio. The advantage over the previous methods is that it is possible to include different DSP blocks like filters. Therefore replay attacks can be performed much more efficiently.

GNU Radio comes with a graphical user interface called companion. The major benefit of the companion is that it does not require any programming knowledge. A companion graph consists of different blocks for various steps in the signal processing. Figure 15.11 shows the used blocks to receive a signal with the HackRF and displays it on an FFT. Between the source block and the visualization is a band-pass filter embedded which cuts the signal between the defined frequencies.

Figure 15.12 shows the amplified signal spike of the garage door opener with band-pass filter embedded.

Now because the signal is located and isolated from the noise, it can be sent via GNU Radio. Similar to the source block, the gr-osmosdr library contains a sink to transmit signals with the HackRF. Therefore the signal was saved in a file, which

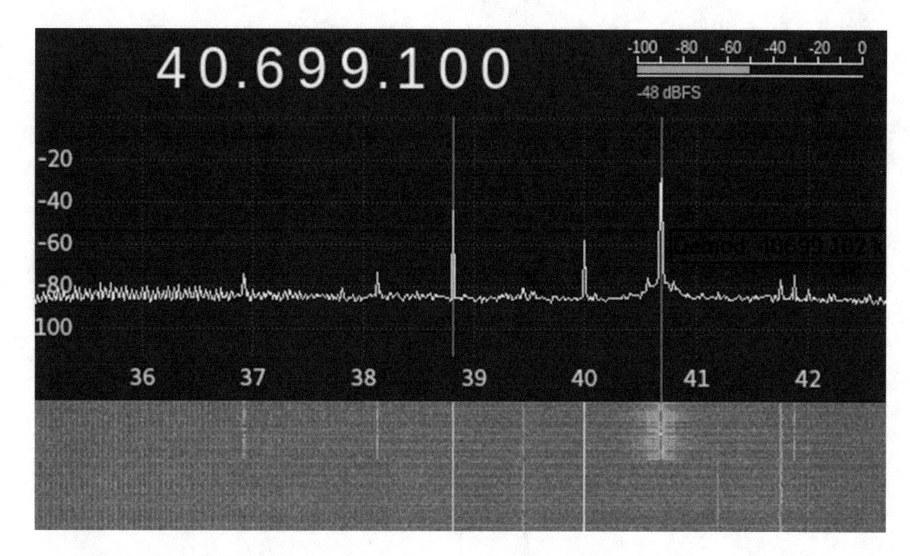

Fig. 15.9 GQRX showing signal of garage door opener at 40.69 MHz

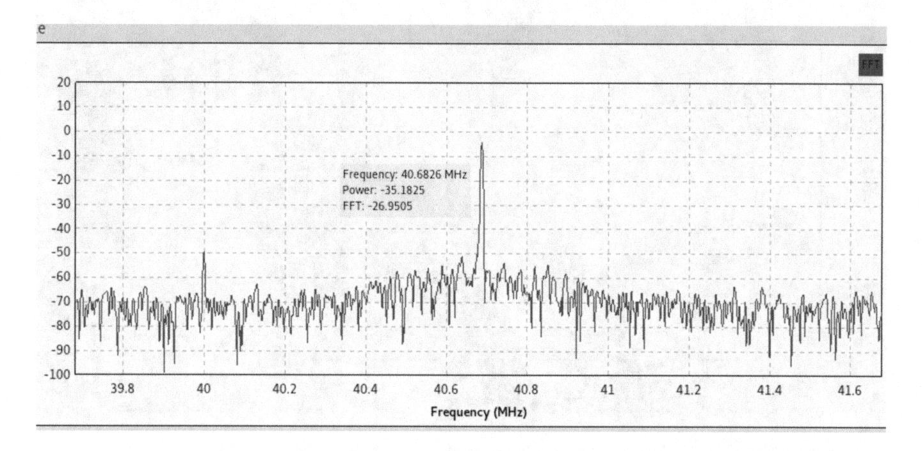

Fig. 15.10 Osmocom_fft spectrum analyzer

is also shown in Fig. 15.11. This basic form of replay attack already worked and therefore opened the garage door.

Another tool which is capable of executing replay attacks is the hackrf_transfer tool. It is part of the hackrf package from the Debian package manager. Figure 15.13 shows the commands for recording and replaying the signal on the command line. With the parameter "-f" the frequency can be specified. The "-x" parameter of the transmit signal defines the gain of the HackRF.

To further analyze the signal sent by the remote, the recorded signal from osmocom_fft was used. Therefore the file was opened in inspectrum, which is shown

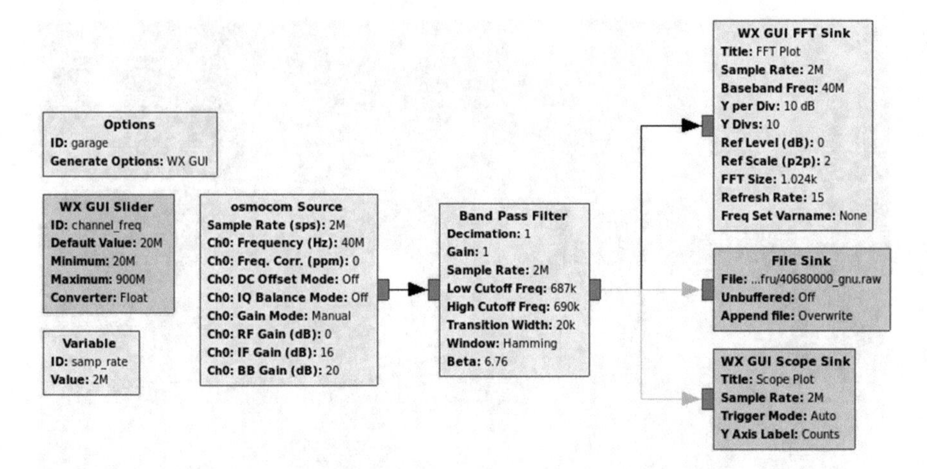

Fig. 15.11 GNU Radio companion with Osmocom source and band-pass filter

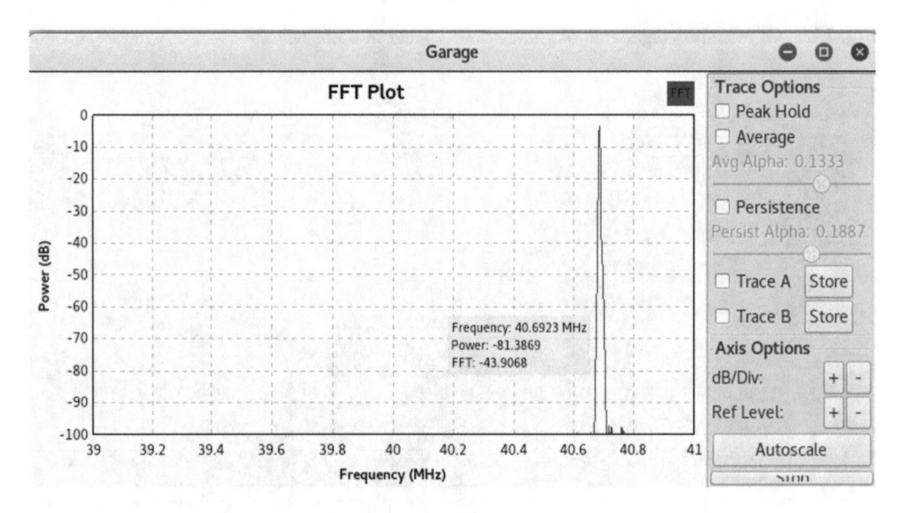

Fig. 15.12 GNU Radio companion spectrum analyzer with band-pass enabled

```
hackrf_transfer -r 386840000.raw -f 386840000
hackrf_transfer -t 386840000.raw -f 386840000 -x 20
```

Fig. 15.13 Command for executing a replay attack with the HackRF

in Fig. 15.14. One can easily identify the OOK modulation used. The same analysis would also have been possible with the use of the recorded signal from GQRX.

Also the encoding technique is obvious in this example. After the long pulse, the signal starts with ten long pulses, which most likely describe a binary “1”, followed by short pulses, which represent a binary “0”. With this knowledge, it was now

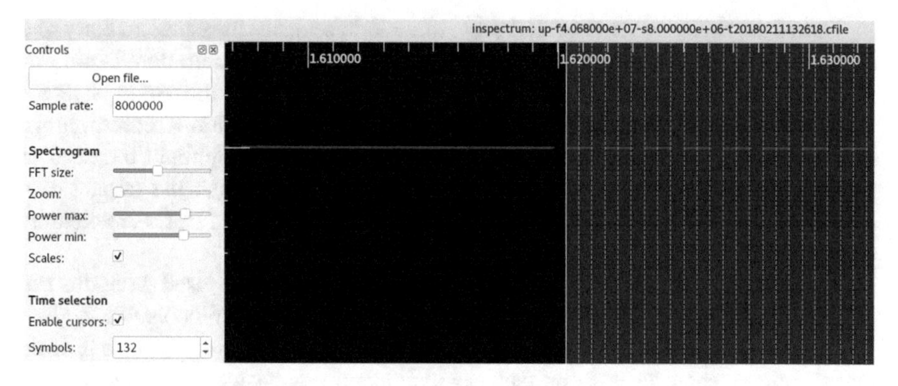

Fig. 15.14 Recorded file from osmocom_fft opened in inspectrum

```
1111 1111 1100 0010 0110 1011 1010 0111 1111 1000 0011 - up
1111 1111 1100 0010 0110 1011 1010 0111 1111 0001 0011 - down
```

Fig. 15.15 Bit representation of the captured signal from the remote control

possible to decode the signals. Listing 15.15 shows the bit representation of both possible signals from the remote control.

Both signals start with a similar pattern which starts with a long set of "1" bits. This looks like a preamble, which basically tells the device that a signal is starting. After the preamble, a fixed pattern is present. This is probably the encoding of the garage itself. The last 8 bits indicate the pressed button on the remote.

With this pattern, it is now also possible to send the signal with the YARD Stick One and the RFCat software.

This case study should show the shortcomings of an engineering process, where security was not included sufficiently. To stop attacks like the one shown, the engineers could have integrated some kind of cryptographic countermeasures, like rolling codes in combination with a short validity period. The validity period protects again DoS and replay attacks in combination.

Depending on the used technology, the investigation process differs. As mentioned in the beginning, the approach shown in this case study would be applicable in scenarios where a proprietary technology is in place, or a black box test is conducted.

15.5 Conclusion

As shown in this chapter, wireless technologies are already well established in the industrial field. Many different technologies are available for different purposes, especially in the field level with PLCs and in the sensor level. Because industrial machines have a long lifetime, wireless security needs to play a major part in the engineering process.

Therefore, this chapter provided ways to include security measures in early and later stages of such processes. Particularly the practice of Penetration Testing of wireless systems was analyzed in detail. The chapter proposes the use of SDR for this process. As an example for a proprietary protocol investigation, a remote garage door was used. The vulnerabilities found in this device also highlight the need of security measures early in every engineering process. Although the impact of a hacked garage door may be minor, attacks on wireless connected PLCs or actuators on industrial machines may also raise safety issues.

As mentioned in Sect. 15.2.4, also common technologies used in industrial settings have been affected by multiple vulnerabilities. Even protocols like ZigBee which have different security measures in place were already affected. This is often due to implementation bugs or faults, which again emphasize the application of Penetration Testing during the lifespan of devices. To conclude the chapter, no device can be considered secure, just because there is no applicable attack at the moment.

References

Airspy. (2018). Airspy low cost high performance sdr. https://airspy.com/. Accessed 15 August 2018.

Akerberg, J., Gidlund, M., & Bjoerkman, M. (2011). Future research challenges in wireless sensor and actuator networks targeting industrial automation. In *2011 9th IEEE International Conference on Industrial Informatics* (pp. 410–415). https://doi.org/10.1109/INDIN.2011.6034912.

Alliance, W. F. (2006). Wpa2™ security now mandatory for wi-fi certified™ products. https://www.wi-fi.org/news-events/newsroom/wpa2-security-now-mandatory-for-wi-fi-certified-products. Accessed 04 March 2019.

Alliance, W. F. (2018). Wi-fi alliance® introduces wi-fi certified wpa3™ security. https://www.wi-fi.org/news-events/newsroom/wi-fi-alliance-introduces-wi-fi-certified-wpa3-security. Accessed: 04 March 2019.

Alrabady, A. I., & Mahmud, S. M. (2005). Analysis of attacks against the security of keyless-entry systems for vehicles and suggestions for improved designs. *IEEE Transactions on Vehicular Technology, 54*(1), 41–50. https://doi.org/10.1109/TVT.2004.838829.

Analog Devices Inc. (2018a). Adalm-pluto. http://www.analog.com/en/design-center/evaluation-hardware-and-software/evaluation-boards-kits/ADALM-PLUTO.html. Accessed 11 August 2018.

Analog Devices Inc. (2018b). Adalm-pluto sdr: Unboxing and initial testing. https://www.rtl-sdr.com/adalm-pluto-sdr-unboxing-and-initial-testing/. Accessed 11 August 2018.

Analog Devices Inc. (2018c). Why "pluto". https://wiki.analog.com/university/tools/pluto/users/name. Accessed 11 August 2018.

atlas0fd00m. (2018). Rfcat. https://github.com/atlas0fd00m/rfcat. Accessed 15 August 2018.

Audacity. (2018). Audacity. https://www.audacityteam.org/. Accessed 15 August 2018.

Bluetooth, S. (2016). Bluetooth core specification v5. 0. San Jose, CA: Bluetooth SIG.

Bowers, B. (2012). Zigbee wireless security: A new age penetration testers toolkit.

Caro, D., et al. (2014). Wireless networks for industrial automation. ISA.

Chapman, A. (2014). Hacking into internet connected light bulbs. Context 4.

Chernyshev, M. (2013). Verification of primitive sub ghz rf replay attack techniques based on visual signal analysis.

Clancy, T. C., & Goergen, N. (2008). Security in cognitive radio networks: Threats and mitigation. In *2008 3rd International Conference on Cognitive Radio Oriented Wireless Networks and Communications (CrownCom 2008)* (pp. 1–8). https://doi.org/10.1109/CROWNCOM.2008.4562534.

Cryptic Apps. (2018). Hopper v4. https://www.hopperapp.com/. Accessed 19 December 2018.

Ettus. (2018) Usrp b200. https://www.ettus.com/product/details/UB200-KIT. Accessed 11 August 2018.

Fadlullah, Z. M., Nishiyama, H., Kato, N., & Fouda, M. M. (2013). Intrusion detection system (ids) for combating attacks against cognitive radio networks. *IEEE Network, 27*(3), 51–56. https://doi.org/10.1109/MNET.2013.6523809.

FCC. (2018). Fcc - what we do. https://www.fcc.gov/about-fcc/what-we-do. Accessed 19 August 2018.

Fragkiadakis, A. G., Tragos, E. Z., & Askoxylakis, I. G. (2013). A survey on security threats and detection techniques in cognitive radio networks. *IEEE Communications Surveys Tutorials, 15*(1), 428–445. https://doi.org/10.1109/SURV.2011.122211.00162.

Frotzscher, A., Wetzker, U., Bauer, M., Rentschler, M., Beyer, M., Elspass, S., et al. (2014). Requirements and current solutions of wireless communication in industrial automation. In *2014 IEEE International Conference on Communications Workshops (ICC)* (pp. 67–72). https://doi.org/10.1109/ICCW.2014.6881174.

Github. (2018). Github. https://github.com/. Accessed 20 August 2018.

GQRX. (2018). Gqrx sdr. http://gqrx.dk/. Accessed 12 August 2018.

Great Scott Gadgets. (2016). Hackrf one. https://greatscottgadgets.com/hackrf/. Accessed 01 March 2018.

Haataja, K. M. J., & Hypponen, K. (2008). Man-in-the-middle attacks on bluetooth: A comparative analysis, a novel attack, and countermeasures. In *2008 3rd International Symposium on Communications, Control and Signal Processing* (pp. 1096–1102). https://doi.org/10.1109/ISCCSP.2008.4537388.

Hayes, K. (2016). Penetration testing vs red teaming. https://blog.rapid7.com/2016/06/23/penetration-testing-vs-red-teaming-the-age-old-debate-of-pirates-vs-ninja-continues/. Accessed. 26 July 2018.

Heinaearo, K. (2015). Cyber attacking tactical radio networks. In *2015 International Conference on Military Communications and Information Systems (ICMCIS)* (pp. 1–6). https://doi.org/10.1109/ICMCIS.2015.7158684.

Hex-Rays. (2018). Ida:about. https://www.hex-rays.com/products/ida/index.shtml. Accessed 19 December 2018.

Howard, M., & Lipner, S. (2006). *The security development lifecycle*, Vol. 8. Redmond: Microsoft Press.

Kamkar, S. (2018). Rolljam. https://www.wired.com/2015/08/hackers-tiny-device-unlocks-cars-opens-garages/. Accessed 06 July 2018.

Kar, G., Mustafa, H., Wang, Y., Chen, Y., Xu, W., Gruteser, M., et al. (2014). Detection of on-road vehicles emanating gps interference. In *Proceedings of the 2014 ACM SIGSAC Conference on Computer and Communications Security, CCS '14* (pp. 621–632). New York, NY: ACM. https://doi.org/10.1145/2660267.2660336.

Klostermeier, G., & Deeg, M. (2016). Security of modern wireless input. https://www.youtube.com/watch?v=JPaiqJIMFCU.

Li, C., Raghunathan, A., & Jha, N. K. (2011). Hijacking an insulin pump: Security attacks and defenses for a diabetes therapy system. In *13th IEEE International Conference on e-Health Networking Applications and Services (Healthcom), 2011* (pp. 150–156). Piscataway: IEEE.

Li, Z., Xu, W., Miller, R., & Trappe, W. (2006). Securing wireless systems via lower layer enforcements. In *Proceedings of the 5th ACM Workshop on Wireless Security, WiSe '06* (pp. 33–42). New York, NY: ACM. https://doi.org/10.1145/1161289.1161297.

Liechti, M., Lenders, V., & Giustiniano, D. (2015). Jamming mitigation by randomized bandwidth hopping. In *Proceedings of the 11th ACM Conference on Emerging Networking Experiments and Technologies, CoNEXT '15* (pp. 11:1–11:13). New York, NY: ACM. https://doi.org/10.1145/2716281.2836096

Machado, R. G., & Wyglinski, A. M. (2015). Software-defined radio: Bridging the analog-digital divide. *Proceedings of the IEEE 103*(3), 409–423. https://doi.org/10.1109/JPROC.2015.2399173.

Miek. (2018). Inspectrum. https://github.com/miek/inspectrum. Accessed 12 August 2018.

NooElec. (2018). Nooelec nesdr mini sdr and dvb-t usb stick. http://www.nooelec.com/store/sdr/sdr-receivers/nesdr-mini-rtl2832-r820t.html. Accessed 01 March 2018.

Nuand. (2018). Bladerf. https://www.nuand.com/. Accessed 01 March 2018.

Ossmann, M. (2016). Rapid radio reversing. Tech. rep.

Ossmann, M., & Spill, D. (2017). What's on the wireless? Automating rf signal identification. Tech. rep.

Owor, R. S., Dajani, K., Okonkwo, Z., & Hamilton, J. (2007). An elliptical cryptographic algorithm for rf wireless devices. In *Proceedings of the 39th Conference on Winter Simulation: 40 Years! The Best is Yet to Come, WSC '07* (pp. 1424–1429). Piscataway, NJ: IEEE Press.

Pancake. (2018). Radare. https://www.radare.org/r/. Accessed 19 August 2018.

Picod, J., Lebrun, A., & Demay, J. (2014). Bringing software defined radio to the penetration testing community. In *Black Hat USA Conference*

Pohl, J., & Noack, A. (2018). Universal radio hacker: A suite for analyzing and attacking stateful wireless protocols. In *12th USENIX Workshop on Offensive Technologies (WOOT 18)*. Baltimore, MD: USENIX Association.

Punal, O., Aguiar, A., & Gross, J. (2012). In vanets we trust?: Characterizing rf jamming in vehicular networks. In *Proceedings of the Ninth ACM International Workshop on Vehicular Inter-networking, Systems, and Applications, VANET '12* (pp. 83–92). New York, NY: ACM. https://doi.org/10.1145/2307888.2307903.

Qu, Y., & Chan, P. (2016). Assessing vulnerabilities in bluetooth low energy (ble) wireless network based iot systems. In *2016 IEEE 2nd International Conference on Big Data Security on Cloud (BigDataSecurity), IEEE International Conference on High Performance and Smart Computing (HPSC), and IEEE International Conference on Intelligent Data and Security (IDS)* (pp. 42–48). https://doi.org/10.1109/BigDataSecurity-HPSC-IDS.2016.63.

Reaves, B., & Morris, T. (2012). Analysis and mitigation of vulnerabilities in short-range wireless communications for industrial control systems. *International Journal of Critical Infrastructure Protection, 5*(3–4), 154–174.

Ronen, E., Shamir, A., Weingarten, A., & O'Flynn, C. (2017). Iot goes nuclear: Creating a zigbee chain reaction. In *2017 IEEE Symposium on Security and Privacy (SP)* (pp. 195–212). https://doi.org/10.1109/SP.2017.14.

RTL-SDR. (2018). About rtl-sdr. https://www.rtl-sdr.com/about-rtl-sdr/. Accessed 10 August 2018.

Sastry, N., & Wagner, D. (2004). Security considerations for ieee 802.15.4 networks. In *Proceedings of the 3rd ACM Workshop on Wireless Security, WiSe '04* (pp. 32–42). New York, NY: ACM. https://doi.org/10.1145/1023646.1023654.

Shafagh, H., & Hithnawi, A. (2014). Poster: Come closer: Proximity-based authentication for the internet of things. In *Proceedings of the 20th Annual International Conference on Mobile Computing and Networking, MobiCom '14* (pp. 421–424). New York, NY: ACM. https://doi.org/10.1145/2639108.2642904.

Shodan. (2018). Shodan. https://www.shodan.io/. Accessed 29 July 2018.

Stolnikov, D. (2018). osmocom gnu radio blocks. https://osmocom.org/projects/gr-osmosdr/wiki/GrOsmoSDR. Accessed 12 August 2018.

The GNU Radio Foundation (2018) What is gnu radio? https://www.gnuradio.org/about/. Accessed 12 August 2018.

The Penetration Testing Standard. (2014). The penetration testing standard. http://www.pentest-standard.org. Accessed 26 July 2018.

Vanhoef, M. (2017). Key reinstallation attacks. https://www.krackattacks.com/. Accessed 04 March 2019.

Vanhoef, M., & Piessens, F. (2017). Key reinstallation attacks: Forcing nonce reuse in WPA2. In *Proceedings of the 24th ACM Conference on Computer and Communications Security (CCS)*. New York, NY: ACM.

Wright, J., & Cache, J. (2015). *Hacking exposed wireless: Wireless security secrets & solutions* (3rd ed.). New York: McGraw-Hill Education Group.

Chapter 16
Secure and Safe IIoT Systems via Machine and Deep Learning Approaches

Aris S. Lalos, Athanasios P. Kalogeras, Christos Koulamas, Christos Tselios, Christos Alexakos, and Dimitrios Serpanos

Abstract This chapter reviews security and engineering system safety challenges for Internet of Things (IoT) applications in industrial environments. On the one hand, security concerns arise from the expanding attack surface of long-running technical systems due to the increasing connectivity on all levels of the industrial automation pyramid. On the other hand, safety concerns magnify the consequences of traditional security attacks. Based on the thorough analysis of potential security and safety issues of IoT systems, the chapter surveys machine learning and deep learning (ML/DL) methods that can be applied to counter the security and safety threats that emerge in this context. In particular, the chapter explores how ML/DL methods can be leveraged in the engineering phase for designing more secure and safe IoT-enabled long-running technical systems. However, the peculiarities of IoT environments (e.g., resource-constrained devices with limited memory, energy, and computational capabilities) still represent a barrier to the adoption of these methods. Thus, this chapter also discusses the limitations of ML/DL methods for IoT security and how they might be overcome in future work by pursuing the suggested research directions.

Keywords Machine learning · Deep learning · Security threats in IoT

16.1 Introduction

The Internet of Things is envisioned as a multitude of heterogeneous devices densely interconnected and communicating with the objective of accomplishing a diverse range of objectives, often collaboratively. The term "Internet of Things" was used

A. S. Lalos (✉) · A. P. Kalogeras · C. Koulamas · C. Alexakos · D. Serpanos
Industrial Systems Institute, ATHENA Research Center, Patras, Greece
e-mail: Lalos@isi.gr; Kalogeras@isi.gr; Koulamas@isi.gr; Alexakos@isi.gr; Serpanos@isi.gr

C. Tselios
Citrix Systems, Patras, Greece
e-mail: christos.tselios@citrix.com

S. Biffl et al. (eds.), *Security and Quality in Cyber-Physical Systems Engineering*,
https://doi.org/10.1007/978-3-030-25312-7_16

for the first time in Mr. Kevin Ashton's presentation in 1999,[1] while a significant milestone from the perspective of the IoT was the period between the years 2008 and 2009, when, according to the Cisco estimation, the number of devices (in general) connected to the Internet exceeds the number of the world's population (Evans 2011). The advent of IoT is accompanied by a number of developments: miniaturization of devices and sensors, increasing mobility of devices, wearable devices, ubiquitous robotics and growing automation of all functions of IoT, presenting numerous benefits in a diverse number of applications ranging from smart homes, smart health, and energy management to connected cars and smart farming.

As a term, Industrial IoT (IIOT) has been introduced to describe the application of IoT in the industry, namely, the utilization of disruptive elements such as sensors, actuators, control systems, machine-to-machine communication interfaces, and enhanced security mechanisms to improve industrial systems and shape the futuristic Smart Factory concept. The proliferation of IoT in industrial environments and value chains will allow companies, manufactures, and workers to operate in a more efficient manner and will have a great impact in several fields, such as automation, industrial manufacturing, logistics, business processes, process management, and transportation (Schmidt et al. 2015). Along with the overall expansion of the core manufacturing process, the digital transformation advancements and the constantly rising node interconnectivity allow new applications to emerge, mostly related to (i) process automation and optimization, (ii) optimized resource consumption, and (iii) autonomous system generation and security intensification. It is already identified that IIoT radically changes the product life cycle, thus providing a new way of doing business in general and highly affecting the overall competitiveness of any organization. As mentioned in (Schmidt et al. 2015), IIoT will integrate products and processes in such a way that will eventually shift the productivity line effectiveness from mass production to mass customization. This simply translates to more modular and configurable products, tailor-made according to specific customer requirements (Jazdi 2014). In a nutshell, IIoT will transform manufacturing as we know it through innovative and highly agile products and services that can become partially independent, responsive, and interactive, track their activity in real time, and optimize the whole value chain into providing relevant status information throughout their life cycle.

The imminent adoption of the emerging IIoT paradigm will provide a significant boost also to the concept of Industry 4.0, a convoluted technological system that has been gaining significant traction over the last few years. Industry 4.0 can be seen as a superordinate term for describing a novel industrial paradigm which aims to combine among others cyber-physical manufacturing systems (CPMS), omnipresent and time-sensitive networks, robotics, big data analytics, and edge computing paradigms. The adoption of these technological pillars is crucial for the development of a highly

[1]"I could be wrong, but I'm fairly sure the phrase 'Internet of Things' started life as the title of a presentation I made at Procter & Gamble (P&G) in 1999," Kevin Ashton, RFID Journal, 22 June 2009.

intelligent manufacturing process that will incorporate machines, sensors, production modules, and incomplete products, all enhanced with the ability to independently exchange information, trigger actions, and control each other, thus creating a fully automated, optimized, and independent manufacturing environment (Weyer et al. 2015). The Industrial IoT is a key element of Industry 4.0, bringing together modern sensor technology, fog–cloud computing platforms, and AI to create intelligent, self-optimizing industrial equipment and facilities.

The aforementioned advancements can be definitely perceived as a big blessing; however, big challenges also arise related to the dynamic management and security mechanisms of Industrial IoT (IIoT) components across heterogeneous objects, transmission technologies, and networking architectures. Another major area of concern is privacy with regard to personal information that will potentially reside on networks, also a likely target for cyber criminals. Finally, it should be mentioned that IoT allows the virtual world to interact with the physical world, and this brings big safety issues. Machine and deep learning (ML/DL) have advanced considerably over the last few years (Jordan and Mitchell 2015; Goodfellow et al. 2016), and machine intelligence has transitioned from laboratory curiosity to practical machinery in several important applications. The ability to monitor IoT devices intelligently provides a significant solution to new or zero-day attacks. ML and DL are powerful methods of data exploration for learning about "normal" and "abnormal" behavior according to how IoT components and devices perform within the IoT environment. Consequently, these methods are important in transforming the security of IoT systems from merely facilitating secure communication between devices to security-based intelligence systems.

The goal of this chapter is to provide a comprehensive survey of ML methods and recent advances in DL methods that can be used to develop enhanced security methods for modern IoT and IIoT systems that are used in smart manufacturing environments. IoT security threats, either inherent or newly introduced, are presented, and various potential IoT system areas of the attack surface and the possible threats and vulnerabilities are discussed. A thorough discussion of the opportunities and challenges involved in applying ML/DL to IoT security is offered. The presented solutions and challenges are expected to provide a novel insight at a key area with renewed research interest, where high potential for novel improvements is feasible in the near future.

The rest of this chapter is organized as follows: Sect. 16.2 provides a review of different layered architecture for IoT and IIoT systems. A comprehensive discussion on the potential vulnerabilities and areas of the attack surface of IoT systems is provided in Sect. 16.3. Section 16.4 presents an in-depth review of the machine learning (ML) and recent advances in deep learning (DL) methods that have been applied for identifying IoT security threats and vulnerabilities. IoT physical world applications and Safety Challenges are presented in Sect. 16.5, and conclusions are drawn in Sect. 16.6.

16.2 IoT and IIoT Layered Architecture Review

The cornerstone for the successful design and deployment of IoT infrastructure and relevant IoT applications is the efficient combination of cutting-edge technological achievements in the areas of networks, hardware, and informatics (Atzori et al. 2010). Only hierarchical, modular, loosely coupled, flexible, and scalable system architectures can manage and coordinate this complex system of different components, networks, data, and software. From the architectural perspective, the first approaches of IoT ecosystems deploy the service-oriented architecture (SOA) as the inspiration for designing and implementing their IoT solutions (Xu et al. 2014). SOA key idea is the fact that each system exposes its independent functionalities in terms of web services, which can be invoked by other systems over computer networks. IoT consists of devices (systems) that are connected through networking. Thus, SOA is considered appropriate to support IoT at the early years (Atzori et al. 2010; Miorandi et al. 2012).

The evolution of IoT brought new challenges such as utilization of limited computational resources, low power consumption, networked devices distributed in a large geographical area, real-time and latency sensitivity, collection and processing of large amount of data, new business models, and social requirements. Although the multi-layer SOA architecture provided a workable solution for IoT, these new challenges forced researchers to seek out alternatives to the SOA. After a decade of IoT existence, there is no widely accepted reference architecture that is established as a standardized design approach for IoT. Closer to SOA, most of researchers' opinions about conventional IoT architecture (Mashal et al. 2015; Mainetti et al. 2011; Wu et al. 2010) follow a three-layer approach which comprises:

1. The *perception (or sensing) layer* being the physical layer, consisting of smart objects/devices such as sensors and actuators that are able for sensing and gathering information about the environment as well as interacting with it and its elements.
2. The *network layer* realizing the connection and communication of the smart objects, network devices, and servers. Furthermore, the network layer is responsible for the transmission and processing of sensor data.
3. The *application layer* consisting of applications that deliver IoT-based services to the end users, including smart homes, smart energy, smart health, and smart cities.

As the three-layer architecture, due to its simplicity, was a popular solution, researchers identify that the complexity of orchestrating the large number of smart devices, as well as the size of associated information, cannot be handled efficiently at the network or application layer. The solution was the introduction of a layer between them, usually named as *middleware layer*, thus defining a four-layer architecture. This layer is responsible for service management and storage of data, as well as for decision-making based on the results of information processing. Such a paradigm is the IoT reference architecture proposed by ITU-T (International Telecommunications Union-Telecommunication Standardization Sector) (ITY-T 2012), where the service and application support layer (middleware layer) provides generic services, such as data processing or data storage, and application-specific services, which cater for the requirements of diversified applications.

The four-layer model provides the flexibility in designing IoT applications, overcoming the most of technical challenges. But the IoT applications are more complex than the classic computer applications regarding their target users. Due to their nature, IoT applications involve many collaborative devices satisfying the needs of various stakeholders as end users (Evans 2011), meaning that different user requirements must be met by a distributed network of heterogeneous nodes. Organizations from both private and public sectors, or even individual citizens, are some examples of potential end users of an IoT application, i.e., a smart city paradigm. The diversity of business requirements and the social impact of the IoT applications lead to the specification of another layer on top of the application layer, separating the data analysis and machine and deep learning from the business models that provide this data to the users. The commonly known business layer has to do with the conversion of the data received by application level to meaningful services to the different groups of users (Wu et al. 2010; Sethi and Sarangi 2017; Aazam et al. 2014; Khan et al. 2012). Furthermore, data analytics provide insights with practical and useful knowledge to the users. Furthermore, data access management and users' privacy are some of the most important features of this layer.

The evolution of the layered IoT architecture was unavoidable, and the addition of new layers permitted both the inclusion of all the factors that affect the operation of IoT applications and the development of technologies and tools to deal with the modern challenges. Figure 16.1 shows the evolution of IoT reference architectures, ending with the five-layered architecture. At this point, it should be mentioned

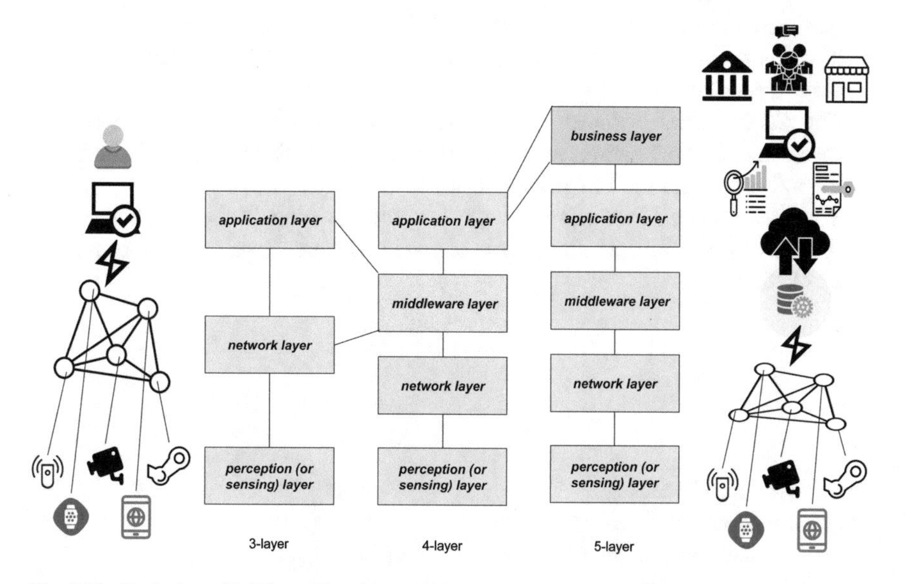

Fig. 16.1 Evolution of IoT layered architecture

that IoT is one of the technologies that were rapidly integrated by the industry into its products. Today, following academia's paradigm and the concept of the Fourth Industrial Revolution (4IR), the industry took large steps toward the well-known Industrial Internet of Things (IIoT) by establishing large and complex IIoT applications in various deployment areas (i.e., cities, energy grid, buildings, manufacturing, etc.). Although the industry is favorable to work with standards, still there is a lack of standardization relevant to the architectural design of IIoT applications. Nevertheless, significant consortiums, consisting of the key industry players, were created worldwide in order to define such standards. The Industrial Internet Consortium (IIC)[2] (USA) and the Industry 4.0 platform[3] (Europe) are two of the mainstream initiatives toward standardization of IIoT systems, supplemented by further initiatives such as Japan's Society 5.0[4] and Made in China 2025.[5] As early results, each of the first two initiatives has proposed IIoT Architecture reference models providing a guidance by specifications for the development of system and application architectures.

The Industry 4.0 platform introduced the Reference Architectural Model Industry 4.0 (RAMI 4.0) (Adolphs et al. 2016). RAMI 4.0 is recognized as a DIN standard (DIN SPEC 91345) and an international pre-standard (IEC PAS 63088). RAMI 4.0 is based on a three-dimensional model covering all the industrial aspects from the industrial hierarchy to the product life cycle. Its three dimensions are (a) the hierarchy defining the functional areas of the IIoT applications selecting from smart product, smart factory, and connected world; (b) architecture, which provides the system architecture; and finally (c) the product life cycle, which covers development, production, and maintenance aspects. Focusing on the architecture dimension, RAMI 4.0 defines six layers:

- The *Asset Layer* representing the physical layer including devices and their hardware parts as well as the human factor.
- The *Integration Layer* defining the provision of informational data and asset control services.
- The *Communication Layer* applying standardized communication between the assets and the applications at the higher layer, always following the formality of the information at the Integration Layer.
- The *Information Layer* dealing with the preprocessing of the information and the generation of events. In the case of events, the asset control services may be invoked.

[2]https://www.iiconsortium.org/.

[3]https://www.plattform-i40.de/.

[4]https://www8.cao.go.jp/cstp/english/society5_0/index.html.

[5]http://english.gov.cn/2016special/madeinchina2025/.

- The *Functional Layer* receiving preprocessed information from the Information Layer and implementing rules and decision-making logic. Furthermore, Functional Layer is the only remote access point to the data as in the layers below the data is protected for ensuring information integrity.
- The *Business Layer* being the layer where the functions of the Functional Layer are integrated to the business processes.

In the USA, the Industrial Internet Consortium (IIC) proposed the Industrial Internet Reference Architecture (IIRA) (Lin et al. 2017a). Contrary to the RAMI 4.0, which is specialized in the manufacturing business processes, IIRA deals with a wider range of IIoT applications, from transportation to energy. IIRA also follows a three-dimensional model, but with a different approach to RAMI 4.0. Its three dimensions are the (a) *product life cycle*; (b) *the industrial sectors* that define the area of deployment, and (c) *a four-level layer consisting of viewpoints*, each one associated with particular stakeholders and their concerns. The business viewpoint deals with business-oriented aspects, such as business value, expected return on investment, cost of maintenance, and product liability. The realization of the key capabilities defined by the business viewpoint is the main concern of the usage viewpoint. The next viewpoint, the functional viewpoint, deals with system functional components, interfaces, and interactions. The last viewpoint, the implementation viewpoint is concerned with the technologies and system components required, implementing the functional requirements defined at the functional viewpoint. IIRA is a general reference model, which doesn't define a specific architecture but proposes some architectural patterns that can be used to deal with functional requirements of an IIoT application. These three patterns are (a) three-tier architecture pattern, (b) gateway-mediated edge connectivity and management architecture pattern, and (c) layered databus pattern.

The existence of two different reference architectures led to the collaboration of the two involved consortiums toward the publication of mapping and alignment guidelines between RAMI 4.0 and IIRA (Lin et al. 2017b). From the architectural perspective, this effort focuses on the mapping of the functional blocks that can be defined in the IIRA model with the layers of architecture dimension of RAMI 4.0. In the context of this model alignment, Fig. 16.2 presents a mapping of the functional blocks of the IIoT architecture with the aforementioned IoT five-layer architecture. The adaptation of a layered architecture for the design of an IIoT infrastructure assists the engineers to clarify both the technologies that will be used at each layer and the implementation of the provided operations/services. Due to the difference of the functionalities and of the technologies/standards used at each layer, the abstraction of an IIoT infrastructure to independent layers allows the examination of security vulnerabilities and safety challenges separately for each layer. The following sections deal with the challenges of IoT security threats and vulnerabilities classifying them in the basic layers of an IoT architecture.

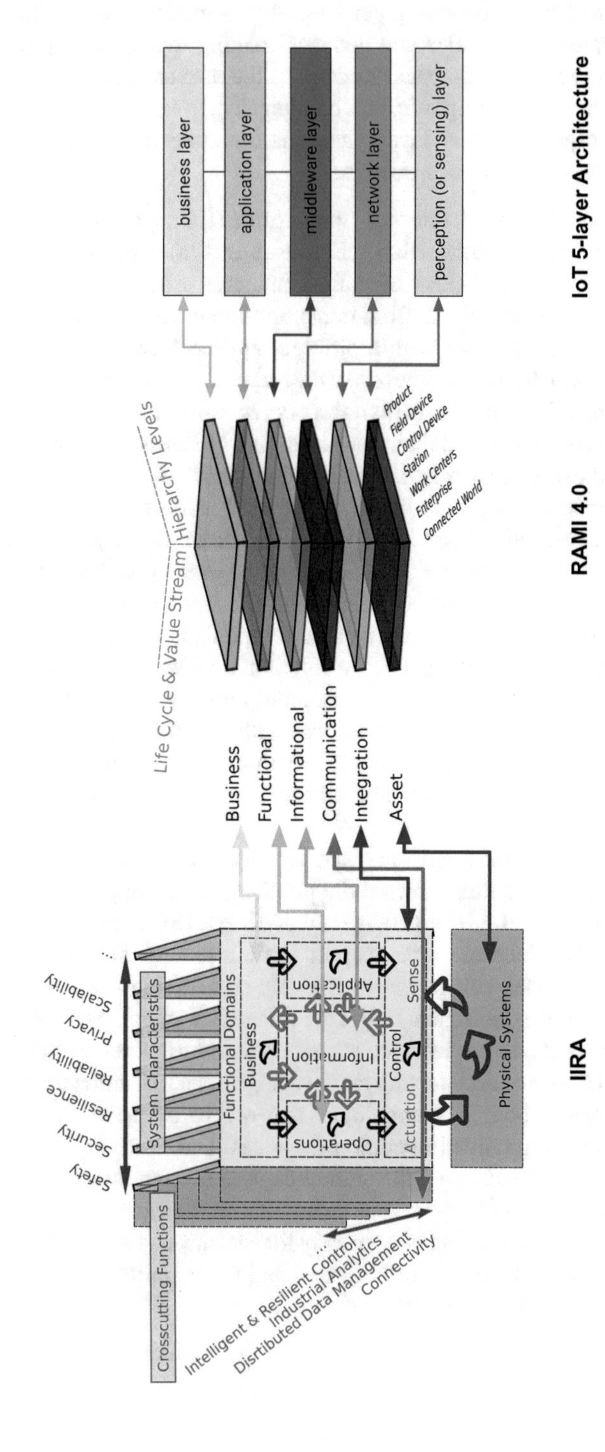

Fig. 16.2 Mapping between IIRA, RAMI 4.0, and IoT five-layered architecture. Partially taken from (Lin et al. 2017b)

16.3 IoT Security Threats and Vulnerabilities

As with any IT system, the principal information security requirements of availability, integrity, confidentiality, authentication/authorization, and non-repudiation constitute critical requirements of IoT-based systems as well. However, the specific characteristics of IoT system components define a well-differentiated domain, requiring thus unique approaches in identifying threats and vulnerabilities (Xu et al. 2014), as well as in detecting and responding to relevant attacks, in order to guarantee the trustworthiness of modern IoT-based systems. Important specificities of IoT systems are related to the typical involvement of (1) a large number of resource-constrained, wirelessly networked, miniaturized embedded devices and (2) distributed and/or centralized Big-Data processing infrastructures introducing significant security challenges. In such systems, these challenges become even harder to be addressed, due to the criticality of supported applications, considering also the Industrial IoT (IIoT) and its applications in critical infrastructures, as well as other systems in avionics, automotive, and medical equipment domains, where safety, reliability, and resilience are of the highest priority. There are already various existing studies and proposals in the literature to identify the peculiarities of IoT security threats (Humayed et al. 2017; Mena et al. 2018; Chen et al. 2018). Among the most representative efforts to structure the typically extensive threat taxonomies, in a way tailored to the IoT specifics, are these of the ENISA agency (ENISA Report 2017, 2018a,b,c) and the OWASP-IoT project (OWASP 2018), which are also referenced by the IIC Security Framework Architecture document (IIC 2016). According to (ENISA Report 2017, 2018a,b,c), there are 8–9 high-level threat groups and a large number of identified threats, depending on the case, while in (OWASP 2018) there are 18 identified areas of the attack surface and a multitude of possible vulnerabilities.

For a smart manufacturing application context case, the different threats identified by ENISA are grouped under the following high-level threat categories (ENISA Report 2018a):

- **Nefarious Activity/Abuse:** It classifies the most widely known threats, such as the *Denial of Service (DoS), malware, manipulation of hardware and software, manipulation of information, personal data abuse, brute force*, and other *targeted* attacks.
- **Eavesdropping/Interception/Hijacking:** This group contains main network-related threats, including the *man-in-the-middle* attacks or *session hijacking*, which involve eavesdropping and actively relaying of messages accompanied possibly by modifications or deletion of the transmitted data. It also contains *protocol hijacking* and *network reconnaissance*, which mainly lead to information leakage, including information related to passwords or network structure.
- **Physical Attacks:** It includes threats related to *device modifications*, such as tampering physically unsecured ports, and *device destruction* or theft (i.e., the attacker's goal is typically sabotage) attacks.

- **Unintentional Damage:** Unintentional changes of data or configuration or erroneous use and administration of devices and systems, as well as damages caused by a third party, such as a maintenance subcontractor or a manufacturer software update, are all considered as threats of this category.
- **Failures or Malfunctions:** This category describes the threats of a general device failure, either at the sensor/actuator or at the control system level. It also contains malfunctions due to various uncategorized *software vulnerabilities*, e.g., due to *weak or default passwords, software bugs*, and *configuration errors*, as well as failures due to *services* which the system depends on.
- **Outages:** This group includes the loss of availability of communication links, or power supply, as well as of higher-level needed support services.
- **Disaster:** *Natural disasters* (floods, landslides, etc.), as well as other *environmental disasters* related to the immediate IoT equipment environment, fall under this threat group.
- **Legal:** It refers to threats related to violation of *rules and regulations* or *breach of legislation* and *abuse of personal data*, as well as to threats related to *failures* to meet *contractual requirements*, all leading to possible financial losses either direct (fines) or indirect (reputation).

On a different perspective, the OWASP-IoT approach starts from the definition of the set of areas of the attack surface, for which then the various vulnerabilities are enumerated. The attack surface list is rather elaborate and includes (OWASP 2018):

- **Ecosystem (General):** interoperability standards, security enrollment, system decommissioning, lost access procedures, and other system-wide vulnerabilities
- **Device Memory:** Leakage of sensitive data (various types of credentials)
- **Device Physical Interfaces:** Firmware extraction, command interfaces, privilege escalations, tamper resistance, removable storage media, debug ports, and device ID exposure
- **Device Web Interface:** Code injection, broken authentication, sensitive data exposure, broken access control, security misconfigurations, cross-site scripting, insecure deserialization, vulnerable components, insufficient logging and monitoring, credential management
- **Device Firmware:** Sensitive data exposure, backdoor accounts, hardcoded credentials, encryption implementation, vulnerable services due to old software versions, security API exposure, firmware downgrades
- **Device Network Services:** Information disclosure, command interfaces, injection, DoS, unencrypted channels, poor encryption implementations, existence of development/test services, OTA update blocks, replay, no payload verification, no integrity checks, credential management
- **Administrative Interface:** Common web interface vulnerabilities, credential management, security/encryption options, logging options, two-factor authentication, insecure direct object references, inability to wipe device

- **Local Data Storage:** Unencrypted or weakly encrypted data, discovered keys, no integrity checks, static keys
- **Cloud Web Interface:** Common web interface vulnerabilities, credential management, transport encryption, two-factor authentication
- **Third-Party Backend APIs:** Device information leakage, location leakage
- **Update Mechanism:** Unencrypted updates, not signed, verified or authenticated updates, malicious updates, missing update mechanisms, no manual update mechanisms
- **Mobile Application:** Implicit trusts, username enumeration, account lockout, default credentials, weak passwords, insecure data storage, transport encryption, insecure password recovery, two-factor authentication
- **Vendor Backend APIs:** Inherent trusts, weak authentication and access controls, injection attacks, hidden services
- **Ecosystem Communication:** Heath checks, heartbeats, de-provisioning, updates
- **Network Traffic:** Protocol fuzzing, wireless medium, range
- **Authentication/Authorization:** Data disclosure or reuse, multiple schemes, weak authentication
- **Privacy:** Data disclosure
- **Hardware (Sensors):** Sensing environment manipulation, physical tampering and damage

Other taxonomies may follow a threat classification based on a purpose/target threat model, as in (Humayed et al. 2017), where the five threat classes are criminal, financial, political, privacy, and physical threats, followed by a detailed enumeration of application domain-specific, physical, cyber, and cyber-physical vulnerabilities. Alternatively, they follow a layered approach, as in (Chen et al. 2018), where the attack threats are categorized on a four-layer basis:

- **Application Layer:** Code injection, buffer overflow, sensitive data permission/manipulation
- **Middleware Layer:** Flooding attack, cloud malware injection, signature wrapping attack, web browser attack, SQL injection attack
- **Network Layer:** Traffic analysis, sniffing attack, DoS, Sybil, sinkhole, replay, man-in-the-middle attacks
- **Perception Layer:** Unauthorized tag access, tag cloning, eavesdropping, RF jamming, spoofing attack

Finally, as the overall IoT architecture contains also typical web components and interfaces, detailed classifications that apply to the wider web environment may also get into the picture (WASC 2012).

Attempting to organize the broad set of threats and areas of the attack surface under the structural view presented in the previous section, Table 16.1 can be constructed.

Table 16.1 Classification of vulnerabilities and threats in modern IoT and IIoT systems

Vulnerabilities, threats	Physical	Cyber	
Attack surface		Passive	Active
Physical device	Modifications Destruction Tampering Theft Failure Malfunction Power outage Link outage Environmental disasters Natural disasters	HW/SW failure Personal data leakage Unauthorized tag Access	DoS Malware False data injection HW/SW manipulation Info. manipulation Personal data abuse Brute force attacks Tag clonning
Network service	Failure Malfunction Environmental disasters Natural disasters Power outage Link outage	Network Reconnaissance Traffic analysis Eavesdropping Sniffing	DoS Man in the middle Session hijacking Protocol hijacking False data injection Sybil Sinkhole Replay Spoofing RF jamming
Cloud, web and application service	Failure Malfunction Environmental disasters Natural disasters Power outage Link outage	HW/SW failure Personal data Leakage	DoS Malware HW/SW manipulation Info. manipulation Personal data abuse Brute force & Targeted attacks Code injection Buffer overflow Signature wrapping Web browser attack SQL injection attack

16.4 Detailed Review of ML and DL Methods for Securing IoT Systems

Pervasive sensors continuously collecting massive amounts of information have rendered data-driven learning increasingly important. Learning algorithms focus on the construction of schemes that progress automatically through experience (Jordan and Mitchell 2015). Machine and deep learning approaches have been widely applied in a surprising number of applications including medical, financial, and automotive industry, and recently they are finding their way into the manufacturing industry, providing from increased production capacity to more efficient plant operation and everything in between (Sharp et al. 2018).

Machine-learning algorithms are usually classified into the following learning categories: supervised, unsupervised, semi-supervised, active, and reinforcement, as it is shown also in Fig. 16.3. Supervised algorithms are used for learning a function that maps an input to an output, based on several input-output pairs known as training data. Supervised learning approaches are applied for solving classification and regression problems, where the output variable is either a category (e.g., "threat" or "no threat") or a real value. Unsupervised learning algorithms model the underlying structure or distribution of data to learn more about the data without using any corresponding output variables. The unsupervised learning problems are further grouped into clustering and association problems. In the clustering case the goal is to discover inherent groupings in the data, while in the association case the focus is on finding rules that describe large portions of data, for example, learning temporal state-based specifications for electric power systems to accurately differentiate between disturbances, normal control operations, and cyber-attacks (Pan et al. 2015). Active learning emphasizes on learning from limited amount of training samples, based on the experience of users that play the role of "omniscient" to label the selected data (Yang et al. 2018). It is naturally suited for the design of intrusion detection systems, provided that the labeling process for intrusion detection is either a very time-consuming process or even impossible for cases that intrusion never happened before. Active learning boosts the power of machine learning by exploiting the experience of a domain expert, significantly decreasing the labeling efforts and increasing at the same time the reliability of a supervised learning model for intrusion detection. Finally, in reinforcement learning uses a software agent that learns an optimal policy of actions over the set of states in an environment. Depending on the performed action, the environment sends a reward to the agent, while each agent tries to maximize its rewards over time by choosing action that results in higher rewards. This approach has been widely adopted to obtain optimal or near-optimal, integrated maintenance and production control policies for deteriorating, stochastic production/inventory systems (Xanthopoulos et al. 2018).

Deep learning is a subcategory of machine learning that focuses on learning data representations. Most deep learning approaches are based on artificial neural networks and more specifically they use a cascade of multiple layers of nonlinear processing units for extracting informative features. Successive layers use the output from the previous layer as input. DL approaches can be also classified into supervised and unsupervised schemes, and their main characteristic is their ability to learn multiple machine and deep levels that correspond to different levels of abstraction. In the following part of this section, we discuss both ML and DL approaches, to provide readers with in-depth review of both of them, and we focus on applications in securing Industrial IoT systems.

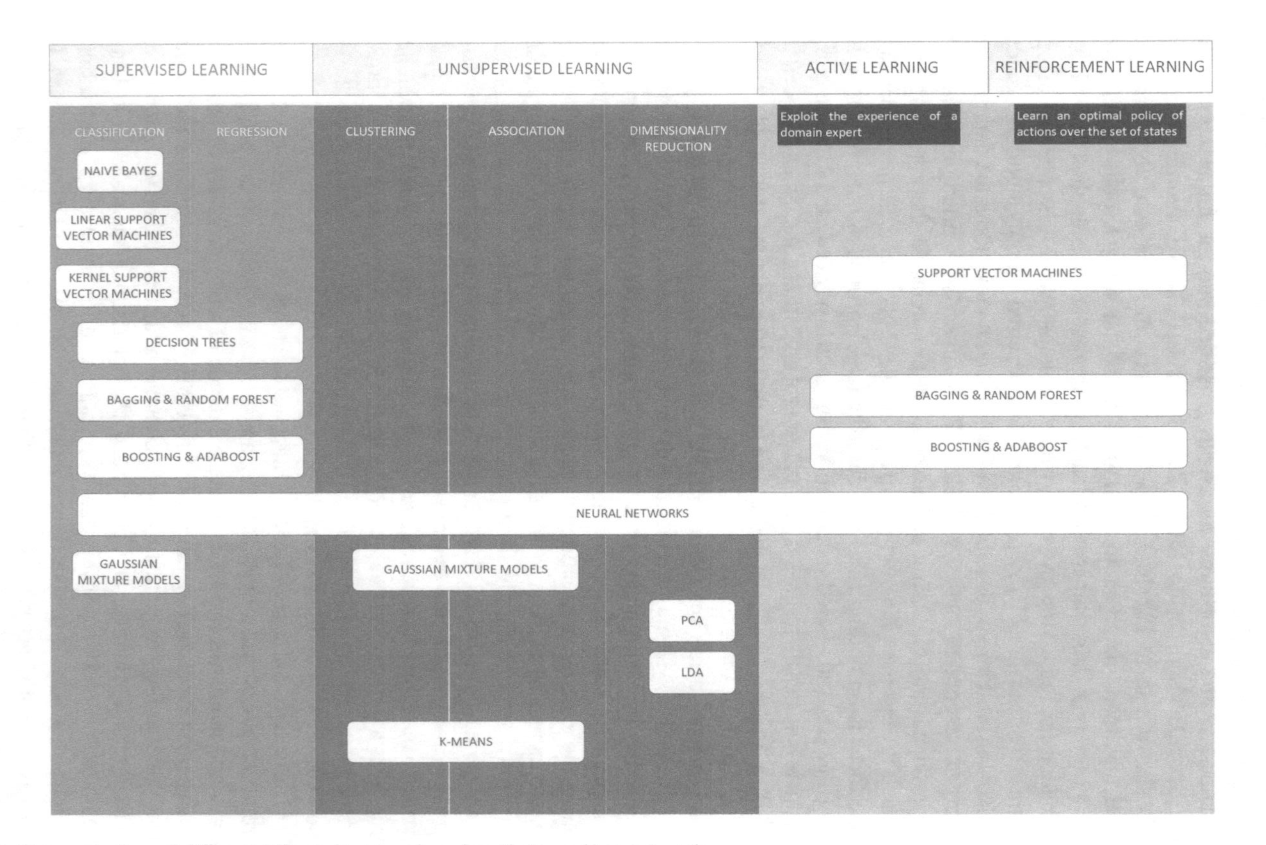

Fig. 16.3 Categorization of different ML applications based on their working principle

16.4.1 Machine Learning (ML) Methods for IIoT Security

This subsection focuses on the presentation of the most common ML approaches including decision trees, support vector machines, Bayesian algorithms, k-nearest neighbors, and random forests. More specifically, we will briefly describe their strengths and weakness and the threats that are usually detected in IIoT security challenges.

Decision trees (DTs) are used for solving classification problems by sorting samples according to some indicative feature values. Each vertex (node) in a tree represents a feature, and each edge (branch) denotes a value that is assigned to the vertex corresponding to the sample that needs to be classified. The samples are then classified starting from the origin vertex and with respect to their feature values. The identification of the optimal feature is based on different metrics including information gain (Quinlan 1986) and Gini index (Du and Zhan 2002). Despite their wide adoption in different security applications, including intrusion detection (Kim et al. 2014) and detection of suspicious traffic sources (Alharbi et al. 2017), they usually involve a massive construction of trees with several decision nodes, increasing significantly the computation and storage requirements.

Support vector machines (SVMs) classify samples by assigning them to points in space and creating a splitting hyperplane between two or more classes, such that the distance between the hyperplane and the most adjacent points of each class is maximized and thus the separate classes are divided by a clear gap that is as wide as possible. Although they are fairly robust against overfitting, especially in high-dimensional space, it is trickier to be tuned due to the importance of selecting the right parameters and do not scale well to larger datasets. SVMs have been employed to improve the effectiveness of prediction and diagnosis of induction motor faults particularly during the maintenance judgment process (Gangsar and Tiwari 2017), to detect attacks in a smart grid (Ozay et al. 2016) or as a tool to exploit IoT device security (Lerman et al. 2015).

Bayesian Methods: Bayes' theorem describes the probability of an event based on previous information related to the event. **Naive Bayes (NB)** is a well-known ML technique for constructing classifiers that calculate the posterior probability of an event and use the Bayes' theorem to evaluate the probability that a particular feature set of unlabeled samples fits a specific label, assuming independence among features. For example, NB can be used for classifying network traffic as normal or abnormal, using as features the connection duration, the connection protocol (e.g., TCP, UDP), and the connection status flag. These features are considered independent although in practice there are dependencies. The aforementioned schemes can be easily implemented and have been applied both in binary and multiclass problems, though they completely ignore interactions among features, which in many complex tasks contribute in increasing the discrimination power of a classification model (Ng and Jordan 2001). They have been successfully applied for identifying nefarious activity/abuse in smart manufacturing systems, including malware attacks (Ye et al. 2017).

k-Nearest neighbor classifiers categorize data patterns (e.g., node behaviors) in normal and malicious, on the basis of the votes provided by a selected number (i.e., k) of its nearest neighbors. Euclidean distance between features representing different patterns is usually adopted for identifying neighborhoods. Determining the optimal value of k may become a challenging and time-consuming process, though this method has been successfully applied in several network intrusion and anomaly detection schemes (Syarif and Gata 2017; Su 2011).

Random forests are supervised learning algorithms that use several decision trees (DTs) to combine decisions and thus acquire precise and more reliable classification results. They are composed of several trees which are constructed randomly and they are trained to vote for an output class. The class with the most votes is selected as the final classification output. The number of required trees depends on the size of the training dataset, and the construction of several trees may be impractical in several real-time applications. Several works suggest their application for DDoS detection (Doshi et al. 2018) and for detection of unauthorized IoT devices (Meidan et al. 2017).

Other ML approaches, such as **association rule, ensemble learning, bugging and boosting**, and **k-means clustering** algorithms, have been also applied for the detection of intrusion, anomalies, and malware, as well as for improving the efficiency of protection of data, without reducing the quality of anonymization (Xie et al. 2017).

16.4.2 Deep Learning (DL) Methods for IIoT Security

Recently, several researchers, system engineers, and software developers have shown increasing interest in the application of DL approaches for addressing security threats and vulnerabilities in modern IoT systems. This phenomenon is mainly attributed to their superior performance over traditional ML schemes, especially when both methods utilize large datasets. DL approaches are capable of learning data representations with several levels of abstraction by using computational architectures with several nonlinear processing layers. This is also the reason why they are known as hierarchical learning methods. Most modern deep learning methods are based on neural networks (NNs) (please refer to Fig. 16.4), while learning can be supervised, alternatively known as discriminative (e.g., convolutional and recurrent NN), unsupervised (generative learning, e.g., generative adversarial networks), or semi-supervised (e.g., auto-encoders, deep belief networks, restricted Boltzmann machines). In the remaining part of this section, we briefly review the working principles of the aforementioned DL schemes and their potential application for identifying different IoT security threats.

convolutional neural networks (CNNs) focus on reducing the connection between layers by exploiting sparse interactions, parameter sharing, and translation invariant characteristics. They consist of two different types of layers: (1) the convolutional layer where data parameters are convolved with multiple filters of equal size and (2) the pooling layer, where different approaches for subsampling the output

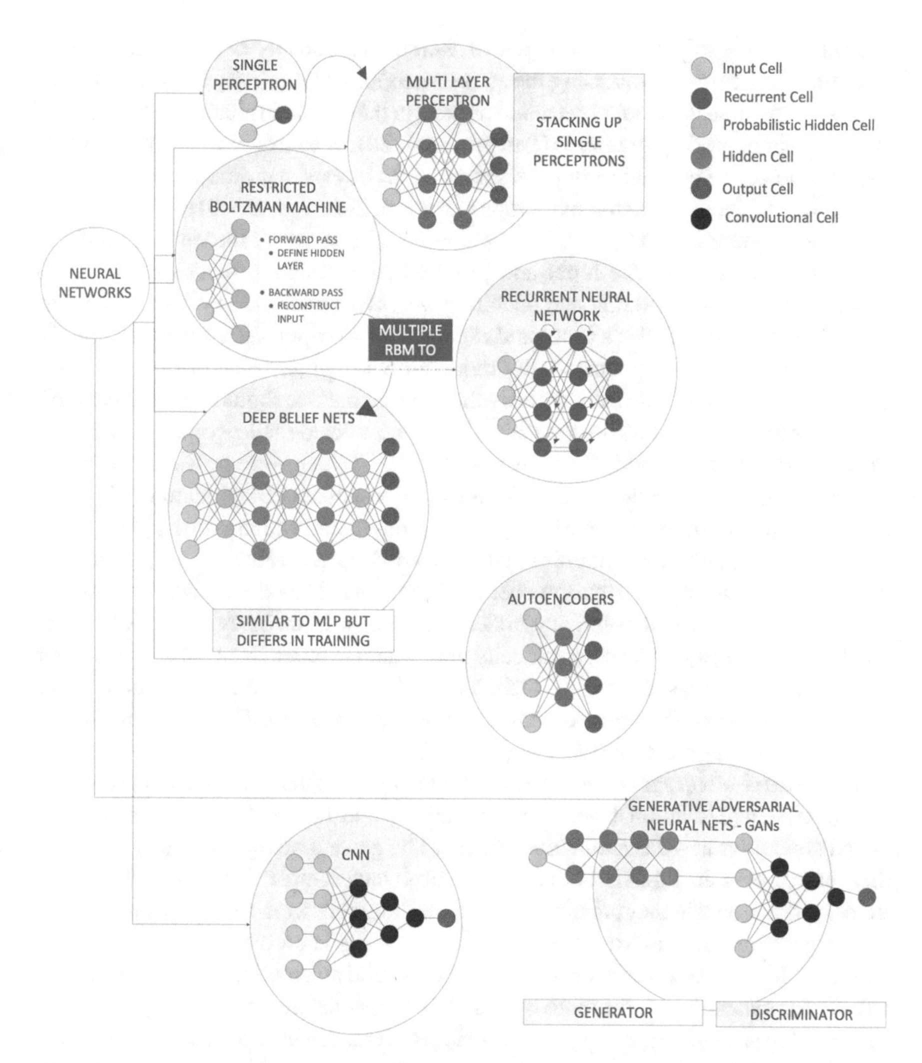

Fig. 16.4 Different DL NN-based working principles for detecting threats in IIoT systems. Red nodes indicate the classification output (e.g., normal, malicious behavior, etc.)

and decreasing the size of subsequent layers are applied. Their benefits compared to traditional NN are increased scalability and reduced training complexity, while their wide adoption is attributed to their ability to automatically learn features from raw data. Still their complexity is quite high, making their integration to resource-constrained devices a very challenging task. CNNs have been successfully utilized for Malware detection (McLaughlin et al. 2017)and for also breaking cryptographic implementations (Maghrebi et al. 2016).

Recurrent Neural Networks (RNNs) have been utilized in applications where the data is available sequentially (e.g. speech, video, sensor measurements). RNNs are created by applying the same set of weights recursively over a differentiable graph-like structure by traversing the structure topologically. They are very efficient in processing data in an adaptive manner, though their main limitation is the issue of vanishing or exploding gradients (Pascanu et al. 2013). RNNs have been previously used for detecting anomalies in time-series-based threats, e.g., monitoring network traffic flow to detect potential malicious behaviors (Torres et al. 2016).

An auto-encoder (AE) is a NN composed of two parts, the encoder and the decoder, which obtains the input and provides an abstraction (code) as an output and vice versa. The encoding and decoding weights are selected by minimizing the error between the encoder's input and the decoder's output. AEs are important for feature extraction and dimensionality reduction without any data prior knowledge, though in order to operate satisfactorily the training dataset should be representative of the testing dataset, while they also consume considerable computation time. Previous studies have used AEs to extract features, which were proven informative for detecting impersonation attacks in Wi-Fi environments (Aminanto et al. 2018) and cyber-attacks in fog computing systems (Abeshu and Chilamkurti 2018).

Restricted Boltzmann machines (RBMs) are deep generative models utilized for learning a probability distribution over the input data. They are undirected models, while there is no link between any nodes in the same layer. They consist of visible and hidden layers and they hierarchically understand features from data. Again, their complexity is increased, making their integration to resource-constrained devices a challenging task. The most common applications that use RBMs are related to network anomaly detection (Fiore et al. 2013).

Generative adversarial networks (GANs) have recently emerged as a promising DL approach. GANs are based on the training and use of two different models called generative and discriminative models. The generative model goal is to learn a distribution over the input dataset and generate a data sample and the discriminative model prediction whether the input is from the dataset or from the generative model. GANs generate samples very fast, though its training is hard and usually unstable. Despite this drawback, GANs have been used to build an architecture for securing an IoT system cyberspace (Hiromoto et al. 2017). GANs have a potential application in IoT security, since they are capable of learning different attack scenarios and generate samples similar to a zero-day attack (e.g., variations of existing attacks), providing security approaches that are robust against unknown attacks (Zenati et al. 2018). All the aforementioned ML and DL approaches provide solutions for detecting threats on how IoT devices interact with each other and with the environment, using the data collected by different heterogeneous devices that can be integrated in dynamic environments. Table 16.2 summarizes the various security/vulnerability threats that are detected using aforementioned ML and DL approaches.

In Fig. 16.4 we present different DL NN-based architectures for detecting threats in IIoT systems. Red nodes indicate the classification output (e.g., normal, malicious behavior, etc.) and (light) green nodes correspond to (probabilistic) hidden layer, while blue nodes denote recurrent cells and purple nodes correspond to convolutional cells.

Table 16.2 Summary of studies on ML and DL for securing IoT and IIoT

Reference	Method	Threats detected or security application	Areas of the attack surface			
			Physical device	Network service	Cloud service	Web service
Kim et al. (2014)	DT	Intrusion detection	✓	✓	–	–
Alharbi et al. (2017)	DT	Denial of service	✓	✓	✓	–
Gangsar and Tiwari (2017)	SVM	Fault prediction	✓	–	–	–
Ozay et al. (2016)	SVM	False data injection	–	–	✓	✓
Lerman et al. (2015)	SVM	Attacks to masked Advanced encryption Schemes (AES)	–	–	–	✓
Ye et al. (2017)	NB	Malware attack	✓	–	–	✓
Syarif and Gata (2017)	kNN	Intrusion detection	✓	✓	–	–
Su (2011)	kNN	Denial of service	✓	✓	✓	–
Doshi et al. (2018)	RF	Denial of service	✓	✓	✓	–
Meidan et al. (2017)	RF	Unauthorized access	–	–	–	✓
Maghrebi et al. (2016)	CNN	Masked AES attacks	–	–	✓	✓
McLaughlin et al. (2017)	CNN	Malware attacks	✓	–	–	✓
Torres et al. (2016)	RNN	Malicious behavior	–	–	✓	✓
Aminanto et al. (2018)	AE	Anomaly-based IDS	–	✓	–	–
Abeshu and Chilamkurti (2018)	AE	Fog cyber-attacks	–	✓	✓	–
Fiore et al. (2013)	RBM	Network Anomaly Detection	–	✓	–	–
Hiromoto et al. (2017)	GAN	Vulnerabilities to malicious supply chain risk	✓	–	–	✓

16.5 Achieving Safety Using ML and DL Approaches

Learning from large volumes of data using powerful algorithms, as those presented above, brings significant benefits in securing IIoT systems, though questions about safety still need to be carefully examined. Although workhorse machine and deep learning tools are expected to have intelligence that in many cases surpasses human abilities or something in between, they are still technological components that have to be engineered with safety in mind (Conn 2015). The term "safety" is widely used in a large number of diverse engineering disciplines, indicating the absence of system failures or the absence of dangerous conditions. Authors in (Maller and Hansson 2008) introduce a decision-theoretic definition of safety, making a link to the minimization or reduction of risk and uncertainty to undesirable states which can be considered as harmful. This generic definition applies to many different domains and systems and indicates that the cost of undesirable states is expected to be quite high in a human sense for events that are harmful and that safety is achieved by minimizing the probability of both expected and unexpected harms.

In IIoT, safety is related to (1) the ability of reasoning about the behavior of the IIoT devices and more specifically that of the actuators and (2) the ability of identifying and preventing unintended and unexpected failures or harmful events. Those are very hard challenges, since they require the system(s) to be able to identify "normal" behaviors and at the same time develop device interaction approaches, mechanisms that enforce safety properties. More importantly, they usually become even harder to be addressed, due to their heterogeneity, the lack of standardization, and the ineffectiveness of traditional defense mechanisms, including firewalls and antivirus software.

The strategies that could be applied for ensuring safety are strongly related to the specific application, though the authors in (Maller and Hansson 2008) have analyzed different strategies across different domains suggesting four main categories of safety approaches. The first approach known as safe design suggests the exclusion instead of the control of a hazard (e.g., excluding hydrogen from the buoyant material of a dirigible airship ensures safety). The second one suggests using multiplicative or additive reserves, known also as safety factors/margins. A safety factor in mechanical engineering is a ratio between the maximum load that does not lead to failure and the optimal load that the system was designed to support, while the corresponding safety margin is determined as the difference between the two. The third category is known as the "safe fail strategy" according to which a system remains safe, even when it fails in its intended operation (e.g., dead man's switches on trains, safety valves on boilers, etc.). The fourth final category suggests including measures, known as procedural safeguards, that are beyond the ones designed in the core functionality of the system, such us audit diagrams, posted warnings, etc. In the rest of this section, we provide details about deploying these strategies using machine and deep learning methods.

Inherently Safe Design One of the major goals in the ML context is to provide robust approaches that address the uncertainties when the training set has not sampled from the test distribution. Training dataset may have biases and patterns, which are unknown to the users, will not be present at the test, and might lead to unsafe or undesirable operations. Recent approaches including gradient boosting and deep neural networks are capable of exploiting the biases, thus achieving higher accuracies; however, making safe predictions of unknown shifts in data, incorrect patterns, or harmful rules seems to remain a safety challenge (Caruana et al. 2015).

These models are usually complex introducing difficulties in understanding their behavior in such shifts or whether their outcome will be unsafe. Therefore, the most widely adopted best practices to introduce inherently safe design is by deploying models that can be interpreted by humans and by excluding features which are not casually related to the outcome (Freitas 2014; Rudin 2014; Athey and Imbens 2015; Welling 2015). The use of interpretable models, features, or processing approaches that are capable of identifying and excluding irregular patterns are expected to enhance safety. Moreover, the successful identification of variables that are causally linked to the outcome could lead to exclusion of behaviors which are not part of the true "physics" of the system, ensuring that any undesirable operation can be avoided. At this point it should be noted that post hoc interpretation and repair of complex uninterpretable models is not the decision rule of a decision-making process, and therefore it does not assure safety via inherently safe design.

Safe Fail Technique It is used in ML for rejecting options, which are not confident (Varshney et al. 2013). More specifically, the model reports whether it cannot provide a reliable output, thus avoiding any unsafe or undesirable output. In cases that the output of a model is the reject option, then the user intervenes checks and test sample and provides a manual prediction. This actually means that there is an assumption that a distance from the decision boundary is inversely related to confidence. This assumption is valid in parts of feature space with high probability density and large number of training samples, since the decision boundary is located where there is a large overlap in likelihood functions, though parts of the feature space with low density may not contain any training samples at all, introducing uncertainties in the decision boundaries. In this case, the distance from the decision boundary is fairly meaningless and the typical rule for triggering the reject option should be avoided (Attenberg et al. 2015). For a rare combination of features in a test sample, a safe fail strategy is to manually examine the test sample. At this point, it should be noted that manual intervention options are suitable for applications with long time scales, while when working in ms scale, only options similar to dead man's switches that stop operations in a reasonable manner are applicable.

Procedural Safeguards Two relevant directions in ML and DL that can be deployed for increasing safety are user experience and openness. Despite the fact that many decision-making systems in several IIoT applications are based on ML and DL systems, the operators and the designers of these systems are usually nonspecialists in the ML and DL domains. However, the definition of the training data and the setup of the evaluation procedures have certain constraints that could lead to undesirable outcomes if they are not done correctly. User experience design can certainly guide and warn nonspecialists to address the aforementioned issues properly, increasing significantly safety. In addition, it's worth mentioning that most ML and DL approaches are open source, allowing their wide deployment and for the possibility also of the public audit, facilitating the identification of safety hazards and potential harms via the examination of the source code. Of course, one should also take into account that the source software is not sufficient, since these approaches are driven by data. Opening therefore data, making them available to be freely used, reused, and redistributed by anyone, is a widely adopted procedural safeguard for increasing safety (Shaw 2015; Kapoor et al. 2015).

16.6 Future Challenges, Discussion, and Conclusion

The latest advancements in learning approaches facilitated the development of machine and deep learning methods for addressing different security threats and vulnerabilities. However, there are still challenges that need to be addressed for satisfying complex requirements related to the physical devices, the collected data wireless transmission technologies, and the mobile and cloud architectures, which are described in detail in the following part of this section.

Availability of Security-Related Datasets One of the major challenges that should be addressed in IIoT systems using ML and DL approaches is the extraction and generation of realistic and high-quality training data that contain various possible attacks. A vital future research approach toward this direction is the use of crowdsourcing methods for generating datasets related to IoT threats and attacks. This approach could lead to the inclusion of all the potential attacks in rich training datasets that could be used for benchmarking the accuracy of new algorithms. At this point, however, it should be also noted that generating collaborative IoT threat dataset that will be continuously updated with new attacks is a challenging task mainly due to the large diversity in the technical characteristics of the various IoT devices. More importantly many privacy concerns also arise, since sensitive and critical information may be shared publicly especially when we focus on industrial and medical IoT devices.

Learning to Secure IoT with Low-Quality Data IoT and IIoT systems deploy a large number of heterogeneous connected devices with memory, power, and computational constraints that usually affect also the data quality (e.g., data with missing entries, outliers, noise). Therefore, learning to secure IoT systems requires effective algorithms capable of handling and learning from noisy and low-quality data. Toward this direction, there is clear need for multimodal and effective ML and DL models that are capable of handling heterogeneous data and with contaminated/noisy data segments.

Lifelong Learning for Learning IoT Threats IoT and IIoT systems represent dynamic systems where several new devices either join or leave the system for satisfying the need of various application with evolving needs. Due to their dynamic nature, distinguishing between normal and abnormal behaviors cannot be predefined, thus becoming a challenging task. To address this issue, frequent updates of the security models are required in order to track and understand the system modifications. Therefore, lifelong learning is a significant attribute that should be supported in long-term real-world applications, and it is directed toward the construction of a model that can perform the retraining process repeatedly for the learning of new emerging patterns related to each behavior. The model should be able to continuously adapt to and learn from new environments.

Implementation of ML and DL at the Edge Edge computing is an essential solution that immigrates IoT service solutions to the network edge, minimizing delays, realizing real-time processing performance, improving energy efficiency, and enhancing the scalability of lightweight IoT devices. Thus, the implementation of DL and ML approaches at the edge for IoT security are expected to offer an effective framework for data processing with reduced network traffic load. Though there are still significant challenges that need to be addressed for allowing the deployment of ML and DL approaches on edge devices, exploiting the benefits of edge computing. The design of ML and DL approaches that can process scalable data representations compatible with adaptive data transmission protocols is an interesting and important direction for improving the performance of transparent computing. In addition, the programming language of the framework should take into account the heterogeneity of hardware and the capacity of the resources in the workflow. Thus an appropriate ML and DL end-to-end framework that will take into account hardware and software reconfigurations is still a challenging problem. Finally, distributed security solutions is still an open direction of research, meaning that future security solutions should not only exploit the capability of edge servers for building more secure IoT devices but also be able to guarantee the security of the distributed and sometimes resource-constrained edge servers.

Data Security and Privacy Concerns Because of the everyday and pervasive nature of IoT scenarios, security and privacy concerns take a broader dimension, demanding for cross and multidisciplinary approaches through efforts from different areas in order to bring citizens into the loop. Nowadays, when talking about the strong development of the IoT, most estimates provide very impressive data on the number of interconnected devices in the coming years. Consequently, many security and privacy approaches in IoT are proposed from a device perspective with the aim of addressing these concerns in a broader environment. However, the IoT ecosystem is not only composed of communication-enabled devices but of a huge amount of heterogeneous smart objects, middleware, and services, where security and privacy requirements from different actors (citizens, companies, or regulatory bodies) need to be reconciled. Given the degree of heterogeneity, one of the most significant challenges is to build a secure, privacy-aware but still interoperable IoT framework. Therefore, there is a strong need to move toward a holistic security and privacy approach by addressing the IoT ecosystem as a whole, beyond such device-centric vision.

Yet, industry has already realized that the true value of IoT is not on the physical interconnected devices per se but on the massive datasets and crude, unrefined information they contain and consequently how this hidden commodity can be efficiently processed in a fast and meaningful manner. Through IoT, individuals will produce an unprecedented amount of raw information about their daily routine which can be exploited manifold by operators and malicious eavesdroppers alike. This clearly violates user privacy, especially when considering that the user has willingly purchased the device which now may handle his personal data over to third-party data silos to be further processed. Consequently, it is essential for users to demand and legislative authorities to enforce a certain move toward data-centric security schemes that will penalize paradox and improper data usage. Users also need to be empowered with mechanisms to control how data from their devices are shared, to whom, and under what circumstances. Machine and deep learning can be used for properly identifying the once again fuzzy lines between using data analysis for benevolent service optimization or arbitrary behavior mapping which can be later sold to the highest bidder.

This chapter presented the working principles together with the strength and weakness of several machine and deep learning approaches, focusing on the identification and mitigation of modern IoT security threats and vulnerabilities. Therefore, it is expected to serve as a useful manual encouraging researchers to advance the security of IoT systems either by addressing device or end-to-end security challenges.

Acknowledgements We acknowledge support of this work by the project "I3T—Innovative Application of Industrial Internet of Things (IIoT) in Smart Environments" (MIS 5002434) which is implemented under the "Action for the Strategic Development on the Research and Technological Sector," funded by the Operational Programme "Competitiveness, Entrepreneurship and Innovation" (NSRF 2014–2020) and co-financed by Greece and the European Union (European Regional Development Fund).

The views and opinions expressed are those of the authors and do not necessary reflect the official position of Citrix Systems Inc.

References

Aazam, M., Khan, I., Alsaffar, A. A., & Huh, E. (2014). Cloud of things: Integrating internet of things and cloud computing and the issues involved. In *Proceedings of 2014 11th International Bhurban Conference on Applied Sciences Technology (IBCAST) Islamabad, Pakistan, 14th–18th January, 2014* (pp. 414–419). https://doi.org/10.1109/IBCAST.2014.6778179.

Abeshu, A., & Chilamkurti, N. (2018). Deep learning: The frontier for distributed attack detection in fog-to-things computing. *IEEE Communications Magazine, 56*(2), 169–175. ISSN 0163-6804. https://doi.org/10.1109/MCOM.2018.1700332.

Adolphs, P., Cabot, J., & Wimmer, M. (2016). *Structure of the Administration Shell: Continuation of the Development of the Reference Model for the Industrie 4.0 Component*. Platform Industrie 4.0. https://www.plattform-i40.de/I40/Redaktion/EN/Downloads/Publikation/structure-of-the-administration-shell.pdf.

Alharbi, S., Rodriguez, P., Maharaja, R., Iyer, P., Subaschandrabose, N., & Ye, Z. (2017). Secure the internet of things with challenge response authentication in fog computing. In *2017 IEEE 36th*

International Performance Computing and Communications Conference (IPCCC) (pp. 1–2). https://doi.org/10.1109/PCCC.2017.8280489.
Aminanto, M. E., Choi, R., Tanuwidjaja, H. C., Yoo, P. D., & Kim, K. (2018). Deep abstraction and weighted feature selection for wi-fi impersonation detection. *IEEE Transactions on Information Forensics and Security, 13*(3), 621–636. ISSN 1556-6013. https://doi.org/10.1109/TIFS.2017.2762828.
Athey, S., & Imbens, G. (2015). Machine learning methods for estimating heterogeneous causal effects.
Attenberg, J., Ipeirotis, P., & Provost, F. (2015). Beat the machine: Challenging humans to find a predictive model's "unknown unknowns". *Journal of Data and Information Quality, 6*(1), 1:1–1:17. ISSN 1936-1955. https://doi.org/10.1145/2700832.
Atzori, L., Iera, A., & Morabito, G. (2010). The internet of things: A survey. *Computer Networks, 54*(15), 2787–2805. ISSN 1389-1286. https://doi.org/10.1016/j.comnet.2010.05.010.
Caruana, R., Lou, Y., Gehrke, J., Koch, P., Sturm, M., & Elhadad, N. (2015). Intelligible models for healthcare: Predicting pneumonia risk and hospital 30-day readmission. In *Proceedings of the 21th ACM SIGKDD International Conference on Knowledge Discovery and Data Mining*, KDD '15 (pp. 1721–1730), New York, NY: ACM. ISBN 978-1-4503-3664-2. https://doi.org/10.1145/2783258.2788613.
Chen, K., Zhang, S., Li, Z., Zhang, Y., Deng, Q., Ray, S., et al. (2018). Internet-of-things security and vulnerabilities: Taxonomy, challenges, and practice. *Journal of Hardware and Systems Security, 2*(2), 97–110. ISSN 2509-3428. https://doi.org/10.1007/s41635-017-0029-7.
Conn, A. (2015). *The AI wars: The battle of the human minds to keep artificial intelligence safe*. Needham: Industrial Internet Consortium. http://futureoflife.org/2015/12/17/the-ai-wars-the-battle-of-the-human-minds-to-keep-artificial-intelligence-safe.
Doshi, R., Apthorpe, N., & Feamster, N. (2018). Machine learning ddos detection for consumer internet of things devices. In *2018 IEEE Security and Privacy Workshops (SPW)* (pp. 29–35). https://doi.org/10.1109/SPW.2018.00013.
Du, W., & Zhan, Z. (2002). Building decision tree classifier on private data. In *Proceedings of the IEEE International Conference on Privacy, Security and Data Mining*, CRPIT '14 (Vol. 14, pp. 1–8), Darlinghurst: Australian Computer Society, ISBN 0-909-92592-5. http://dl.acm.org/citation.cfm?id=850782.850784.
ENISA Report. (2017). Baseline Security Recommendations for IoT. https://www.enisa.europa.eu/publications/baseline-security-recommendations-for-iot.
ENISA Report. (2018a). Good Practices for Security of Internet of Things, https://www.enisa.europa.eu/publications/good-practices-for-security-of-iot.
ENISA Report. (2018b) Hardware Threat Landscape and Good Practice Guide, https://www.enisa.europa.eu/publications/hardware-threat-landscape.
ENISA Report. (2018c). Ad-hoc and sensor networking for M2M Communications, https://www.enisa.europa.eu/publications/m2m-communications-threat-landscape.
Evans, D. (2011). *The internet of things—how the next evolution of the internet is changing everything*. White Paper. San Jose: CISCO.
Fiore, U., Palmieri, F., Castiglione, A., & De Santis, A. (2013). Network anomaly detection with the restricted boltzmann machine. *Neurocomputing, 122*, 13–23. ISSN 0925-2312. https://doi.org/10.1016/j.neucom.2012.11.050.
Freitas, A. A. (2014). Comprehensible classification models: A position paper. *SIGKDD Explorations Newsletter, 15*(1), 1–10. ISSN 1931-0145. https://doi.org/10.1145/2594473.2594475.
Gangsar, P., & Tiwari, R. (2017). Comparative investigation of vibration and current monitoring for prediction of mechanical and electrical faults in induction motor based on multiclass-support vector machine algorithms. *Mechanical Systems and Signal Processing, 94*, 464–481. ISSN 0888-3270. https://doi.org/10.1016/j.ymssp.2017.03.016.
Goodfellow, I., Bengio, Y., & Courville, A. (2016). *Deep learning*. Cambridge: The MIT Press. ISBN 0262035618, 9780262035613.

Hiromoto, R. E., Haney, M., & Vakanski, A. (2017). A secure architecture for iot with supply chain risk management. In *2017 9th IEEE International Conference on Intelligent Data Acquisition and Advanced Computing Systems: Technology and Applications (IDAACS)* (Vol. 1, pp. 431–435). https://doi.org/10.1109/IDAACS.2017.8095118.

Humayed, A., Lin, J., Li, F., & Luo, B. (2017). Cyber-physical systems security—a survey. *IEEE Internet of Things Journal, 4* (6), 1802–1831. ISSN 2327-4662. https://doi.org/10.1109/JIOT.2017.2703172.

IIC. Industrial Internet of Things Volume G4: Security Framework (2016). https://www.iiconsortium.org/IISF.htm.

ITY-T. Overview of Internet of Things (2012)

Jazdi, N. (2014). Cyber physical systems in the context of industry 4.0. In *2014 IEEE International Conference on Automation, Quality and Testing, Robotics* (pp. 1–4). https://doi.org/10.1109/AQTR.2014.6857843.

Jordan, M. I., & Mitchell, T. M. (2015). Machine learning: Trends, perspectives, and prospects. *Science, 349*(6245), 255–260. ISSN 0036-8075. https://doi.org/10.1126/science.aaa8415.

Kapoor, S., Mojsilovic, A., Strattner, J. N., & Varshney, K. R. (2015). From open data ecosystems to systems of innovation: A journey to realize the promise of open data. In *Proceedings of the Data for Good Exchange Conference*, New York, NY, USA.

Khan, R., Khan, S. U., Zaheer, R., & Khan, S. (2012). Future internet: The internet of things architecture, possible applications and key challenges. In *2012 10th International Conference on Frontiers of Information Technology* (pp. 257–260). https://doi.org/10.1109/FIT.2012.53.

Kim, G., Lee, S., & Kim, S. (2014). A novel hybrid intrusion detection method integrating anomaly detection with misuse detection. *Expert Systems with Applications, 41*(4), 1690–1700. ISSN 0957-4174. https://doi.org/10.1016/j.eswa.2013.08.066.

Lerman, L., Bontempi, G., & Markowitch, O. (2015). A machine learning approach against a masked AES. *Journal of Cryptographic Engineering, 5*(2), 123–139. ISSN 2190-8516. https://doi.org/10.1007/s13389-014-0089-3.

Lin, S.-W., Crawford, M., Miller, B., Durand, J., & Bleakley, G. (2017a). *The industrial internet of things volume G1: reference architecture*. Needham: Industrial Internet Consortium. https://www.iiconsortium.org/IIC_PUB_G1_V1.80_2017-01-31.pdf.

Lin, S.-W., Murphy, B., Clauer, E., Loewen, U., & Bleakley, G. (2017b). *Architecture alignment and interoperability*. Industrial Internet Consortium and Plattform Industrie 4.0 Joint Whitepaper. http://www.iiconsortium.org/pdf/JTG2_Whitepaper_final_20171205.pdf.

Maghrebi, H., Portigliatti, T., & Prouff, E. (2016). Breaking cryptographic implementations using deep learning techniques. In *IACR Cryptology ePrint Archive*.

Mainetti, L., Patrono, L., & Vilei, A. (2011). Evolution of wireless sensor networks towards the internet of things: A survey. In *SoftCOM 2011, 19th International Conference on Software, Telecommunications and Computer Networks* (pp. 1–6).

Maller, N., & Hansson, S. O. (2008). Principles of engineering safety: Risk and uncertainty reduction. *Reliability Engineering & System Safety, 93*(6), 798–805. ISSN 0951-8320. https://doi.org/10.1016/j.ress.2007.03.031.

Mashal, I., Alsaryrah, O., Chung, T.-Y., Yang, C.-Z., Kuo, W.-H., & Agrawal, D. P. (2015). Choices for interaction with things on internet and underlying issues. *Ad Hoc Networks, 28*, 68–90. ISSN 1570-8705. https://doi.org/10.1016/j.adhoc.2014.12.006.

McLaughlin, N., Martinez del Rincon, J., Kang, B., Yerima, S., Miller, P., Sezer, S., Safaei, Y., Trickel, E., Zhao, Z., Doupé, A., & Joon Ahn, G. (2017). Deep android malware detection. In *Proceedings of the Seventh ACM on Conference on Data and Application Security and Privacy*, CODASPY '17 (pp. 301–308). New York, NY: ACM. ISBN 978-1-4503-4523-1. https://doi.org/10.1145/3029806.3029823.

Meidan, Y., Bohadana, M., Shabtai, A., Ochoa, M., Tippenhauer, N. O., Guarnizo, J. D., & Elovici, Y. (2017). Detection of unauthorized iot devices using machine learning techniques. *CoRR*, abs/1709.04647. http://arxiv.org/abs/1709.04647.

Mena, D. M., Papapanagiotou, I., & Yang, B. (2018). Internet of things: Survey on security. *Information Security Journal: A Global Perspective, 27*(3), 162–182. https://doi.org/10.1080/19393555.2018.1458258.

Miorandi, D., Sicari, S., Pellegrini, F. D., & Chlamtac, I. (2012). Internet of things: Vision, applications and research challenges. *Ad Hoc Networks, 10*(7), 1497–1516. ISSN 1570-8705. https://doi.org/10.1016/j.adhoc.2012.02.016.

Ng, A. Y., & Jordan, M. I. (2001). On discriminative vs. generative classifiers: A comparison of logistic regression and naive bayes. In *Proceedings of the 14th International Conference on Neural Information Processing Systems: Natural and Synthetic*, NIPS'01 (pp. 841–848). Cambridge, MA: MIT Press. http://dl.acm.org/citation.cfm?id=2980539.2980648.

OWASP. The free and open software security community (2018). http://www.owasp.org/index.php/OWASP_Internet_of_Things_Project.

Ozay, M., Esnaola, I., Yarman Vural, F. T., Kulkarni, S. R., & Poor, H. V. (2016). Machine learning methods for attack detection in the smart grid. *IEEE Transactions on Neural Networks and Learning Systems, 27*(8), 1773–1786. ISSN 2162-237X. https://doi.org/10.1109/TNNLS.2015.2404803.

Pan, S., Morris, T., & Adhikari, U. (2015). Developing a hybrid intrusion detection system using data mining for power systems. *IEEE Transactions on Smart Grid, 6*(6), 3104–3113. ISSN 1949-3053. https://doi.org/10.1109/TSG.2015.2409775.

Pascanu, R., Mikolov, T., & Bengio, Y. (2013). On the difficulty of training recurrent neural networks. In *Proceedings of the 30th International Conference on International Conference on Machine Learning*, ICML'13 (Vol. 28, pp. 1310–1318). JMLR.org. http://dl.acm.org/citation.cfm?id=3042817.3043083.

Quinlan, J. R. (1986). Induction of decision trees. *Machine Learning, 1*(1), 81–106. ISSN 0885-6125. https://doi.org/10.1023/A:1022643204877.

Rudin, C. (2014). Algorithms for interpretable machine learning. In *Proceedings of the 20th ACM SIGKDD International Conference on Knowledge Discovery and Data Mining*, KDD '14 (pp. 1519–1519). ACM: New York, NY. ISBN 978-1-4503-2956-9. https://doi.org/10.1145/2623330.2630823.

Schmidt, R., Möhring, M., Härting, R.-C., Reichstein, C., Neumaier, P., & Jozinović, P. (2015). Industry 4.0 – potentials for creating smart products: Empirical research results. In W. Abramowicz (Ed.), *Business information systems* (pp. 16–27). Cham: Springer. ISBN 978-3-319-19027-3.

Sethi, P., & Sarangi, S. R. (2017). Internet of things: Architectures, protocols, and applications. *Journal of Electrical and Computer Engineering, 2017*, 9324035:1–9324035:25.

Sharp, M., Ak, R., & Hedberg, T. (2018). A survey of the advancing use and development of machine learning in smart manufacturing. *Journal of Manufacturing Systems, 48*, 170–179. ISSN 0278-6125. https://doi.org/10.1016/j.jmsy.2018.02.004. Special Issue on Smart Manufacturing.

Shaw, E. (2015). *Improving service and communication with open data*. Data Smart City solutions. https://datasmart.ash.harvard.edu/news/article/improving-service-and-communication-with-open-data-702.

Su, M.-Y. (2011). Real-time anomaly detection systems for denial-of-service attacks by weighted k-nearest-neighbor classifiers. *Expert Systems with Applications, 38*(4), 3492–3498. ISSN 0957-4174. https://doi.org/10.1016/j.eswa.2010.08.137.

Syarif, A. R., & Gata, W. (2017). Intrusion detection system using hybrid binary pso and k-nearest neighborhood algorithm. In *2017 11th International Conference on Information Communication Technology and System (ICTS)* (pp. 181–186). https://doi.org/10.1109/ICTS.2017.8265667.

Torres, P., Catania, C., Garcia, S., & Garino, C. G. (2016). An analysis of recurrent neural networks for botnet detection behavior. In *2016 IEEE Biennial Congress of Argentina (ARGENCON)* (pp. 1–6). https://doi.org/10.1109/ARGENCON.2016.7585247.

Varshney, K. R., Prenger, R. J., Marlatt, T. L., Chen, B. Y., & Hanley, W. G. (2013). Practical ensemble classification error bounds for different operating points. *IEEE Transactions on Knowledge and Data Engineering, 25*(11), 2590–2601. ISSN 1041-4347. https://doi.org/10.1109/TKDE.2012.219.

WASC. Threat Classification v2.0 (2012). http://projects.webappsec.org/w/page/13246978/Threat%20Classification.

Welling, M. (2015). Are ml and statistics complementary. IMS-ISBA Meeting on Data Science in the Next 50 Years.

Weyer, S., Schmitt, M., Ohmer, M., & Gorecky, D. (2015). Towards industry 4.0 – standardization as the crucial challenge for highly modular, multi-vendor production systems. *IFAC-PapersOnLine, 48*(3), 579–584. ISSN 2405-8963. https://doi.org/10.1016/j.ifacol.2015.06.143. 15th IFAC Symposium onInformation Control Problems inManufacturing.

Wu, M., Lu, T.-J., Ling, F.-Y., Sun, J., & Du, H.-Y. (2010). Research on the architecture of internet of things. In *2010 3rd International Conference on Advanced Computer Theory and Engineering(ICACTE)* (Vol. 5, pp. V5–484–V5–487). https://doi.org/10.1109/ICACTE.2010.5579493.

Xanthopoulos, A. S., Kiatipis, A., Koulouriotis, D. E., & Stieger, S. (2018). Reinforcement learning-based and parametric production-maintenance control policies for a deteriorating manufacturing system. *IEEE Access, 6*, 576–588. ISSN 2169-3536. https://doi.org/10.1109/ACCESS.2017.2771827.

Xie, M., Huang, M., Bai, Y., & Hu, Z. (2017). The anonymization protection algorithm based on fuzzy clustering for the ego of data in the internet of things. *Journal of Electrical and Computer Engineering, Hindawi, 1* (1), 1–10. Article ID 2970673.

Xu, L. D., He, W., & Li, S. (2014) Internet of things in industries: A survey. *IEEE Transactions on Industrial Informatics, 10*(4), 2233–2243. ISSN 1551-3203. https://doi.org/10.1109/TII.2014.2300753.

Yang, K., Ren, J., Zhu, Y., & Zhang, W. (2018). Active learning for wireless iot intrusion detection. *IEEE Wireless Communications, 25*(6), 19–25. ISSN 1536-1284. https://doi.org/10.1109/MWC.2017.1800079.

Ye, Y., Li, T., Adjeroh, D., & Iyengar, S. S. (2017). A survey on malware detection using data mining techniques. *ACM Computing Surveys, 50*(3), 41:1–41:40. ISSN 0360-0300. https://doi.org/10.1145/3073559.

Zenati, H., Foo, C. S., Lecouat, B., Manek, G., & Chandrasekhar, V. R. (2018) Efficient GAN-based anomaly detection. *CoRR*, abs/1802.06222. http://arxiv.org/abs/1802.06222.

Chapter 17
Revisiting Practical Byzantine Fault Tolerance Through Blockchain Technologies

Nicholas Stifter, Aljosha Judmayer, and Edgar Weippl

Abstract The connection between Byzantine fault tolerance and cryptocurrencies, such as Bitcoin, may not be apparent immediately. Byzantine fault tolerance is intimately linked to engineering and design challenges of developing long-running and safety-critical technical systems. Its origins can be traced back to the question of how to deal with faulty sensors in distributed systems and the fundamental insight that majority voting schemes may be insufficient to guarantee correctness if arbitrary, or so-called Byzantine failures, can occur. However, achieving resilience against Byzantine failures has its price, both in terms of the redundancy required within a system and the incurred communication overhead. Together with the complexity of correctly implementing Byzantine fault-tolerant (BFT) protocols, it may help to explain why BFT systems have not yet been widely deployed in practice, even though practical designs exist for almost 20 years. On the other hand, asking anyone about Bitcoin or blockchain 10 years ago would have only raised quizzical looks. Since then, the ecosphere surrounding blockchain technologies has grown from the pseudonymously published proposal for a peer-to-peer electronic cash system into a multi-billion-dollar industry. At the heart of this success story lies not only the technical innovations presented by Bitcoin but a colorful and diverse community that has succeeded in bridging gaps and bringing together various disciplines from academia and industry alike. Bitcoin reinvigorated interest in the topic of BFT as it was arguably the first system that achieved a practical form of Byzantine fault tolerance with a large and changing number of participants. Research into the fundamental principles and mechanisms behind the underlying

N. Stifter (✉) · E. Weippl
Christian Doppler Laboratory for Security and Quality Improvement in the Production System Lifecycle (CDL-SQI), Institute of Information Systems Engineering, Technische Universität Wien, Vienna, Austria

SBA Research, Vienna, Austria
e-mail: nstifter@sba-research.org; eweippl@sba-research.org

A. Judmayer
SBA Research, Vienna, Austria
e-mail: ajudmayer@sba-research.org

S. Biffl et al. (eds.), *Security and Quality in Cyber-Physical Systems Engineering*,
https://doi.org/10.1007/978-3-030-25312-7_17

blockchain technology of Bitcoin has since helped advance the field and state of the art regarding BFT protocols. This chapter will outline how these modern blockchain technologies relate to the field of Byzantine fault tolerance and outline advantages and disadvantages in their design decisions and fundamental assumptions. Thereby, we highlight that Byzantine fault tolerance should be considered a practical and fundamental building block for modern long-running and safety critical systems and that the principles, mechanisms, and blockchain technologies themselves could help improve the security and quality of such systems.

Keywords Blockchain · Byzantine fault tolerance · Distributed ledger technologies · Bitcoin · Distributed systems

17.1 Introduction

Currently, the term "blockchain" is hardly associated with long-running, software-intensive, and production- and other, so-called, cyber-physical systems. Instead, many people will likely recall news and articles that cover topics such as the high volatility and speculative nature of cryptocurrencies, security breaches, and technical failures that have led to large financial losses[1] or promises of potential applications of blockchains that are reminiscent of the "peak of inflated expectations" found in the Gartner hype cycle for emerging technologies.[2]

However, beyond this hype, academia and industry alike have started to take a closer look at the technical foundations and possible applications of blockchain and distributed ledger technologies (DLT). Many of their fundamental concepts and building blocks are actually well established and researched technologies, such as cryptographic hash functions, Merkle trees, elliptic-curve cryptography, or moderately hard proof-of-work puzzles (Dwork and Naor 1992; Back 2002; Jakobsson and Juels 1999). The novel and particularly effective interplay between these components within the Bitcoin protocol, as well as the addition of game-theoretic incentives, facilitated the breakthrough which established Bitcoin (Nakamoto 2008) as the first viable cryptographic currency that could operate in a peer-to-peer environment without having to rely on a trusted third party.

It is precisely this seeming ability to avoid any (single) trusted third parties that renders blockchain protocols highly interesting for a variety of use-cases that reach well beyond the realm of virtual currencies. Hereby, the system as a whole exhibits a certain resilience against malicious activities from participants as long as their number and capabilities are reasonably bounded. Essentially, Bitcoin addresses the decades-old problem of *Byzantine fault tolerance* from a different and mostly practically oriented angle.

[1]See https://www.nytimes.com/2016/06/18/business/dealbook/hacker-may-have-removed-more-than-50-million-from-experimental-cybercurrency-project.html.

[2]cf. https://www.gartner.com/en/research/methodologies/gartner-hype-cycle.

Byzantine fault tolerance is of particular importance in the context of critical information infrastructures and other systems where both availability and resilience against faults is essential (Veronese et al. 2009; Esteves-Verissimo et al. 2017). However, widespread adoption of BFT has, at least in part, been hampered by the high resource requirements of early solutions, which may have contributed to the stigma that such protocols are largely impractical, although this issue has been addressed almost 20 years ago (Castro and Liskov 1999). Advancements in both the capacity and cost of technology, as well as new and efficient BFT protocols themselves, have rendered this overhead small enough and that Byzantine fault tolerance should not only be considered for an application in the most critical infrastructure but as a general design philosophy for any system with multiple distinct components that form complex interactions.

Through the current interest and research on blockchain and distributed ledger technologies, the topic of Byzantine fault tolerance is again being drawn into focus. Especially in the context of private or restricted environments, where not every participant should be able to partake in the consensus protocol and be allowed to propose updates to the underlying ledger or shared data structure, classical BFT protocols offer inherent advantages in both security and performance over proof-of-work-based blockchain designs such as Bitcoin.

The rest of this chapter is organized as follows: First, we give an introduction to the research field of Byzantine fault tolerance by outlining its history (Sect. 17.2). Second, we address the topic of blockchain technologies and how they originate from the development of the Bitcoin peer-to-peer cryptocurrency system (Sects 17.3–17.6). An outlook on future challenges and opportunities in this research field is given in (Sect. 17.7). We then outline potential use-cases for blockchain technologies that reach beyond cryptographic currencies (Sect. 17.8). In (Sect. 17.9) the potential application of BFT and blockchain technologies to production system engineering is discussed in more detail before the chapter is concluded (Sect. 17.9).

17.2 Byzantine Fault Tolerance

The origin of the term *Byzantine* failure traces back to the seminal work of Lamport et al. that introduces and addresses an agreement problem called the *Byzantine generals problem* (Lamport et al. 1982). In prior work, the same set of authors had first identified that ensuring consistency in the presence of *arbitrary* failures within distributed systems is more difficult than one would intuitively expect (Pease et al. 1980). Generally speaking, the terms Byzantine and arbitrary failure are used interchangeably, even though the former more explicitly considers the possibility of adversarial behavior. If a system is allowed to exhibit arbitrary failures, it follows that there can also exist execution traces where the sequence and type of failures are indiscernible to that of any adversarial strategy. A clear distinction between the two failure models is usually not made.

Initial research on the Byzantine generals problem, or more generally how to reach agreement, i.e., *consensus*, among a set of processes in the presence of faults, was spawned from practical engineering challenges at the time. Improvements in both microprocessors and networking capabilities had led to a consideration for their application in safety critical systems such as the SIFT fault-tolerant aircraft control system (Wensley et al. 1978). However, thorough analysis of a concrete design problem, namely, clock synchronization among multiple clocks, revealed that synchronization algorithms become impossible for three clocks if one of them is faulty and can drift arbitrarily. The generalization of this problem, that is, reaching agreement upon a vector of values where each value is the private input of a participant in the agreement protocol and the agreed-upon vector must either contain the private input of each participant or that the particular participant was faulty, is referred to as *interactive consistency*.

Pease et al. (1980) were able to show that even in a synchronous system model, i.e., where there is an a priori known upper bound Δ on computation and message transmission times, and a fully connected graph of reliable, point-to-point communication channels without message authentication, interactive consistency requires $3f + 1$ participants to arrive at a solution, where f denotes the maximum number of faulty participants that can exhibit arbitrary failures.

A few years later it was proven in (Fischer et al. 1985), what is now referred to as the *FLP impossibility result*, that deterministic consensus becomes impossible in an asynchronous system if only a single process is allowed to fail in the crash-stop model, even if communication between processes is reliable. The result, however, does not extend to consensus protocols exhibiting only *probabilistic* guarantees for liveness or correctness. Hence, so-called *randomized* Byzantine consensus algorithms, first presented by Ben-Or (1983) and Rabin (1983), which instead eventually terminate with probability $P(1)$ or have a non-zero probability for disagreement, are hereby not affected.

Nevertheless, at the time, the takeaway from these first results was that systems for reaching consensus, in particular in the presence of *Byzantine failures* while in principle feasible, were largely impractical for most real-world scenarios (Castro and Liskov 2002). For instance, the papers presenting the Byzantine generals problem and interactive consistency contain accompanying solutions where the distributed algorithms have an exponential message complexity in the number of participating processes. Together with the additional computational overhead, as well as large number of additional replicas that are required to tolerate Byzantine failures over the more benign crash failures, early BFT consensus protocols were simply too prohibitive for most use-cases.

It would take over a decade until publications such as *Practical Byzantine Fault Tolerance* (PBFT) by Castro and Liskov (Castro and Liskov 1999) showed that Byzantine fault-tolerant consensus algorithms could indeed be rendered practicable under realistic system assumptions. Nevertheless, while research on the topic of BFT consensus was ongoing (Cachin et al. 2000; Clement et al. 2009; Guerraoui et al. 2010; Veronese et al. 2013), it remained a comparatively isolated topic area, given the broad range of potential applications. In part, this may be attributed to the fact that consensus protocols are often discussed in the context of *state machine replication* (Lamport 1984; Schneider 1990) and achieving active replication for services such as databases. For these scenarios, all replicas, i.e., participants, may be under the control of a single entity and achieving only the more benign crash-fault tolerance can often be a tenable system model. In particular, Lamport's crash-fault-tolerant *Paxos* consensus algorithm (Lamport 1998) and derivations thereof have found their way into practical applications (Chandra et al. 2007).

However, even in such scenarios where crash-fault tolerance may have previously been considered acceptable, it can still be advantageous to gain the additional resilience of BFT. In particular, because these previously isolated systems are increasingly becoming interconnected, e.g., by operating in cloud environments (Vukolić 2010), it appears sensible to be able to tolerate Byzantine failures. Even before the advent of Bitcoin and blockchain technologies, calls from the scientific community had become louder that Byzantine fault-tolerant protocols could increasingly meet a wide range of practical demands and should hence be adopted (Clement et al. 2008; Liu et al. 2016). The recent hype surrounding blockchain and distributed ledger technologies has seemingly provided a crucial stepping stone in this regard and could help achieve more widespread adoption of BFT protocols. Demand for private or consortium blockchains, as well as a quest for achieving better scalability in terms of transaction throughput and resource consumption, has put modern BFT consensus protocols at the heart of many new ledger designs (Vukolić 2015). Further, research and newly found insights into the fundamental principles and mechanisms of Bitcoin and similar proof-of-work-based blockchains have resulted in hybrid system models and promising new approaches for BFT protocols (e.g., Miller et al. 2016; Abraham et al. 2018; Gilad et al. 2017; Pass and Shi 2018). Together, these advancements may facilitate the deployment of BFT protocols in various systems as part of the process of exploring and familiarizing oneself how blockchain technologies could be meaningfully integrated.

17.3 What is Blockchain?

Nowadays, blockchain is all too often encountered as a marketing buzzword or fuzzy umbrella term whose intended meaning is best translated to *"technologies that are loosely related to Bitcoin"*. *Bitcoin* is a proposal and subsequent implementation of a *"peer-to-peer electronic cash system"*, whose novel approach promises to solve the distributed *double spending problem* (Jarecki and Odlyzko 1997; Hoepman 2007) without having to rely on a trusted third party.

However, beyond the hype the underlying concepts and technologies have managed to spark the interest of the scientific community and led to a plethora of research efforts and publications from various different disciplines, such as cryptography, IT security, distributed and fault-tolerant computing, formal methods, game theory, economics, and legal sciences. This serves to highlight the interdisciplinary nature of cryptocurrencies and blockchain technologies, as a new area of research is beginning to take shape.

Interestingly, the term *blockchain* itself was not directly introduced by the pseudonymous author or authors going by the name *Satoshi Nakamoto* in the original Bitcoin white paper (Nakamoto 2008); instead only the words *blocks* and *chains* are mentioned. As part of Bitcoin's underlying data structure, transactions are grouped into blocks which are linked or *chained* together using hash pointers (Narayanan et al. 2016). The combination of these words was subsequently used early on within the Bitcoin community when referring to certain concepts of this so-called cryptographic currency or simply *cryptocurrency*.

As a result, two common spellings can be encountered throughout the literature, namely, *blockchain* and *block chain*. Although the latter variant was actually used by Satoshi Nakamoto in a comment within the original source code,[3] the former, i.e., blockchain, has established itself as the de facto standard in both the community and academic literature.

Generally speaking, blockchain or blockchain technologies may be used to refer to the mechanisms and principles by which Bitcoin and similar systems are able to achieve some form of *decentralized agreement* upon a shared ledger. On the other hand, blockchain may also specifically refer to the underlying data structure of such systems. Currently, there is no broad agreement on the exact meaning of the term, and definitions are evolving as research in this field is ongoing.

17.4 The Early Days of Cryptocurrencies

In the early 1980s, around the same time early research on BFT consensus was being established, David Chaum presented the cryptographic concept of *blind signatures* (Chaum 1983) together with a use-case in the form of an untraceable (electronic) payment system. It was arguably the first step toward the development of research on (anonymous) electronic cash systems, and the heavy reliance on cryptography to instill upon such systems new desirable properties would eventually lead to the term cryptographic currency or simply *cryptocurrency*. However, while Chaum's proposal presented a significant improvement toward preserving the privacy of users, it still suffered from the drawback that a single (trusted) authority was necessary to issue currency units and prevent their *double spending*. Unfortunately, despite various commercialization efforts (Pitta 1999), this early concept failed to reach a broad audience. Nevertheless, the seed had been planted that would inspire further research toward electronic cash systems that could better satisfy desirable properties of traditional physical money.

What followed was a new generation of cryptocurrencies such as Wei Dai's *b-money* (Dai 1998), Nick Szabo's *bit gold* (Szabo 2005), Hal Finney's reusable

[3]https://github.com/trottier/original-bitcoin/blob/master/src/main.h#L795-L803.

proofs-of-work (RPOW) (Finney 2004), and Adam Back's *Hashcash* (Back 2002). While these second-generation systems still could not entirely avoid the necessity for a trusted third party, they started to incorporate an interesting cryptographic primitive as a new approach for controlling the issuance of new currency units, referred to as proof-of-work (PoW). The underlying concept of proof-of-work, namely, to require the solution to a *moderately hard* but easy to verify computation as some form of pricing mechanism, was originally devised as a means for combating junk mail by Dwork and Naor (1992).

In the context of the presented research for both BFT fault tolerance and cryptocurrencies, Bitcoin was able to provide an interesting and novel approach that appeared to be practical.

17.5 The Decentralization of Trust

Bitcoin is the first cryptographic currency that does not have to rely on a trusted third party to solve the double-spending problem. It achieves this by combining clever incentive engineering and well-studied cryptographic primitives in a novel way, such that participants are able to establish (eventual) agreement on the state changes of the underlying transaction ledger (Bonneau et al. 2015).

When Bitcoin was first presented by Satoshi Nakamoto, both the publication and subsequent release of a prototype implementation (Nakamoto 2008) garnered relatively little attention, in particular from academia.

Interestingly, the original Bitcoin white paper did not relate its proposed solution to the distributed double-spending problem to previous research on Byzantine fault tolerance or consensus,[4] thereby rendering it less likely for readers to immediately make a connection to this field of research. Similarly, from the perspective of a cryptographer, at a first glance Bitcoin did not introduce any fundamentally new concept beyond a novel application of proof-of-work.

Furthermore, it can be argued that despite its reliance on well-discussed primitives such as cryptographic hash functions, (Elliptic curve) digital signatures, and Merkle trees, the presented concept behind Bitcoin nevertheless left room for skepticism, in particular because the author(s) did not provide formalizations of the claimed properties and security guarantees of the system.

Irrespective of this initial obscurity to much of the scientific community, Bitcoin as a system continuously gained real-world adoption and quickly outgrew its hobbyist cradle, both in valuation and ability to be effectively mined on consumer hardware (Taylor 2013). In retrospect, one may argue that Bitcoin was able to

[4]Satoshi Nakamoto did claim that Bitcoin's fundamental mechanism is a solution to the Byzantine generals problem in the cryptography mailing list; see http://www.metzdowd.com/pipermail/cryptography/2008-November/014849.html.

effectively bridge various research fields precisely because it avoided placing itself into a single category early on.

The first peer-reviewed publications related to Bitcoin were published in 2011 (e.g., Reid 2011), and most of the early works covering this topic area had a focus on double-spending attacks, network properties, and the privacy guarantees that could be achieved in such systems (Androulaki et al. 2012; Ron and Shamir 2013; Meiklejohn et al. 2013).

In 2014 Miller and LaViola made a first step toward the formalization of Bitcoin's consensus mechanism in a synchronous system model by considering its applicability for solving a single instance of (eventual) binary consensus (Miller and LaViola 2014). The following year, Garay et al. presented the first formal analysis and description of Bitcoin's underlying protocol and consensus approach, referring to it as the "Bitcoin Backbone Protocol" (Garay et al. 2015). Their initial system model assumes a static set of participants, i.e., nodes, where the ratio of computational power among them remains the same, a fully connected network that supports synchronous message communication and constant mining difficulty. Formalization efforts of the Bitcoin protocol and its underlying consensus mechanism, generally referred to as *Nakamoto consensus*, are ongoing (Stifter et al. 2018; Garay and Kiayias 2018), extending, for instance, to models of weaker (partial) synchrony (Pass et al. 2017) and chains of variable difficulty (Garay et al. 2017).

This novel (Byzantine fault tolerant) consensus approach and the practical demonstration of its feasibility are significant scientific contributions of Bitcoin. Nakamoto consensus allows for so-called *permissionless* participation (Vukolić 2015) because it only requires a very weak form of identity in the shape of computational resources. Arguably, decentralization poses the requirement that (consensus) participants are readily able to join and leave the system at will. Classical BFT consensus assumes a *static* set of participants that are *a priori determined*, because allowing for so-called dynamic group membership has proven to be difficult to solve for the Byzantine failure case, demanding strong system assumptions (Kihlstrom et al. 1998) that are unrealistic to achieve in a peer-to-peer electronic cash system over the Internet.

Part of the problem lies in preventing an adversary from simply generating multiple identities to perform a so-called Sybil attack. The concept of Sybil attacks is first introduced in Douceur (2002) and addresses the problem that an adversary in a peer-to-peer environment can cheaply (in terms of utilized resources) generate multiple identities with which to participate and thereby undermine any redundancy requirements that are employed to mitigate faulty or malicious behavior. Interestingly, a few years before the Bitcoin white paper was released, Aspnes et al. proposed the utilization of *moderately hard puzzles* as a way to expose Byzantine impostors (Aspnes et al. 2005) and address the Sybil attack. In such a model an adversary that is bounded in its computational resources also becomes bounded in the number of identities it can generate over a given period of time, thereby rendering Byzantine consensus solvable as long as a sufficiently large fraction of the overall computational power used to create identities is controlled by honest participants. Individual participants are able to join and leave the consensus process by either commencing or seizing the generation of puzzle solutions; however certain assumptions may still need to hold to provide meaningful guarantees.

Analogously, Bitcoin also leverages on proof-of-work as a core component of its consensus mechanism which acts as a weak form of identity by designating a round leader eligible for proposing the next state updates to the underlying ledger. In a sense, the group membership problem is hereby avoided and a system where participants can potentially anonymously partake in becomes possible.

Another important contribution Nakamoto consensus makes is its scalability in terms of the possible number of consensus participants (Vukolić 2015). Traditional BFT consensus protocols have so far achieved a message communication complexity that is at best quadratic in the number of participants, i.e., $O(n^2)$ (Miller et al. 2016). This generally limits the number of consensus nodes to less than one hundred active participants if the protocol is to remain practicable. Given an efficient peer-to-peer gossip mechanism and initial setup, Nakamoto consensus is able to achieve a communication complexity that lies in $O(n)$ (Garay et al. 2015).

It is these aspects of Nakamoto consensus that facilitate decentralization and permit a permissionless, peer-to-peer consensus setting.

17.6 From Bitcoin to Blockchain

With a continuous influx of new users and developers interested in Bitcoin, questions about its design and also what other applications could potentially benefit from the underlying technology were increasingly being discussed and explored.

In 2011, *Namecoin* was developed as the first successful fork and extension of the (open source) Bitcoin protocol code. Namecoin extends the concept of a cryptocurrency by adding a decentralized key-value store to allow for such use-cases as providing a decentralized domain name service (Schwarz 2011). It was followed by a growing number of alternative implementations with a variety of different goals in mind.

Alternative cryptocurrencies, in short *altcoins*, is a broad term encompassing the nowadays hundreds of cryptocurrency designs[5] that loosely follow Bitcoin's principles or its backbone protocol. Needless to say, not all altcoins have been successful, some of which only existed for a short period of time. Many of these projects are variations on the parametrization of the Bitcoin protocol with few actual modifications to the underlying code (Palmer et al. n.d; Litecoin.org n.d.). Some, however, have incorporated more profound changes and even provide entirely new code bases (King and Nadal 2012; Ben Sasson et al. 2014; Schwartz et al. 2014), where their applicability as a *cryptocurrency* may only play a secondary role, i.e., as part of a decentralized smart contract and application platform such as Ethereum (Buterin 2014a).

[5]See https://coinmarketcap.com/.

The difficulty in drawing a clear distinction between a cryptocurrency and an alternative application based on blockchain technology becomes apparent when we consider the core principles behind Nakamoto consensus: Participants compete to solve a proof-of-work of certain difficulty over their proposed state changes to the underlying ledger, referred to as *mining*. Furthermore, each puzzle input also explicitly includes the reference to a previous solution, in order to establish a causal relationship between them. As an agreement mechanism, the longest consecutive chain[6] of such puzzle solutions, starting from a pre-agreed-upon *genesis block*, is considered valid, and its current head will be referenced by honest participants when searching for new puzzle solutions.

The security of this approach also depends on the game theoretic aspect that consensus participants, so-called miners, are incentivized by being rewarded cryptocurrency units for finding valid puzzle solutions and extending the longest chain. Since this property is an intrinsic and natural byproduct of a cryptocurrency system such as Bitcoin, it is not easily replaceable in other application scenarios without potentially affecting the security guarantees of the underlying system. Therefore, as a prudent approach, many projects resort to adopting all properties of a blockchain-based cryptocurrency and add their additional application-specific components on top of them.

If we recall the previously outlined *decentralization* properties that Bitcoin's Nakamoto consensus provides, an interesting question that arises is how modifications to the underlying consensus affect the resulting system. In a distributed environment, the utilization of an authenticated data structure such as a blockchain can have its merits beyond an immediate application as part of a permissionless cryptocurrency. Depending on the application scenario, it may not actually be necessary, or even desirable, to allow such permissionless access to the underlying consensus mechanism. In particular, the required continuous resource consumption of proof-of-work renders Nakamoto consensus both impractical and insecure for small-scale deployment, as the provided security guarantees only hold under the assumption that the majority of computational power is controlled by honest participants.

Furthermore, Nakamoto consensus achieves decentralization at the cost of rendering transaction scalability seemingly more difficult to achieve than what traditional BFT consensus approaches are able to offer (Vukolić 2015). Therefore, so-called *permissioned* blockchains with alternative Byzantine fault-tolerant consensus mechanisms are increasingly being considered for corporate application scenarios (Dinh et al. 2017; Vukolić 2015). However, applying those technologies to a different use-case or system, while at the same time preserving desirable characteristics of blockchain technologies, has turned out to be a nontrivial task (Cachin and Vukolić 2017).

More recently, *hybrid* consensus models have emerged that aim toward bringing together properties from both permissioned and permissionless systems (Pass and

[6]More precisely, it is the chain with the most cumulative proof-of-work.

Shi 2017b; Luu et al. 2016; Pass and Shi 2018). The quest for addressing resource consumption in Nakamoto consensus has furthermore led to promising research and results on the topic of so-called *proof-of-stake* (PoS) consensus protocols (Kiayias et al. 2016; Bentov et al. 2016; Micali 2016). In such PoS systems, *virtual resources* in the form of cryptocurrency units are *staked* instead of requiring actual computational effort while retaining most of the desirable properties of PoW-based Nakamoto consensus.

Overall, we can conclude that the rise in popularity of Bitcoin and its derivatives has also led to an increased and renewed interest in the underlying technologies and core components behind blockchain and DLT, e.g., BFT consensus, that render such systems possible.

17.7 Future Challenges and Opportunities in Blockchain Research

Albeit academia's initial slow reaction to Bitcoin and blockchain technologies, the pace of new publications, and research has continuously increased over the last few years. With a growing understanding of the fundamental principles behind blockchain technologies, the focus is now shifted toward both new application domains and potential improvements. State-of-the-art findings and insights are increasingly being adopted and considered in new system proposals and improvements, such as Ethereum's incorporation of a variant of GHOST (Sompolinsky and Zohar 2013) as part of its design.

Blockchain and distributed ledger technologies have many different aspects and can therefore be viewed from various angles, including the *financial* and *economic* perspective, *legal* perspective, *political* and *sociological* perspective, as well as *technical* and *socio-technical* perspectives. These very different viewpoints can be separated even further; for example, the technical aspects can be divided into the following non-exhaustive list of fields: *cryptography*, *distributed computing*, *game theory*, *data science*, and *software and language security*. Because of these many different viewpoints and the broad potential applicability of these technologies, it is not only helpful but necessary to strive for interdisciplinary collaboration.

As the adoption and use of DLTs is steadily increasing, new challenges and limitations of the underlying technologies are increasingly becoming apparent (Croman et al. 2016; Vukolić 2015). In particular, concerns about future scalability and performance are a current driving force behind new research and discussions. Furthermore, many open questions on governance, the handling of human and technological failures, and other life cycle events of blockchain technologies are no longer just hypothetical (Buterin 2016) but have been rendered current and pressing issues by real-world events (De Filippi and Loveluck 2016). We outline some of these open questions in more detail.

17.7.1 Scalability

Bitcoin-like cryptocurrencies that are based on proof-of-work blockchains have certain drawbacks when it comes to scalability. Due to network latencies and structure and the very nature of the computationally expensive proof-of-work, there are certain performance limitations. The Bitcoin network is currently capable of handling around 7–10 transactions per second (Vukolić 2015; Decker and Wattenhofer 2013; Croman et al. 2016). Compared to traditional payment networks or BFT protocols, this is a relatively small number. For example, PayPal is capable of handling a few hundred transactions per second (Kiayias and Panagiotakos 2015), whereas VISA can process up to several thousand transactions per second (Kiayias and Panagiotakos 2015; Croman et al. 2016). It is well known that there are certain tradeoffs between the security and performance of PoW-based cryptocurrencies (Bamert et al. 2013; Kiayias and Panagiotakos 2015; Sompolinsky and Zohar 2013; Gervais et al. 2016). Optimizing the performance of decentralized blockchains while still being able to provide accurate estimates and formal proofs on the security impact of any changes is an ongoing topic of research. Several different approaches have been proposed that aim to minimize intrusive changes to existing protocols, such as *Bitcoin-NG* (Eyal et al. 2016). Others propose switching to entirely different underlying consensus mechanisms (Vukolić 2015; Vukolic 2016). Hybrid system models (Pass and Shi 2017a) that aim to consolidate advantages of both approaches are also being discussed. So-called layer two scaling solutions are another possibility to increase scalability by shifting some of the transaction load off-chain, i.e., in direct payment and state channels (Poon and Dryja 2016; Dziembowski et al. 2017; Coleman et al. 2018). For a general summary of possible directions, see (Croman et al. 2016).

17.7.2 Resource Consumption

All proof-of-work-based schemes rely on the existence of a limited resource that nodes are required to draw upon if they want to generate PoWs. In Bitcoin, this resource is a combination of energy, hardware, and network capacity. If there were a proof-of-work that did not rely on a limited resource, and instead could be claimed in unbounded quantities by anybody, Sibyl attacks would again become possible. It is actually the PoW that allows mining participants to remain "anonymous" and not have to reveal any previous information about themselves when participating in Nakamoto consensus. In a non-anonymous setting, this problem can be partially addressed by determining a set of nodes that are responsible for maintaining consensus on the blockchain's state; however in this case, a certain degree of trust needs to be placed in those nodes. The question of how to solve Byzantine fault tolerance in a dynamic membership setting is however still part of ongoing research.

The question that arises is whether there are provably secure yet practical and scalable schemes that permit a virtualization of the required PoW resources while still providing protection against Sibyl attacks in the permissionless model. Such a scheme would mean that instead of being forced to waste physical resources such as energy and computing hardware, one could only simply rely on virtual counterparts. One of the first approaches toward virtualizing such PoW resources, namely, *proof-of-stake* (PoS), was first introduced in cryptocurrencies such as *Peercoin* (King and Nadal 2012). The general idea behind proof-of-stake is to allow participants to lock up or *stake* part of their cryptocurrency units, which, in relation to the number of units staked by other miners, gives them a certain probability at which they can mine, or *mint*, a new block. Several difficulties and attacks with regard to proof-of-stake cryptocurrencies have been initially pointed out (Bentov et al. 2014) and until recently, concepts and presented protocols often lacked formal models and security proofs. This situation however has been amended by recent works such as *Ouroboros* (Kiayias et al. 2017) and *Snow White* (Bentov et al. 2016), which both present provably secure proof-of-stake blockchain protocols.

Another approach that could help improve the security of proof-of-stake protocols which is, for instance, being pursued by the Ethereum foundation is to integrate or leverage economic incentives and game theory in the PoS consensus process. The proposed protocol is designed to render (certain types of) malicious behavior detectable and consequently *punishes* such behavior by destroying locked-up funds or potential block rewards of the perpetrator (Buterin 2014b; Buterin and Griffith 2017). Research toward understanding and leveraging on game theoretic incentives that influence behavior of protocol participants in the realm of cryptocurrencies has been dubbed *cryptoeconomics*. In the context of traditional BFT protocols, this concept has also been explored in, e.g., the BAR (Byzantine, altruistic, rational) model (Aiyer et al. 2015; Li et al. 2006) or by reevaluating known possibility and impossibility results of distributed protocols, such as consensus, when a subset of participant is modeled as rational actors that follow certain optimization strategies (Groce et al. 2012).

17.7.3 Centralization vs. Decentralization

Studies on the mining landscape of Bitcoin, as well as other cryptocurrencies, show that there is a potential trend toward mining pool centralization in PoW-based systems (Judmayer et al. 2017). The question is, how decentralized should a cryptographic currency ecosystem be, and what methods can be used to enforce certain levels of decentralization? Which single points of failure are acceptable and which are not—for example, powerful exchanges, mining pools, and influential developers?

In the case of blockchain technologies that are based on Byzantine fault-tolerant systems, the question is how to compose and maintain a set of trusted nodes for consensus and who decides which nodes are allowed to participate. If the set of consensus nodes is small and static, resilience against Byzantine failures is more readily achievable; however the system is strongly centralized. The question of how to achieve Byzantine fault tolerance in a dynamic group membership setting which could potentially allow for more decentralization remains part of ongoing research.

17.8 Blockchain Use-Cases Beyond Cryptocurrencies

So far, we have primarily outlined the characteristics and technical challenges of blockchain technologies without expanding upon the potential use-cases that reach beyond the realm of cryptographic currencies. The following examples showcase problem domains and scenarios where an application of blockchain and distributed ledger technologies can be both promising and warranted, given that their engineering goals, challenges, desirable properties of the resulting systems, and also threat models have various overlaps and similarities to those encountered in the cryptocurrency space.

17.8.1 Trusted Timestamping and Data Provenance

The concept of trusted timestamping is not new and it has a wide range of useful applications such as providing tamper-resistant proofs of existence, for instance, for intellectual properties such as patent applications, or to document and commit to a particular state or information (e.g., a Merkle tree root which was derived from the relevant system data or a Git commit hash). In case of a system breach where the adversary may have tampered with data, such cryptographic commitments can later serve as vital references to determine data integrity.

However, this scheme of course requires that the commitment itself is safe from manipulation and ideally spread across multiple systems and media. Public proof-of-work (PoW)-based blockchains, such as Bitcoin, present an ideal platform to record these commitments as part of regular transaction data (requiring the committing party to only pay the appropriate transaction fee). The security and manipulation resistance of such blockchains stems from the sequential chaining of moderately hard puzzles which renders it (exponentially) increasingly unlikely for an adversary to be able to change any recorded transactions with respect to the length of newly mined blocks.

The advantage of PoW-based constructions over basic signature schemes with one or multiple trusted third parties is that, unless a severe flaw is found within the cryptographic hash function, no private keys or trapdoors exist that efficiently allow for equivocation. That is, if an adversary were to gain access to the private keys used in a signature-based timestamping scheme, they could readily forge backdated commitments with very little resource requirements, whereas in a PoW-based model they would have to recompute sequential PoWs which impose a highly prohibitive constraint both in terms of available time and computational resources. Blockchain-based timestamping has been described both in the scientific community (e.g.,

Gipp et al. 2015; Szalachowski 2018) and is employed in commercial products (cf., https://guardtime.com/ which is partnered with Lockheed Martin to secure systems engineering processes).

Permissioned blockchain systems that are based on classical BFT protocols can also offer advantages in combination with signature schemes for timestamping, especially if they employ the use of append only authenticated data structures such as hash chains. As long as the signing keys for timestamping are not reused in the BFT consensus mechanism and are therefore independent, an adversary would have to compromise both systems to effectively and fully conceal its malicious activity.

17.8.2 PKI and Digital Identities

An interesting proposition is the utilization of blockchain technologies to record identity information or serve as the basis for public key infrastructures (PKI). A general problem with most identity systems is the establishment of trusted infrastructure that secures and links cryptographic public keys to identities. Blockchain technologies could help augment traditional approaches such as certificate authorities (CAs). In particular, more recent developments in this area, such as certificate transparency, already embrace authenticated data structures as a means of identifying manipulation attempts. In the context of production systems, blockchain-based public key infrastructure could, for instance, help provide more robust mechanisms for establishing (and revoking) digital identities that are used for aspects such as access control or rights management, both in the development process and the operational design of the system. Another interesting application lies in the area of supply chain management where blockchain-based identity systems may help improve provenance. Proposals from the scientific community that leverage blockchain technologies, for instance, regarding certificate transparency, already exist (Wang et al. 2019) and there are currently concerted development efforts under way for establishing both standards and working systems for blockchain-based identity systems (e.g., https://identity.foundation/). However, many of these use-cases raise several important questions regarding user privacy and compliance with legislation and regulations, such as the EU General Data Protection Regulation (GDPR), and leave many open research questions and challenges.

17.8.3 Smart Contracts and Trusted Execution Environments

The currently established term "smart contract" is an unfortunate misnomer when it comes to succinctly describing its principal purpose or functionality, as it easily draws upon associations to legal contracts. A smart contract can be best thought of as program code that is executed in some distributed trusted execution environment. More specifically, the execution environment is generally a distributed or decentralized platform that offers both replication and, more importantly, Byzantine fault tolerance to ensure the correct execution and integrity of the smart contract code and its data storage. This is in contrast to the prevalent approach of implementing *Trusted Execution Environments* (TEEs) within computer hardware, such as Intel's SGX platform (McKeen et al. 2016; Costan and Devadas 2016), where the hardware manufacturer still acts as a single trusted third party to ensure the correctness and integrity of code execution. Permissionless blockchain-based smart contract platforms such as Ethereum (Buterin 2014a), but also permissioned counterparts such as the various incarnations of the Hyperledger platform (Cachin 2016), offer a unique trusted execution environment for program code, where the correctness and agreement upon the result of computations can be publicly verified and is secured by Byzantine agreement. As generalized computing platforms,[7] the previously mentioned use-cases can readily be implemented within smart contracts, thereby allowing the contract owner to leverage the security and availability of the base platform to provide such services without having to deploy another blockchain where the desired functionality then has to be integrated.

17.9 BFT and Blockchain Technologies in Production System Engineering

In this section we address how Byzantine fault tolerance and blockchain technologies can contribute to address the fundamental challenges and research questions that have been outlined in Chap. 1 regarding the engineering process of software-intensive technical systems.

Recapitulating the general system assumptions and challenges, this engineering process is often conducted by multiple teams and possibly subcontractors which have to collaborate and exchange engineering artifacts and other data. These collaborating actors may not extend mutual trust toward each other and there may even be incentives for participants to act dishonestly, such as attempting to gain access to inside knowledge of competitors, manipulate data and engineering artifacts, or otherwise disrupt the engineering process. This challenging collaborative environment calls for novel security methods where confidentiality, integrity, and availability can be guaranteed, as well as establishing traceability and accountability for engineering artifacts and the different actors within this environment. In the following, we address aspects among the research questions posed within Chap. 1, in particular regarding questions *RQ3a: Which security concepts mitigate cyber threats targeting the engineering process of complex cyber-physical systems?* and *RQ3b: How can the security of complex cyber-physical systems be enhanced by considering security*

[7]In principle such platforms offer Turing completeness for code; however executions are generally bounded in their complexity by requiring users to pay a certain price for each operation to prevent trivial denial of service attacks.

aspects during the engineering phase?, through domain-related research on block chain and BFT technologies.

17.9.1 Byzantine Fault Tolerance in PSE

The challenge of securing collaborative environments and shared data-stores against adversarial behavior is a long-standing research topic that has been addressed and informed by various research fields such as cryptography, (Byzantine) fault tolerance, and distributed systems. For instance Herlihy and Tygar (1987), address the question of how replicated data can be made more secure, where the notion of security encompasses the two properties of *secrecy* and *integrity*. It is argued that there seemingly is a trade-off between easier security measures for centralized services and a resilience to faults, which can be improved through redundancy. A solution employing threshold cryptography is therein presented where an adversary cannot ascertain or alter the state of a (shared) data object, if it can only compromise fewer than a threshold of repositories. Subbiah and Blough (2005) improve upon this approach by utilizing a more efficient secret sharing scheme and considers collaborative work environments. The topic of intrusion tolerance in collaborative environments is considered in Dutertre et al. (2002) where techniques for Byzantine fault tolerance and secret sharing are applied to *group communication* primitives (Chockler et al. 2001) to render the design more robust against adversaries. Kallahalla et al. (2003) present a protocol for scalable and secure file sharing using untrusted storage. Hereby, the novelty stems from a practical approach of an encrypt-on-disk system where key management and distribution is handled by the participants' clients rather than the storage provider or administrator. The approach helps to protect against data leakage attacks, e.g., by an untrusted administrator or compromised server; allows users to set arbitrary policies for key distribution; and improves scalability by shifting computationally demanding cryptographic operations to the client. In Zhao and Babi (2013) Byzantine fault tolerance in the context of real-time collaborative editing systems is addressed and a comprehensive threat analysis is performed. An interesting insight the paper presents, namely, that the detection of malicious updates to a document can only be done by its publisher or participants because it is application-specific, can be related to the more general result of Doudou et al. (2002) that the detection of Byzantine behavior by a failure detector cannot be entirely independent of the algorithm in which the failure detector is used.

17.9.2 Blockchain Technologies in PSE

We have previously outlined possible use-cases for blockchain and DLT technologies in Sect. 17.8 that reach beyond the realm of cryptocurrencies. In this regard it is not always clear if a scenario will stand to substantially benefit by its adoption. Wang et al. (2017) explores possible application scenarios of blockchain and DLT for construction engineering management and attempts to envision how the technologies may be employed in these settings. Technical details and their feasibility are presented primarily at a conceptual level, as the intention of the publication is to present a possible outlook what these technologies may offer to the problem domain. The desire or necessity to reduce reliance on *trusted third parties* and *tamper-proof documentation of interaction between parties and modification of shared data* are, among others, identified as key aspects where DLT could provide advantages.

A promising research topic related to the challenges of production system engineering is the implementation or improvement of access control mechanisms through blockchain technologies. For instance, Maesa et al. (2017) outlines, based on the example of Bitcoin, how attribute-based access control (ABAC) can be integrated into existing blockchain-based systems. In Paillisse et al. (2019) it is shown how access/policy-based networking for multi-administrative domains can be implemented and managed using blockchain technologies by presenting a prototype implementation that expands upon Group-Based Policy (GBP) and is based on the Hyperledger Fabric (Cachin 2016) blockchain framework.

17.9.3 Discussion

So far, core aspects of BFT and blockchain technologies were outlined, and research in these fields was presented that also addresses challenges which are encountered in production system engineering. Hereby we show that many of the fundamental problems, assumed system models, and security threats (e.g., the existence of multiple distrusting parties that may act maliciously while at the same time have to collaborate to produce some common result, the necessity to provide resistance against manipulation of shared data) between these different fields are closely related. Decades of insights and improvements in developing practical Byzantine fault-tolerant consensus protocols, as well as the hype and subsequent explosion of research in blockchain technologies, can be leveraged when seeking to provide solutions or improvements to the security and quality of production system engineering as well as the long-running software-intensive technical systems that are hereby created.

As a concrete example, consider Chap. 12 of this book, which covers the topic of securing information manipulation in PSE. The therein assumed system architecture, as depicted in Fig. 12.3, may be augmented with concepts and techniques described within this chapter. Instead of relying on a single centralized PSE platform to exchange data, for example, the secure and scalable file sharing approach outlined by Kallahalla et al. (2003), could prove both beneficial and practical. Its design considers key exchange and access management on client devices rather than by a centralized provider. This aspect could further be strengthened by employing blockchain-based access control management where said clients act as nodes in either a public or private DLT network and commit all relevant updates to the access rights to the ledger. The advantage of such an approach is that a tamper-resistant, (eventually) ordered log of events is distributed among the participants such that manipulation by a subset of them is rendered difficult and readily detectable.

Further research in this regard is both necessary and warranted to determine if such an application of modern BFT protocols and blockchain technologies can indeed lead to practical designs and techniques for production system engineering and, more importantly, contribute toward an improvement in their quality and security.

17.10 Conclusion

While blockchain technologies are hardly the answer to life, the universe, and everything, as ideologists or advertising sometimes paint it,[8] the fusion of its underlying principles and methods has opened up new pathways and outlined new possibilities in different areas of research. Bitcoin has created a new class of BFT consensus systems and rekindled research in the field of distributed computing and Byzantine fault tolerance, leading to new and interesting permissioned, permissionless, and hybrid blockchain constructs. It furthermore bootstrapped a vivid and diverse community that is driving the development and practical application of this set of technologies further.

The renewed interest in Byzantine fault tolerance, fueled by the hype surrounding blockchain technologies, may also prove to be highly beneficial to a variety of other problem domains, such as the herein discussed topic of production system engineering. Intriguing design proposals, promising system architectures, and even fully functional prototypes that tolerate Byzantine failures are being rediscovered and reconsidered as both practical and desirable approaches when evaluating if blockchain or DLT could benefit a particular use-case. Many of the examples presented in this chapter highlight that these technologies should not be considered as mutually exclusive and can actually stand to benefit from each other, showcasing that interdisciplinary thinking can lead to novel approaches and solutions with practical applications. It remains to be seen what impact blockchain technologies and cryptocurrencies will ultimately have on society and technology; however it is clear that they have the potential to be more disruptive than just a superficial speculative bubble.

Acknowledgements We thank Georg Merzdovnik as well as the participants of Dagstuhl Seminar 18152 "Blockchains, Smart Contracts and Future Applications" for valuable discussions and insights. This research was funded by Bridge 1 858561 SESC, Bridge 1 864738 PR4DLT (all FFG), the Christian Doppler Laboratory for *Security and Quality Improvement in the Production System Lifecycle (CDL-SQI)*, Institute of Information Systems Engineering, TU Wien, and the competence center SBA-K1 funded by COMET. The financial support by the Christian Doppler

[8] Cf., https://www.theguardian.com/world/2016/jul/07/blockchain-answer-life-universe-everything-bitcoin-technology.

Research Association; the Austrian Federal Ministry for Digital and Economic Affairs; and the National Foundation for Research, Technology, and Development is gratefully acknowledged.

References

Abraham, I., Gueta, G., & Malkhi, D. (2018). Hot-stuff the linear, optimal-resilience, one-message bft devil. arXiv:1803.05069. https://arxiv.org/pdf/1803.05069.pdf

Aiyer, A. S., Alvisi, L., Clement, A., Dahlin, M., Martin, J.-P., & Porth, C. (2005). Bar fault tolerance for cooperative services. In *ACM SIGOPS Operating Systems Review* (Vol. 39, pp. 45–58). New York, NY: ACM. http://www.dcc.fc.up.pt/~Ines/aulas/1314/SDM/papers/BAR%20Fault%20Tolerance%20for%20Cooperative%20Services%20-%20UIUC.pdf

Androulaki, E., Capkun, S., & Karame, G. O. (2012). Two bitcoins at the price of one? Double-spending attacks on fast payments in bitcoin. In *CCS*. http://eprint.iacr.org/2012/248.pdf

Aspnes, J., Jackson, C., & Krishnamurthy, A. (2005). *Exposing Computationally-Challenged Byzantine Impostors*. Department of Computer Science, Yale University, New Haven, CT, Tech. Rep. http://www.cs.yale.edu/homes/aspnes/papers/tr1332.pdf

Back, A. (2002). Hashcash-a denial of service counter-measure. Retrieved March 9, 2016, from http://www.hashcash.org/papers/hashcash.pdf

Bamert, T., Decker, C., Elsen, L., Wattenhofer, R., & Welten, S. (2013). Have a snack, pay with bitcoins. In *2013 IEEE Thirteenth International Conference on IEEE Peer-to-Peer Computing (P2P)* (pp. 1–5). Piscataway, NJ: IEEE. http://www.bheesty.com/cracker/1450709524_17035424cb/p2p2013_093.pdf

Ben-Or, M. (1983). Another advantage of free choice (extended abstract): Completely asynchronous agreement protocols. In *Proceedings of the Second Annual ACM Symposium on Principles of Distributed Computing* (pp. 27–30). New York, NY: ACM. http://homepage.cs.uiowa.edu/~ghosh/BenOr.pdf

Ben Sasson, E., Chiesa, A., Garman, C., Green, M., Miers, I., Tromer, E., et al. (2014). Zerocash: Decentralized anonymous payments from bitcoin. In *2014 IEEE Symposium on Security and Privacy (SP)* (pp. 459–474). Piscataway, NJ: IEEE. http://zerocash-project.org/media/pdf/zerocash-extended-20140518.pdf

Bentov, I., Lee, C., Mizrahi, A., & Rosenfeld, M. (2014). Proof of activity: Extending bitcoin's proof of work via proof of stake [extended abstract] y. *ACM SIGMETRICS Performance Evaluation Review, 42*(3), 34–37. http://eprint.iacr.org/2014/452.pdf

Bentov, I., Pass, R., & Shi, E. (2016). Snow white: Provably secure proofs of stake. Retrieved November 11, 2016, from https://eprint.iacr.org/2016/919.pdf

Bonneau, J., Miller, A., Clark, J., Narayanan, A., Kroll, J. A., & Felten, E. W. (2015). Sok: Research perspectives and challenges for bitcoin and cryptocurrencies. In *IEEE Symposium on Security and Privacy*. http://www.ieee-security.org/TC/SP2015/papers-archived/6949a104.pdf

Buterin, V. (2014a). Ethereum: A next-generation smart contract and decentralized application platform. Retrieved August 22, 2016, from https://github.com/ethereum/wiki/wiki/White-Paper

Buterin, V. (2014b). Slasher: A punitive proof-of-stake algorithm. Retrieved March 24, 2017, from https://blog.ethereum.org/2014/01/15/slasher-a-punitive-proof-of-stake-algorithm/

Buterin, V. (2016). Chain interoperability. Retrieved March 25, 2017, from https://static1.squarespace.com/static/55f73743e4b051cfcc0b02cf/t/5886800ecd0f68de303349b1/1485209617040/Chain+Interoperability.pdf

Buterin, V., & Griffith, V. (2017). Casper the friendly finality gadget. arXiv:1710.09437. Retrieved November 6, 2017, from https://arxiv.org/pdf/1710.09437.pdf

Cachin, C. (2016). Architecture of the hyperledger blockchain fabric. Retrieved August 10, 2016, from https://www.zurich.ibm.com/dccl/papers/cachin_dccl.pdf

Cachin, C., Kursawe, K., & Shoup, V. (2000). Random oracles in constantinople: Practical asynchronous byzantine agreement using cryptography. In *Proceedings of the Nineteenth Annual*

ACM Symposium on Principles of Distributed Computing (pp. 123–132). New York, NY: ACM. https://www.zurich.ibm.com/~cca/papers/abba.pdf

Cachin, C., & Vukolić, M. (2017). Blockchain consensus protocols in the wild. In *31 International Symposium on Distributed Computing*. arXiv preprint arXiv:1707.01873

Castro, M., & Liskov, B. (2002). Practical byzantine fault tolerance and proactive recovery. *ACM Transactions on Computer Systems, 20*, 398–461.

Castro, M., Liskov, B. (1999). Practical byzantine fault tolerance. In *OSDI* (Vol. 99, pp. 173–186). http://pmg.csail.mit.edu/papers/osdi99.pdf

Chandra, T. D., Griesemer, R., & Redstone, J. (2007). Paxos made live: An engineering perspective. In *Proceedings of the Twenty-Sixth Annual ACM Symposium on Principles of Distributed Computing* (pp. 398–407). New York, NY: ACM. https://www.kth.se/polopoly_fs/1.116933!/Menu/general/column-content/attachment/paxoslive.pdf

Chaum, D. (1983). Blind signatures for untraceable payments. In *Advances in cryptology* (pp. 199–203). Berlin: Springer. http://blog.koehntopp.de/uploads/Chaum.BlindSigForPayment.1982.PDF

Chockler, G. V., Keidar, I., & Vitenberg, R. (2001). Group communication specifications: a comprehensive study. *ACM Computing Surveys, 33*(4), 427–469.

Clement, A., Marchetti, M., Wong, E., Alvisi, L., & Dahlin, M. (2008). BFT: the time is now. In *Proceedings of the 2nd Workshop on Large-Scale Distributed Systems and Middleware* (p. 13). New York, NY: ACM.

Clement, A., Wong, E. L., Alvisi, L., Dahlin, M., & Marchetti, M. (2009). Making byzantine fault tolerant systems tolerate byzantine faults. In *NSDI* (Vol. 9, pp. 153–168). http://static.usenix.org/events/nsdi09/tech/full_papers/clement/clement.pdf

Coleman, J., Horne, L., & Xuanji, L. (2018). Counterfactual: Generalized state channels [online]. Retrieved May 18, 2019, from https://l4.ventures/papers/statechannels.pdf

Costan, V., & Devadas, S. (2016). Intel sgx explained. *IACR Cryptology ePrint Archive, 2016*(86), 1–118.

Croman, K., Decker, C., Eyal, I., Gencer, A. E., Juels, A., Kosba, A., et al. (2016). On scaling decentralized blockchains. In *3rd Workshop on Bitcoin and Blockchain Research, Financial Cryptography 16*. http://www.tik.ee.ethz.ch/file/74bc987e6ab4a8478c04950616612f69/main.pdf

Dai, W. (1998). bmoney. Retrieved April 4, 2016, from http://www.weidai.com/bmoney.txt

De Filippi, P., & Loveluck, B. (2016). The invisible politics of bitcoin: governance crisis of a decentralised infrastructure. Retrieved October 18, 2017, from https://halshs.archives-ouvertes.fr/halshs-01380617/document

Decker, C., & Wattenhofer, R. (2013). Information propagation in the bitcoin network. In *2013 IEEE Thirteenth International Conference on Peer-to-Peer Computing (P2P)* (pp. 1–10). Piscataway, NJ: IEEE. http://diyhpl.us/~bryan/papers2/bitcoin/Information%20propagation%20in%20the%20Bitcoin%20network.pdf

Dinh, T. T. A., Wang, J., Chen, G., Liu, R., Ooi, B. C., & Tan, K.-L. (2017). Blockbench: A framework for analyzing private blockchains. In *Proceedings of the 2017 ACM International Conference on Management of Data* (pp. 1085–1100). New York, NY: ACM.

Douceur, J. R. (2002). The sybil attack. In *International Workshop on Peer-to-Peer Systems* (pp. 251–260). Berlin: Springer. http://www.cs.cornell.edu/people/egs/cs6460-spring10/sybil.pdf

Doudou, A., Garbinato, B., & Guerraoui, R. (2002). Encapsulating failure detection: From crash to byzantine failures. In *International Conference on Reliable Software Technologies* (pp. 24–50). Berlin: Springer.

Dutertre, B., Crettaz, V., & Stavridou, V. (2002). Intrusion-tolerant enclaves. In *Proceedings of the 2002 IEEE Symposium on Security and Privacy* (pp. 216–224). Piscataway, NJ: IEEE.

Dwork, C., & Naor, M. (1992). Pricing via processing or combatting junk mail. In *Annual International Cryptology Conference* (pp. 139–147). Berlin: Springer. https://web.cs.dal.ca/~abrodsky/7301/readings/DwNa93.pdf

Dziembowski, S., Eckey, L., Faust, S., & Malinowski, D. (2017). *Perun: Virtual Payment Channels Over Cryptographic Currencies*. Cryptology ePrint Archive, Report 2017/635. Retrieved November 20, 2017, from https://eprint.iacr.org/2017/635.pdf

Esteves-Verissimo, P., Völp, M., Decouchant, J., Rahli, V., & Rocha, F. (2017). Meeting the challenges of critical and extreme dependability and security. In *2017 IEEE 22nd Pacific Rim International Symposium on Dependable Computing (PRDC)* (pp. 92–97). Piscataway, NJ: IEEE.

Eyal, I., Gencer, A. E., Sirer, E. G., & van Renesse, R. (2016). Bitcoin-ng: A scalable blockchain protocol. In *13th USENIX Security Symposium on Networked Systems Design and Implementation (NSDI'16)*. Berkeley, CA: USENIX Association. http://www.usenix.org/system/files/conference/nsdi16/nsdi16-paper-eyal.pdf

F. Reid, M. H. (2011). An analysis of anonymity in the bitcoin system. In *2011 IEEE International Conference on Privacy, Security, Risk, and Trust, and IEEE International Conference on Social Computing*. http://arxiv.org/pdf/1107.4524

Finney, H. (2004). Reusable proofs of work (RPOW). Retrieved April 31, 2016, from http://web.archive.org/web/20071222072154/http://rpow.net/

Fischer, M. J., Lynch, N. A., & Paterson, M. S. (1985). Impossibility of distributed consensus with one faulty process. *Journal of the ACM, 32*, 374–382. http://macs.citadel.edu/rudolphg/csci604/ImpossibilityofConsensus.pdf

Garay, J., & Kiayias, A. (2018). *Sok: A Consensus Taxonomy in the Blockchain Era*. Cryptology ePrint Archive, Report 2018/754. https://eprint.iacr.org/2018/754.pdf

Garay, J., Kiayias, A., & Leonardos, N. (2015). The bitcoin backbone protocol: Analysis and applications. In *Advances in Cryptology-EUROCRYPT 2015* (pp. 281–310). Berlin: Springer. http://courses.cs.washington.edu/courses/cse454/15wi/papers/bitcoin-765.pdf

Garay, J., Kiayias, A., & Leonardos, N. (2017). The bitcoin backbone protocol with chains of variable difficulty. In *Annual International Cryptology Conference* (pp. 291–323). Berlin: Springer.

Gervais, A., Karame, G. O., Wüst, K., Glykantzis, V., Ritzdorf, H., & Capkun, S. (2016). On the security and performance of proof of work blockchains. In *Proceedings of the 2016 ACM SIGSAC* (pp. 3–16). New York, NY: ACM.

Gilad, Y., Hemo, R., Micali, S., Vlachos, G., & Zeldovich, N. (2017). Algorand: Scaling byzantine agreements for cryptocurrencies. In *Proceedings of the 26th Symposium on Operating Systems Principles* pp. 51–68. New York, NY: ACM.

Gipp, B., Meuschke, N., & Gernandt, A. (2015). Decentralized trusted timestamping using the crypto currency bitcoin. preprint arXiv:1502.04015.

Groce, A., Katz, J., Thiruvengadam, A., & Zikas, V. (2012). *Byzantine agreement with a rational adversary* (pp. 561–572). Berlin: Springer. http://cs.ucla.edu/~vzikas/pubs/GKTZ12.pdf

Guerraoui, R., Knežević, N., Quéma, V., & Vukolić, M. (2010). The next 700 bft protocols. In *Proceedings of the 5th European Conference on Computer Systems* (pp. 363–376). New York, NY: ACM. https://infoscience.epfl.ch/record/121590/files/TR-700-2009.pdf

Herlihy, M. P., & Tygar, J. D. (1987). How to make replicated data secure. In *Conference on the Theory and Application of Cryptographic Techniques* (pp. 379–391). Berlin: Springer.

Hoepman, J.-H. (2007). Distributed double spending prevention. In *Security Protocols Workshop* (pp. 152–165). Berlin: Springer. http://www.cs.kun.nl/~jhh/publications/double-spending.pdf

Jakobsson, M., & Juels, A. (1999). Proofs of work and bread pudding protocols. In *Secure information networks* (pp. 258–272). Berlin: Springer. https://link.springer.com/content/pdf/10.1007/978-0-387-35568-9_18.pdf

Jarecki, S., & Odlyzko, A. (1997). An efficient micropayment system based on probabilistic polling. In *Financial cryptography* (pp. 173–191). Berlin: Springer. https://www.researchgate.net/profile/Stanislaw_Jarecki/publication/220797099_An_Efficient_Micropayment_System_Based_on_Probabilistic_Polling/links/0f31753c7f02552a9d000000.pdf

Judmayer, A., Zamyatin, A., Stifter, N., Voyiatzis, A. G., & Weippl, E. (2017). Merged mining: Curse or cure? In *Proceedings of the International Workshop on Cryptocurrencies and Blockchain Technology, CBT'17*. https://eprint.iacr.org/2017/791.pdf

Kallahalla, M., Riedel, E., Swaminathan, R., Wang, Q., & Fu, K. (2003). Plutus: scalable secure file sharing on untrusted storage. In *Proceedings of the 2nd USENIX Conference on File and Storage Technologies* (pp. 3). Berkeley, CA: USENIX Association

Kiayias, A., Konstantinou, I., Russell, A., David, B., & Oliynykov, R. (2016). A provably secure proof-of-stake blockchain protocol. Retrieved November 9, 2016, from http://eprint.iacr.org/2016/889.pdf

Kiayias, A., & Panagiotakos, G. (2015). Speed-security tradeoff s in blockchain protocols. Retrieved October 17, 2016, from https://eprint.iacr.org/2015/1019.pdf

Kiayias, A., Russell, A., David, B., & Oliynykov, R. (2017). Ouroboros: A provably secure proof-of-stake blockchain protocol. In *Annual International Cryptology Conference* (pp. 357–388). Berlin: Springer.

Kihlstrom, K. P., Moser, L. E., & Melliar-Smith, P. M. (1998). The securering protocols for securing group communication. In *Proceedings of the Thirty-First Hawaii International Conference on System Sciences* (Vol. 3, pp. 317–326). Piscataway, NJ: IEEE.

King, S., & Nadal, S. (2012). Ppcoin: Peer-to-peer crypto-currency with proof-of-stake. Retrieved January 7, 2017, from https://peercoin.net/assets/paper/peercoin-paper.pdf

Lamport, L. (1984). Using time instead of timeout for fault-tolerant distributed systems. *ACM Transactions on Programming Languages and Systems, 6*, 254–280. http://131.107.65.14/en-us/um/people/lamport/pubs/using-time.pdf

Lamport, L. (1998). The part-time parliament. *ACM Transactions on Computer Systems, 16*, 133–169. https://www.microsoft.com/en-us/research/uploads/prod/2016/12/The-Part-Time-Parliament.pdf

Lamport, L., Shostak, R., & Pease, M. (1982). The byzantine generals problem. *ACM Transactions on Programming Languages and Systems, 4*, 382–401. http://people.cs.uchicago.edu/~shanlu/teaching/33100_wi15/papers/byz.pdf

Li, H. C., Clement, A., Wong, E. L., Napper, J., Roy, I., Alvisi, L., & Dahlin, M. (2006). Bar gossip. In *Proceedings of the 7th Symposium on Operating Systems Design and Implementation* (pp. 191–204). Berkeley, CA: USENIX Association. http://www.cs.utexas.edu/users/dahlin/papers/bar-gossip-apr-2006.pdf

Litecoin.org. (n.d.). Retrieved May 18, 2019, from https://litecoin.org/

Liu, S., Viotti, P., Cachin, C., Quéma, V., & Vukolić, M. (2016). XFT: Practical fault tolerance beyond crashes. In *12th USENIX Symposium on Operating Systems Design and Implementation (OSDI 16)* (pp. 485–500).

Luu, L., Narayanan, V., Zheng, C., Baweja, K., Gilbert, S., & Saxena, P. (2016). A secure sharding protocol for open blockchains. In *Proceedings of the 2016 ACM SIGSAC Conference on Computer and Communications Security* (pp. 17–30). New York, NY: ACM. https://www.comp.nus.edu.sg/~prateeks/papers/Elastico.pdf

Maesa, D. D. F., Mori, P., & Ricci, L. (2017). Blockchain based access control. In *IFIP International Conference on Distributed Applications and Interoperable Systems* (pp. 206–220). Berlin: Springer.

McKeen, F., Alexandrovich, I., Anati, I., Caspi, D., Johnson, S., Leslie-Hurd, R., et al. (2016). Intel® software guard extensions (intel® sgx) support for dynamic memory management inside an enclave. In *Proceedings of the Hardware and Architectural Support for Security and Privacy 2016* (p. 10). New York, NY: ACM.

Meiklejohn, S., Pomarole, M., Jordan, G., Levchenko, K., McCoy, D., Voelker, G. M., & Savage, S. (2013). A fistful of bitcoins: Characterizing payments among men with no names. In *Proceedings of the 2013 Conference on Internet Measurement Conference* (pp. 127–140). New York, NY: ACM. https://cseweb.ucsd.edu/~smeiklejohn/files/imc13.pdf

Micali, S. (2016). Algorand: The efficient and democratic ledger. Retrieved Febraury 9, 2017, from https://arxiv.org/pdf/1607.01341.pdf

Miller, A., & LaViola, J. J. (2014). Anonymous byzantine consensus from moderately-hard puzzles: A model for bitcoin. Retrieved March 9, 2016, from https://socrates1024.s3.amazonaws.com/consensus.pdf

Miller, A., Xia, Y., Croman, K., Shi, E., & Song, D. (2016). The honey badger of bft protocols. In *Proceedings of the 2016 ACM SIGSAC Conference on Computer and Communications Security* (pp. 31–42). New York, NY: ACM. https://eprint.iacr.org/2016/199.pdf

Nakamoto, S. (2008). Bitcoin: A peer-to-peer electronic cash system. Retrieved July 1, 2015, from https://bitcoin.org/bitcoin.pdf

Narayanan, A., Bonneau, J., Felten, E., Miller, A., Miller, A., & Goldfeder, S. (2016). Bitcoin and cryptocurrency technologies. Retrieved March 29, 2016, from https://d28rh4a8wq0iu5.cloudfront.net/bitcointech/readings/princeton_bitcoin_book.pdf

Paillisse, J., Subira, J., Lopez, A., Rodriguez-Natal, A., Ermagan, V., Maino, F., & Cabellos, A. (2019). Distributed access control with blockchain. preprint arXiv:1901.03568.

Palmer, J., Nakamoto S., /u/PowerLemons, Ricks, C. (n.d.). Dogecoin.com [online]. Retrieved May 18, 2019, from https://dogecoin.com/

Pass, R., Seeman, L., & Shelat, A. (2017). Analysis of the blockchain protocol in asynchronous networks. In Annual International Conference on the Theory and Applications of Cryptographic Techniques (pp. 643–673). Berlin: Springer.

Pass, R., & Shi, E. (2017a). Hybrid consensus: Efficient consensus in the permissionless model. In *31st International Symposium on Distributed Computing (DISC 2017)* Merzig-Wadern: Schloss Dagstuhl-Leibniz-Zentrum fuer Informatik.

Pass, R., & Shi, E. (2017b). The sleepy model of consensus. In *International Conference on the Theory and Application of Cryptology and Information Security* (pp. 380–409). Berlin: Springer.

Pass, R., & Shi, E. (2018). Thunderella: Blockchains with optimistic instant confirmation. In *Annual International Conference on the Theory and Applications of Cryptographic Techniques* (pp. 3–33). Berlin: Springer.

Pease, M., Shostak, R., & Lamport, L. (1980). Reaching agreement in the presence of faults. *Journal of the ACM, 27*, 228–234. https://www.microsoft.com/en-us/research/uploads/prod/2016/12/Reaching-Agreement-in-the-Presence-of-Faults.pdf

Pitta, J. (1999). Requiem of a bright idea [online]. Retrieved May 18, 2019, from http://www.forbes.com/forbes/1999/1101/6411390a.html

Poon, J., & Dryja, T. (2016). The bitcoin lightning network. Retrieved July 7, 2016, from https://lightning.network/lightning-network-paper.pdf

Rabin, M. O. (1983). Randomized byzantine generals. In *24th Annual Symposium on Foundations of Computer Science* (pp. 403–409). Piscataway, NJ: IEEE. https://www.cs.princeton.edu/courses/archive/fall05/cos521/byzantin.pdf

Ron, D., & Shamir, A. (2013). Quantitative analysis of the full bitcoin transaction graph. In *International Conference on Financial Cryptography and Data Security* (pp. 6–24). Berlin: Springer.

Schneider, F. B. (1990). Implementing fault-tolerant services using the state machine approach: A tutorial. *ACM Computing Surveys, 22*, 299–319. http://www-users.cselabs.umn.edu/classes/Spring-2014/csci8980-sds/Papers/ProcessReplication/p299-schneider.pdf

Schwartz, D., Youngs, N., & Britto, A. (2014). The ripple protocol consensus algorithm. Retrieved August 8, 2016, from https://ripple.com/files/ripple_consensus_whitepaper.pdf

Schwarz, A. (2011) Squaring the triangle: Secure, decentralized, human-readable names. Retrieved November 12, 2014, from http://www.aaronsw.com/weblog/squarezooko

Sompolinsky, Y., & Zohar, A. (2013). Accelerating bitcoin's transaction processing. Fast money grows on trees, not chains. http://eprint.iacr.org/2013/881.pdf

Stifter, N., Judmayer, A., Schindler, P., Zamyatin, A., & Weippl, E. (2018). *Agreement with Satoshi—on the Formalization of Nakamoto Consensus*. Cryptology ePrint Archive, Report 2018/400. https://eprint.iacr.org/2018/400.pdf

Subbiah, A., & Blough, D. M. (2005). An approach for fault tolerant and secure data storage in collaborative work environments. In *Proceedings of the 2005 ACM workshop on Storage Security and Survivability* (pp. 84–93). New York, NY: ACM.

Szabo, N. (2005). Bit gold. Retrieved April 4, 2016, from http://unenumerated.blogspot.co.at/2005/12/bit-gold.html

Szalachowski, P. (2018). (short paper) towards more reliable bitcoin timestamps. In *2018 Crypto Valley Conference on Blockchain Technology (CVCBT)* (pp. 101–104). Piscataway, NJ: IEEE.
Taylor, M. B. (2013). Bitcoin and the age of bespoke silicon. In *Proceedings of the 2013 International Conference on Compilers, Architectures and Synthesis for Embedded Systems* (p. 16). Piscataway, NJ: IEEE Press. https://cseweb.ucsd.edu/~mbtaylor/papers/bitcoin_taylor_cases_2013.pdf
Veronese, G. S., Correia, M., Bessani, A. N., & Lung, L. C. (2009). Highly-resilient services for critical infrastructures. In *Proceedings of the Embedded Systems and Communications Security Workshop*.
Veronese, G. S., Correia, M., Bessani, A. N., Lung, L. C., & Verissimo, P. (2013). Efficient byzantine fault-tolerance. *IEEE Transactions on Computers, 62*, 16–30. https://www.researchgate.net/profile/Miguel_Correia3/publication/260585535_Efficient_Byzantine_Fault-Tolerance/links/5419615d0cf25ebee9885215.pdf
Vukolić, M. (2010). The byzantine empire in the intercloud. *ACM Sigact News, 41*(3), 105–111.
Vukolić, M. (2015). The quest for scalable blockchain fabric: Proof-of-work vs. bft replication. In *International Workshop on Open Problems in Network Security* (pp. 112–125). Berlin: Springer. http://vukolic.com/iNetSec_2015.pdf
Vukolic, M. (2016). Eventually returning to strong consistency. Retrieved August 10, 2016, from https://pdfs.semanticscholar.org/a6a1/b70305b27c556aac779fb65429db9c2e1ef2.pdf
Wang, J., Wu, P., Wang, X., & Shou, W. (2017). The outlook of blockchain technology for construction engineering management. *Frontiers of Engineering Management, 4*(1), 67–75.
Wang, Z., Lin, J., Cai, Q., Wang, Q., Jing, J., & Zha, D. (2019). Blockchain-based certificate transparency and revocation transparency. In *International Conference on Financial Cryptography and Data Security* (pp 144–162). Berlin: Springer.
Wensley, J. H., Lamport, L., Goldberg, J., Green, M. W., Levitt, K. N., Melliar-Smith, P. M., et al. (1978). Sift: Design and analysis of a fault-tolerant computer for aircraft control. *Proceedings of the IEEE, 66*(10), 1240–1255.
Zhao, W., & Babi, M. (2013). Byzantine fault tolerant collaborative editing. In *IET International Conference on Information and Communications Technologies (IETICT 2013)* (pp. 233–240). IET.

Chapter 18
Conclusion and Outlook on Security and Quality of Complex Cyber-Physical Systems Engineering

Stefan Biffl, Matthias Eckhart, Arndt Lüder, and Edgar Weippl

Abstract Typical assumptions for research in quality and security assurance and improvement for small software-intensive systems may not hold for long-running technical systems, such as critical infrastructure or industrial production systems. Therefore, researchers in quality and security assurance and improvement can benefit from better understanding challenges on quality and security assurance and quality improvement coming from the engineering of *Complex Cyber-Physical Systems* based on the use cases and requirements presented. This chapter summarizes and reflects on the material presented in this book regarding challenges and solutions for Security and Quality of *Complex Cyber-Physical Systems* (C-CPS) Engineering. Contributions in this book consider requirements, risks, and solutions to improve the security and quality of C-CPS. Engineers and project managers will be enabled to identify quality and security challenges they should consider. In addition, the chapter describes measures to assist the involved staff in handling the identified challenges. The chapter discusses the contributions of the chapters to the Research Questions raised in Chap. 1 of this book.

Keywords Complex cyber-physical systems · Engineering process · Multidisciplinary engineering · AutomationML · Information security

S. Biffl (✉)
Institute of Information Systems Engineering, Technische Universität Wien, Vienna, Austria
e-mail: stefan.biffl@tuwien.ac.at

M. Eckhart · E. Weippl
Christian Doppler Laboratory for Security and Quality Improvement in the Production System Lifecycle (CDL-SQI), Institute of Information Systems Engineering, Technische Universität Wien, Vienna, Austria

SBA Research, Vienna, Austria
e-mail: matthias.eckhart@tuwien.ac.at; edgar.weippl@tuwien.ac.at

A. Lüder
Otto-v.-Guericke University/IAF, Magdeburg, Germany
e-mail: arndt.lueder@ovgu.de

S. Biffl et al. (eds.), *Security and Quality in Cyber-Physical Systems Engineering*,
https://doi.org/10.1007/978-3-030-25312-7_18

18.1 Part I: Engineering Complex Cyber-Physical Systems

The improvement of quality and security in engineering *Complex Cyber-Physical Systems* (*C-CPS*) is only possible if the engineering process is understood by all involved stakeholders. This understanding enables identifying possible impacts of the engineering process execution on quality and security properties and makes it possible to develop and evaluate quality and security improvement approaches.

Therefore, the initial research question within this book has been the following.

RQ1a: What are typical characteristics of engineering processes for long-running software-intensive technical systems?

Chapters 2–4 have considered this research question and provided detailed descriptions of practically relevant engineering processes and the challenges to be tackled by engineers using these processes. Nevertheless, the chapters take three viewpoints that represent different types of engineering organization and their engineering processes.

Chapter 2 takes the viewpoint of an *international engineering organization* and discusses challenges emerging from executing engineering processes in this international setting. One possible measure to tackle these challenges is a *readiness check for engineering internationalization* that highlights main questions to answer when setting up an international engineering organization. These questions highlight success critical characteristics of engineering processes, such as workload distribution and process/task descriptions, knowledge management, and information exchange.

Chapter 3 takes the viewpoint of engineering in a *large automotive OEM*. The *New Product and Production system Development Process* (NPPDP) considers the joined engineering of two C-CPS types: cars and car production systems. *Dynamic complexity* is a critical characteristic of these joined engineering processes. This dynamic complexity comes from the diversity of engineering tasks and artifacts, the strong connectivity between these tasks and artifacts, and the uncertainty of required changes within the artifacts. Therefore, *a complexity management framework* can help manage dynamic complexity in a holistic way by providing the required understanding of (a) the engineering processes, (b) the engineering process dependencies, and (c) the engineering data logistics needed.

Chapter 4 provides insights into the *engineering of rail systems* by collecting special challenges that emerge within this field of public transport due to safety and warranty regulations. With a focus on the engineering of signaling systems, the authors present the engineering process and its tool chain. A main element in rail system engineering is the data consistency along the tool chain.

Beyond these more general views on C-CPS engineering, Chaps. 5 and 6 discuss selected special views on C-CPS engineering and derive characteristics of these engineering processes.

Chapter 5 considers interrelations between the engineering and use of a production system by discussing methods to improve the production system engineering based on run-time data. The authors extended model-based engineering to represent and make good use of the interdependencies of models along the complete C-CPS lifecycle.

In contrast, Chap. 6 discusses the important topic of *data-intensive systems* as a special kind of C-CPS, stressing the difficulties to properly test this type of C-CPS and extending test characteristics matrices used. Thereby, the authors show that data management in data-intensive C-CPS is a characteristic with much more facets than usually perceived.

Together, these chapters highlight that *C-CPS engineering processes* can be characterized as *a network of multidisciplinary processes that connect engineering activities, which are optimized independently, by engineering data exchange* (Biffl et al. 2017). The success of the engineering process is strongly related to the quality and security of the *engineering data exchange*.

The second research question considers the support C-CPS engineering can expect from well-proven approaches in (business) informatics.

RQ1b: What are requirement areas from engineering processes for long-running software-intensive technical systems that require informatics contributions?

Chapter 2 derives from the structure and execution of the *readiness check* requirements for IT support considering four requirements clusters. The first cluster is dedicated to the overall engineering organization solution portfolio including requirements like (a) required system-of-systems approaches or (b) adequate data modeling approaches. The second cluster is dedicated to engineering team with (c) dedicated working conditions and (d) security-related issues. The third cluster is related to the engineering artifacts covering requirements related to (e) consistent and scalable data exchange structures, (f) consistency in engineering change management, (g) data quality across different locations, and (h) security of engineering data. Finally, the fourth cluster reflects the used engineering tools with requirements related to (i) interoperability and (j) tool chain design.

Chapter 3 derives from the *complexity management framework* for NPPDP requirements for understanding the engineering processes, their dependencies, and the engineering data logistics. Thereby, the authors highlight requirements to (a) IT systems that enable engineering process modeling and analysis, (b) data logistics in engineering processes impacting engineering quality over data quality assurance, and (c) capabilities for consistent modeling of engineering objects.

Chapter 4 discusses data consistency along the tool chain of rail system engineering and derives requirements for IT systems regarding (a) safe and reliable data exchange, integration, and migration; (b) role-based data security; and (c) support for engineering data reuse.

In the scope of their special views on C-CPS engineering, Chaps. 5 and 6 provide valuable requirements. Chapter 5 stresses requirements for appropriate information modeling and data management as foundation for collecting data from run-time systems and integrating the run-time data with design models to improve the accuracy of the design models. Chapter 6 discusses security issues that testing of data-intensive software systems can help address and identifies new security threats as input to a subsequent security analysis.

In summary, the information engineering processes collected in Part I can be characterized as *interconnected multidisciplinary network of engineering decisions* (Biffl et al. 2017) that strongly rely on

- The available knowledge related to *process structure and execution* rules
- The consistency, quality, and completeness of the complete set of *engineering artifacts* created, exchanged, and used in the engineering network
- The *tool set* applied within the network with its interfaces and use conditions
- The *secure information* creation, exchange, and use within the engineering network

The challenges to information science anticipated in Chap. 1 have been confirmed. They can be summarized as challenges related to

(C1): The *increasing digitalization* of CPS with *multi-model data integration* that enables the assurance of consistency, quality, and completeness of engineering information in an engineering network
(C2): Consistent and fault-free *engineering information exchange* between engineering disciplines along the complete C-CPS engineering lifecycle
(C3): Information and knowledge management in an engineering network
(C4): Secure engineering processes and engineering data management

The chapters in Parts II and III provide contributions from quality improvement and from security improvement to address these challenges in C-CPS engineering.

18.2 Part II: Engineering Quality Improvement

Business informatics aims at improving business processes, in the case of C-CPS these are the processes of engineering organizations. Chaps. 7–10 in Part II focus on the engineering processes of collaborating discipline-specific workgroups in an engineering network to support decision making, quality assurance, and information management. In addition, Chaps. 5 and 6 in Part I play a dual role by providing requirements for quality improvement and by introducing methods for quality improvement.

In the focus of Part II, Chaps. 5–10 address quality improvement challenges identified in Part I, in particular, engineering data/model representation and integration, methods for consistent and fault-free engineering information exchange, and methods for quality assurance. The two research questions of Part II, defined in Chap. 1, consider both the potentials of better quality improvement capabilities for engineering C-CPS and security requirements and issues that quality improvement approaches may introduce or intensify.

RQ2a: How can approaches adapted from business informatics address quality improvement requirements coming from characteristics of C-CPS engineering?

Chapters 5–10 address this research question. Chapters 5, 7, and 10 discuss approaches for better knowledge representation to address challenges of data and model integration as foundation for better information management in an engineering network. Chapters 7 and 8 analyze and improve the process of engineering information exchange between disciplines in a C-CPS engineering network. Chapters 6 and 9 discuss methods for quality assurance, in particular, correctness and consistency of engineering information in an engineering network.

Chapter 5 follows a model-based engineering approach to improve the backflow of information from operation to the engineering phase. The authors introduce the *Temporal Model Framework* to link production system design models to run-time models as foundation for improving the accuracy and completeness of information in the design-time models based on aggregated information from analysis of operation data, such as the confidence interval of the duration for a transport process. The extension of design models with run-time information improves the usefulness of the design models for iterative engineering, keeps the design models relevant along the lifetime of the C-CPS, and collects relevant information for designing similar systems in a C-CPS family.

Chapter 6 discusses the role of data in testing systems and requirements for appropriate testing approaches for data-intensive software. In data-intensive software, the quality of data plays a vital role for the quality of the system, beyond the traditional input data to a system. For example, software functions may depend on a large body of data collected from previous projects or from operation. The quality of this data becomes, therefore, instrumental to the quality of the software. However, an open issue is how to measure and assure the quality of such data and how to provide data with specified quality levels for testing data-intensive software. As C-CPSs are increasingly likely to contain data-intensive software, the quality assurance of data-intensive software and the underlying data becomes an increasing concern.

Chapter 7 analyzes gaps regarding critical engineering knowledge in the exchange between disciplines in a C-CPS engineering network. The authors focus on the information needs of detail engineers regarding knowledge on product and production process requirements that basic engineers know but often fail to document, leading to risks and delays in the C-CPS engineering process. The chapter introduces a method for product/ion-aware analysis of production system engineering (PSE) processes based on the engineering process analysis method (Lüder et al. 2012) and the BPMN 2.0 notation. To address shortcomings in the BPMN 2.0 notation, the authors introduce a notation for representing linked knowledge on products, production processes, and production system resources (PPR) in the BPMN 2.0, the *product/ion-aware data processing map* (PPR DPM). The PPR DPM enables advanced analyses of PPR knowledge needs and gaps as foundation for better information management in a C-CPS engineering network.

Chapter 8 aims at improving the process of engineering information exchange between engineering disciplines in a C-CPS engineering network by enabling a more frequent and efficient synchronization between collaborating workgroups in C-CPS engineering. The authors introduce a method for efficient Engineering Data

Exchange, in particular, for negotiating and representing the data that consumers in an engineering network require, as foundation for configuring an Engineering Data Exchange information system for the efficient and consistent management of information in a C-CPS engineering network. For typical C-CPS engineering networks, which rely on the traditional point-to-point exchange of engineering artifacts, the approach aims at reducing the risk of diverging local data views, effort for rework, and unclear project progress assessment.

Chapter 9 discusses methods for improving the automated testing of software-intensive systems that are used in C-CPS. The authors focus on knowledge representation for the reuse of artifacts for automated testing in similar but different test environments, for example, the efficient generation of test source code in different testing environments. For example, the underlying technology of a system part may change, for example, if a pick-and-place robot is exchanged for another robot with similar functionality but different programming language. In a C-CPS engineering network, the approach can reduce the effort and risk coming from the manual adaptation of testing artifacts.

Chapter 10 discusses an approach for the efficient representation of knowledge on product variants in a family of C-CPSs to identify reusable parts as foundation for better quality assurance. The representation of a product portfolio is the foundation for efficiently identify which C-CPSs are likely to be affected by a defect found in one C-CPS and to manage a suite of test cases and artifacts for the systems in the C-CPS family.

RQ2b: Which interventions of quality improvement approaches are likely to introduce or increase information security requirements and risks?

Chapters 5–10 introduce and discuss approaches that primarily aim at quality improvement in engineering networks. However, these improvement approaches can also be seen as interventions in traditional engineering networks, interventions that change the way of working and that may be relevant for information security concerns. As a cross-cutting quality concern, effective and efficient security cannot be added later, but needs to be designed into an engineering network, its tools, and the C-CPS operational environment. Surprisingly, in traditional C-CPS engineering, security is often an add-on consideration to engineering and quality improvement, possibly due to the discipline-specific focus in C-CPS engineering.

The quality improvement contributions build on the foundations of better knowledge representation and the integration and centralization of previously scattered, often isolated, knowledge. These foundations allow developing more effective and efficient approaches for engineering and quality management by providing better access to integrated knowledge. The knowledge can be used to improve understanding of engineers and managers regarding (a) characteristics of the engineering network, (b) C-CPS characteristics, (c) C-CPS operational characteristics, and (d) the reuse in a C-CPS family of similar systems.

However, these advantages for legitimate users can become security risks if accessible to external attackers, who may use the integrated knowledge for gaining intelligence on the roles and information in the engineering process, the C-CPS, and the planned C-CPS operation functionality. Attackers can extract this intelligence to inform the competition. Attackers can use this intelligence for planning attacks on the engineering network and the C-CPS. Further, attackers can use this intelligence to manipulate engineering artifacts (a) to lower the quality of the C-CPS or its operation as well as (b) to make the C-CPS more vulnerable against future attacks. These risk scenarios emphasize security requirements for capabilities to protect engineering processes and data against adversarial access.

In addition, there are inside attackers, legitimate users, who misuse their access privileges. Therefore, security requirements have to include capabilities (a) to trace access and changes to relevant engineering knowledge and (b) to trace exported data in a legally binding way. Therefore, the C-CPS engineering examples in Part II point out potential targets that IT security should defend, as input to the security methods introduced in Part III of this book.

Overall, readers gain from these chapters' insights into a variety of approaches to improve engineering quality for addressing requirements raised in Part I of this book for making better informed decisions in the design and improvement of C-CPS engineering processes. Further, Part II discusses a range of security requirements and threats as input to the security methods introduced in Part III of this book.

18.3 Part III: Engineering Security Improvement

System integrators, who seek to improve information security in the context of C-CPSs engineering face a twofold challenge: First, it has to be ensured that the engineering process is adequately protected against adversarial access and manipulations. Failure to do so may result in information leaks through industrial cyber espionage attacks or the manipulation of engineering artifacts in order to make the engineered systems inherently vulnerable. Second, security needs to become a *first-class citizen* in the engineering of C-CPSs in order to achieve industrial systems that are built to be secure. The consequences of security negligence in the industrial sector have already manifested themselves and, considering the long lifecycle of C-CPSs, will remain an issue in the years to come. Thus, security must be established as an integral part of (green-field) engineering projects to counter this trend. The two research questions of Part III, defined in Chap. 1, relate to both of the aforementioned issues.

RQ3a: Which security concepts mitigate cyber threats targeting the engineering process of complex cyber-physical systems?

The first two chapters of Part III, that is, Chaps. 11 and 12, address this research question.

Chapter 11 discusses measures to mitigate security risks pertaining to central data repositories used for the exchange of engineering information, specifically in the context of AutomationML. The chapter first gives an introduction to access control and clarifies its relevance for engineering networks. After that, the authors introduce a stepwise approach that can be used to analyze the security of engineering data exchange platforms. The main finding relevant to this research question is that general IT security measures (according to established cybersecurity standards) must be supplemented by cryptographic measures to protect the content of engineering artifacts. In this way, the risk of potential confidentiality breaches caused by external threat actors can be significantly mitigated.

Chapter 12 deals with security concepts that system integrators can adopt in order to thwart insider threats (e.g., from legitimate partners involved in engineering projects). After motivating the need for mechanisms to secure engineering data against exfiltration and manipulation, the chapter provides background information on *fingerprinting* mechanisms and *expert-in-the-loop* systems. On this basis, the authors extend the *doctor in the loop* paradigm (cf. for instance, Kieseberg et al. 2016), to embed audit and control capabilities into engineering environments. Furthermore, the authors propose two measures to mitigate the risk of engineering data exfiltration, namely, (i) incorporating read access into the transaction mechanism of engineering databases, and (ii) applying database steganography through slack space. Their proposed approach also includes ex-post protection by means of *fingerprinting*, enabling system integrators to track and prove data leaks. This reactive measure enables taking legal action against parties, which leak data but cannot be easily controlled or monitored.

Both chapters provide insights into mitigating security risks pertaining to the C-CPSs engineering process. In particular, these two chapters show that general IT security measures (e.g., implemented according to the ISO/IEC 27000-series) constitute merely the foundation for protecting engineering environments. Securing engineering artifacts on a content level via cryptographic methods and integrating data manipulation and exfiltration detection mechanisms into engineering databases are further measures that must be taken to reduce security risks emanating from malicious acts of external threat actors and insiders.

RQ3b: How can the security of complex cyber-physical systems be enhanced by considering security aspects during the engineering phase?

Chapters 13–17 address this research question.

Chapter 13 provides valuable background information on potential security issues of CPSs and discusses how they can be addressed by introducing the *security-by-design* principle to the Operational Technology (OT) domain. Furthermore, the chapter refers to multiple streams of research in CPSs security and attempts to show the current state of knowledge in this area. The authors of this chapter identified run-time security monitoring as a key component of an effective defense, which must be designed in line with the system to be engineered. The reason for this is that the implementation of measures for detecting and preventing cyber-attacks against CPSs is driven by the system's specification as well as by the considered threat model. Thus, developing run-time security monitoring techniques right from the beginning of the CPS's lifecycle provides the means to tailor them to the system's characteristics (e.g., the design of control loops), which may result in fine-tuned implementations that yield a high detection performance.

Chapter 14 reviews literature on the concept of digital twins in the context of information security. The authors first give an overview of the *digital twin* concept by describing its origins, types of digital twins, its use cases in the manufacturing domain, and its connection to the *digital thread* concept, which aims to establish a consistent data link over the systems' lifecycle. Then, the authors expand on the security-related use cases of digital twins that have been initially presented by Eckhart and Ekelhart (2018a, b) and discuss other relevant use cases covered in the literature. The chapter shows that the following security-related digital twin use cases, which can be applied in engineering activities, are already discussed in the literature: (1) supporting the design of more secure CPSs, (2) detecting hardware and software misconfigurations, (3) conducting security testing, and (4) monitoring security and legal compliance. Finally, the authors identify research gaps to address in future research. For instance, use cases applicable to the engineering phase, which are in need of further investigation, are the assessment of security risks, improving the resilience of CPSs, and performing security testing in an automated manner.

Chapter 15 discusses the security risks associated with the use of *Radio Frequency* (RF) technologies in industrial facilities and describes mitigation strategies that vendors and system integrators can implement during engineering activities. First, the chapter provides a brief introduction to wireless protocols and their application in an industrial setting. Then, the chapter shows how typical attacks against RF-based systems can be executed. Based on this knowledge, the authors demonstrate how threat modeling and penetration testing can be embedded in the engineering lifecycle as a proactive measure to mitigate these attacks. In particular, the authors describe how the STRIDE model can be applied in conjunction with the DREAD model to assess security risks pertaining to RF-based systems. Furthermore, they show how *Software Defined Radio* can be leveraged as a penetration testing tool. As the authors correctly point out, conducting threat modeling activities and penetration testing of wireless systems during development may reveal vulnerabilities at an early stage of the system's lifecycle, allowing engineers to fix these weaknesses efficiently at a lower cost.

Chapter 16 provides a review of security and safety challenges concerning *Industrial Internet of Things* (IIoT) systems and surveys *machine learning* and *deep learning* (ML/DL) methods that can be applied to address these challenges. The authors of this chapter highlight the specificities of IIoT systems, such as resource constraints, the prevalence of wireless communication, and, in particular, their distributed and embedded nature. These characteristics need to be considered in the threat model and represent a determining factor in mitigating security risks, especially in the context of ML/DL. The results of their literature review indicate that current papers on detecting cyber-attacks against IIoT systems pay particular attention to DL methods, given their promising performance in such applications. In the context of the engineering lifecycle, the authors suggest that a close collaboration between C-CPSs engineers (i.e., domain experts) and ML/DL specialists is required for proper feature engineering, as it affects the accuracy of ML/DL models.

Chapter 17 examines the connection between *blockchain* technologies and *Byzantine fault tolerance* and discusses their application for addressing technical challenges that emerge when engineering safety-critical systems. After providing an introduction to *Byzantine fault tolerance* and *blockchain*, the chapter delves into the topic of decentralizing trust in the context of *Bitcoin's Nakamoto consensus*. In this context, the chapter presents use cases of *blockchain* technologies beyond cryptocurrencies and thereby establishes a strong link to the book's theme. For instance, trusted timestamping, digital identities, and trusted execution environments are *blockchain*-related use cases that may be implemented to improve the security of C-CPSs. In addition to the discussed concepts for improving the security of C-CPSs, the chapter addresses RQ3a. In particular, the authors describe how Byzantine fault tolerance and *blockchain* technologies can be leveraged to protect the engineering process of C-CPSs by referring to previous research relevant to the subject. The findings of this chapter suggest that *blockchain* technologies do not only represent a promising advance for achieving *Byzantine fault tolerance*, but also for securing engineering processes and C-CPSs themselves.

To sum up, the chapters examine several security concepts that can be implemented as part of C-CPS engineering for the purpose of enhancing the security of C-CPSs. The specific security concepts covered in these chapters are not closely related to each other and therefore provide insights from different perspectives. More specifically, the considered security-improving methods range over intrusion detection, digital twins, wireless communication, machine and deep learning methods, and *blockchain* technologies. Altogether, these chapters only discuss a fraction of the topics relevant to improving the security of C-CPSs. However, the chapters still represent a valuable contribution to the scientific progress in this area, as they discuss a variety of security challenges pertaining to C-CPSs and, more importantly, provide pointers to address them.

Summing up all chapters, this book discusses challenges and solutions for improving quality and security of and within the engineering of C-CPSs. As visualized in Table 18.1, all challenge fields identified in Part I have been addressed and the research questions named in Chap. 1 have been tackled.

In total, the chapters discuss a broad range of methodologies to improve quality and security of C-CPS by improving and securing the C-CPS engineering process. Even if the provided methods cover a wide area, there are still several open issues for research identified within this book in relation to the research questions.

The most relevant open issues are related to the multimodel nature of C-CPS engineering and sufficient data integration in C-CPS engineering as well as the knowledge protection within engineering data exchange, to name only two open issues.

Finally, it can be stated that this book has reached its target of providing a review on the state of the art in its field and to open up existing solutions and open issues for further research. This will enable researchers and practitioners to take better informed decisions when dealing with C-CPS engineering process quality and security.

Table 18.1 Mapping of C-CPS challenge fields Cx to chapter contributions

Chapters	Challenge			
	C1 Multimodel data integration	C2 Information exchange	C3 Information/knowledge management	C4 Secure engineering
5	x		x	
6	x			
7	x	x	x	
8		x		
9	x		x	
10	x		x	
11	x	x		x
12	x			x
13			x	x
14	x		x	x
15			x	x
16				x
17				x

Acknowledgments The financial support by the Christian Doppler Research Association, the Austrian Federal Ministry for Digital and Economic Affairs and the National Foundation for Research, Technology and Development is gratefully acknowledged.

References

Biffl, S., Gerhard, D., & Lüder, A. (2017). *Multi-disciplinary engineering for cyber-physical production systems*. Cham: Springer.

Eckhart, M., & Ekelhart, A. (2018a, May). Towards security-aware virtual environments for digital twins. In *Proceedings of the 4th ACM workshop on cyber-physical system security* (pp. 61–72). New York: ACM.

Eckhart, M., & Ekelhart, A. (2018b). Securing cyber-physical systems through digital twins. *ERCIM NEWS, 115*, 22–23, European Research Consortium for Informatics and Mathematics.

Kieseberg, P., Malle, B., Frühwirt, P., Weippl, E., & Holzinger, A. (2016). A tamper-proof audit and control system for the doctor in the loop. *Brain Informatics, 3*(4), 269–279.

Lüder, A., Foehr, M., Köhlein, A., & Böhm, B. (2012). Application of engineering processes analysis to evaluate benefits of mechatronic engineering. In *Emerging Technologies & Factory Automation (ETFA), 2012 IEEE 17th Conference* (pp. 1–4). IEEE.

Printed in the United States
By Bookmasters